HANDBOOK OF NEUROCHEMISTRY

VOLUME IV

CONTROL MECHANISMS IN
THE NERVOUS SYSTEM

HANDBOOK OF NEUROCHEMISTRY

Edited by Abel Lajtha

HANDBOOK OF NEUROCHEMISTRY

Edited by Abel Lajtha

*New York State Research Institute
for Neurochemistry and Drug Addiction
Ward's Island
New York, New York*

VOLUME IV

CONTROL MECHANISMS
IN THE NERVOUS SYSTEM

℗ PLENUM PRESS • NEW YORK–LONDON • 1970

Library of Congress Catalog Card Number 68-28097

SBN 306-37704-7

ISBN 978-1-4615-7165-0   ISBN 978-1-4615-7163-6 (eBook)
DOI 10.1007/978-1-4615-7163-6
© 1970 Plenum Press, New York
Softcover reprint of the hardcover 1st edition 1970
A Division of Plenum Publishing Corporation
227 West 17th Street, New York, New York 10011

United Kingdom edition published by Plenum Press, London
A Division of Plenum Publishing Company, Ltd.
Donington House, 30 Norfolk Street, London W.C. 2, England

Contributors to this volume:

R. Wayne Albers   Laboratory of Neurochemistry, National Institute of Neurological Diseases and Stroke, National Institutes of Health, Bethesda, Maryland (page 13)

Dorothea Aures   Psychopharmacology Research Laboratories, Veterans Administration Hospital, Sepulveda, California; and the Department of Pharmacology, College of Medicine, University of California, Irvine, California (page 165)

H. S. Bachelard   Department of Biochemistry, Institute of Psychiatry, British Postgraduate Medical Federation, University of London, England (page 1)

Arvid Carlsson   Research Division of the Cleveland Clinic Foundation, Cleveland, Ohio; and the Department of Pharmacology, University of Göteborg, Sweden (page 251)

William G. Clark   Psychopharmacology Research Laboratories; Veterans Administration Hospital, Sepulveda, California; and the Department of Biological Chemistry, University of California, Los Angeles, California (page 165)

Lynwood G. Clemens   Department of Zoology, Michigan State University, East Lansing, Michigan (page 429)

E. Costa   Laboratory of Preclinical Pharmacology, Division of Special Mental Health Research, National Institute of Mental Health, Saint Elizabeth's Hospital, Washington, D.C. (page 45)

David R. Curtis   Department of Physiology, Australian National University, Canberra, Australia (page 115)

Jacques Glowinski   Section de Neuropharmacologie Biochimique du Laboratoire de Neurophysiologie Générale du College de France (page 91)

Roger A. Gorski   Department of Anatomy and Brain Research Institute, UCLA School of Medicine, Los Angeles, California (page 429)

Jack Peter Green   Department of Pharmacology, Mount Sinai School of Medicine, New York, New York (page 221)

Rolf Håkanson — Department of Pharmacology, University of Lund, Lund, Sweden (page 165)

Leslie L. Iversen — Department of Pharmacology, University of Cambridge, Cambridge, England (page 197)

Graham A. R. Johnston — Department of Physiology, Australian National University, Canberra, Australia (page 115)

Robert Katzman — The Saul R. Korey Department of Neurology, Albert Einstein College of Medicine, Bronx, New York (page 313)

Erling Mellerup — Psychochemical Laboratory, University Clinic of Psychiatry, Rigshospitalet, Copenhagen, Denmark (page 361)

K. D. Neame — Department of Physiology, University of Liverpool, Liverpool, England (page 329)

N. H. Neff — Laboratory of Preclinical Pharmacology, Division of Special Mental Health Research, National Institute of Mental Health, Saint Elizabeths Hospital, Washington, D.C. (page 45)

J. H. Quastel — Kinsmen Laboratories of Neurological Research, Faculty of Medicine, University of British Columbia, Vancouver, B.C., Canada (page 285)

Irvine H. Page — Research Division of the Cleveland Clinic Foundation, Cleveland, Ohio; and the Department of Pharmacology, University of Göteborg, Sweden (page 251)

Lincoln T. Potter — Biophysics Department, University College, London, England (page 263)

Ole J. Rafaelsen — Psychochemical Laboratory, University Clinic of Psychiatry, Rigshospitalet, Copenhagen, Denmark (page 361)

Max Reiss — Neuroendocrine Research Unit, Willowbrook State School, Staten Island, New York (page 463)

Howard Sachs   Roche Institute of Molecular Biology, Nutley, New Jersey (page 373)

George J. Siegel   Departments of Neurology and Physiology, Mount Sinai School of Medicine of The City University of New York, New York (page 13)

Leonhard S. Wolfe   The Donner Laboratory of Experimental Neurochemistry, Montreal Neurological Institute and the Department of Neurology and Neurosurgery, McGill University, Montreal, Canada (page 149)

R. J. Wurtman   Department of Nutrition and Food Science, Massachusetts Institute of Technology, Cambridge, Massachusetts (page 451)

G. Zetler   Department of Pharmacology, Medizinsche Akademie Lübeck, Lübeck, Germany (page 135)

PREFACE

The explosive accumulation of new knowledge in the biological sciences in the last decades has advanced our understanding of the basic mechanisms that underlie most biological phenomena. These advances, however, have not been uniform but have varied considerably among the different biological problems. In some cases, e.g., biochemical genetics, radical advances have been made which have changed our ideas and our approaches. In other cases, even with work which has yielded much detailed new knowledge, our understanding of basic mechanisms remains very inadequate.

Among the lines of work that have not yet led to dramatic conceptual advances is the problem of control of biological activities. This problem is, of course, basic both to any full understanding of life as a whole, and to any real understanding of its most minute phenomena. Indeed, the myriad of biological activities that we can observe by direct or indirect means are all under the sway of most exquisitely precise mechanisms. Any malfunctioning of these mechanisms has serious consequences, not only for the particular function itself, but for all the related and interlinked activities.

It is quite possible that the failure to have made definitive progress towards the elucidation of the basic mechanisms of control may be a consequence of the very nature of the knowledge that we are seeking, in the sense that, to formulate in exact terms the uniquely precise mechanisms of control, it is necessary to know first much more than we know at present about the phenomena being controlled.

Under the circumstances, the editor setting out to gather the information available on control mechanisms has a very difficult job indeed. On the one hand he might bring together as much information as might be relevant to the question of control; on the other, he might limit himself to reviewing in depth mechanisms already known to be related to control. In the first case, the result might be a prolix and diffuse enumeration of tenuously related subjects with a loss of emphasis on the really significant points. In the second case he might very well overlook what may eventually prove to be crucial. In the present volume, Dr. Abel Lajtha has skillfully avoided these potential drawbacks and has put together a group of chapters which cover their subject thoroughly but concisely, and in which proper emphasis is not lost. He has done this by avoiding any overall hypothesis and by staying within the limits of positive knowledge. He has divided this body of knowledge into fairly homogeneous subjects, which he has succeeded in having reviewed by authors highly knowledgeable in their respective fields. Specifically, he has treated the transmitters and their metabolism, the enzymes presumably involved in transmission and other neural activities, the hormones related to control, and, finally, some biological phenomena which are necessarily related to control.

The final result of his careful editing is to have made available under a single cover a wealth of information that would require considerable effort to gather from other sources. Thus it will be possible for the reader to obtain, rapidly and economically, specific information both on the various problems reviewed and on their significance to the overall question of biological control of the nervous tissue.

J. Folch-Pi, M.D.

August 1969

CONTENTS

Chapter 5

Chapter 6

Chapter 9

Metabolism of Catecholamines 197
by Leslie L. Iverson

Chapter 10

Histamine 221
by Jack Peter Green

Chapter 13

Amine Oxidases 285

by J. H. Quastel

Chapter 1

CONTROL OF
CARBOHYDRATE METABOLISM

H. S. Bachelard

Department of Biochemistry, Institute of Psychiatry
British Postgraduate Medical Federation
University of London, England

I. INTRODUCTION

The brain as a tissue is distinguished from other tissues of the body by its relatively high metabolic rate, its almost complete dependence on carbohydrates for its sources of energy, and its capacity to respond rapidly and transiently to a requirement for increased energy expenditure when its electrically excitable cells are caused to fire. The tissue comprises about 3 % of the total body weight, yet it utilizes something of the order of 25 % of the total bodily consumption of glucose. The metabolic rate may be as high as 20 times the average for the body as a whole.

Evidence for the peculiar emphasis on carbohydrate as the source of energy has been derived from many studies *in vivo* and *in vitro*.[1] Generally, glucose has been confirmed to be the only satisfactory metabolic support for the brain *in vivo*; no alternative substrate has been found to support brain function adequately under normal conditions. However, under certain conditions, in particular as a result of high rates of fat mobilization during fasting, the brain can utilize β-hydroxybutyrate in partial replacement of glucose in man and in the rat.[71,72] Other substrates that apparently support brain function are considered to be converted to glucose elsewhere in the body before reaching the brain. Similar conclusions have come from studies on isolated cell-containing systems. The rapid metabolism of carbohydrate is also apparent from the high rates of glucose and glycogen depletion during ischemia.[2,3]

The pathways through which carbohydrate is metabolized in the brain are apparently the same as those operating elsewhere in the body, except that no gluconeogenesis has been demonstrated; glucose-6-phosphatase and fructose-1,6-diphosphatase[4] have not been detected in the tissue. Much

1

evidence has accumulated to support the view that, in adult mammalian brain, the contribution of the hexose monophosphate shunt is negligible; i.e., that carbohydrate metabolism is almost exclusively via the glycolytic route.[5,6] The overall rate of glucose utilization in adult mammalian brain *in vivo* is of the order of 20 μmoles/g tissue/hr.[1] All of the individual glycolytic enzymes, as estimated *in vitro*, are capable of rates greatly in excess of that (Table I). It should be noted that such rates, measured under conditions (e.g.,

TABLE I

Rates of Glycolytic Enzymes of the Brain (measured under optimal conditions)

Enzyme	Species	Rate (μmole/g/hr)	Reference
Hexokinase	Mouse	660	8
	Guinea pig	1,350	9
Glucose-phosphate	Mouse	3,300	8
isomerase	Human	3,000	10
Phosphofructokinase	Mouse	550	8
Aldolase	Mouse	310	8
	Human	1,500	10
Glyceraldehyde-3-phosphate	Mouse	3,360	8
dehydrogenase	Rat	1,450	11
Phosphoglycerate kinase	Mouse	10,000	8
	Rat	1,600	12
Phosphoglycerate mutase	Mouse	2,340	8
	Rat	2,000	12
Enolase	Mouse	1,800	8
	Ox	800	13
Pyruvate kinase	Mouse	7,080	8
	Rat	4,500	12
Lactate dehydrogenase	Mouse	3,500	8
	Rat	3,400	12

of pH and substrate concentration) designed to achieve maximum activities, do not necessarily reflect the potential maximum rates likely to be generated under the conditions which might prevail in the immediate cellular environment. However, the individual enzymes must be capable of reacting at greatly elevated rates *in vivo* since the rates of glycolysis and respiration can be increased up to tenfold[1] by stimulation *in vivo* and *in vitro* for brief periods as discussed below. From this arises one of the major features of interest— how the potentially high enzymatic rates are controlled or limited to the level of the normal glycolytic rate.

A further characteristic is the very low carbohydrate store. The brain has relatively little fuel reserve and relies on using very rapidly the glucose brought to it in the bloodstream. The glycogen levels are quite low for a tissue so dependent on its sources of energy and it is possible that there may be little or no intracellular glucose.[7]

Neural tissues are unique in being electrically excitable; one major use of energy is to promote the ion transport mechanisms essential in maintaining physiological function. The measured resting rates of glucose utilization, respiration, and lactate production are considerably increased by electrical stimulation and also by the addition of K^+ salts to the external medium, in experiments *in vitro*.[1] It has therefore been concluded that, since *in situ* the brain exhibits continuous electrical activity, some form of stimulation, particularly by the application of electrical pulses, is necessary for displaying many of its normal activities *in vitro*. The increased rates observed after electrical or K^+ stimulation are considered to be due to increased glycolysis rather than to increased participation of the shunt pathway.[14] Such changes may involve factors causing changes in cell membrane permeability. When a nerve cell fires during the continual electrical activity occurring in the tissue, a portion of the cellular K^+ moves out and Na^+ moves in. During the subsequent recovery process of repolarization, ATP is utilized in transporting these cations back to their previous sites and the tissue is therefore involved in reestablishing electrochemical gradients. Osmotic work is performed at the expense of ATP and the process is of direct relevance to considerations of the mechanisms which regulate the maintenance of ATP levels.

General aspects of brain carbohydrate and energy metabolism have recently been comprehensively reviewed.[7,15-17]

From the described characteristics of the tissue, it follows that any discussion of control mechanisms of carbohydrate metabolism in neural tissues must be in terms of the accessibility of glucose to the tissue, the involvement of ion movements and changes in membrane potentials, and the properties of the enzymes involved in the utilization of glucose and glycogen by glycolysis. The intriguing problems rest not only with the mechanisms by which the potentially high enzymatic rates are suppressed but also by what means this suppression may be rapidly relieved when the situation demands.

The likely relevant factors have been studied at all levels—*in vivo*, by the use of isolated cell-containing preparations, and in cell-free systems where the kinetic properties and inhibition characteristics of individual enzymes have been investigated.

II. ELECTRICAL ACTIVITY AND ION MOVEMENTS

When electrical impulses are applied to respiring cerebral cortex slices, about 6 mμequiv of cations/g of tissue are exchanged per impulse between tissue and medium. Cell firing may proceed at rates up to hundreds of impulses/sec so the rates of cation movement, sustained for brief periods, can be as high as 4–8 μequiv/min/g.[18,19] The increased intracellular Na^+ is extruded during the recovery period as ATP is utilized by cerebral Na^+, K^+-ATPase.[20] The acceleration in carbohydrate utilization caused by either electrical stimulation or by increased external K^+ occurs only if Na^+ is present in the external medium, thus providing further evidence for a relationship to Na^+ transport.[21] Whittam[22,23] has suggested that the cation transport mechanisms, in particular the Na^+, K^+-ATPase, may act as a partial control of respiration in the brain.

In cell-free systems, cations specifically affect certain individual glycolytic enzymes. Na^+ has been shown to inhibit pyruvate kinase[24] which catalyzes the phosphorylation of ADP to form ATP at the expense of phosphoenolpyruvate. K^+ stimulates pyruvate kinase and affects the activity of phosphofructokinase as discussed below.

The effect of glutamate in increasing glycolysis and respiration is also associated with cation movement, with depletion of energy-rich phosphates and with depolarization[25-27] and so provides further indications of correlations between energy metabolism, membrane permeability, and ion transport.

III. GLUCOSE TRANSPORT

Glucose phosphorylation may proceed very rapidly in cerebral tissues; hexokinase (particulate and cytoplasmic)[9] must under normal conditions be proceeding at rates of only a few percent of that of which it is capable. There is no evidence for any accumulation of glucose under these circumstances, which leads to the possibility that inaccessibility of glucose may itself be a limiting factor. The uptake of glucose by the brain has been clearly shown to be a saturable carrier-mediated form of active transport from studies *in vivo* and *in vitro*.[28-32] It is not known whether glucose transport itself is affected by factors which stimulate glucose utilization. It does seem likely, however, that changes which occur in membrane permeability during electrical activity, and in ion dispositions might be also reflected in changes in rates of glucose uptake. So while precise evidence is so far lacking, the possibility that glucose uptake is a control point in carbohydrate metabolism (as it seems to be in heart and muscle)[33,34] cannot be excluded.

IV. GLYCOLYSIS

A. Glycogen Metabolism

Although little is apparently known of the specific enzymes concerned with branching and debranching reactions in glycogen metabolism in the brain (i.e., enzymes analogous to the amylo-1,4 $\longrightarrow$ 1,6-transglucosylase and amylo-1,6-glucosidase of other tissues), the general pathways of glycogen synthesis and breakdown are similar to those operating in liver or muscle.

One important feature common to considerations of control mechanisms in all glycogen-metabolizing tissues is that the pathway of degradation involving phosphorylase is different from the synthetic route via uridine diphosphate glucose.

Under conditions of elevated glucose, glycogen metabolism may be directed more toward synthesis, as glucose has been shown[35] to be a competitive inhibitor of the phosphorylase step in glycogenolysis. Also, any tendency for glucose-6-phosphate to accumulate would be reflected in increased synthesis as the ester activates UDPG-glycogen transglucosylase.[36,37] The most likely point for control of glycogen metabolism would appear to be phosphorylase. Like the enzyme of other tissues, it occurs mainly in an inactive form; formation of the active enzyme involves adrenaline, cyclic AMP, and phosphorylase kinase.[38-41] After removal of the blood supply to the brain, glycogen and glucose are rapidly lost. Over 85% of the glucose disappears within minutes,[2,3] but delays have been noted in the disappearance of glycogen which is suggestive of a control feature in operation. The delay of 1 min noted in the conversion of inactive to active phosphorylase[41] during ischemia does not account entirely for the delay of over 2 min before glycogen starts to disappear during postmortem autolysis. Most of the glucose has been utilized by this stage. Such considerations led to the suggestion[42,43] that the delay in glycogen breakdown may be due not only to the time required for phosphorylase activation, but also to that required for sufficient AMP to accumulate to provide maximum phosphorylase rates. Thus phosphorylase would appear to be involved in controlling rates of glycogen utilization and the transglucosylase (glycogen synthetase) may have a role in the regulation of glycogen synthesis. As mentioned above, the synthetase is activated by glucose-6-phosphate and its optimal rate is the lowest of those studied in the synthetic pathway. It may be the pacemaker step in synthesis. However, it must be noted that the rate (25 μmoles/g/hr)[37] is considerably in excess of the overall rates (1–2 μmoles/g/hr) observed for glycogen synthesis *in vivo*[44] and *in vitro*.[45]

B. Glycolytic Enzymes

Overall glycolytic rates measured *in vitro* are conditioned partly by the concentration of inorganic phosphate in the medium. The rate of lactate formation is proportional to the phosphate concentration up to 15–20 mM.[1] Inorganic phosphate, in addition to being an essential requirement together with ADP for oxidative phosphorylation, is also involved with the activity of certain individual glycolytic enzymes. It is an activator or deinhibitor for phosphofructokinase,[46] a substrate for glyceraldehyde-3-phosphate dehydrogenase, and may be important in relieving hexokinase inhibition by glucose-6-phosphate.[47,48] Cerebral tissues normally contain 3.5–5 μmoles of inorganic phosphate per gram, concentrations below those required for maximal rates of respiration or glycolysis,[1] so the rates are likely to be limited by the phosphate concentration.

Control of rates of individual glycolytic enzymes may also be exerted by limiting amounts of other substrates or coenzymes such as NAD^+ for dehydrogenase reactions and ADP for two kinase reactions (pyruvate kinase and phosphoglycerate kinase). Competition between enzymes in the same sequence requiring the same substrate or coenzyme may also result in limited overall rates.

1. Enzymatic Control Points

When the blood supply to the brain has been cut off, the brain is converted to a "closed system"—no exogenous fuel (glucose or O_2) is available and the tissue rapidly metabolizes, by anaerobic glycolysis, its stores of carbohydrate for the few minutes before these become depleted. Examination[3] of the levels of the substrates of the glycolytic pathway during short time intervals after ischemia showed that within half a minute glucose and the hexose monophosphates (G-6-P and F-6-P) had decreased in concentration while subsequent intermediates had increased. Figure 1 shows the levels of the substrates in anesthetized and nonanesthetized 10-day-old mouse brain 25 sec after removal of the blood supply. Anesthetics have little effect on zero time levels but apparently slow down the rate of postmortem glycolysis.[49] Peak fluxes were considered[3] to occur at 25 sec in anesthetized and 6 sec in nonanesthetized animals. The levels in the latter group are included in Fig. 1.

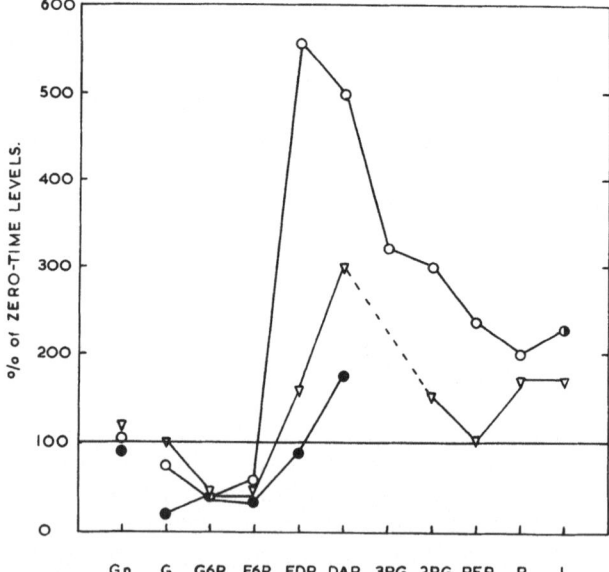

Fig. 1. Effect of ischemia on substrates of glycolysis in the brains of 10-day-old mice. Gn, glycogen; G, glucose; G6P, glucose-6-phosphate; F6P, fructose-6-phosphate; FDP, fructose-1,6-diphosphate; DAP, dihydroxyacetone-phosphate; 3PG, 3-phosphoglycerate; 2PG, 2-phosphoglycerate; PEP, phosphoenolypyruvate; P, pyruvate; L, lactate. ● Unanesthetized mice, 25 sec ischemia. ○ Anesthetized mice, 25 sec ischemia. ▽ Unanesthetized mice 6 sec ischemia. The values were taken from Lowry et al.[3]

Lowry[3] concluded that the control points in glycolysis in the brain must therefore occur before the stage of formation of fructose diphosphate, i.e., at the hexokinase and particularly the phosphofructokinase steps. Kinetic studies on the individual enzymes also provided supporting evidence for control points at hexokinase and phosphofructokinase.[8]

2. Phosphofructokinase

This enzyme undergoes a peculiarly complex series of interrelated inhibitions and deinhibitions involving its substrates, its products, and other compounds in a fashion which emphasizes its role in regulation. The activity is inhibited by an excess of one of its substrates, ATP; at noninhibitory levels of ATP, the Michaelis constants for the two substrates, ATP (0.1 mM) and F-6-P (0.04 mM) were found each to be independent of the other.[46] Free ATP is considerably more inhibitory than Mg-ATP; as Mg^{2+} itself was shown to be slightly inhibitory, the relationship between the concentrations of Mg^{2+} and ATP and the enzymatic activity is a complex one. Citrate, also inhibitory, acts synergistically with ATP[46,50]; F-6-P decreases the inhibition of both ATP and citrate. The inhibition due to ATP is also decreased by inorganic phosphate, K^+, NH_4^+, 5'-AMP and 3',5'-cyclic AMP. These properties have led Lowry to suggest that the enzyme possesses at least 7, and possibly as many as 12, separate sites for substrates, inhibitors, and activators or deinhibitors.[46]

The citrate inhibition of a glycolytic enzyme provides an interesting possibility of regulatory contact between the tricarboxylic cycle and glycolysis.[50] Possibly under normal circumstances phosphofructokinase activity may be regulated by citrate inhibition without the involvement of other factors,[43] but in situations when the rate of ATP utilization exceeds the rate of ATP generation, so that AMP and phosphate levels tend to increase, the changes in metabolite levels would be expected to result in increases in enzyme activity. The NH_4^+ released during electrical activity would also be expected to result in increased phosphofructokinase activity, by deinhibition.

3. Hexokinase

Inhibition of phosphofructokinase causes an accumulation of glucose-6-phosphate which could provide a coupled regulation of hexokinase activity. Glucose-6-phosphate, which inhibits hexokinase by competition with ATP[54], is by itself unlikely to suppress the enzymatic activity to the extent required. Glycolytic rates are only 1–3% of the maximal potential rate of total hexokinase activity and about 5–6% of the potential rate of the cytoplasmic enzyme. Levels of glucose-6-phosphate equivalent to those which occur *in vivo* (about 0.1 mM)[3] produce inhibition of hexokinase measured under optimum conditions from well below 50%[48,52] to over 70%.[53]

Partial relief by inorganic phosphate of the inhibition due to G-6-P has been reported[47,48,53] although one group of investigators did not observe the effect.[8] ADP also inhibits hexokinase competitively, but is less potent in its action than is G-6-P.[53-55] Approximately 90–95% inhibition of the cytoplasmic enzymatic activity is required to bring it down to the level of the general glycolytic rate. The total inhibition due to G-6-P plus ADP, when added in concentrations approximating to those which occur *in vivo*, is unlikely to be sufficient as based on enzymatic rates measured under maximally activating conditions of Mg^{2+} and ATP. However, it must be remembered that the calculations tend to be based on mean levels of enzymes, substrates and inhibitors estimated in extracts from a heterogeneous mixture of cell types. The amounts of glucose available to the enzyme are likely to be limiting, as discussed above, and the level of inhibition of a lowered substrate-limited rate could perhaps be sufficient if the inhibitory products are sufficiently concentrated at the precise sites of phosphorylation. The requisite degree of control may be approached[53a] if the inhibition due to free ATP as well as to G-6-P and ADP are considered in terms of the likely intracellular concentrations of these and of Mg^{2+}

V. HORMONES AND DRUGS

Marked decreases in cerebral glycogen content have been noted as a result of insulin-induced hypoglycemia;[56-58] this is probably due to the hypoglycemia rather than to a direct effect of insulin on the brain. Any direct involvement of insulin in brain function *in vivo* is difficult to assess as the extent of penetration into the brain of the hormone, especially if it is derived from exogenous sources, appears to be very limited.[59,60] Studies *in vitro* on the effect of insulin on glucose transport into isolated samples of a variety of neural tissues have yielded conflicting results. (See Ref. 60 for a discussion on this topic.) *In vivo*, no effect of insulin on glucose transport into dog brain could be demonstrated[30] although more recently it has been suggested[61] that human brain may be sensitive.

While adrenaline stimulates the adenyl cyclase activity of brain extracts, particularly of the cerebellum,[40] no effect of the catecholamine on glycogen metabolism of the brain under normal conditions has been clearly demonstrated.[56,62] However, stress,[63] in which increased adrenaline levels might be expected, causes lowered glycogen contents in the median eminence and anterior pituitary of the rat, and the apparent activation of mouse brain phosphorylase due to ethanol was concluded to be due to increased permeability of the brain to adrenaline.[64] Rates of brain glycogen catabolism have also been shown to be stimulated by methamphetamine[65] and convulsant drugs.[58,66,67]

Certain hormonal effects have been implicated in brain carbohydrate metabolism at the hexokinase step. Attempts (see Ref. 68) to confirm an earlier report[69] that pituitary extracts affect the hexokinase reaction, and that the effect was modified by insulin, have proved inconclusive. However, the studies did provide evidence that erythrocyte extracts contain an activator and plasma constituents an inhibitor of brain hexokinase.[68,70] The inhibition due to the plasma factor was modified by insulin.[68] Whether the observed effects were due to hormonal constituents could not be decided. No direct effect on brain hexokinase activity could be elicited by the addition of adrenocortical extracts or purified hormones even though the inhibitory activity appeared in plasma under conditions which correlated with those known to stimulate the activity of the pituitary-adrenocortical system.

The possibility that hormones may play a direct role in the regulation of carbohydrate metabolism in the brain is a fascinating one; at present insufficient evidence is available and any evaluation of the possibility must await the outcome of further study.

VI. CONCLUSIONS

Although the relationships cannot yet be defined in precise quantitative terms, regulation of carbohydrate metabolism in the brain may be envisaged in terms of the levels of inorganic ions, phosphate, and the phosphate esters of adenosine and the hexoses in a complex interrelated fashion with feedback control involving glycolysis, the tricarboxylic acid cycle, and the electron transport system.

As pointed out by Racker,[48] the involvement of orthophosphate and adenosine phosphates in the activities of the glycolytic enzymes together with the competition between the pathways of glycolysis and oxidative phosphorylation for phosphate and ADP, can provide a basis for an understanding of the regulation of carbohydrate metabolism as manifested by the Pasteur effect.

VII. REFERENCES

1. H. McIlwain, *Biochemistry and the Central Nervous System*, 3rd ed., Churchill, London (1966).
2. S. E. Kerr and M. Ghantus, The carbohydrate metabolism of brain. III: On the origin of lactic acid, *J. Biol. Chem.* **117**:217–225 (1937).
3. O. H. Lowry, J. V. Passonneau, F. X. Hasselberger, and D. W. Schulz, Effect of ischemia on known substrates and cofactors of the glycolytic pathway in brain, *J. Biol. Chem.* **239**:18–30 (1964).
4. H. A. Krebs and M. Woodford, Fructose 1,6-diphosphate in striated muscle, *Biochem. J.* **94**:436–445 (1965).
5. W. Sacks, Cerebral metabolism of isotopic glucose in normal human subjects, *J. Appl. Physiol.* **10**:37–44 (1957).
6. F. C. G. Hoskin, Effect of inhibitors on the metabolism of specifically labelled glucose by brain, *Biochim. Biophys. Acta* **40**:309–313 (1960).
7. H. S. Bachelard and H. McIlwain, in *Comprehensive Biochemistry* (M. Florkin and E. H. Stotz, eds.), Vol. 17, Chapter 6, Elsevier, Amsterdam., pps. 191–218 (1969).
8. O. H. Lowry and J. V. Passonneau, The relationship between substrates and enzymes of glycolysis in brain, *J. Biol. Chem.* **239**:31–42 (1964).

9. H. S. Bachelard, The sub-cellular distribution and properties of hexokinases in the guinea-pig cerebral cortex, *Biochem. J.* **104**:286–292 (1967).

10. N. Robinson and B. M. Phillips, Glycolytic enzymes in human brain, *Biochem. J.* **92**:254–259 (1964).

11. R. H. Laatsch, Glycerol phosphate dehydrogenase activity of developing rat central nervous system, *J. Neurochem.* **9**:487–492 (1962).

12. R. von Fellenberg, H. Eppenberger, R. Richterich, and H. Aebi, The glycolytic enzymes from the liver, kidney, skeletal muscle, heart muscle and brain of the rat and mouse, *Biochem. Z.* **336**:334–350 (1962).

13. T. Wood, The purification of enolase from cerebral tissue, *Biochem. J.* **91**:453–460 (1964).

14. J. J. O'Neill, S. H. Simon, and W. W. Shreeve, Alternate glycolytic pathways in brain, A comparison between the action of artificial electron acceptors and electrical stimulation, *J. Neurochem.* **12**:797–802 (1965).

15. H. S. Bachelard, in *Handbook of Neurochemistry* (A. Lajtha, ed.), Vol. 1, Part 1, Chapter 2, Plenum Press, New York, pp. 25–31 (1969).

16. R. Balázs, in *Handbook of Neurochemistry* (A. Lajtha, ed.), Vol. 3, Chapter 1, Plenum Press, New York (1969).

17. R. V. Coxon, in *Handbook of Neurochemistry* (A. Lajtha, ed.), Vol. 3, Chapter 2, Plenum Press, New York (1969).

18. H. S. Bachelard, W. J. Campbell, and H. McIlwain, The sodium and other ions of mammalian cerebral tissues, maintained and electrically stimulated *in vitro*, *Biochem. J.* **84**:225–232 (1962).

19. H. McIlwain, Chemical Exploration of the Brain: A Study of Cerebral Excitability and Ion Movement, Elsevier, Amsterdam (1963).

20. A. Schwartz, H. S. Bachelard, and H. McIlwain, The sodium-stimulated adenosine-triphosphatase activity and other properties of cerebral microsomal fractions and sub-fractions, *Biochem. J.* **84**:626–637 (1962).

21. M. B. R. Gore and H. McIlwain, Effects of some inorganic salts on the metabolic response of sections of mammalian cerebral cortex to electrical stimulation, *J. Physiol.* **117**:471–483 (1952).

22. R. Whittam, Active cation transport as a pace-maker of respiration, *Nature (London)* **191**:603–604 (1961).

23. R. Whittam, The dependence of the respiration of brain cortex on active cation transport, *Biochem. J.* **82**:205–212 (1962).

24. M. F. Utter, The role of carbon dioxide fixation in carbohydrate utilization and synthesis, *Ann. N.Y. Acad. Sci.* **72**:451–461 (1959).

25. R. J. Woodman and H. McIlwain, Glutamic acid, other amino acids and related compounds as substrates for cerebral tissues: their effect on tissue phosphates, *Biochem. J.* **81**:83–93 (1961).

26. H. F. Bradford and H. McIlwain, Ionic basis for the depolarization of cerebral tissues by excitatory acidic amino acids, *J. Neurochem.* **13**:1163–1177 (1963).

27. H. H. Hillman and H. McIlwain, Membrane potentials in mammalian cerebral tissues *in vitro*: dependence on ionic environment, *J· Physiol.* **157**:263–278 (1961).

28. P. Joanny, J. Corriol, A. Kleinzeller, and H. H. Hillman, Transport of monosaccharides into slices of guinea-pig brain cortex, *Abstr. 1st Intern. Meeting, I.S.N.*, p. 110, Strasbourg (1967).

29. R. A. Fishman, Carrier transport of glucose between blood and cerebrospinal fluid, *Am. J. Physiol.* **206**:836–844 (1964).

30. C. Crone, Facilitated transfer of glucose from blood into brain tissue, *J. Physiol.* **181**:103–113 (1965).

31. J. C. Gilbert, Mechanism of sugar transport in brain slices, *Nature (London)* **205**:87–88 (1965).

32. P. G. Le Fevre and A. A. Peters, Evidence of mediated transfer of monosaccharides from blood to brain in rodents, *J. Neurochem.* **13**:35–46 (1966).
33. P. J. Randle and G. H. Smith, Regulation of glucose uptake by muscle. 2. The effects of insulin, anaerobiosis and cell poisons on the penetration of isolated rat diaphragm by sugars, *Biochem. J.* **70**:501–508 (1958).
34. H. E. Morgan, M. J. Henderson, D. M. Regen, and C. R. Park, Regulation of glucose uptake in muscle. I. The effects of insulin and anoxia on glucose transport and phosphorylation in the isolated, perfused heart of normal rats, *J. Biol. Chem.* **236**:253–261 (1961).
35. G. T. Cori and C. F. Cori, The kinetics of the enzymatic synthesis of glycogen from glucose-1-phosphate, *J. Biol. Chem.* **135**:733–756 (1940).
36. B. M. Breckenridge and E. J. Crawford, Glycogen synthesis from uridine diphosphate glucose in brain, *J. Biol. Chem.* **235**:3054–3057 (1960).
37. D. K. Basu and B. K. Bachhawat, Purification of uridine diphosphoglucose-glycogen transglucosylase from sheep brain, *Biochim. Biophys. Acta* **50**:123–128 (1961).
38. G. T. Cori, S. P. Colowick, and C. F. Cori, The formation of glucose-1-phosphoric acid in extracts of mammalian tissues and of yeast, *J. Biol. Chem.* **123**:375–380 (1938).
39. T. W. Rall and E. W. Sutherland, Adenyl cyclase. II. The enzymatically catalyzed formation of 3′5′-phosphate and inorganic pyrophosphate from adenosine triphosphate, *J. Biol. Chem.* **237**:1228–1232 (1962).
40. L. M. Klainer, Y.-M. Chi, S. L. Freidberg, T. W. Rall, and E. W. Sutherland, Adenyl cyclase. IV. The effects of neurohormones on the formation of adenosine 3′5′-phosphate by preparations from brain and other tissues, *J. Biol. Chem.* **237**:1239–1243 (1962).
41. B. M. Breckenridge and J. H. Norman, Glycogen phosphorylase in brain, *J. Neurochem.* **9**:383–392 (1962).
42. O. H. Lowry, D. W. Schulz, and J. V. Passonneau, The kinetics of glycogen phosphorylases from brain and muscle, *J. Biol. Chem.* **242**:271–280 (1967).
43. O. H. Lowry, in *Nerve as a Tissue* (K. Rodahl and B. Issekutz, eds.), pp. 163–174, Harper & Row, New York (1966).
44. J. Krivanek, Changes of brain glycogen in the spreading EEG-depression of LEAO, *J. Neurochem.* **2**:337–343 (1958).
45. F. N. Le Baron, The resynthesis of glycogen by guinea pig cerebral cortex slices, *Biochem. J.* **61**:80–85 (1955).
46. O. H. Lowry and J. V. Passonneau, Kinetic evidence for multiple binding sites on phosphofructokinase, *J. Biol. Chem.* **241**:2268–2279 (1966).
47. H. Tiedemann and J. Born, On the mechanism of the Pasteur reaction. The influence of phosphate ions on the activity of the structurally bound hexokinase, *Z. Naturforsch.* **14B**:477–478 (1959).
48. K. Uyeda and E. Racker, Regulatory mechanisms in carbohydrate metabolism. VII. Hexokinase and phosphofructokinase, *J. Biol. Chem.* **240**:4682–4688 (1965).
49. F. N. Minard and R. V. Davis, Effect of chlorpromazine, ether and phenobarbital on the active phosphate level of rat brain: an improved extraction technique for acid-soluble phosphates, *Nature (London)* **193**:277–278 (1962).
50. J. V. Passonneau and O. H. Lowry, P-fructokinase and the control of the citric acid cycle, *Biochem. Biophys. Res. Commun.* **13**:372–379 (1963).
51. N. D. Goldberg, J. V. Passonneau, and O. H. Lowry, Effects of changes in brain metabolism on the levels of citric acid cycle intermediates, *J. Biol. Chem.* **241**:3997–4003 (1966).
52. R. K. Crane and A. Sols, The non-competitive inhibition of brain hexokinase by glucose-6-phosphate and related compounds, *J. Biol. Chem.* **210**:597–606 (1954).
53. E. A. Newsholme, F. S. Rolleston, and K. Taylor, Factors affecting the glucose-6-phosphate inhibition of the hexokinase from cerebral cortex tissue of the guinea pig, *Biochem. J.* **106**:193–201 (1968).

53a H. S. Bachelard and P. S. G. Goldfarb, Adenine nucleotides and magnesium ions in relation to control of mammalian cerebral-cortex hexokinase. *Biochem. J.* **112**:579–586 (1969).

54. H. J. Fromm and V. Zewe, Kinetic studies of the brain hexokinase reaction, *J. Biol. Chem.* **237**:1661–1667 (1962).

55. A. Sols and R. K. Crane, The inhibition of brain hexokinase by adenosine diphosphate and sulfhydryl reagents, *J. Biol. Chem.* **206**:925–936 (1954).

56. S. E. Kerr and M. Ghantus, The carbohydrate metabolism of brain. II : The effect of varying the carbohydrate and insulin supply on the glycogen, free sugar and lactic acid in mammalian brain, *J. Biol. Chem.* **116**:9–20 (1936).

57. M. R. A. Chance and D. C. Yaxley, Central nervous function and changes in brain metabolite concentration. I : Glycogen and lactate in convulsing mice, *J. Exptl. Biol.* **27**:311–323 (1950).

58. S. H. Carter and W. E. Stone, Effects of convulsants on brain glycogen in the mouse, *J. Neurochem.* **7**:16–19 (1961).

59. N. Haugaard, M. Vaughan, E. S. Haugaard, and W. C. Stadie, Studies of radioactive injected labelled insulin, *J. Biol. Chem.* **208**:549–563 (1954).

60. O. J. Rafaelson, Studies on a direct effect of insulin on the central nervous system : a review, *Metabolism* **10**:99–114 (1961).

61. W. J. H. Butterfield, M. E. Abrams, R. A. Sells, G. Sterky, and M. J. Whichelow, Insulin sensitivity of the human brain, *Lancet* (i) 557–560 (1966).

62. R. V. Coxon, E. C. Gordon-Smith, and J. R. Henderson, The incorporation of isotopic carbon (^{14}C) into the cerebral glycogen of rabbits, *Biochem. J.* **97**: 776–781 (1965).

63. D. Jacobowitz and B. H. Marks, Effect of stress on glycogen and phosphorylase in the rat anterior pituitary, *Endocrinology* **75**:86–88 (1964).

64. C.-J. Estler and H. P. T. Ammon, Phosphorylase activity and glycogen content of the brain under the influence of ethanol and adrenaline, *J. Neurochem.* **12**:871–876 (1965).

65. C.-J. Estler and H. P. T. Ammon, The influence of propanolol on the methamphetamine-induced changes of cerebral function and metabolism, *J. Neurochem.* **14**:799–805 (1967).

66. F. N. Minard, C. H. Kang, and I. K. Mushahwar, The effect of periodic convulsions induced by 1,1-dimethylhydrazine on the glycogen of rat brain, *J. Neurochem.* **12**:279–286 (1965).

67. J. R. Klein and N. S. Olsen, Effect of convulsive activity upon the concentration of brain glucose, glycogen, lactate and phosphates, *J. Biol. Chem.* **167**:747–756 (1947).

68. H. Weil-Malherbe and D. Bone, Activators and inhibitors of hexokinase in human blood, *J. Mental Sci.* **97**:635–662 (1951).

69. S. P. Colowick, G. T. Cori, and M. W. Slein, The effect of adrenal cortex and anterior pituitary extracts and insulin on the hexokinase reaction, *J. Biol. Chem.* **168**:583–596 (1947).

70. J. Stern, Inhibitors and activators of brain hexokinase, *Biochem. J.* **58**:536–542 (1954).

71. O. E. Owen, A. P. Morgan, H. G. Kemp, J. M. Sullivan, M. G. Herrera, and C. F. Cahill, Brain metabolism during fasting, *J. Clin. Invest.* **46**:1589 (1967).

72. A. L. Smith, H. S. Satterthwaite, and L. Sokoloff, Induction of brain D(-)β-hydroxybutyrate dehydrogenase activity by fasting, *Science* **163**:79 (1969).

Chapter 2

NUCLEOSIDE TRIPHOSPHATE PHOSPHOHYDROLASES

George J. Siegel*

Departments of Neurology and Physiology
Mount Sinai School of Medicine
of The City University of New York
New York, New York

and

R. Wayne Albers

Laboratory of Neurochemistry
National Institute of Neurological
Diseases and Stroke
National Institutes of Health
Bethesda, Maryland

I. INTRODUCTION

The pivotal role of ATP in energy metabolism is manifest in the frequency with which biological energy transducers are represented at the enzyme level as ATPases. Although some endergonic work processes may be coupled directly to electron transport or to electrochemical fluxes, energy transfer through ATP is apparently the most common mode.

This review is principally concerned with the mechanisms which couple ATP hydrolysis to the performance of mechanical or electro-osmotic work. It does not attempt to be comprehensive: citations are often confined to other reviews and to papers which include literature surveys. No attempt is made to limit the discussion to enzymes of the nervous system since the number of such ATPases is increasing as their characteristics become better understood. We emphasize data which illustrate experimental attacks on the problems relating ATPases to the molecular mechanisms of various processes. This is done with the thought that approaches which prove useful in one case may often suggest tests of analogous questions in others, even though the mechanisms of the individual enzymes may be quite diverse.

* Portions of this work were performed while in the Laboratory of Neurochemistry, National Institute of Neurological Diseases and Stroke, National Institutes of Health, Bethesda, Maryland.

A. Classification

The ATPases have been classified according to their association with subcellular structures and also according to their substrate and activator specificities. Many ATPases are so susceptible to modification that they do not conform to simple classification based on specificities. Thus myosin ATPase is often described as a Ca-ATPase,[1] although both EDTA and Mg^{2+} activate this enzyme and influence its specificity for nucleotides.[2] As the relationships between ATPases and function become established, these provide the more rational basis for classification.

B. Nonfunctional and Pseudo-ATPases

There may be instances of ATPases which function in a regulatory capacity, simply to hydrolyze excess ATP,[3] although this has not been demonstrated. ATPase may arise artifactually, e.g., through the combined action of an ATP-phosphotransferase and a phosphatase. However, most phosphotransferases have a remarkable ability to exclude water from the transferase site and therefore do not, in themselves, exhibit such activity. Thus the ratio of ATPase to transferase activities in crystalline hexokinase is 10^{-6}. [4]

A pseudo-ATPase system consisting of diglyceride kinase and phosphatidic acid phosphatase may have a physiological role in the mechanism responsible for the acetylcholine stimulation of phosphate turnover in microsomal phospholipids.[5] The rapid turnover of phosphoprotein phosphate in electrically stimulated brain slices[6] seems to be an example of the combined action of a phosphoprotein kinase and a phosphatase.[7]

Several synthetases can catalyze ATP hydrolysis in the presence of a second substrate. Thus carbamyl phosphate synthetase in the presence of bicarbonate catalyzes the hydrolysis at 10% of the rate of carbamyl phosphate synthesis :[8]

$$\text{enzyme} + \text{ATP} + HCO_3^- \leftrightarrow \text{enzyme-}(CO_3PO_3H) + \text{ADP}$$

$$\rightarrow \text{enzyme} + CO_2 + PO_4^{3-} + \text{ADP}$$

In contrast, the succinate-dependent GTPase activity of succinyl thiokinase is only 0.04% as rapid as the synthesis of succinyl CoA in the complete system.[9]

II. ATPase ACTIVITIES RELATED TO CONTRACTILE PROTEINS

Much of the prime data concerning the conversion of biochemical energy into useful work have come from studies of muscle. The hypothesis that the mechanical force of contraction derives from the interaction and hydrolysis of ATP with contractile proteins was proposed by Engelhardt and co-workers in 1939[10,11] although experiments which substantiate this directly were devised only recently.[12] The papers which introduced the concept of "high energy" chemical bonds were contemporary with Engelhardt's.[13,14] Lipmann stated the problem of joining these two hypotheses: "No decision can be taken from where $\sim$ph is taken off to operate contraction. Hydrolysis through pyrophosphatase (ATPase), in fact, completely dissipates the potential energy accumulated in the energy-rich bond.... In cyclic processes of utilization, we think that to know how inorganic phosphate is recovered would frequently imply knowing the mechanism because complete utilization of the potential energy in $\sim$ph converts it necessarily into inorganic phosphate.... It is an obvious deduction that the energy-rich phosphate should link up to the contracting protein. There is, however, no experimental indication even to encourage such a view ...".[13]

Some confusion may arise from the multiplicity of ATPases associated with muscle. In addition to the activities deriving from the contractile proteins, there are distinct ATPases associated with the plasma membrane, the sarcoplasmic reticulum, and mitochondria.

The biochemistry of the contractile proteins is frequently reviewed[15-18] and only those features which are relevant to the more general problem of energy transduction will be outlined here.

A. Myosin

1. General Properties of Myosin ATPase

Myosin (myosin A, myosin L) is quantitatively the major component of muscle. Native myosin is soluble only at high ionic strengths; therefore most studies are performed in 0.5 M KCl or greater. Under such conditions the activity is Ca^{2+} dependent and inhibited by Mg^{2+}. Although myosin Mg-ATPase is low relative to the Ca-ATPase activity, there are indications that the former may be more closely related to the contractile mechanism.[17] The properties of the enzyme are also modified by ionic strength and by numerous reagents. In an attempt to rationalize these effects, it has been noted that a number of agents alter the pH activity relationship and stimulate the Mg-ATPase activity in a manner similar to the effect of actin. This has been discussed in terms of a conformational transition of myosin between "native" and "activated" states.[19,20]

There is some uncertainty about the weight and substructure of the molecule.[17] Recent measurements indicate a weight somewhat less than 500,000.[21] However, extensive studies by Harrington et al. were interpreted in terms of a super coil of three 200,000 mol. wt. subunits.[22] Several investigators have now reported the association of a low molecular weight component with the 200,000 mol. wt. subunit.[23-25]

Arguments against the three-stranded model are based on evidence that there exists only one site per 4–500,000 mol. wt. for ATP,[26] Mg^{2+},[27] and phosphorylation.[28]

2. Relation of Myosin ATPase to Contraction

There is a close correlation in a wide variety of muscle types between speed of contraction and myosin ATPase activity.[29] This holds for the ATPase activated by Ca^{2+}, or actin $+ Mg.^{2+}$ In the latter case, the source of actin is not a factor. Myosins from red (slow) and white (fast) muscles of the same species have different turnover numbers,[30] different activation responses to sulfhydryl reagents,[31] and different pH stabilities.[32]

3. Studies on the Mechanism of the Myosin ATPase Reaction

Many hypothetical mechanisms for this reaction have been advanced; most of these are predicated on considerations of the contraction mechanism.

A predominant issue in studies of ATPase mechanisms is the manner in which free energy is transferred from ATP to the energy transducer, in this case, the contractile proteins. If this transfer is to be efficient, the ΔF of reaction 1 must be small.

(1) ATP + enzyme ↔ activated enzyme + products

This requirement remains valid whatever phase of the contraction process may be synchronous with the reaction, since by definition, the activation step does not produce work.

4. Proposals Involving Energy Transfer Through the Michaelis Complex

Activation could occur through the formation of a noncovalent complex of ATP with myosin:

(2) ATP + enzyme ↔ enzyme(ATP), or equivalently

(3) ATP + enzyme(ADP) ↔ ADP + enzyme(ATP)

In these models, the potential energy transfer is usually ascribed to electrostatic interactions. Morales proposed that interaction with ATP and a polycationic peptide chain might produce contraction during complex formation.[33]

In Davies' model,[34] the nucleotide contributes charge which overrides other forces and produces extension of a peptide chain. In either case, hydrolysis of ATP is assumed to allow the products to dissociate and the protein to return to its "relaxed" configuration. Relevant thermodynamic data concerning the binding of ATP to myosin are not available. A large negative enthalpy associated with a small ΔF of binding would support these models.

5. Evidence for the Phosphorylation of Myosin

In many reactions of intermediary metabolism, energy transfer accompanies the transfer of a phosphoryl group from ATP to an acceptor molecule, e.g., the ATP-creatine transphosphorylation:

(4) MgATP + creatine $\leftrightarrow$ MgADP + creatine phosphate, K ca. 10

Since, by the Engelhardt hypothesis, enzyme and energy acceptor are identical, the analogous activation of myosin would occur through the formation of phosphoryl myosin. Attempts to isolate myosin-^{32}P after incubation of myosin with AT^{32}P have failed.[35,36]

Frequently the demonstration of an ATP-ADP exchange is considered to be evidence of an intermediary phosphoryl enzyme:

(5) ATP + enzyme $\leftrightarrow$ ADP + phosphoryl enzyme

Myofibrils and actomyosin are reported to catalyze this exchange. However, myosin can be extracted from the myofibrils without reducing the exchange activity. Purified myosin A and some actomyosin preparations do not possess exchange activity.[37]

Less direct evidence has arisen from measurements of the incorporation of ^{18}0 from water into phosphate during ATP hydrolysis. Phosphate produced from ATP by the myosin Ca-ATPase reaction contains one equilibrated oxygen per phosphate:[38]

$$(6)\ ATP + H_2{}^{18}O \rightarrow ADP + {}^-O-\overset{\displaystyle O}{\overset{\displaystyle \|}{\underset{\displaystyle |}{P}}}-O^- + H^+$$
$$^{18}OH$$

This amount of incorporation demonstrates that O—P cleavage occurs exclusively, but does not bear on the question of enzyme intermediates since either reaction 7 or reactions 8 + 9 would produce the same equilibration:

(7) enzyme(ATP) + H$_2$^{18}O $\rightarrow$ enzyme + ADP + H^{18}OPO$_3$$^{2-}$ + H$^+$;

(8) enzyme(ATP) $\leftrightarrow$ P-enzyme(ADP) + H$^+$,

(9) P-enzyme(ADP) + H$_2$^{18}O $\rightarrow$ enzyme + ADP + H^{18}OPO$_3$$^{2-}$

In contrast to the myosin Ca-ATPase reaction, the reaction catalyzed by myosin Mg-ATPase equilibrates more than one ^{18}O into the phosphate cleaved from ATP. In this case, there is no isotope incorporation into ADP and relatively little into exogenous phosphate.[39] Thus on the average, more than one O—P bond cleavage occurs per mole of ATP hydrolyzed. The additional incoproration must occur when the phosphate is at a chemical potential intermediate between ATP and orthophosphate or separated from these by high activation energy barriers. One mechanism for the isotope exchange could involve oxygen exchange into the enzyme followed by formation of covalent enzyme-^{18}O—P and enzyme—O bond cleavage. Another possibility is the enzymatic catalysis of a reversible dehydration to the metaphosphate level:

(10) enzyme (PO$_4$H)$^{2-}$ $\leftrightarrow$ enzyme (PO$_3$)$^-$ + OH$^-$, which need not involve covalent bonds.

A line of evidence implicating phosphoryl myosin has been developed from the observation of Weber and Hasselbach that myosin and actomyosin ATPases are both characterized by an initial rapid burst of phosphate release.[40-42] The extent of this initial burst is dependent upon the type of enzyme preparation and the conditions of assay. Tonomura[43] considers that an anionic enzyme site is phosphorylated and that ADP is also bound in the initial phase to the extent of one mole per mole myosin:

(11) enzyme + ATP $\leftrightarrow$ enzyme(ATP) $\to$ P-enzyme(ADP) $\xrightarrow{\text{denaturation}}$ protein + phosphate + ADP

Sartorelli *et al.* have confirmed the quantitative relationship between initial burst phosphate and myosin protein.[36] When the $H_2{}^{18}O$ is added with the denaturant, essentially none of the ^{18}O equilibrates into the phosphate released during the initial phase. However, if the $H_2{}^{18}O$ is present throughout the initial phase, a large excess of ^{18}O appears in the phosphate. This seems to demonstrate that (a) an enzyme-O—P linkage is formed, (b) enzyme-O cleavage occurs during the acid denaturation, and (c) a P—O bond is reversibly cleaved during the lifetime of the enzyme-P. Alternatively, no covalent bond may be formed between phosphate and myosin: the initial burst phosphate is sequestered in some manner before denaturation and the ^{18}O exchange proceeds by reaction 10.

6. Evidence that "Phosphoryl Myosin" may be an Acyl Phosphate

Equating initial burst phosphate with phosphoryl myosin leads to the requirement that no hydrogen ion be released during the enzyme phosphorylation. This suggests that an anionic residue such as a carboxyl group may be the phosphate acceptor.[43] Thus at $pH > 8$,

(12) enzyme-COO^{-1} + ATP^{-4} $\to$ enzyme-$COOPO_3{}^{-2}$ + ADP^{-3}

Tonomura *et al.*[40] concluded that the pH stability of the intermediate is similar to that of acetyl phosphate. As mentioned previously, the ^{18}O data are consistent with an enzyme-O—P linkage and appear to rule out N—P, S—P, etc.[36]

B. Actin

1. Interaction of ATP with Actin

The actin monomer (G-actin) has a molecular weight of about 60,000. At low ionic strength, G-actin combines reversibly with ATP and Ca^{2+}. At higher ionic strengths the bound ATP is hydrolyzed and a helical polymer is formed (F-actin) which contains more tightly bound ADP.[44] Thus with each cycle of polymerization and depolymerization a stoichiometric amount of ATP is hydrolyzed:

(13) G-actin (ATP) + H_2O $\to$ F-actin (ADP) + phosphate;

(14) F-actin (ADP) + ATP → G-actin (ATP) + ADP

Presumably a similar cycle is responsible for the hydrolysis of ATP catalyzed by F-actin which is observed when it is subjected to sonic vibration.[45]

2. Actomyosin ATPase

Actomyosin (myosin B) is formed by the combination of one or two molecules of actin with one molecule of myosin. The actin combining site is near the myosin ATPase site. The combination requires 10^{-7} M Ca^{2+}. At low ionic strength, actomyosin displays nearly equal Mg- and Ca-ATPase activities; the molar specific activities of actomyosin ATPase and myosin Ca-ATPase are similar.

When excess ATP is added to actomyosin at sufficiently low ionic strength, a gel contraction phenomenon occurs which is termed "super precipitation" or "syneresis." Szent-Györgyi and Prior[44] find that a large fraction of bound actomyosin ADP exchanges with medium ATP during syneresis.

Tonomura et al. consider that the combination of actin with myosin accelerates the rate of decomposition of "phosphoryl myosin."[28] They have investigated the proposition that the bound ADP of actin might act as a phosphoryl acceptor and concluded that it does not.[46]

After the onset of ATP hydrolysis by myofibrils and before the ^{18}O exchange begins, Benson et al. have detected a lag period of 3 to 4 min.[39] After this period the relative rates of ATP hydrolysis and ^{18}O exchange remain constant. They suggest that this may reflect an ATP-induced conformational change in myofibrillar structure.

3. Tropomyosin and Troponin

When the bound Ca^{2+} of myofibrils or actomyosin is removed by chelation with EGTA, the Mg-ATPase is inhibited. The Ca^{2+} requirement and the EGTA sensitivity of the reaction disappear when the preparations are extracted at very low ionic strength. The extraction removes two accessory proteins: tropomyosin and troponin.[17] Tropomyosin probably acts chiefly to bind troponin to actin.[47] Troponin in turn binds 5 moles of Ca^{2+} per 100,000 grams.[48] In the absence of Ca^{2+}, i.e., in the presence of EGTA, troponin inhibits the Mg-ATPase of actomyosin. Removal of tropomyosin from actomyosin increases the Ca-ATPase activity. However, tropomyosin does not inhibit the Ca-ATPase activity of myosin A[48] Troponin has been resolved into two factors.[49]

4. The Relationship of ATP Hydrolysis to Muscle Contraction

Although a comprehensive discussion of the physiology of the contraction process is outside the scope of this review, some recent attempts to correlate ATP consumption with heat production bear on the molecular events of a single contraction cycle and the chemical basis of energy transfer.

About 0.2 μmole of ATP is hydrolyzed per gram of wet muscle during the rising phase of a single isotonic contraction of frog sartorius muscle.[50] Assuming that myosin constitutes 5 % of muscle protein, its molar concentration is 10^{-4} or 0.1 μmole per gram wet muscle. Despite this rather close correspondence, one cannot conclude that a single contraction of this muscle corresponds to a single cycle of the total actomyosin ATPase. The fraction of the total contractile protein activated by a single "maximal" twitch is not known. Moreover, part of the ATP consumption must arise from the Ca^{2+} pump, even during the rising phase. The amount of ATP consumed by the Ca^{2+} pump is nonlinearly related to the rate of activation since less Ca^{2+} is released per activation at higher frequencies.[50]

A body of morphological and physical studies have been interpreted to support a "sliding filament" theory of contraction.[51] Actin and myosin filaments appear to interdigitate and produce contraction by the interaction of myosin crossbridges. Other interpretations have been made in terms of contractions of the individual fibrils.[33] Whatever the model, it is evident that free energy is transferred from ATP to the proteins which in turn produce a force parallel to the fibrils.

A more detailed concept requires assumptions about the locus of the effective ATPase activity within the fibrillar structure. The properties of actomyosin ATPase and myofibrillar ATPase are similar and in most respects consistent with the proposition that they are modifications of the intrinsic ATPase of myosin. Even so, the possible relevance of ATPase activity associated with a cyclic transformation of actin, and hence primarily an "actin-ATPase," cannot be neglected.[44]

The sliding filament model proposes that multiple crossbridges are formed and broken during a contraction. The number of contacts made by a filament subunit would be inversely proportional to the distance from the margin of the overlap area for those units outside the initial area of overlap. The number of contacts at any instant is presumably much less than the number of myosin bridges because the periodicity of actin filaments is slightly less than that of the myosin filaments; if exact registry is necessary for an effective interaction, as few as 5 % of the units may be in effective contact at a given instant. Huxley has suggested that the range of movement associated with a single interaction might be about 8 Å.[51]

5. Other ATPase Activities Associated with Mobility

Cilia and flagellae are characterized by a remarkable uniformity of internal structure whether associated with protozoan or mammalian cells. This structure consists of eleven fibers, two located centrally and nine peripherally within the cilia. The protein of the outer fibers has been isolated from *Tetrahymena pyriformis* and has similarities to actin. Another protein, dynein, has ATPase activity and interacts with the outer fibers. The specific activity of dynein ATPase is comparable to that of myosin ATPase.[52] However, the basic subunit appears to be a globular protein of 10^5 mol. wt.[53]

From studies of the kinetics of ATP activation of the beat frequency of sea urchin spermatozoa, Brokaw has estimated that one molecule of ATP is hydrolyzed per dynein molecule per beat cycle.[54]

Proteins having some of the characteristics of actomyosin have been isolated from diverse cell types: slime mold, sea urchin eggs, fibroblasts. These proteins often interact with ATP and exhibit some ATPase activity, in most cases of the order of milliunits per milligram.[55]

III. ATPase ACTIVITY OF MITOCHONDRIA

Intact mitochondria have little ATPase activity. Procedures which uncouple phosphorylation stimulate ATP hydrolysis. The inner membrane system of mitochondria is the site of the principal ATPase activity as well as of oxidative phosphorylation. The solubilized ATPase prepared from these membranes consists of 80–100 Å diameter particles (F_1 coupling factor) which appear identical with the particles seen attached to the inner membranes in negatively stained electron micrographs.[56]

The F_1 ATPase has a molecular weight of 284,000, but at low temperatures it dissociates into inactive 26,000 mol. wt. subunits.[57] The solubilized enzyme differs from the activity of uncoupled mitochondria in several respects. It is not inhibited by oligomycin while the dinitrophenol activation and nucleotide specificity are less. The association of the F_1 particles with the inner membrane[56] and an additional protein factor[58] are required for oligomycin sensitivity. The higher nucleotide specificity of mitochondria is evidently imposed by a carrier-mediated exchange of adenine nucleotides across the inner mitochondrial membrane (the atractyloside-sensitive factor).[59]

Mitochondria also catalyze ATP-ADP and ATP-Pi exchange reactions.[60] The ATP-ADP exchange enzyme of Colomb et al.[61] does not catalyze Pi → ATP exchange. This enzyme has a number of characteristics similar to the phosphoryl transferase isolated from mitochondria by Beyer.[62] The Beyer protein can also accept phosphate generated by oxidative phosphorylation of submitochondrial particles. However, it is not clear from kinetic studies that these enzymes function directly in oxidative phosphorylation.[62]

Boyer et al. conclude that phosphorylated intermediates which directly involve electron carriers are inconsistent with the small effects of the respiratory state on the rate of phosphate exchange reactions.[60] The high energy intermediate generated by electron transport is, in their view, more plausibly conceived as an energized state of the membrane. The energized membrane might be attained and stabilized equivalently by the redox state of electron transport carriers or by reaction with ATP. If the inner membrane is a mosaic of identical units, highly cooperative unit interactions are to be expected.[63] The chemical activity of a particular unit would be determined by the predominant state of neighboring units as well as by ligands at the particular unit. Several workers have related relatively large-scale morphological changes of the inner membrane cristae to the respiratory state of the preparation.[63-65]

IV. ATPases ASSOCIATED WITH ACTIVE TRANSPORT

Energy-dependent processes which can convey ions or molecules through membranes to a higher electrochemical potential are termed active transport processes.[6,66,67] Several instances of active ion transport are linked with ATPase activity.[68] In each case the terminal phosphate of ATP is hydrolyzed by an enzyme which is specifically stimulated by the ion, or ions, transported.

A. Na-K-ATPase

In most cells, sodium ion transport is clearly dependent on cellular energy production and ATP is a specific, immediate, and sufficient energy source.[69-71] The experimental support for this is reviewed elsewhere.[72-75] A possible enzymatic link to transport was suggested by Skou who observed that crab nerve membranes contain $(Na^+ + K^+)$-stimulated ATPase activity.[76] A similar activity occurs in erythrocyte membranes and the kinetic characteristics compare favorably with those of active Na^+ transport out of red cell ghosts.[77,78] ATP hydrolysis and Na^+ transport both require Na^+ within the cell and K^+ in the external medium. High external Na^+ concentration inhibits both systems[79-82] and cardioactive steroids are effective inhibitors only in the external medium.[78,83,84]

1. Distribution of the Enzyme

Na-K-ATPase has been found in a large variety of tissues from organisms ranging from bacteria to man.[68,74,85-87]

The enzyme is intimately associated with cell membrane fragments. These usually sediment in the so-called "microsomal" subcellular fractions,[88,89] although more intact plasma membranes containing the enzyme can be prepared.[90] The distribution of Na-K-ATPase in brain is similar to that of acetylcholinesterase,[91] being enriched in gray matter[85,92] and in subcellular fractions containing nerve-ending membranes,[93-97] particularly of the cholinergic variety.[98] Synaptic vesicles, however, contain little activity.[93,99] The activity in rat brain increases with neonatal maturation.[100] Highest specific activity is found in the cerebral cortex.[101]

In preparations from muscle,[102-104] the relationship of the enzyme to specific membrane elements (i.e., sarcolemma, transverse tubules, reticulum) has not been defined partly because these tend to sediment in the same fractions.

Histochemical evidence is also conflicting as to the enzyme's presence in portions of the sarcoplasmic reticulum.[105,106] The demonstration of

Na-K-ATPase by histochemical means has proved difficult.[107] It is complicated by the inhibitory effects of heavy metals[108] and tissue fixatives,[109] the occurrence of Pb^{2+}-catalyzed ATP hydrolysis[110] and the difficulty in discerning small $(Na^+ + K^+)$-dependent increments in lead phosphate precipitates.[108]

2. Physical Properties of the Enzyme

General experience has been that chemical or physical fractionations of the particulate preparations inactivate rather than solubilize the enzyme. There is evidence that some element of structural integrity among the phospholipid as well as protein components, is essential for catalytic activity.[111–113]

Stable dispersions in detergents have been prepared and the molecular weight of one such preparation has been estimated as 670,000.[114] This compares with estimates of 250,000,[115] 500,000,[116] and 1,000,000,[117] obtained by radiation inactivation methods.

3. Characteristics of the Enzyme Reaction

Magnesium and sodium ions are absolute requirements while potassium and other univalent cations $(NH_4^+ > Rb^+ > Cs^+ > Li^+)$ stimulate the activity to a degree dependent on the sodium ion concentration. Maximal activation generally occurs in 50–100 mM Na^+, 10–20 mM K^+, 3 mM Mg^{2+} and 3 mM ATP at physiological pH and ionic strength. Excesses of any of the cations are inhibitory.[72,118] Ca^{2+}, Cu^{2+} and F^- are potent inhibitors; the inhibition by F^- is irreversible.[118–121]

The enzyme is distinguished from nonspecific Mg-ATPase in that it is specific for ATP and is inhibited by low concentrations of cardioactive steroids. Na-K-ATPase is also characterized by a relatively high activation energy.[122,123]

Activations of Na-K-ATPase by the various ligands or effectors (Na^+, K^+, Mg^{2+}, ATP) may modify the enzyme response both to successive additions of the same (homotropic) and other (heterotropic) effectors.[123,124] It is presumed that a specific ligand-binding site may be modified through enzyme conformational changes which are induced by binding another ligand molecule to a topographically distinct site.[125]

There may be two or more cooperatively interacting binding sites each for Na^+ and K^+.[68,123,124,126] This is consistent with estimates of the sodium pump efficiency which indicate that two or three sodium ions are exchanged per molecule of ATP hydrolyzed.[127]

Observations that optimal activity is usually obtained at equimolar Mg^{2+} and ATP have led to the suggestion that the enzyme substrate is a Mg-ATP complex.[128] However, cooperativity and varying optimal ratios have been observed with regard to Mg^{2+} and ATP.[97,123] These and other considerations discussed below with regard to conformational changes indicate the occurrence of a Mg-enzyme complex.

4. Evidence for Phosphorylated Intermediate

There is strong evidence that the hydrolysis of ATP proceeds through the formation of a phosphorylated enzyme intermediate.[129,130] Preparations of Na-K-ATPase from a variety of sources incorporate ^{32}P from $(\gamma\text{-}^{32}P)$-ATP into an acid-precipitable protein. The phosphopeptides obtained from these sources have similar electrophoretic mobilities.[131] The steady-state levels of the phosphoryl enzyme bear a fairly constant relationship to the enzyme-specific activity despite wide variations in purity of the preparations.[68,131]

Formation of the phosphorylated enzyme is dependent on Mg^{2+} and Na^+:

(15) $ATP + enzyme \xrightarrow{Mg^{2+}, Na^+} P\text{-enzyme} + ADP$

Potassium and other univalent cations accelerate the turnover of the P-enzyme, and are, therefore, thought to activate the dephosphorylation:[129–133]

(16) $P\text{-enzyme} + H_2O \xrightarrow{K^+} enzyme + Pi$

5. Transphosphorylation

The demonstration that Na-K-ATPase preparations from electric organ catalyze a Mg^{2+}-dependent, Na^+-stimulated ATP-ADP exchange[134,135] constitutes further evidence for the participation of a phosphorylated intermediate in the enzyme reaction. This Na^+-dependent exchange has also been demonstrated in beef brain Na-K-ATPase preparations.[136] It is observed only under conditions of low Mg^{2+} concentrations or in the presence of certain agents which inhibit Na^+-dependent ATP hydrolysis. ATPase inhibitors which activate the exchange are N-ethyl maleimide (NEM), BAL-arsenite, and oligomycin.[135,137]

Interaction of these agents with the enzyme can be shown to produce parallel effects on hydrolysis, phosphorylation, and ATP-ADP exchange. As the hydrolytic reaction is progressively inhibited, the ability of K^+ to dephosphorylate the enzyme is lost and the Na^+-dependent exchange rate increases.[137] These findings indicate that at least a portion of the phosphoryl enzyme has sufficient free energy to be in reversible equilibrium with ATP and ADP (reaction 17), and is not subject to hydrolysis (reaction 16):

(17) $ATP + enzyme \xleftrightarrow{Mg^{2+}, Na^+} P \sim enzyme + ADP$

The associated sodium-*independent* exchange activity that occurs in most Na-K-ATPase preparations[72] has been separated from Na-K-ATPase.[138]

6. Chemistry of the Phosphorylation Site

Electrophoretic studies of the ^{32}P-tagged peptide fragments indicate the presence of at least one diacidic and one dibasic residue[139] and one sulfhydryl group, which is not reactive with NEM.[140] Phospholipids have been excluded as intermediates.[141]

The nature of the phosphate linkage to the peptide is unknown. Some evidence suggests the participation of a serine residue in the catalysis, but not necessarily its identity with the active site.[68,142] Recently, the recovery of a peptide fragment with electrophoretic mobility similar to that of a glutamyl hydroxamate has been considered as supporting evidence for an acylphosphate linkage with the ω-carbon of glutamate.[143] But the low yield (2% of the expected) leaves this matter unsettled.

Other arguments for the formation of an acylphosphate are based upon the relative stability of the phosphate complex at low pH, and its instability in the presence of acylphosphatase or hydroxylamine.[132] The stability of acetylphosphate, which is used as the reference, is, in fact, greatest at pH 5–6[144] while this phosphoryl ATPase is most stable at about pH 2. The observation of instability in the presence of hydroxylamine is, of course, not equivalent to the demonstration of acylhydroxamate formation.

The presence of contaminating ammonium ions may account for the effect of hydroxylamine on ^{32}P-labeling of the native enzyme simply by stimulating the enzyme turnover.[145,146] There is a slow catalysis of the decomposition of denatured phosphoenzyme by hydroxylamine[147] but at a rate insufficient to account for its effect on the native enzyme.[148]

7. Potassium-Activated Phosphatase

Phosphate incorporation into protein from (^{32}P)-acetylphosphate has been observed[149,150] and pepsin digestion of a protein labeled by acetylphosphate liberates phosphopeptides with electrophoretic mobilities similar to those obtained with ATP as substrate.[151] It is possible that similar, if not identical, phosphorylated intermediates are involved in the K^+-activated phosphatase activities (acting upon acetylphosphate, and carbamylphosphate). These have a number of properties in common with Na-K-ATPase, including inhibition by ouabain[152–156] and, it has been suggested that the K^+-activated phosphatase may be a manifestation of that portion of Na-K-ATPase involved in reaction 16. However, data pertaining to the effect of K^+ on ^{32}P labeling by (^{32}P)-acetylphosphate is at present conflicting.[149,150]

8. Evidence Against a Phosphorylated Intermediate

Some authors have observed that at low temperatures K^+ fails to stimulate hydrolysis while it does activate dephosphorylation.[72,157] To account for this apparent dissociation between the K^+ effects, Skou has suggested that hydrolysis proceeds through formation of a Michaelis complex with ATP in the simultaneous presence of Na^+ and K^+ but that the protein phosphorylation occurs "abnormally" when Na^+ occupies K^+ sites on the enzyme.[73] A different view has been offered by Kamazawa et al.[157]

In other preparations, however, the relative extent of K^+ activation of hydrolysis is variably reduced at $0°C$[133] but may still amount to a six- to tenfold increase.[137] A temperature-dependent quantitative dissociation between the effects of K^+ on hydrolysis and on dephosphorylation may be alternatively interpreted as reflecting a rate-limiting transition in the non-phosphorylated enzyme (reaction 21, below).[68] Direct evidence for the nature of the rate-limiting step is, however, not at hand.

9. Inhibition by Cardioactive Steroids

Cardioactive steroids (CS) and their glycosides are potent inhibitors of Na-K-ATPase and the sodium pump,[158] but their mode of inhibition has been obscure.[159] K^+ has been known to antagonize the actions of CS. Reversible competition of CS with K^+ in reaction 16 has been suggested.[84,126,160] However, it is now clear that CS inhibit Na-K-ATPase by combining irreversibly with the enzyme.[161-163] The *rate*, but not the *extent* of CS binding to the enzyme is decreased by K^+.[163] Thus, as originally suggested by Glynn,[158] undoubtedly some of the anomalies in kinetic studies of CS inhibition have arisen from the progressive inhibition of the enzyme during its assay.

The rate of the binding reaction between CS and Na-K-ATPase is markedly influenced by Mg^{2+}, Na^+, nucleotides, and orthophosphate as well as by K^+.[163,164] What is perhaps most remarkable is that the CS-enzyme complex will incorporate orthophosphate into acid-stable linkage.[163] While this phenomenon requires further characterization, the ^{32}P-enzyme and the radioactive peptide fragments obtained therefrom appear identical in their electrophoretic patterns to those obtained after labeling with $AT^{32}P$.[165] CS do not form covalent linkages with either the enzyme or the phosphate since they are extracted from the enzyme by methanol and by acid denaturation, while the ^{32}P incorporation is stable to both of these treatments. It seems most probable that the CS-enzyme interaction produces a major conformational change in the enzyme which alters the reactivity of the phosphorylation site and the free energy content of the phosphoryl enzyme.

10. Evidence for Conformational Changes of the Enzyme

a. *Phosphorylated Enzyme.* While conformational changes of the enzyme have not been demonstrated by physical means, certain of the kinetic studies support the likelihood of their importance.[123,124]

As mentioned above, a sodium-stimulated nucleotide exchange (reaction 17) is observed in low Mg^{2+} (0.15 mM) but not in concentrations that are optimal for ATP hydrolysis (3 mM). As might be expected, the maximal phosphorylation of the enzyme occurs at a much lower Mg^{2+} concentration than that required for maximal hydrolysis.[166]

These facts are consistent with the occurrence of a magnesium-induced transition (reaction 19) of the phosphoryl enzyme from a form of high energy character ($E_1 \sim P$) capable of participating in the exchange to a form ($E_2 - P$) of lower free energy content. By this interpretation, K^+ activation (reaction 20), resulting in dephosphorylation, involves the lower energy intermediate.[137] (Reaction 17 is rewritten as 18 in accordance with the distinction between $E_1 \sim P$ and $E_2 - P$).

(18) $ATP + E_1 \xrightarrow{\text{Mg}^{2+}, \text{Na}^+} E_1 \sim P + ADP$

(19) $E_1 \sim P \xrightarrow{\text{Mg}^{2+}} E_2 - P$

(20) $E_2 - P + H_2O \xrightarrow{\text{K}^+} E_2 + Pi$

The observed modifications of the enzyme by NEM or BAL-arsenite are consistent with these reactions. These agents do not inhibit phosphorylation but do inhibit the K^+-activated dephosphorylation (reaction 20) to the same extent to which they activate the exchange (reaction 18). Their action is explicable as a stabilization of $E_1 \sim P$ at the expense of $E_2 - P$, the latter apparently predominating in the native enzyme under high Mg^{2+} concentrations.[135,137]

The nature of the transition to a lower energy level is not known but it may involve either a chemical bond transfer or an enzyme conformation change, or both. However, since electrophoresis of the P-peptide from NEM-treated enzyme has shown no dissimilarity from that of the native enzyme,[131] it is considered likely that the stabilization of $E_1 \sim P$ by the sulfhydryl reactive agents results from constraints placed upon the conformation of the enzyme.

b. Nonphosphorylated Enzyme. The considerations presented so far refer to possible transitions of the phosphoryl enzyme. If these occur, then the dephosphorylated enzyme must undergo transitions in the reverse direction to regenerate E_1, the form reactive with Na^{2+} and ATP in reaction 18. Hence, reaction 21 is required to complete the cycle:

(21) $E_2 \leftrightarrow E_1$

The sum of reactions 18–21 is the net hydrolysis of ATP. The existence of multiple nonphosphorylated forms in equilibrium is consistent with the observations that agents which are thought to stabilize $E_1 \sim P$ also increase the affinity of the enzyme for Na^+ as measured by the extent of phosphorylation. Since reaction 21 is the reverse of 19 with respect to enzyme conformation, both effects are likely to derive from the same conformational restraints: the increased Na^+ affinity may be another consequence of the higher proportion of enzyme in the E_1 form.[137]

The observation that CS binding rates are changed by ADP and other effectors which do not phosphorylate the enzyme is additional evidence that the nonphosphorylated enzyme is also subject to conformational transition.[163,164]

B. Correlation of the Properties of Na-K-ATPase with the Requirements of the Sodium Pump

Properties of a carrier which are essential for its mediation of active transport include: (a) that the affinity or availability of the binding sites on the carrier for a specific ligand must change as the carrier traverses the membrane, and (b) that the passage of the carrier in one direction must be accompanied by release of free energy.[167,168]

The postulated enzyme transitions described in reactions 18 to 21 may be directly related to the mechanism of active sodium transport. E_1 and $E_1 \sim P$ may represent carrier and activated carrier, respectively, both with relatively high Na^+ affinity and oriented inward from the cell membrane. $E_2 - P$ and E_2 would then constitute forms of the deenergized carrier with relatively high K^+ affinity at the external surface. If so, an exchange of bound potassium for sodium at the inner, and of bound sodium for potassium ions at the outer membrane surfaces would follow the transitions:

$$E_2 \rightarrow E_1 \text{ and } E_1 \sim P \rightarrow E_2 - P, \text{ respectively}$$

The direction of the transitions, and hence of the cation fluxes, would be mainly determined by the large energy difference between $E_1 \sim P$ and $E_2 - P$.[163]

The available information about Na-K-ATPase and the sodium pump is almost entirely kinetic in nature. None of this is direct evidence for testing the validity of a number of alternative models which attempt to define more explicitly the physical or chemical correlates of the pump activity.[129,169,170] However, the intimate association of the enzyme with the actual ion transport mechanism is substantiated by the recent observations of Glynn et al.[171]

C. Metabolic Regulation and Cation Transport

Active cation transport may be a major energy-consuming function of specialized tissues such as brain and kidney. Indications that energy production may be stimulated by increased transport activity are discernible in early observations that the respiration of brain slices incubated in low Ca^{2+} media is stimulated by high K^+.[172,173] This stimulation was later found to be inhibited by ouabain.[174] Electrical pulsation of brain slices also results in similar cation-dependent respiratory stimulation accompanied by acceleration of active cation fluxes.[175-177]

Whittam and his co-workers have studied the dependence of respiration on transport activity. When incubated in low Ca^{2+} media, either deprived of Na^+ or containing ouabain, brain and kidney slices lose their ability to take up and retain K^+ in parallel with a reduction in oxygen consumption.[178-180] Under similar conditions, the Na-K-ATPase activity and ADP levels of homogenates are reduced comparably.[181-183]

These authors have estimated that 25 to 40 % of the respiration of kidney and brain in vitro is devoted to cation transport and that, to this extent, respiration may be controlled by the rate of transport. Mitochondrial respiration is known to be sensitive to the cellular concentration of ADP.[184] Thus, the transport control of respiration is thought to be mediated through the cellular level of ADP which in turn is partly regulated by the activity of Na-K-ATPase.

In erythrocytes, studies of the influences of Na:K ratios and ouabain on lactate production have indicated that 15 to 20% of glycolysis is dependent on Na-K-ATPase activity. This control may be mediated through the stimulatory effects of ADP or Pi on glycolytic enzymes.[185–187]

While it has been presumed that the action of ouabain is restricted to Na-K-ATPase, it should be noted that ouabain can modify the effects of Ca^{2+} in brain. For example, elevated Ca^{2+} inhibits brain slice respiration. But in the presence of ouabain, increased Ca^{2+} produces first stimulation, which eventually is followed by depression of oxygen uptake.[174,188–190] Pertinent to this observation are the facts that Ca^{2+} is a known uncoupler of mitochondrial respiration[191] and that Ca^{2+} mobility in brain slices[192] and Ca^{2+} uptake by brain mitochondria[193] are increased by ouabain. These data are consistent with the suggestion that some ouabain effects on respiration may be related to an alteration of Ca^{2+} content in mitochondria.[189] It is not known whether this alteration may ensue indirectly from an action of ouabain on Na-K-ATPase or more directly through an action on a hypothetical Ca^{2+} translocation or binding system.

D. Ca-ATPase of Sarcoplasmic Reticulum

The work of Marsh[194] and Bendall[195] first showed that extracts of muscle could reverse the ATP-induced syneresis of actomyosin and contraction of myofibrils and the associated actomyosin-ATP splitting activity. Evidence accrued from subsequent investigations that the active component in the extracts consists of vesicular elements or "grana" derived from the sarcoplasmic reticulum (SR).[196] The grana contain an ATPase, and, in addition, they exhibit an ATP-dependent capacity for rapidly sequestering calcium ions to produce an intravesicular calcium concentration as much as 1000-fold greater than in the external media.[197,198]

The observations that minute amounts of calcium are required for superprecipitation of actomyosin[199] and that chelating agents exert relaxing effects in proportion to their capacities for complexing calcium ions, support the proposal[197] that these elements of the sarcoplasmic reticulum exert relaxation effects in vitro by lowering calcium available to the actomyosin.[200] The dissociation constant for calcium from actomyosin is 10^{-7} M and the SR grana are able to lower the free calcium in solution to 10^{-8} M (See below). Their effects in vitro are generally consistent with the calculated requirements for regulation of myofibril activity by calcium in intact muscle.[201]

The reader is referred elsewhere for detailed accounts of the role of the sarcoplasmic reticulum,[202] and the calcium pump in muscle physiology.[203,204]

1. Isolation and Structure

Relaxing grana have been isolated from both skeletal and cardiac muscle by differential centrifugation.[205–207] They are found as a component of microsome fractions largely free of nuclei and mitochondria.

Prior to the recognition of the physiological significance of the grana, Kielley and Meyerhoff[208] isolated a lipoprotein-bound Mg-ATPase from sarcoplasm which was later related to sarcoplasmic granules.[209] Lecithinase treatment was found to inactivate this enzyme in proportion to the extent of phospholipid hydrolysis.

The SR grana have high lecithin and phospholipid contents.[210] Lecithinase inactivates the relaxing effect and ATPase activity of the SR grana[211] and the rate of loss of the two activities is similar.[210]

Both activities may be restored by addition of lecithin, lysolecithin, or phosphatidic acid.[210] In these experiments, disruption of membrane structure by the phospholipase treatment is not evident in electron micrographs. Deoxycholate visibly disrupts the grana membranes and destroys calcium transport ability, although the ATPase activity is preserved and solubilized.[212] Removal of the detergent permits spontaneous regeneration of the vesicle membrane continuity and restores the calcium transport activity.

Electrophoretic separation of the solubilized enzyme protein components leads to loss of ATPase activity which may be restored, however, by reversing the polarity and recombining the separated proteins. Thus, it appears that ATPase activity of the grana depends on interaction of multiple protein units with phospholipid and that calcium transport has the additional requirement of membrane structural integrity.[212] Sulfhydryl[213] and imidazole[214] groups have been implicated as functional determinants.

2. Characteristics of the Enzyme Reaction

The ATPase activity of the grana preparations consists of a Mg^{2+}-dependent portion (basal splitting) and a $(Mg^{2+} + Ca^{2+})$-dependent portion (extra splitting or Ca-ATPase).[198] These activities appear to be related to different enzymes which can be distinguished by their activation energies,[215,216] and pH optima.[216] The sulfhydryl reactive agents, salyrgan and N-ethylmaleimide, inhibit the extra splitting induced by calcium but not the basal splitting.[198,203] On the other hand, the Ca-ATPase and the calcium transport activities are similar by these criteria.

Parallel activations of Ca-ATPase and calcium uptake by the grana occur with increasing magnesium concentrations up to the optimal level of 2–5mM. Mn^{2+} or Zn^{2+}, but not Ba^{2+}, may substitute for Mg^{2+}.[217] In the presence of optimal Mg^{2+} levels, Ca^{2+} activation of ATP hydrolysis and uptake occurs in the range of 10^{-8} to 10^{-5} M while higher calcium concentrations are inhibitory.[217–219] Although Sr^{2+} may be taken up by the grana,[220] its presence does not interfere with calcium uptake.[217]

While ATP is the optimal substrate, both the Ca-ATPase and transport activities are supported at lesser rates by other nucleosidetriphosphates.[221,222]

The optimal enzyme activity is seen at physiological pH[223] and ionic strengths. Higher salt concentrations inhibit the net calcium uptake without inhibiting the ATPase.[205,218,224] This is associated with an increased release of calcium from the grana[225] and may be related to competition of cations with Ca^{2+} for a fixed number of cation binding sites within the vesicles (see below).

The calcium uptake and ATPase activities of the relaxing grana are not inhibited by concentrations of azide, oligomycin, or 2,4-dinitrophenol which inhibit mitochondrial calcium uptake, thus providing a convenient means of selectively reducing contaminating mitochondrial activity.[59,205,207]

As in the case of Na-K-ATPase, the usual tissue fixatives are inhibitory.[226] Although histochemical methods have shown ATPase activity in the terminal cisternae of the triads,[105,106] the lack of decisive biochemical controls coupled with inhibition of activity by the procedure itself, complicate the distinct localization of Ca-ATPase.

3. Relation of the Enzyme to Ca^{2+} Transport

In the presence of oxalate (which forms an insoluble calcium salt permitting larger quantities of calcium to be sequestered with concommittantly larger values of total ATP hydrolysis), the calcium uptake and ATP extra splitting that commence upon addition of calcium continue at a ratio of about 2:1 until the external concentration is reduced to 10^{-8} to 10^{-7} M calcium, when both activities cease. Calcium oxalate is accumulated until the ionic activity product inside the vesicles is about 500 times that found outside. Thus, it appears that the external Ca^{2+} activates both the ATPase and pumping activities and is itself transported against a concentration gradient, a process which requires an estimated 3.6 Kcal/mole Ca^{2+} at 5 mM ATP.[219,221] This value, however, may be lower at lower ATP concentrations[223] and, in addition, may depend on whether the calcium flux is in a steady state.[220,227] There is evidence that the intravesicular as well as extravesicular Ca^{2+} levels regulate the transport and ATPase activities.[227]

It has been estimated that the free calcium ions within the vesicles must be significantly less than the total calcium uptake on the basis of solubility and ionic activity products,[215,220] and Ebashi has proposed that the calcium is bound to specific sites on the vesicular membranes.[228] This is supported by the findings of Caravalho[229] that about 80% of the sequestered calcium may bind to nondiffusable anions within the vesicles, displacing Mg^{2+}, K^+, and H^+. The effect of ATP is to increase the apparent affinity for calcium 1,000-fold but the quantity of total bound cation equivalents remains constant. These studies do not, however, distinguish between a process in which the availability of internal calcium is increased by active transport and one in which the binding affinity is primarily increased through the action of ATP. It should be noted that these possibilities are not mutually exclusive.

4. Evidence for a Phosphorylated Intermediate in Calcium Transport

a. Transphosphorylation. The grana preparations catalyze a terminal phosphate exchange between NTP and NDP. This transphosphorylation exhibits characteristics similar to those of the Ca-ATPase: activation by Ca^{2+} in parallel with transport and hydrolysis, inhibition by salyrgan and oleic acid[198] and similar magnesium and hydrogen dependency.[196] The exchange does not show nucleotide specificity, in common with the Ca-ATPase, and it can be differentiated from myokinase and other nucleoside diphosphokinases.[230]

Assuming the transphosphorylation to be catalyzed by the same enzyme, it has been proposed that ATP transfers phosphate reversibly to a component on the outer surface of the vesicle, producing an increase in affinity of the component for Ca^{2+}. When the calcium complex passes to the inner surface, the phosphate is split off and the calcium affinity is reduced.[198,230]

b. Phosphorylation of SR grana. Yamamoto and Tonomura[216] have recently obtained an acid-stable ^{32}P-labeled material from SR grana incubated with (γ-^{32}P)-ATP. The formation of the labeled constituent is dependent on Ca^{2+} and inhibited by *N*-ethylmaleimide in common with the Ca-ATPase. The extent of ^{32}P-labeling is proportional to the rate of hydrolysis.

E. Other ATPases Possibly Related to Transport

It is pertinent to mention briefly several examples of ATPases possibly related to transport but which have characterizations that so far are not sufficient to permit any conclusion.

1. Calcium ATPases

Microsome and nerve-ending fractions from brain have been observed to bind calcium in the presence of ATP and magnesium, thus resembling in certain respects the SR grana.[231–233] In addition, similar preparations, including synaptic vesicle fractions, exhibit calcium-stimulated ATPase activity which may be related to the calcium sequestration.[99,234–236] The phenomenon is not specific for neuronal tissue, however, since calcium binding has been found also in fractions from liver and kidney and (Mg + Ca)-ATPase has been reported in erythrocytes.[237]

While calcium transport is probably a general cellular function, special regulatory roles may exist apropos membrane excitability[238] and synaptic transmission,[239] as in muscle contractility, all of which are sensitive to calcium ions.

2. Anion-Stimulated ATPase

Kasbekar and Durbin[240,241] have separated a membrane-bound ATPase from fractions of frog gastric mucosa. This enzyme (or enzymes) requires magnesium and is stimulated by bicarbonate and halide ions. Thiocyanate, which inhibits gastric acid secretion, also inhibits the ATPase while sodium, potassium, and ouabain have no influence on the activity. Since gastric acid secretion is dependent on metabolism, the possibility that this enzyme participates in HCl secretion has been suggested. Enzymatic activity with similar properties has been obtained in particulate fractions from yeast.[242]

V. GTPase ASSOCIATED WITH PROTEIN SYNTHESIS

It is well known that GTP is a specific requirement for protein synthesis.[243] Recently some insight has been gained into the nature of its action following observations that combinations of ribosomes with dialyzed supernatant fractions catalyze the specific hydrolysis of GTP to GDP plus Pi. The hydrolysis of GTP proceeds in conjunction with mRNA-directed incorporation of amino acids from aminoacyl-sRNA into peptides with a stoichiometric relationship of one peptide bond synthesized per mole of GTP hydrolyzed.[244,245]

The specific GTPase activity, in common with peptide synthesis, is dependent on the presence of ribosomes and magnesium and is stimulated by mRNA, sRNA, and NH_4^+ or K^+ while inhibited by Na^+, Li^+, and GDP. In bacterial systems, Ca^{2+} or Mn^{2+} may replace Mg^{2+}.[244,246]

Supernatant factors from mammalian reticulocytes and *Escherischia coli* required for amino acid polymerization have been partially separated into at least two major components which are not necessarily homogeneous. One component (designated TF-1 or translocase in reticulocytes and F-1 or initiator factors in *E. coli*) catalyzes the binding of aminoacyl-sRNA to complete ribosomes (50 S plus 30 S subunits) in the presence of mRNA, GTP, and magnesium concentrations of about 4 mM. The addition of the second component (termed TF-2 or peptide synthetase in reticulocytes and F-11 or T plus G factors in *E. coli*) then leads to peptide bond formation.[247–249] GTP is not required for peptide bond synthesis directly, as shown by experiments in which ribosomal 50 S subunits support synthesis of *N*-formylmethionyl puromycin in the absence of GTP and intiator factors.[250]

The enzymatic, GTP-dependent binding associated with GTP hydrolysis apparently involves a specific ribosome site or conformation. Nonenzymatic binding which can occur in high magnesium concentrations does not lead to peptide bond formation unless the intiator factors plus GTP are added.[247–249] Also, a GTP analogue which is not hydrolyzed may replace GTP in support of bonding but not of peptide bond synthesis.[251] Finally, there is evidence for a GTP binding step without phosphate cleavage occurring prior to an interaction with the ribosome.[252]

There is general agreement that the initiator factors and GTP are necessary for the translocation of aminoacyl-sRNA from an "acceptor" to a "donor" (peptidyl) site on the ribosome, or fixation to the donor sites.[248,249] This process, it has been speculated, may be linked to the advancement of mRNA in positioning successive codons at the ribosome sites.[244]

VI. REFERENCES

1. M. Dixon and E. C. Webb, *Enzymes*, 2nd ed., Academic Press, New York (1964).
2. W. W. Kielley, in *The Enzymes*, 2nd ed. (P. D. Boyer, H. Lardy, and K. Myrbäch, eds.), Vol. 5, pp. 159–168, Academic Press, New York (1961).
3. S. Gatt and E. Racker, Regulatory mechanisms in carbohydrate metabolism. I. Crabtree effect in reconstructed systems, *J. Biol. Chem.* **234**:1015–1023 (1959).
4. K. A. Trayser and S. P. Colowick, Properties of crystalline hexokinase from yeast. II. Studies on ATP-enzyme interaction, *Arch. Biochem. Biophys.* **94**:161–168 (1961).
5. L. E. Hokin and M. R. Hokin, Biological transport, *Ann. Rev. Biochem.* **32**:553–578 (1963).
6. P. J. Heald, Phosphoprotein metabolism and ion transport in nervous tissue: a suggested connexion, *Nature (London)* **193**:451–454 (1962).
7. D. A. Hems and R. Rodknight, Properties of phosphate bound to cerebral microsomes during ATPase activity, *Biochem. J.* **101**:516–523 (1966).
8. P. M. Anderson and A. Meister, Bicarbonate-dependent cleavage of ATP and other reactions catalysed by *E. coli* carbamyl phosphate synthetase, *Biochemistry* **5**:3157–3163 (1966).
9. S. Cha, C. M. Cha, and R. E. Parks, Jr., Succinic thiokinase. IV. Improved method of purification, arsenolysis of GTPase activity, and some other properties of the enzyme. *J. Biol. Chem.* **242**: 2577–2581 (1967).
10. W. A. Engelhardt and W. V. Lyubimova, Myosine and adenosinetriphosphatase, *Nature (London)* **144**:668–669 (1939).
11. W. A. Engelhardt, Enzymatic and mechanical properties of muscle proteins, *Yale J. Biol. Med.* **15**:21–38 (1942), trans. by P. Talalay.
12. D. F. Cain and R. E. Davies, Breakdown of ATP during a single contraction of working muscle, *Biochem Biophys. Res. Comm.* **8**:361–366 (1962).
13. F. Lipmann, Metabolic generation and utilization of phosphate bond energy, *Adv. Enzymol.* **1**:99–162 (1941).
14. H. M. Kalckar, The nature of energetic coupling in biological synthesis, *Chem. Rev.* **28**:71–178 (1941).
15. J. Gergely, ed., *Biochemistry of Muscle Contraction*, Little, Brown, Boston (1962).
16. W. W. Kielley, Biochemistry of muscle, *Ann. Rev. Biochem.* **33**:403–430 (1964).
17. S. V. Perry, The structure and interactions of myosin, *Progress Biophys. and Mol. Biol.* **17**:327–381 (1967).
18. D. R. Wilkie, Muscle, *Ann. Rev. Physiol.* **28**:17–38 (1966).
19. J. C. Warren, L. Stowring, and M. F. Morales, The effect of structure-disrupting ions on the activity of myosin and other enzymes, *J. Biol. Chem.* **241**:309–316 (1966).
20. T. Sekine, in *Molecular Biology of Muscular Contraction* (S. Ebashi, F. O. Oosawa, T. Sekine, and Y. Tonomura, eds.), pp. 33–44, Elsevier, New York (1965).
21. Y. Tonomura, P. Appel, and M. F. Morales, On the molecular weight of myosin. II. *Biochemistry* **5**:515–521 (1966).
22. D. M. Young, W. F. Harrington, and W. W. Kielley, The dissociation and reassociation of the subunit polypeptide chains of myosin, *J. Biol. Chem.* **237**:3116–3122 (1962).
23. P. Dreizen, D. J. Hartshorne, and A. Stracher, The subunit structure of myosin. I. Polydispersity in 5 M guanidine, *J. Biol. Chem.* **241**:443–448 (1966).
24. E. Gaetjens, K. Bárány, G. Bailin, H. Oppenheimer, and M. Bárány, Studies on the low molecular weight components in rabbit skeletal myosin, *Arch. Biochem. Biophys.* **123**:82–96 (1968).
25. S. Seifter and P. M. Gallop, in *The Proteins* (H. Neurath, ed.), 2nd ed., Vol. 4, pp. 153–458, Academic Press, New York (1966).
26. L. B. Nanninga and W. F. H. H. Mommaerts, Kinetic constants of the interaction between myosin and ATP, *Proc. Nat. Acad. Sci. U.S.A.* **46**:1166–1173 (1960).

27. T. Nihei, M. Morris, and A. L. Jacobson, Activation and inhibition of myosin B ATPase by Mg^{2+} and Ca^{2+} at low concentration of KCl, *Arch. Biochem. Biophys.* **113**:45–51 (1966).

28. T. Kanazawa and Y. Tonomura, The pre-steady state of the myosin-adenosine triphosphate system. I. Initial rapid liberation of inorganic phosphate, *J. Biochem.* **57**:604–615 (1965).

29. M. Bárány, ATPase activity of myosin correlated with speed of muscle shortening, *J. Gen. Physiol.* **50**:No. 6, Part 2, 197–218 (1967).

30. M. Bárány, K. Bárány, T. Reckard, and A. Volpe, Myosin of fast and slow muscles of the rabbit, *Arch. Biochem. Biophys.* **109**:185–191 (1965).

31. F. A. Sreter, J. C. Seidel, J. Gergely, Studies on myosin from red and white skeletal muscles of the rabbit. I. ATPase activity, *J. Biol. Chem.* **241**:5772–5776 (1966).

32. J. C. Seidel, Studies on myosin from red and white skeletal muscles of the rabbit. II. Inactivation of myosin from red muscles under mild alkaline conditions, *J. Biol. Chem.* **242**:5623–5629 (1967).

33. M. F. Morales, in *Enzymes: Units of Biological Structure and Function* (O. H. Gaebler, ed.), pp. 325–336, Academic Press, New York (1956).

34. R. E. Davies, A molecular theory of muscle contraction: calcium-dependent contractions with hydrogen bond formation plus ATP-dependent extensions of the myosin-actin cross bridges, *Nature (London)* **199**:1068–1074 (1963).

35. M. E. Dempsey, P. D. Boyer, and E. S. Benson, Characteristics of an orthophosphate oxygen exchange catalysed by myosin, actomyosin, and muscle fibers, *J. Biol. Chem.* **238**:2708–2715 (1963).

36. L. Sartorelli, H. J. Fromm, R. W. Benson, and P. D. Boyer, Direct and ^{18}O-exchange measurements relevant to possible activated or phosphorylated states of myosin, *Biochemistry* **5**:2877–2884 (1966).

37. D. E. Koshland, Jr., Z. Budenstein, and A. Kowalsky, Mechanism of hydrolysis of ATP catalysed by purified muscle proteins, *J. Biol. Chem.* **211**:279–287 (1954).

38. H. M. Levy and D. E. Koshland, Jr., Mechanism of hydrolysis of ATP by muscle proteins and its relation to muscular contraction, *J. Biol. Chem.* **234**:1102–1107 (1959).

39. R. W. Benson, M. E. Dempsey, and E. S. Benson, Kinetic relationships between phosphate-oxygen exchange and hydrolytic cleavage of ATP catalysed by myofibrils, *J. Biol. Chem.* **242**:1612–1616 (1967).

40. Y. Tonomura, S. Kitagawa, and J. Yoshimura, The initial phase of myosin A-ATPase and the possible phosphorylation of myosin A, *J. Biol. Chem.* **237**:3660–3666 (1962).

41. K. Imamura, T. Kanazawa, M. Tad, and Y. Tonomura, The pre-steady state of the myosin-adenosine triphosphate system, III. Properties of the intermediate, *J. Biochem.* **57**:627–636 (1965).

42. W. J. Bowen, L. C. Stewart, and H. L. Martin, Studies on the fast initial rate of ATPase of actomyosin and glycerolo-treated muscle, *J. Biol. Chem.* **238**:2926–2931 (1963).

43. Y. Tonomura, The reaction of myosin with ATP, *7th International Congress of Biochemistry*, Tokyo, Abstracts II, pp. 317–318 (1967).

44. A. G. Szent-Györgyi and G. Prior, Exchange of ADP bound to actin in superprecipitated actomyosin and contracted myofibrils, *J. Mol. Biol.* **15**:515–538 (1966).

45. F. Oosawa, S. Asabura, S. Higashi, M. Kasai, S. Kobayashi, E. Nakano, T. Ohnishi, and M. Taniguchi, in *Molecular Biology of Muscular Contraction* (S. Ebashi, F. Oosawa, T. Sekine, and Y. Tonomura, eds.), pp. 56–76, Elsevier, New York (1965).

46. T. Tokiwa and Y. Tonomura, The pre-steady state of the myosin-adenosine triphosphate system. II. Initial absorption and liberation of H^+, *J. Biochem.* **57**:616–626 (1965).

47. S. Ebashi and A. Kodama, Interaction of troponin with F-actin in the presence of tropomyosin, *J. Biochem.* **59**:425–426 (1966).

48. S. Ebashi, F. Ebashi, a. Kodama, Troponin as the Ca^{++}-receptive protein in the contractile system, *J. Biochem.* **62**:137–138 (1967).

49. D. J. Hartshorne and H. Mueller, Fractionation of troponin into two distinct proteins: A = activating, B = inhibiting, *Biochem. Biophys. Res. Comm.* **31**:647–683 (1968).

50. R. E. Davies, M. J. Rushmerick, and R. E. Larson, ATP, activation, and the heat of shortening of muscle, *Nature (London)* **214**:148–151 (1967).

51. H. E. Huxley, in *Muscle* (W. M. Paul, E. E. Daniel, C. M. Kay, and G. Monckton, eds.), pp. 3–28, Pergamon Press, New York (1965).

52. I. A. Gibbons, Studies on the ATPase activity of ^{14}S and ^{30}S dynein from cilia of *Tetrahymena*, *J. Biol. Chem.* **241**:5590–5596 (1966).

53. I. R. Gibbons, Proteins associated with movement in cilia, *7th International Congress of Biochemistry*, Tokyo, Abstracts II, p. 333 (1967).

54. C. S. Brokaw, ATP usage by flagella, *Science* **156**:76–78 (1967).

55. H. Hoffman-Berling, in *Comparative Biochemistry* (M. Florkin and H. S. Mason, eds.), Vol. 2, pp. 341–370, Academic Press, New York (1960).

56. E. Racker and L. Horstman, Partial resolution of the enzymes catalyzing oxidative phosphorylation, XIII. Structure and function of submitochondrial particles completely resolved with respect to coupling factor I, *J. Biol. Chem.* **242**:2547–2551 (1967).

57. M. E. Pullman and G. Schatz, Mitochondrial oxidations and energy coupling, *Ann. Rev. Biochem.* **36**:539–610 (1967).

58. D. H. MacLennan, J. M. Smoly, and A. Tzagoloff, Studies on the mitochondrial adenosine triphosphatase system. I. Restoration of adenosine triphosphate-dependent reactions in salt-extracted submitochondrial particles, *J. Biol. Chem.* **243**:1589–1597 (1968).

59. G. A. Vigers and F. D. Ziegler, Azide inhibition of mitochondrial ATPase, *Biochem. Biophys, Res. Comm.* **30**:83–88 (1968).

60. P. D. Boyer, L. L. Bieber, R. A. Mitchell, and G. Szabolcsi, The apparent independence of the phosphorylation and water formation reactions of oxidative phosphorylation, *J. Biol. Chem.* **241**:5384–5390 (1966).

61. M. G. Colomb, J. G. Laturaze, and P. V. Vignais, Isolation of a phosphorylated intermediate involved in the ADP-ATP exchange reaction, *Biochem. Biophys. Res. Comm.* **24**:909–915 (1966).

62. R. E. Beyer, Isolation and reactions of a phosphorylated transferase from beef mitochondria, *Arch. Biochem. Biophys.* **125**:884–894 (1968).

63. D. E. Green, M. Asai, R. A. Harris, and J. T. Penniston, Conformational basis of energy transformations in membrane systems. III. Configurational changes in the mitochondrial inner membrane induced by changes in functional states, *Arch. Biochem. Biophys.* **125**:684–705 (1968).

64. C. R. Hackenbrock, Ultrastructural bases for metabolically linked mechanical activity in mitochondria. II. Electron transport linked ultrastructural transformations in mitochondria, *J. Cell. Biol.* **37**:345–369 (1968).

65. D. W. Deamer, K. Utsumi, and L. Packer, Oscillatory states of mitochondria. III. Ultrastructure of trapped conformational states, *Arch. Biochem. Biophys.* **121**:641–651 (1967).

66. B. Anderson and H. H. Ussing, in *Comparative Biochemistry* (M. Florkin and H. Mason, eds.), Vol. 2, pp. 371–402, Academic Press, New York (1960).

67. W. Wilbrandt, Aktiver Transport durch Grenzflachen, *Klin. Wochenschr.* **41**:138–147 (1963).

68. R. W. Albers, Biochemical aspects of active transport, *Ann. Rev. Biochem.* **36**:727–756 (1967).

69. J. F. Hoffman, The link between metabolism and the active transport of Na^+ in human red cell ghosts, *Fed. Proc.* **19**:127 (1960).

70. P. C. Caldwell, A. L. Hodgkin, R. D. Keynes, and T. I. Shaw, The effects of injecting energy rich phosphate compounds on the active transport of ions in the giant axons of *Loligo*, *J. Physiol.* **152**:561–590 (1960).

71. C. M. Connelly, in *Biophysical Science—A Study Program* (J. L. Oncley, ed.), pp. 475–484, Wiley, New York (1959).

72. J. C. Skou, Enzymatic basis for active transport of Na$^+$ and K$^+$ across cell membranes, *Physiol. Rev.* **45**:596–617 (1965).

73. J. D. Judah and K. Ahmed, The biochemistry of sodium transport, *Biol. Rev.* **39**:160–193 (1964).

74. A. I. Katz and F. H. Epstein, Physiologic role of sodium-potassium-activated adenosine triphosphatase in the transport of cations across biologic membranes, *New Eng. J. Med.* **278**:253–261 (1968).

75. W. Brodsky and I. L. Schwartz, experimental results.

76. J. C. Skou, The influence of some cations on an adenosine triphosphatase from peripheral nerve, *Biochim. Biophys. Acta* **23**:394–401 (1957).

77. R. L. Post, C. R. Merritt, C. R. Kinsolving, and C. D. Albright, Membrane adenosine triphosphatase as a participant in the active transport of sodium and potassium in the human erythrocyte, *J. Biol. Chem.* **235**:1796–1802 (1960).

78. E. T. Dunham and I. M. Glynn, Adenosinetriphosphatase activity and the active movements of alkali metal ions, *J. Physiol.* **156**:274–293 (1961).

79. R. A. Sjodin and L. A. Beauge, The ion selectivity and concentration dependence of cation coupled active sodium transport in squid giant axons, *Currents in Modern Biology* **1**:105–115 (1967).

80. P. J. Garrahan and I. M. Glynn, The sensitivity of the sodium pump to external sodium, *J. Physiol.* **192**:175–188 (1967).

81. H. J. Schatzman, The role of Na$^+$ + K$^+$-activated membrane adenosine triphosphatase, *Biochim. Biophys. Acta* **94**:89–96 (1965).

82. R. Whittam and M. E. Ager, Vectorial aspects of adenosinetriphosphatase activity in erythrocyte membranes, *Biochem. J.* **93**:337–348 (1964).

83. P. C. Caldwell and R. D. Keynes, Effect of ouabain on efflux of sodium from squid giant axon, *J. Physiol.* **148**:8 P (1959).

84. I. M. Glynn, The action of cardiac glycosides on ion movements, *Pharmacol. Rev.* **16**:381–407 (1964).

85. S. L. Bonting, K. A. Simon, and N. M. Hawkins, Studies on sodium-potassium-activated adenosine triphosphatase. I. Quantitative distribution in several tissues of the cat, *Arch. Biochem. Biophys.* **95**:416–423 (1961).

86. J. C. M. Hafkenscheid and S. L. Bonting, Studies on (Na$^+$ + K$^+$)-activated ATPase. XIX. Occurrence and properties of a (Na$^+$ + K$^+$)-activated ATPase in *Escherichia coli*, *Biochim. Biophys. Acta* **151**:204–211 (1968).

87. E. Heinz, Transport through biological membranes, *Ann. Rev. Physiol.* **29**:21–58 (1967).

88. D. F. Wallach and D. Ullrey, Studies on the surface and cytoplasmic membranes of Ehrlich ascites carcinoma cells. II. Alkali-cation-activated adenosine triphosphate hydrolysis in a microsomal membrane fraction, *Biochim. Biophys. Acta* **88**:620–629 (1964).

89. M. Barclay, R. K. Barclay, and E. S. Essner, Plasma membranes of rat liver: isolation of lipoprotein macromolecules, *Science* **156**:665–667 (1967).

90. P. Emmelot and C. J. Bos, Studies on plasma membranes. III. Mg^{2+}-ATPase, (Na$^+$ − K$^+$ − Mg^{2+})-ATPase and 5^1-nucleotidase activity of plasma membranes isolated from rat liver, *Biochim. Biophys. Acta* **120**:369–382 (1966).

91. E. Lewin and H. H. Hess, Intralaminar distribution of Na$^+$ − K$^+$ adenosine triphosphatase in rat cortex, *J. Neurochem.* **11**:473–481 (1964).

92. F. J. Samaha, Studies on Na$^+$ − K$^+$-stimulated ATPase of human brain, *J. Neurochem.* **14**:333–341 (1967).

93. R. W. Albers, G. R. de Lores Arnaiz, and E. DeRobertis, Sodium-potassium-activated ATPase and potassium-activated *p*-nitrophenylphosphatase: a comparison of their subcellular localizations in rat brain, *Proc. Nat. Acad. Sci. U.S.A.* **53**:557–564 (1965).

94. R. J. A. Hosie, The localization of adenosine triphosphatases in morphologically characterized subcellular fractions of guinea-pig brain, Biochem. J. 96:404–412 (1965).

95. H. F. Bradford, E. K. Brownlow, and D. B. Gammack, The distribution of cation-stimulated adenosine triphosphatase in subcellular fractions from bovine cerebral cortex, J. Neurochem. 13:1283–1297 (1966).

96. M. Kurokawa, T. Sakamoto, and M. Kato, Distribution of sodium-plus-potassium-stimulated adenosine-triphosphatase activity in isolated nerve-ending particles, Biochem. J. 97:833–844 (1965).

97. W. Schoner, C. von Silberg, R. Kramer, and W. Seubert, On the mechanism of Na^+-and-K^+-stimulated hydrolysis of adenosine triphosphate 1. Purification and properties of a Na^+- and K^+-activated ATPase from ox brain, European J. Biochem. 1:334–343 (1967).

98. E. DeRobertis, M. Alberici, G. R. de Lores Arnaiz, and J. M. Azcurra, Isolation of different types of synaptic membranes from the brain cortex, Life Sciences 5:577–582 (1966).

99. M. Germain, and P. Proulx, Adenosine triphosphatase activity in synaptic vesicles of rat brain, Biochem. Pharm. 14:1815–1819 (1965).

100. F. E. Samson, Jr. and D. J. Quinn, $Na^+ - K^+$-activated ATPase in rat brain development, J. Neurochem. 14:421–427 (1967).

101. S. Fahn and L. J. Cote, Regional distribution of sodium-potassium activated adenosine triphosphatase in the brain of the rhesus monkey, J. Neurochem. 15:433–436 (1968).

102. H. Matsui and A. Schwartz, Purification and properties of a highly active ouabain-sensitive Na^+, K^+-dependent adenosine triphosphatase from cardiac tissue, Biochim. Biophys. Acta 128:380–390 (1966).

103. F. J. Samaha and J. Gergely, Studies on the Na^+- and K^+-activated ATPase in human striated muscle, Arch. Biochem. Biophys. 114:481–487 (1966).

104. Y. Tashima, T. Nakao, K. Nagano, N. Mizuno, and M. Nakao, Partial purification and characterization of sodium- and potassium-dependent adenosine triphosphatase from rat-heart muscle, Biochim. Biophys. Acta 117:54–62 (1966).

105. L. W. Tice and A. G. Engel, Cytochemistry of phosphatase of the sarcoplasmic reticulum. II. In situ localization of the Mg-dependent enzyme, J. Cell Biol. 31:489–499 (1966).

106. F. Giacomelli, C. Bibbiani, E. Bergamini, and C. Pellegrino, Two ATPases in the sarcoplasmic reticulum of skeletal muscle fibers, Nature (London) 213:679–682 (1967).

107. J. Tormey, Significance of the histochemical demonstration of ATPase in epithelia noted for active transport, Nature 210:820–822 (1966).

108. V. T. Marchesi and G. E. Palade, The localization of Mg-Na-K-activated adenosine triphosphatase on red cell ghost membranes, J. Cell Biol. 35:385–404 (1967).

109. J. S. Gordon and R. M. Torack, Inhibition of cerebral adenosine triphosphatase activity by various aldehyde fixatives, J. Neurochem. 14:1155–1160 (1967).

110. H. L. Moses and A. S. Rosenthal, On the significance of lead-catalysed hydrolysis of nucleoside phosphates in histochemical systems, J. Histochem. Cytochem. 15:354–355(1967).

111. J. Somogyi, The effect of proteases on the $(Na^+ + K^+)$-activated adenosine triphosphatase system of rat brain, Biochim. Biophys. Acta 151:421–428 (1968).

112. K. Ahmed and J. D. Judah, Preparation of lipoproteins containing cation-dependent ATPase, Biochim. Biophys. Acta 93:603–613 (1964).

113. R. Tanaka and K. P. Strickland, Role of phospholipid in the activation of Na^+, K^+-activated adenosine triphosphatase of beef brain, Arch. Biochem. Biophys. 111:583–592 (1965).

114. F. Medzihradsky, M. H. Kline, and K. E. Hokin, Studies on the characterization of the sodium-potassium transport adenosine triphosphatase. I. Solubilization, stabilization, and estimation of apparent molecular weight, Arch. Biochem. Biophys. 121:311–316 (1967).

115. G. R. Kepner and R. I. Macey, Molecular weight estimation of membrane bound ATPase by in vacuo radiation inactivation, Biochem. Biophys. Res. Comm. 30:582–587 (1968).

116. M. Nakao, K. Nagano, T. Nahao, N. Mizuno, and Y. Tashima, Molecular weight of Na, K-ATPase approximated by the radiation inactivation method, *Biochem. Biophys. Res. Com.* **29**:588–592 (1967).

117. G. R. Kepner and R. I. Macey, Red cell membrane ATPase: radiation inactivation estimates of "size," *Biochem. Biophys. Res. Com.* **23**:202–207 (1966).

118. J. C. Skou, Further investigations on a Mg^{++} + Na^+-activated adenosine triphosphatase, possibly related to the active, linked transport of Na^+ and K^+ across the nerve membrane, *Biochim. Biophys. Acta* **42**:6–23 (1960).

119. R. A. Peters, M. Shorthouse, and J. M. Walshe, The effect of Cu^{++} on the membrane $(Na^+ + K^+)$-ATPase and its relation to the initiation of convulsions, *Biochem. J.* **96**:47P (1965).

120. L. J. Opit, H. Potter, and J. S. Charnock, The effect of anions on $(Na^+ + K^+)$-activated ATPase, *Biochim. Biophys. Acta* **120**:159–161 (1966).

121. H. Yoshida K. Nagai, M. Kamei, and Y. Nakagawa, Irreversible inactivation of (Na + K)-dependent ATPase and K^+-dependent ATPase and K^+-dependent phosphatase by fluoride, *Biochim. Biophys. Acta* **150**:162–164 (1968).

122. N. Gruener and Y. Avi-Dor, Temperature-dependence of activation and inhibition of rat-brain adenosine triphosphatase activated by sodium and potassium ions, *Biochem. J.* **100**:762–767 (1966).

123. J. D. Robinson, Kinetic studies on a brain microsomal adenosine triphosphatase. Evidence suggesting conformational changes, *Biochemistry* **6**:3250–3258 (1967).

124. R. F. Squires, On the interactions of Na^+, K^+, Mg^{++} and ATP with the Na^+, K^+, Mg^{++} and ATP with the Na^+ plus K^+-activated ATPase from rat brain, *Biochem. Biophys. Res. Comm.* **19**:27–32 (1965).

125. D. E. Atkinson, Regulation of enzyme activity, *Ann. Rev. Biochem.* **35**:85–124 (1966).

126. K. Ahmed, J. D. Judah, and P. G. Scholefield, Interaction of Na^+ and K^+ with cation-dependent ATPase system from rat brain, *Biochim. Biophys. Acta* **120**:351–360 (1966).

127. P. J. Garrahan and I. M. Glynn, The stoichiometry of the sodium pump, *J. Physiol.* **192**:217–235 (1967).

128. K. P. Wheeler and R. Whittam, Structural and enzymic aspects of the hydrolysis of adenosine triphosphate by membranes of kidney cortex and erythrocytes, *Biochem. J.* **93**:349–363 (1964).

129. R. W. Albers, S. Fahn, and G. J. Koval, The role of sodium ions in the activation of electrophorus electric organ ATPase, *Proc. Nat. Acad. Sci. U.S.A.* **50**:474–481 (1963).

130. R. L. Post, A. K. Sen, and A. S. Rosenthal, A phosphorylated intermediate in adenosine triphosphate-dependent sodium and potassium transport across kidney membranes, *J. Biol. Chem.* **240**:1437–1445 (1965).

131. H. Bader, R. L. Post, and G. H. Bond, Comparison of sources of a phosphorylated intermediate in transport ATPase, *Biochim. Biophys. Acta* **150**:41–46 (1968).

132. L. E. Hokin, P. S. Sastry, P. R. Galsworthy, and A. Yoda, Evidence that a phosphorylated intermediate in a brain transport adenosine triphosphatase is an acyl phosphate, *Proc. Nat. Acad. Sci. U.S.A.* **54**:177–184 (1965).

133. K. Nagano, N. Mizuni, M. Fujita, Y. Tashima, T. Nakao, and M. Nakao, On the possible role of the phorphorylated intermediate in the reaction mechanism of $(Na^+ + K^+)$-ATPase, *Biochim. Biophys. Acta* **143**:239–248 (1967).

134. S. Fahn, G. J. Koval, and R. W. Albers, Sodium-potassium-activated adenosine triphosphatase of *Electrophorus* electric organ. I. An associated sodium-activated transphosphorylation, *J. Biol. Chem.* **241**:1882–1889 (1966).

135. S. Fahn, M. R. Hurley, G. J. Koval, and R. W. Albers, Sodium-potassium-activated adenosine triphosphatase of *Electrophorus* electric organ. II. Effects of N-ethylmaleimide and other sulfhydryl reagents, *J. Biol. Chem.* **241**:1890–1895 (1966).

136. W. L. Stahl, Sodium-stimulated [^{14}C] adenosine diphosphate-adenosine triphosphate exchange activity in brain microsomes, *J. Neurochem.* **15**:511–518 (1968).

137. G. J. Siegel and R. W. Albers, Sodium-potassium-activated adenosine triphosphatase of *Electrophorus* electric organ. IV. Modification of responses to sodium and potassium by arsenite *plus* 2,3-dimercaptopropanol, *J. Biol. Chem.* **242**:4972–4979 (1967).

138. W. L. Stahl, A. Sattin, and H. McIlwain, Separation of adenosine diphosphate-adenosine triphosphate-exchange activity from the cerebral microsomal sodium-plus-potassium ion-stimulated adenosine triphosphatase, *Biochem. J.* **99**:404–412 (1966).

139. D. H. Jean and H. Bader, Dissociation constants of the pronase fragment of the phosphorylated (Na$^+$ + K$^+$)-dependent ATPase, *Biochem. Biophys. Res. Comm.* **27**:650–654 (1967).

140. J. Bader, R. L. Post, and D. H. Jean, Further characterization of a phosphorylated intermediate in (Na$^+$ + K$^+$)-dependent ATPase, *Biochim. Biophys. Acta* **143**:229–238 (1967).

141. I. M. Glynn, C. W. Slayman, J. Eichberg, and R. M. C. Dawson, The adenosine-triphosphatase system responsible for cation transport in electric organ: exclusion of phospholipids as intermediates, *Biochem. J.* **94**:692–699 (1965).

142. L. E. Hokin and A. Yoda, Inhibition by diisopropylfluorophosphate of a kidney transport adenosine triphosphatase by phosphorylation of a serine residue, *Proc. Nat. Acad. Sci. U.S.A.* **52**:454–461 (1964).

143. A. Kahlenberg, P. R. Galsworthy, and L. W. Hokin, Sodium-potassium adenosine triphosphatase: acyl phosphate "intermediate" shown to be L-glutamyl-γ-phosphate, *Science* **157**:434–436 (1967).

144. F. Lipmann and L. C. Tuttle, Acetyl phosphate: chemistry, determination, and synthesis, *J. Biol. Chem.* **153**:571–582 (1944).

145. J. S. Charnock, L. J. Opit, and H. A. Potter, Hydroxylamine and a ^{32}P-labelled intermediate in sodium-plus-potassium ion-activated adenosine triphosphatase, *Biochem. J.* **104**:17c–19c (1967).

146. C. F. Chignell and E. Titus, The effect of hydroxylamine and *N*-methylhydroxylamine on beef brain microsomal (Na$^+$ + K$^+$)-ATPase, *Biochim. Biophys. Acta* **159**:345–351 (1968).

147. D. A. Hems and R. Rodnight, Properties of phosphate bound to cerebral microsomes during adenosine-triphosphatase activity, *Biochem. J.* **101**:516–523 (1966).

148. S. Fahn, G. J. Koval, and R. W. Albers, Sodium-potassium-activated adenosine triphosphatase of *Electrophorus* electric organ. V. Phosphorylation by AT^{32}P, *J. Biol. Chem.* **243**:1993–2002 (1968).

149. G. H. Bond, H. Bader, and R. L. Post, Acetyl phosphate as substrate for (Na$^+$ + K$^+$)-ATPase, *Fed. Proc.* **25**:567 (1966).

150. G. Sachs, J. D. Rose, and B. I. Hirschowitz, Acetyl phosphatase in brain microsomes: a partial reaction of (Na$^+$ + K$^+$)-ATPase, *Arch. Biochem. Biophys.* **119**:277–281 (1967).

151. Y. Israel and E. Titus, A comparison of microsomal (Na$^+$ + K$^+$)-ATPase with K$^+$-acetylphosphatase, *Biochim. Biophys. Acta* **139**:450–459 (1967).

152. H. Bader and A. K. Sen, (K$^+$)-dependent acyl phosphatase as part of the (Na$^+$ + K$^+$)-dependent ATPase of cell membranes, *Biochim. Biophys. Acta* **118**:116–123 (1966).

153. P. Emmelot and C. J. Bos, Studies on plasma membranes. II. K$^+$-dependent *p*-nitrophenyl phosphatase activity of plasma membranes isolated from rat liver, *Biochim. Biophys. Acta* **121**:375–385 (1966).

154. M. Fujita, T. Nakao, Y. Tashima, N. Mizuno, K. Nagano, and M. Nakao, Potassium-ion stimulated *p*-nitrophenyl-phosphatase activity occurring in a highly specific adenosine triphosphatase preparation from rabbit brain, *Biochim. Biophys. Acta* **117**:42–53 (1966).

155. A. E. Rega, P. J. Garrahan, and M. I. Pouchan, Effects of ATP and Na$^+$ on K$^+$-activated phosphatase from red blood cell membranes, *Biochim. Biophys. Acta* **150**:742–743 (1968).

156. K. Nagai, F. Izumi, and H. Yoshida, Studies on potassium dependent phosphatase; its distribution and properties, *J. Biochem.* **59**:295–303 (1966).

157. T. Kanazawa, M. Saito, and Y. Tonomura, Properties of a phosphorylated protein as a reaction intermediate of $Na^+ - K^+$ sensitive ATPase, *J. Biochem.* **61**:555–566 (1967).

158. I. M. Glynn, The action of cardiac glycosides on sodium and potassium movements in human red cells. *J. Physiol.* **136**:148–173 (1957).

159. J. F. Hoffman, The red cell membrane and the transport of sodium and potassium, *Am. J. Med.* **41**:666–680 (1966).

160. J. S. Charnock and R. L. Post, Evidence of the mechanism of ouabain inhibition of cation activated adenosine triphosphatase, *Nature (London)* **199**:910–911 (1963).

161. H. Matsui and A. Schwartz, ATP-dependent binding of ^3H-digoxin to a Na^+, K^+ ATPase from cardiac muscle, *Fed. Proc.* **26**:398 (1967).

162. G. J. Siegel, G. J. Koval, and R. W. Albers, Irreversible ouabain interaction with $(Na^+ + K^+)$-ATPase: modification by cations, nucleotides and Pi, *Neurology* **18**:296 (1968).

163. R. W. Albers, G. J. Koval, and G. J. Siegel, Studies on the interaction of ouabain and other cardioactive steroids with sodium-potassium-activated adenosine triphosphatase, *Molecular Pharmacol.* **4**:324–336 (1968).

164. A. Schwartz, H. Matsui, and A. H. Laughter, Tritiated digoxin binding to $(Na^+ + K^+)$-activated adenosine triphosphatase: possible allosteric site, *Science* **160**:323–325 (1968).

165. G. J. Siegel, G. J. Koval, and R. W. Albers, Sodium-potassium-activated adenosine triphosphatase VI. Characterization of the phosphoprotein formed from orthophosphate in the presence of ouabain, *J. Biol. Chem.* **244**:3264–3269 (1969).

166. F. J. Samaha and R. W. Albers, Unpublished experiments (1968).

167. T. Rosenberg and W. Wilbrandt, Carrier transport uphill. I. General, *J. Theoret. Biol.* **5**:288–305 (1963).

168. Y. Inui and H. N. Christensen, Discrimination of single transport systems: the Na^+-sensitive transport of neutral amino acids in the Ehrlich cell, *J. Gen Physiol.* **50**:203–224 (1966).

169. O. Jardetzky, Simple allosteric model for membrane pumps, *Nature* **211**:969–970 (1966).

170. L. J. Opit and J. S. Charnock, A molecular model for a sodium pump, *Nature* **208**:471–474 (1965).

171. I. M. Glynn and U. Luthi, The relation between ouabain-sensitive K^+ efflux and the hypothetical dephosphorylation step in the "transport ATPase" system, *J. Gen Physiol.* **51**:Part 2, 385–391 (1968).

172. C. A. Ashford and K. C. Dixon, The effect of potassium on the glucolysis of brain tissue with reference to the Pasteur effect, *Biochem. J.* **29**:157–168 (1935).

173. F. Dikens and G. D. Greville, The metabolism of normal and tumor tissue. XIII. Neutral salt effects, *Biochem. J.* **29**:1468–1483 (1935).

174. R. Whittam, D. M. Blond, and M. Ruscak, The influence of ions on the metabolism of brain-cortex slices, *Biochem. J.* **96**:47p–48p (1965).

175. H. S. Bachelard, W. J. Campbell, and H. McIlwain, The sodium and other ions of mammalian cerebral tissues, maintained and electrically stimulated *in vitro*, *Biochem. J.* **84**:225–232 (1962).

176. J. C. Keesey and H. Walgren, Movements of radioactive sodium in cerebral-cortex slices in response to electrical stimulation, *Biochem. J.* **95**:301–310 (1965).

177. H. McIlwain, *Chemical Exploration of the Brain*, pp. 55–66, Elsevier, New York (1963).

178. R. Whittam, Active cation transport as a pace-maker of respiration, *Nature (London)* **191**:603–604 (1961).

179. R. Whittam, The dependence of the respiration of brain cortex on active cation transport, *Biochem. J.* **82**:205–212 (1962).

180. R. Whittam and J. S. Willis, Ion movements and oxygen consumption in kidney cortex slices, *J. Physiol.* **168**:158–177 (1963).

181. R. Whittam and D. M. Blond: *Biochem. J.* **92**:147–158 (1964).

182. D. M. Blond and R. Whittam, The regulation of kidney respiration by sodium and potassium ions, *Biochem. J.* **92**:158–167 (1964).

183. D. Blond and R. Whittam, Effects of sodium and potassium ions on oxidative phosphorylation in relation to respiratory control by a cell-membrane adenosine triphosphatase, *Biochem. J.* **97**:523–531 (1965).

184. B. Chance and B. Hess, Metabolic control mechanisms. IV. The effect of glucose upon the steady state of respiratory enzymes in the ascites cell, *J. Biol. Chem.* **234**:2421–2427 (1959).

185. R. Whittam and M. E. Ager, The connection between active cation transport and metabolism in erythrocytes, *Biochem. J.* **97**:214–227 (1965).

186. S. Minakami and H. Yoshikawa, Studies on erythrocyte glycolysis. III. The effects of active cation transport, pH and inorganic phosphate concentration on erythrocyte glycolysis, *J. Biochem.* **59**:145–150 (1966).

187. J. S. Wiley, Effect of ouabain on deoxynucleoside metabolism in hereditary spherocytic human erythrocytes, *Biochim. Biophys. Acta* **135**:1071–1074 (1967).

188. A. Schwartz, The effect of ouabain on potassium content, phosphoprotein metabolism and oxygen consumption of guinea pig cerebral tissue, *Biochem. Pharmacol.* **11**:389–391 (1962).

189. R. S. Bourke and D. B. Tower, Fluid compartmentation and electrolytes of cat cerebral cortex *in vitro*. II. Sodium, potassium and chloride of mature cerebral cortex, *J. Neurochem.* **13**:1099–1117 (1966).

190. P. D. Swanson and K. Ullis, Ouabain-induced changes in sodium and potassium content respiration of cerebral cortex slices: dependence on medium calcium concentration and effects of protamine, *J. Pharmacol. Exp. Therap.* **153**:321–328 (1966).

191. B. Chance, The energy-linked reaction of calcium with mitochondria, *J. Biol. Chem.* **240**:2729–2747 (1965).

192. H. Fujisawa, K. Kajikawa, Y. Ohi, Y. Hashimoto, and H. Yoshida, Movement of radioactive calcium in brain slices and the influence of protoveratrine, ouabain, potassium chloride, and cocaine on the movement, *Japan. J. Pharmacol.* **15**:327–334 (1965).

193. D. B. Tower, Ouabain and the distribution of calcium and magnesium in cerebral tissue *in vitro*, *Exptl. Brain Res.* **6**:273–283 (1968).

194. B. B. Marsh, The effects of adenosine triphosphate on the fibre volume of a muscle homogenate, *Biochim. Biophys. Acta* **9**:247–260 (1952).

195. J. R. Bendall, Effect of the "Marsh Factor" on the shortening of muscle fiber models in the presence of adenosine triphosphate, *Nature (London)* **170**:1058–1060 (1952).

196. S. Ebashi and F. Lipmann, Adenosine triphosphate-linked concentration of calcium ions in a particulate fraction of rabbit muscle, *J. Cell Biology* **14**:389–400 (1962).

197. S. Ebashi, Calcium binding activity of vesicular relaxing factor, *J. Biochem.* **50**:236–244 (1961).

198. W. Hasselbach and M. Makinose, ATP and active transport, *Biochem. Biophys. Res. Comm.* **7**:132–136 (1962).

199. A. Weber and S. Winicur, The role of calcium in the superprecipitation of actomyosin, *J. Biol. Chem.* **236**:3198–3202 (1961).

200. A. Weber and R. Herz, in *Biochemistry of Muscle Contraction* (J. Gergely, ed.), pp. 222–231, Little, Brown, Boston (1964).

201. A. Weber, R. Herz, and I. Reiss, Role of calcium in contraction and relaxation of muscle, *Fed. Proc.* **23**:896–900 (1964).

202. D. S. Smith, The organization and function of the sarcoplasmic reticulum and T-system of muscle cells, *Prog. Biophys. Mol. Biol.* **16**:109–137 (1966).

203. W. Hasselbach, Relaxing factor and the relaxation of muscle, *Prog. Biophys.* **14**:167–222 (1964).

204. A. Weber, in *Current Topics in Bioenergetics* (D. R. Sanadi, ed.), Vol. 1, pp. 203–254, Academic Press, New York (1966).

205. B. Fanburg and J. Gergely, Studies on adenosine triphosphate-supported calcium accumulation by cardiac subcellular particles, *J. Biol. Chem.* **240**:2721–2728 (1965).

206. F. N. Briggs, E. W. Gertz, and M. L. Hess, Calcium uptake by cardiac vesicles: inhibition by amytal and reversal by ouabain, *Biochem. Zeit.* **345**:122–131 (1966).

207. F. J. Samaha and J. Gergely, Ca^{++} uptake and ATPase of human sarcoplasmic reticulum, *J. Clin. Inves.* **44**:1425–1431 (1965).

208. W. W. Kielley and O. Meyerhoff, Studies on adenosine triphosphatase of muscle. IV. The lipoprotein nature of the magnesium-activated adenosine-triphosphatase, *J. Biol. Chem.* **183**:391–401 (1950).

209. S. V. Perry, The adenosine triphosphatase activity of lipoprotein granules isolated from skeletal muscle, *Biochim. Biophys. Acta* **8**:499–509 (1952).

210. A. Martonosi, J. Donley, and R. A. Halpin, Sarcoplasmic reticulum. III. The role of phospholipids in the adenosine triphosphatase activity and Ca^{++} transport, *J. Biol. Chem.* **243**:61–70 (1968).

211. S. Ebashi, A granule-bound relaxation factor in skeletal muscle, *Arch. Biochem. Biophys.* **76**:410–423 (1958).

212. A. Martonosi, Sarcoplasmic reticulum. IV. Solubilization of microsomal adenosine triphosphatase, *J. Biol. Chem.* **243**:71–81 (1968).

213. W. Hasselbach and K. Seraydarian, The role of sulfhydryl groups in calcium transport through the sarcoplasmic membranes of skeletal muscle, *Biochem. Zeit.* **345**:159–172 (1966).

214. B. P. Yu, E. J. Masaro, and F. D. DeMartinis, Imidazole and sequestration of calcium ions by sarcoplasmic reticulum, *Nature (London)* **216**:822–824 (1967).

215. G. Inesi and S. Watanabe, Temperature dependence of ATP hydrolysis and calcium uptake by fragmented sarcoplasmic membranes, *Arch. Biochem. Biophys.* **121**:665–671 (1967).

216. T. Yamamoto and Y. Tonomura, Reaction mechanism of the Ca^{++}-dependent ATPase of sarcoplasmic reticulum from skeletal muscle, *J. Biochem.* **64**:137–145 (1968).

217. A. Martonosi and R. Feretos, Sarcoplasmic reticulum. I. The uptake of Ca^{++} by sarcoplasmic reticulum fragments, *J. Biol. Chem.* **239**:648–658 (1964).

218. A. Martonosi and R. Feretos, Sarcoplasmic reticulum. II. Correlation between adenosine triphosphatase activity and Ca^{++} uptake, *J. Biol. Chem.* **239**:659–668 (1964).

219. W. Hasselbach and M. Makinose, Uber den Mechanismus des Calcium Transportes durch die Membranen des Sarcoplasmatischen Reticulums, *Biochem. Zeit.* **339**:94–111 (1963).

220. A. Weber, R. Herz, and I. Reiss, Study of the kinetics of calcium transport by isolated fragmented sarcoplasmic reticulum, *Biochem. Zeit.* **345**:329–369 (1966).

221. M. Makinose and R. The, Calcium-Akkumulation und Nucleosidtriphosphat-Spaltung durch die Vesikel des Sarcoplasmatischen Reticulum, *Biochem. Zeit.* **343**:383–393 (1965).

222. M. E. Carsten and W. F. H. M. Mommaerts, The accumulation of calcium ions by sarcotubular vesicles, *J. Gen. Physiol.* **48**:183–197 (1964).

223. F. Ebashi and I. Yamanouchi, Calcium accumulation and adenosine triphosphatase of the relaxing factor, *J. Biochem.* **55**:504–509 (1964).

224. J. C. Fratantoni and A. Askari, Effect of monovalent cations on the adenosinetriphosphatase of a skeletal muscle microsomal preparation, *Biochim. Biophys. Acta* **99**:259–269 (1965).

225. K. S. Lee, K. Tanaka, and D. H. Yu, Studies on the adenosine triphosphatase, calcium uptake and relaxing activity of the microsomal granules from skeletal muscle, *J. Physiol.* **179**:456–478 (1965).

226. J. R. Sommer and W. Hasselbach, The effect of glutaraldehyde and formaldehyde on the calcium pump of the sarcoplasmic reticulum, *J. Cell. Biol.* **34**:902–905 (1967).

227. A. Weber, Regulation of Ca^{++} transport by isolated reticulum, *7th International Congress of Biochemistry*, Tokyo, Abstracts II, pp. 323–324 (1967).

228. S. Ebashi, in *Molecular Biology of Muscular Contraction* (S. Ebashi, F. Oosawa, T. Sekine, and Y. Tonomura, eds.), pp. 197–206, Elsevier, New York (1965).

229. A. P. Caravalho and B. Leo, Effects of ATP on the interaction of Ca^{++}, Mg^{++} and K^+ with fragmented sarcoplasmic reticulum isolated from rabbit skeletal muscle, *J. Gen. Physiol.* **50**:1327–1352 (1967).

230. M. Makinose, Die Nucleosidtriphosphat-Nucleosid-diphosphat-Transphosphorylase-Aktivitat der Vesikel des sarkoplasmatischen Reticulums, *Biochem. Zeit.* **345**:80–86 (1966).

231. M. Otsuka, I. Ohtsuki, and S. Ebashi, ATP-dependent Ca^{++} binding of brain microsomes, *J. Biochem.* **58**:188–190 (1965).

232. H. Yoshida, K. Kadota, and H. Fujisawa, Adenosine triphosphate dependent calcium binding of microsomes and nerve endings, *Nature (London)* **212**:291–292 (1966).

233. J. D. Robinson and W. D. Lust, Adenosine triphosphate-dependent calcium accumulation by brain microsomes, *Arch. Biochem. Biophys.* **125**:286–294 (1968).

234. Y. Nakamaru, M. Kosakai, and K. Konishi, Some properties of brain microsome adenosine triphosphatases activated by magnesium and calcium, *Arch. Biochem. Biophys.* **120**:15–21 (1967).

235. Y. Nakamaru, Mg-ATPase activated by a low concentration of Ca^{++} in brain microsomes, *J. Biochem.* **63**:626–631 (1968).

236. K. Kadota, S. Mori, and R. Imaizumi, The properties of ATPase of synaptic vesicle fraction, *J. Biochem.* **61**:424–432 (1967).

237. P. Wins and E. Schoffeniels, Studies on red cell ghost ATPase systems: properties of a $(Mg^{++} + Ca^{++})$-dependent ATPase, *Biochim. Biophys. Acta* **120**:341–350 (1966).

238. I. Tasaki, A. Watanabe, and L. Lerman, Role of divalent cations in excitation of squid giant axons, *Am. J. Physiol.* **213**:1465–1474 (1967).

239. B. Katz and R. Miledi, The effect of calcium on acetylcholine release from motor nerve terminals, *Proc. Roy. Soc. Ser. B* **161**:496–503 (1964).

240. R. P. Durbin and D. K. Kasbekar, Adenosine triphosphate and active transport by the stomach, *Fed. Proc.* **24**:1377–1381 (1965).

241. D. K. Kasbekar, Studies of resting isolated frog gastric mucosa, *Proc. Soc. Exp. Biol. Med.* **125**:267–271 (1967).

242. C. Ling, An anion stimulated particulate ATP phosphohydrolase (ATPase) from bakers' yeast, *Life Sci.* **6**:1077–1084 (1967).

243. E. B. Keller and P. C. Zamecnik, The effect of guanosine diphosphate and triphosphate on the incorporation of labelled amino acids into proteins, *J. Biol. Chem.* **221**:45–59 (1956).

244. T. W. Conway and F. Lipmann, Characterization of a ribosome-linked guanosine triphosphatase in *Escherichia coli* extracts, *Proc. Nat. Acad. Sci. U.S.A.* **52**:1462–1469 (1964).

245. Y. Nishizuka and F. Lipmann, Comparison of guanosine triphosphate split and polypeptide synthesis with a purified *E. coli* system, *Proc. Nat. Acad. Sci. U.S.A.* **55**:212–219 (1966).

246. J. Gordon and F. Lipmann, Role of divalent cations in poly U-directed phenylalanine polymerization, *J. Mol. Biol.* **23**:23–33 (1967).

247. J. M. Ravel, Demonstration of a guanosine-triphosphate-dependent enzymatic binding of aminoacyl-ribonucleic acid to *Escherichia coli* ribosomes, *Proc. Nat. Acad. Sci. U.S.A.* **57**:1811–1816 (1967).

248. J. Lucas-Lenard and F. Lipmann, Initiation of polyphenylalanine synthesis by N-acetylphenylalanine-sRNA, *Proc. Nat. Acad. Sci. U.S.A.* **57**:1050–1057 (1967).

249. R. Heintz, H. McAllister, R. Arlinghaus, and R. Schweet, Formation and function of the active ribosome complex, *Cold Spring Harbor Symp. Quant. Biol.* **31**:633–639 (1966).

250. R. W. Monro, Catalysis of peptide bond formation by 50 S ribosomal subunits from *Escherichia coli*, *J. Mol. Biol.* **26**:147–151 (1967).

251. J. S. Anderson, J. E. Dahlberg, M. S. Bretscher, M. Revel, and B. F. C. Clark, GTP-stimulated binding of initiator-tRNA to ribosomes directed by f2 bacteriophage RNA, *Nature* **216**:1072–1076 (1967).

252. J. Gordon, A stepwise reaction yielding a complex between a supernatant fraction from *E. coli*, guanosine 5^1-triphosphate, and aminoacyl-sRNA, *Proc. Nat. Acad. Sci. U.S.A.* **59**:179–183 (1968).

Chapter 3

ESTIMATION OF TURNOVER RATES TO STUDY THE METABOLIC REGULATION OF THE STEADY-STATE LEVEL OF NEURONAL MONOAMINES

E. Costa and N. H. Neff

Laboratory of Preclinical Pharmacology
Division of Special Mental Health Research
National Institute of Mental Health
Saint Elizabeths Hospital, Washington, D.C.

I. INTRODUCTION

In the recent history of neurochemistry, probably no group of chemicals has made an impact on the growth of this discipline greater than that made by serotonin (5-HT) and catecholamines. A number of reports[1-5] published in the last 5 years have contributed to the formulation of new concepts concerning their neurobiological role. This chapter deals with the theoretical intricacies and the practical problems encountered in designing methods to calculate the turnover rates of neuronal monoamines *in vivo*. Usually, these methods describe monoamine stores in terms of synthesis, diffusion, and transport across neuronal membranes, because these processes converge and integrate to maintain the dynamic equilibrium of neuronal monoamine stores. These new attempts to measure turnover inevitably involve oversimplifications which appear justifiable only because they may serve as a framework to obtain more detailed information.

The conclusion that monoamine tissue levels are relatively stable in the face of changing rates of neuronal depolarization and of transmitter release is an undisputed meeting point of the attempts to interpret the physiological, biochemical, and pharmacological role of brain catecholamines and serotonin. A major problem, however, is precisely that of explaining the mechanism which maintains the constancy of brain monoamine levels. Associated with this challenge is the most trying imposition of revealing the nature of the process that interlocks depolarization of neuronal membranes with the turnover rates of monoamine stores. The necessity of determining how karyoplasmic processes are related to phenomena occurring in excitable neuronal membranes has been on the mind of neurobiologists who are unhappy with our present trend to explain brain function in terms of conventional synaptic mechanisms. Since we are attempting to understand the function of neuronal membranes and karyoplasmic constituents as a converging dynamic inter-action, it is important to keep in mind that the biochemical link between these topographically distinct structures may have to operate with a time dimension which mirrors the time constant of neuronal depolarization.

Currently, it is believed that a relationship between turnover rates and functional output of neurons is a factor contributing to maintain monoamine levels constant when neurons respond to physiological stimuli.[6] Ideally, it is hoped that functional output of a neuronal population can be estimated by calculating instant changes of monoamine turnover rates in small samples of tissue taken from unanesthetized freely moving animals. This approach may define in dynamic terms neuronal participation in a given brain func-tion. The available methods of measuring monoamine turnover rates have facilitated some understanding of the dynamics of monoamine stores, but they must be developed further to be adequate to the exacting requirements imposed by physiological experimentation.

II. MONOAMINE TURNOVER MODELS

A. Conceptual Considerations

Our inability to express several morphological features of nerve endings in terms of the dynamic compartments which participate in synaptic function is a major difficulty besetting the calculation of turnover rates of monoamine stores in living animals.[7a,b] Provisionally, this difficulty has been circum-vented by defining the morphology of nerve endings as a system of idealized metabolic compartments with kinetic parameters reflecting relationships known to exist between anatomical compartments. Forced as we are to depart from morphological considerations, it becomes convenient to repre-sent monoaminergic nerve endings with a hypothetical model, describing only those kinetic properties of the biological system which we can formulate mathematically. Turnover rate measurements, as a study of the relationships between the elements of a system, involve an analysis of the manner in which these elements react as a unit. In these models, certain intrinsic properties of nerve terminals may remain unknown without impairing our attempts to use the model to calculate turnover rates. Frequently, in this abstract repre-sentation of reality, each component of the mathematical model corresponds to one and only one component in the biological system, but not all of the components of the biological system are represented in the model. This kind of representation is usually termed homomorphic and is opposed to an iso-morphic representation of reality where the model corresponds one to one with all parameters of the biological system.

Several reports have recently appeared to reinforce the belief that the rate of monoamine turnover depends upon the functional state of the neuronal population under study.[8-10] This relationship implies that the formation of new amine molecules, although continuous, may not proceed at a constant rate. It is proposed that the monoamine levels remain constant at any time because influx and efflux of materials from these idealized metabolic compartments are equal. Provisionally, we accept the view that monoamines are stored in an open single compartment where newly synthesized molecules rapidly mix with the molecules stored in the compartment.[11] These suppositions do not have to be true in all cases to be useful.

Classically, the term turnover refers to the process of renewal of a substance in the body or in a tissue. Renewal may be accomplished by different ways: (a) a substance may be synthesized in a given compartment, or (b) a substance may be synthesized elsewhere and arrive in the compartment by means of the circulation. The term turnover rate as applied to an open system implies that a steady state exists where synthesis and transport of an endogenous substance into the metabolic compartment equals the rate of breakdown and exit. The turnover rate of brain monoamine stores seems to equal its synthesis rate, as only an insignificant quantity of the amines crosses the blood–brain barrier.[12] However, if a large quantity of amines were to originate from the circulation, the turnover rate would be greater than the synthesis rate. Hypothetically, if one represents monoamine stores as an open two-compartment system, it is possible to describe a situation where the synthesis rate is higher than the turnover rate.[13] However, available data fail to support such a model.[11]

Implicit in the kinetic definition of a nerve terminal as an open single compartment is the postulate that newly formed molecules mix with those in the storage compartment with a time constant greater than the fractional rate constant of the compartment. When the turnover rate measurements are carried out to study the effects of drugs on monoamine stores, one cannot accept axiomatically that the drug has not changed the rate of the mixing process. The present understanding of the action of several drugs on the transport of monoamines across either neuronal or cytoplasmic membranes actually makes such an interaction quite probable.[14] The practical consequences of this type of drug action cannot be overlooked when one keeps in mind that 60 to 90% of the total cell substance is made up of membranes together with associated enzymes and other special components.[15]

B. Morphological Considerations

Present electron microscopic technology may only partially reveal the morphological features involved in the molecular flow pertinent to monoamine turnover.[16] Thus, any attempt to build an isomorphic model of nerve endings may at best result in a homomorphic representation. Efforts to bridge morphology with biochemistry are hampered by our limited understanding of the following:

1. Localization of Enzymes for Monoamine Biosynthesis

Subcellular fractionation by gradient centrifugation methods of brain and other tissue samples containing monoaminergic neurons has contributed conflicting information on the localization of the enzymes involved in mono-amine biosynthesis.[17,18] The most discordant data concern the localization of the oxygenases which catalyze the 5-hydroxylation of tryptophan and 3-hydroxylation of tyrosine. These two processes are believed to limit the rate of 5-HT and catecholamine formation. Available data indicate that these enzymes are localized in synaptosomes according to a bimodal distribution, variable portions of the enzyme being soluble or bound to vesicles.[19] Since the distribution of the enzyme in adrenal medullary granules changes quantitatively with the composition of the medium used to attain their separation,[2] tyrosine hydroxylase may establish ionic bonds with synaptic vesicle membranes and these bonds may be broken by homogenization in various media. It is conceivable that the extent of binding depends on pH and ionic composition of the surroundings. Since enzyme kinetics differ when enzymes are part of membranes,[21] enzyme binding to membranes may be a valid mechanism to control the rate of monoamine biosynthesis. If this view were supported by experiments, it would follow that *in vitro* measurements of tyrosine hydroxylase activity in tissue homogenates may not reflect the *in vivo* activity of the enzyme.

2. Formation and Migration of Synaptic Vesicles

The axonal length and the eccentricity of the neuronal soma give rise to special problems with regard to molecular biosynthesis, internal transport of cellular organelles, and elimination of wastes. The concept of axonal flow reflects an attempt to formulate, in biochemical terms, the cellular dynamics imposed by the specialized morphology of neurons. The occurrence of several vesicles in front of constricted nerve fibers and at the blind end of severed axons has suggested that vesicles are made in the soma and are transported distally.[22] The transport appears to be faster than the translational movement of neuronal proteins. Since the efflux of synaptic vesicles from soma proceeds at a high rate, the possible error on turnover rate estimation caused by vesicle flow may be proportional to the number of mono-aminergic somata present in any given tissue sample. Such an interference may explain the finding that the turnover of norepinephrine (NE) stores in peripheral sympathetic ganglia is several-fold faster than that of stores in terminal varicosities (peripheral sympathetic nerve endings) of axons emanating from the ganglionic somata.[23]

Recently, the concept of a rapid transport of adrenergic vesicles within sympathetic nerve has been challenged, and data have been presented to support a new thesis for the accumulation of NE and dense core vesicles proximal to a nerve ligation.[24] The main point is that ligated nerves contain an abnormally high quantity of neurotubules and granulated vesicles proximal to the ligation, in contrast to normal sympathetic nerves, which contain relatively few organelles. Nerve ligation may cause budding off from enlarged neurotubules to form dense core vesicles which may not be related to catecholamine-containing vesicles. In support of this hypothesis, granulated vesicles are also formed in cholinergic nerve after ligation. Since the accumulation of vesicles proximal to a ligation did not appear related to NE-containing vesicles, the rate of NE accumulation after ligation may not be an accurate measure of vesicle transport.

Preliminary reports indicate that electrical stimulation of sympathetic axons fails to change the vesicle flow rate.[25] However, it remains to be ascertained whether vesicle flow rate is also unaffected by a persistent increased rate of physiological stimuli impinging on the soma of monoaminergic neurons. Provisionally, it seems reasonable to accept that the increase of NE turnover rate described in heart of mice[6] and rats[10] exposed to cold reflects the augmented turnover rate of NE molecules rather than an increased vesicle flow in monoaminergic neurons. Furthermore, the latter possibility seems improbable because the life span of an adrenergic synaptic vesicle has been estimated to be about 2 weeks,[22] while the turnover time of monoamine stores is several hours.[26,27] However, we should not exclude the possibility that small (500 Å) monoaminergic vesicles might form in terminal varicosities from the large (1000 Å–1500 Å) organelles which have been observed in monoaminergic varicosities.[28]

3. Monoamine Compartments in Varicosities

The years that have elapsed since the first description of vesicles in presynaptic terminal varicosities[4] evidently have brought little changes in our understanding of the biochemical mechanisms involved in the release of monoamines elicited by nerve stimulation. This is not surprising if we keep in mind that we fail to understand whether the synaptic vesicles freely move in the sap of terminal varicosities or are kept relatively immobile by an intricate net of neurotubules.[15] If a net were present, it would protect an extravesicular compartment of monoamine transmitter from enzymatic destruction and keep it readily available for the release by nerve impulses. Thus the net of neurotubules would divide the nerve terminals into a multicompartment system. If the vesicles were free they might continuously move in a random fashion, leading to their collision with each other or with fixed points of release in the membrane of the varicosities. That synaptic vesicles freely move in nerve varicosities lacks direct experimental support, but appears consistent with the following observations: (a) the recording of the spontaneous end plate potentials in smooth muscle cells proximal to sympathetic nerve endings reveals that at rest the transmitter efflux from adrenergic terminals follows a Poisson distribution.[29] Such a finding is best explained by randomized movement of synaptic vesicles which release the transmitter on collision with fixed points of release in the membrane; (b) catecholamines, ATP, and a soluble protein are released from the adrenal medullary granules into the extracellular fluid.[30,30a,30b] Since empty granules[31] may be seen in stimulated chromaffin cells, it has been inferred that granules retain their identity after they are emptied of their catecholamine content.[32] Similar findings have been reported for the vesicles in terminal varicosities of rats given reserpine. The biochemical composition of synaptic vesicle membranes[15] indicates that vesicles have a net negative charge which may be instrumental in maintaining vesicle motion. Thus, it seems possible that catecholamines are also released directly from synaptic vesicles of nerve endings into the extraneuronal space; and (c) most of the monoamines stored in the neuron are localized in the vesicles. A study of adrenergic nerves depleted of their catecholamine content by reserpine indicates that the vesicular compartment of amine plays a fundamental physiological role. In sympathetic nerves from reserpine-treated animals, when this NE compartment is depleted and NE uptake by vesicles is blocked, the function of sympathetic nerves is also impaired. Moreover, in rats treated with reserpine, nerve impulses reacquire their capacity to elicit a response from adrenergic receptors when the uptake of NE-^{3}H into vesicles of terminal varicosities is partially recovered.[33]

These three lines of evidence merely indicate that, in nerve endings, neurotubules connecting the vesicles to the external membranes are not necessary to explain adrenergic function. Thus, a multicompartment system is not the only possible model of nerve endings. Actually, such a model lacks valid morphological and physiological support. Hence, an open multicompartment system may be considered only if one remembers that one of the two compartments is very small and contains a fraction of the total NE stored in a tissue.[13]

C. Biochemical Considerations

Many reports suggest that during stimulation of adrenergic neurons, the steady-state concentration of NE is maintained close to control levels by an increased rate of monoamine biosynthesis.[6,34] In addition, similar findings have been reported for the stimulation of serotonergic neurons in the raphé medianus nuclei of the rat brain.[35] Thus, the energy required to maintain the steady-state concentrations of neuronal monoamines increases with the frequency of nerve impulses. In many metabolic reactions involving multienzyme systems, the first enzyme is often inhibited by the product of the last reaction of the sequence. Usually, such inhibition is noncompetitive with the substrate of the first enzyme, that is, the inhibition cannot be reversed by increasing the concentration of the substrate of the first enzyme. Functionally, such enzyme systems are self-regulating. The first enzyme in the sequence is called an allosteric enzyme to indicate that the enzyme has two sites, one specific for its substrate and the second for the allosteric effector. The effector may activate or inhibit the enzyme; allosteric enzymes have been described which can be activated or inhibited by more than one molecular type of effector. Reports described in detail later in this chapter indicate that the multienzyme system concerned with the biosynthesis of catecholamines fits, in part, the pattern of allosteric inhibition as outlined above.

It has been postulated that enzymes change their conformation as they combine with substrates, and thus may exist in two forms in labile equilibrium:[36]

$$E_s \rightleftharpoons E_i$$
$$E + I \rightleftharpoons E_i I$$

Where E_s is the catalytically active form and E_i is the inactive form, the configuration of these two enzymes may differ only slightly. If the affinity of E_s is maximal for a substrate (S) and that of E_i is maximal for the inhibitor (I), then in the presence of excess of inhibitor, the equilibrium is perturbed and the inactive form E_i will prevail. It is now becoming clear that many soluble enzymes are not randomly distributed in the intact cell,[37] but rather organized in specific ways into functionally significant assemblages. For instance, "soluble" enzymes appear to be more or less loosely attached to foundation macromolecules which are an integral part of the membranes.[38] The net result is not only compartmentation of enzymes but formation of complexes with probable functional roles. It may be inferred that aggregation of an enzyme may reduce, enhance, or modify its catalytic activity.[21] The idea that the formation of a protein aggregate may confer new three-dimensional arrangements and thereby new modes of activity to the components of the complex is an attractive one for the allosteric concept. In fact, the three-dimensional structures of proteins are relatively easily perturbed by interaction with specific compounds of low molecular weight. If the controversial reports on the localization of tyrosine hydroxylase (see Section II, A, 1) are interpreted in the light of these biochemical concepts, it may become interesting to study whether the catalytic activity of this soluble enzyme changes in the presence of synaptic vesicles. The possibility that the control of NE synthesis can be formulated by an allosteric inhibition is enhanced by reports that NE may combine with the effector site of tyrosine hydroxylase.[39] Hence NE impairs tyrosine hydroxylase by competing with a tetrahydropteridine (2-amino-4-hydroxy-6,7-dimethyl-5,6,7,8-tetrahydropteridine).[40]

Recent reports show that the tyrosine hydroxylase activity of a constricted axon does not increase when synaptic vesicles accumulate above the constriction, indicating that the enzyme in axons may not be bound to vesicles.[20] Since tyrosine hydroxylase activity of isolated axons cannot be increased by nerve stimulation, depolarization *per se* may not control NE synthesis.[41] Hence, the control involves a dynamic equilibrium between enzyme, substrates, and products; such a pattern is compatible with an allosteric process.

The hydroxylation of tryptophan at the 5-position is postulated as being the rate-limiting event in 5-HT synthesis.[42] *In vitro* evidence for inhibition of this enzyme by 5-HT is not yet available. Other factors may control 5-HT biosynthesis *in vivo*. Diaz et al.[43] have shown that increasing the concentration of O_2 in the inspired air stimulates brain 5-HT synthesis. Perhaps the availability of O_2 limits the rate of 5-HT synthesis *in vivo*. Phenylalanine and catecholamines are potent inhibitors of tryptophan hydroxylase *in vitro*,[42] and thus might influence 5-HT formation during pathological states when they are abnormally high.

Eiduson[44] has suggested that 5-HT levels in the developing chick brain may influence tryptophan hydroxylase formation by end-product repression. Contractor and Jeacock[45] have provided circumstantial evidence that high levels of 5-HT can inhibit aromatic amino acid decarboxylase *in vitro*. These authors suggested that 5-HT might regulate its own synthesis *in vivo* by negative feedback at this step. The mechanisms for regulating indolealkylamine metabolism in brain are not known; however feedback inhibition has not been demonstrated.[45a]

D. Physiological Considerations

Peripheral sympathetic nerves retrieve part of the transmitter released by nerve impulses from receptor sites.[46] Thus, the rate of dissociation of catecholamines from postjunctional receptors can be viewed as controlled by two processes: (a) diffusion into the circulation and (b) retrieval from receptor sites by an amine pump localized on the membrane of the nerve terminal. This pump may also be present in other portions of the neuronal membrane.[47] The NE pump can be saturated,[48] therefore, it has a limited capacity. Only a minimal part of the transmitter released at rest escapes retrieval by the membrane pump.[47] If sympathetic nerves are stimulated at a frequency of 6 impulses per second, about 70 to 80% of the released NE appears to be retrieved.[47] When the rate of stimulation is increased to 10 impulses per second, a greater proportion of the released amine escapes retrieval.[49] Proper functioning of the amine pump involves the binding of the transmitter to the receptor.[50] It has been reported that irreversible blockade of receptors with phenoxybenzamine does not impair the uptake of blood-borne catecholamines but does inhibit the retrieval of the neurally released amine during blockade of receptors.[50] When sympathetic nerves to the colon are stimulated at 10 impulses per second while the receptor is blocked by phenoxybenzamine, initially a large portion of the transmitter released reaches the circulation. However, after a few seconds of stimulation, the amount of NE in the effluent declines because the quantity of transmitter released by each impulse is greatly decreased. At this point, the tissue NE content is only slightly decreased.[50] If stimulation is applied after 30 min of rest, the initial rate of NE release reappears. These results indicate that (a) nerve impulses do not release from the total store of NE but have access to a limited compartment which can be readily exhausted when stimulation occurs in the presence of phenoxybenzamine; (b) during nerve stimulation, the increased synthesis rate that concurs to maintain the steady-state level of NE is not sufficient to cope with the rate of NE release elicited by unphysiological frequencies of nerve stimulation.[51] These interpretations are consistent with the following: (a) when high frequency stimulation is applied, synaptic vesicles adhere persistently to fixed release points on the varicosity membrane;[46] (b) nerve impulses release NE from those synaptic vesicles which are attached to the varicosity membranes;[46] (c) the increase of NE synthesis elicited by nerve stimulation is probably localized to the vesicles involved in NE release.[50] When high frequency stimulation is applied, synthesis of NE may be a process occurring in a small neuronal compartment.[51] Hypothetically, this could be represented by the selected population of synaptic vesicles adhering to the membrane.

If synthesis and reuptake are two integral components of the dynamic equilibrium that controls the stability of monoamine levels, conditions affecting one of these processes may indirectly change the rate of the other.[52] Compensation between these two mechanisms has been invoked to explain the increase of amine turnover elicited by drugs that block amine retrieval by nerve endings. However, the increased synthesis associated with the action of drugs that impair retrieval does not overcome their action on monoaminergic function: the latter remains imprecise, sluggish, and retarded.[46] Thus it is presently believed that in peripheral sympathetic neurons, synthesis and reuptake contribute to assure a high level of probability in the delivery of neuronal message while reuptake limits its time course. A number of the concepts expressed above were developed from studies of peripheral adrenergic function. They should not be extrapolated to the CNS neurons that store NE and 5-HT, unless there is valid evidence that these neurons mirror the function of sympathetic neurons.

III. CATECHOLAMINE TURNOVER

A. Synopsis of Catecholamine Biosynthesis

As reported in detail in another chapter of this book, in 1939, Blaschko[53] postulated that the major pathway for the biosynthesis of catcholamines proceeded as shown by the following sequence of events.

L-tyrosine → L-3,4-dihydroxyphenylalanine →
                   (DOPA)

3,4-dihydroxyphenylethylamine → (−)3,4-dihydroxyphenylethanolamine →
   Dopamine (DM)   Norepinephrine (NE)

(−)3,4-dihydroxyphenylethanol-N-methylamine
     Epinephrine (EPI)

This sequence of events is supported by a number of reports[54–57] and is held as a valid model of the catecholamine biosynthesis in mammals, including man.

It is currently believed that NE is the transmitter substance in peripheral sympathetic nerves[58] and in CNS nerve endings.[59] DM not only is the precursor of NE but is a final neuroeffector[60] in the nigro neostriate pathway[61] and in other pathways of the diencephalon and telencephalon.[62] The cytoplasmic particles of the adrenal medulla are the principal repositories for EPI.[63]

Only recently, many of the enzymes involved in the formation of catecholamines have been described in detail. Tyrosine hydroxylase, which catalyzes the initial step of catecholamine biosynthesis, is the proposed rate-limiting enzyme for the formation of catecholamines.[40] The enzyme, as prepared from bovine adrenal medulla, requires O_2 and a tetrahydropteridine for activity. Bagchi and McGeer[64] studied the properties of tyrosine hydroxylase from rat and bovine caudate nucleus. Enzyme activity was associated with a particulate fraction, but no requirement for a tetrahydropteridine was reported. In bovine splenic nerve homogenate, however, Stjarne and Lishajko[65] reported that the enzymatic hydroxylation of tyrosine occurred in the particulate-free, high-speed supernatant. At present, the information concerning the subcellular localization of tyrosine hydroxylase in adrenal medulla and other tissues suggests that a bimodal distribution of the enzyme may be a more realistic description of its subcellular localization.

Since the hydroxylation of tyrosine is thought to be the rate-limiting step in the formation of catecholamines, it is the most likely site for the regulation of amine synthesis. Indeed, catecholamines can inhibit the formation of DOPA from tyrosine *in vitro*.[39] Apparently, they compete with the pteridine cofactor and not with the enzyme substrate. These data do not rule out the possibility that catecholamines interact with an allosteric site on the enzyme. This possibility is supported by several reports showing that, *in vivo*, an enhancement of catecholamine tissue levels is associated with a reduced catecholamine synthesis rate,[23,66] prompting the suggestion that product inhibition is a controlling mechanism for catecholamine synthesis.[66,67,68]

In 1938, Holtz *et al.*[69] discovered the enzyme 3,4-dihydroxyphenylalanine (DOPA) decarboxylase. Studies with other substrates and purified enzyme preparations have indicated that DOPA decarboxylase acts on many substrates. Lovenberg *et al.*[70] have proposed the name "aromatic amino acid decarboxylase" instead of DOPA decarboxylase because 5-hydroxytryptophan, DOPA, phenylalanine, histidine, tryptophan, and α-methyl-DOPA are all decarboxylated by the same enzyme. Pyridoxal phosphate is a cofactor for decarboxylation. Since the substrate specificity of this soluble enzyme cannot be established from studies with tissue homogenate, perhaps it is still premature to consider that the enzyme, as it operates *in vivo*, lacks specificity. In gradient centrifugation studies, Rodrigues de Lores Arnaiz and DeRobertis[71] have found decarboxylase activity in the brain synaptosome fraction; rigorous homogenization of brain usually results in a soluble enzyme.

The conversion of DM to NE is catalyzed by dopamine-β-hydroxylase. This enzyme has been purified from bovine adrenal particles and was shown to require ascorbate, a dicarboxylic acid such as fumarate, and O_2.[71a] The enzyme has an average molecular weight of approximately 2.9×10^5 and contains approximately 2 μmoles of Cu^{2+} per μmole of enzyme.[72] Evidence has been obtained that the Cu^{2+} in the enzyme undergoes cyclic reduction and oxidation during the overall hydroxylation reaction.[72] In heart, dopamine-β-hydroxylase is associated with the NE storage particles.[73]

Recently, Stjarne[74] reported that dopamine-β-hydroxylase is the only enzyme involved in catecholamine biosynthesis localized in bovine splenic nerve synaptic vesicles. In synaptic vesicles and chromaffin granules, this enzyme is surrounded by a semipermeable membrane,[75] which could play a role as a rate-limiting factor in the biosynthesis of catecholamines.

The final step in the biosynthesis of EPI is the N-methylation of NE to EPI. In 1949, Bulbring first demonstrated that epinephrine was formed from NE in the minced adrenal gland.[76] Subsequently, Axelrod partially purified and described the properties of phenylethanolamine-N-methyltransferase, the enzyme thought to be responsible for this conversion.[77,78] The enzyme is found in the high-speed supernatant, and S-adenosylmethionine is an absolute requirement for this reaction. This enzyme is specific for phenylethanolamine and not for phenylethylamine derivatives. Both dextro and levo compounds are N-methylated, but the levo isomers have a higher affinity. The enzyme activity of adrenal medulla is apparently regulated by glucocorticoids secreted by the adrenal cortex.[79]

B. Synopsis of Catecholamine Metabolism

Catecholamines are metabolized primarily by two enzymes: monoamine oxidase (MAO) and catechol-O-methyltransferase (COMT).

MAO is a term used to describe enzymatic preparations that have the ability to catalyze the following reactions:

$$R\text{-}CH_2\text{-}NH_2 + O_2 + H_2O \xrightarrow{\text{MAO}} R\text{-}CHO + NH_3 + H_2O_2$$

These oxidases can be provisionally separated into two groups by their sensitivity to carbonyl reagents: (a) those insensitive to carbonyl reagents catalyze the oxidative deamination of 5-HT, NE, and DM, and will be termed in this synopsis MAO; (b) those sensitive to carbonyl reagents can be subdivided into diamine oxidases and polyamine oxidases.[80]

MAO is found complexed with the outer membrane of mitochondria.[81,82] Copper was thought to be a metal cofactor of the enzyme.[83] Recent reports show that the purified enzyme contains 0.73 moles of copper in 3×10^5 g of protein, indicating that enzyme-bound copper does not seem to be a requisite for enzyme activity.[84,85] Evidence is also available which suggests that a riboflavinlike material is present in MAO of mitochondria, probably covalently linked to the peptide chain.[85] The enzyme extracted from different organs shows some substrate selectivity.[80] These observations support an earlier suggestion that MAO is not one enzyme, but a mixture of different enzymes.[86] Gorkin has recently shown selective inhibition of enzymatic deamination of biogenic monamines by some MAO inhibitors.[80,87] Harmine is 1000 times more potent as an inhibitor of the deamination of 5-HT than of tyramine or tryptamine.[80] Giachetti and Shore[88] reported that various preparations of rat and guinea pig MAO showed a preference for the levo isomers of the catecholamines, confirming an early report by Blaschko et al.[89]

In vivo, aldehyde products of oxidative deamination are usually further oxidized by aldehyde dehydrogenase to acids. However, some of the aldehyde products are reduced to alcohols. Deaminated products may undergo further catabolism by COMT, yielding as end products either a methoxy acid or methoxy alcohol.[90]

Oxidative deamination is generally assumed to be the primary fate of catechols released from their storage sites within the nerves; the acidic and alcoholic end products leave the neuron and are methoxylated extraneuronally.[91,92] Catecholamines released at the synaptic junction by nerve impulses are lost from neuronal stores as such.[50,93,94] Extraneuronally, either they diffuse into interstitial fluid and are methylated in the meta position by COMT, or they are recaptured by a specialized pump functioning on the excitable neuronal membrane.[93,95]

COMT, first described in detail by Axelrod and Tomchick,[96] is found primarily in the high-speed supernatant and has an absolute requirement for a divalent cation. S-Adenosylmethionine is apparently the methyl donor. The enzyme does not show specificity toward d or l catechol isomers. All of the normally occurring catechols are O-methylated, including NE, DM, and DOPA and their deaminated acidic and alcoholic metabolites. O-Methylation is an important step in the inactivation of circulating catecholamines. Most metabolism occurs in the liver; the activity of other tissues, including that of sympathetic ganglia, is usually low.[97]

IV. METHODS OF ESTIMATING CATECHOLAMINE TURNOVER

A. Isotopic Methods

1. Labeling of Peripheral Stores with DL-NE-³H

The first *in vivo* attempt[98,99] to measure catecholamine turnover rates in peripheral sympathetic nerves was based on the assumption that injected radioactive NE was removed from the circulation by the amine pump in terminal varicosities of adrenergic nerves[100] and in turn mixed with NE stores.[98,99] The only aspect of the problem which is still controversial concerns the rate and mode of mixing of the label with the stored amine.[11,13]

High specific activity (5–7 C/mmole), racemic, tritium NE-7³H (NE-³H) is used most often when studying NE turnover because its high specific activity makes it possible to label NE stores in sympathetic nerves without changing the steady state.[3,95] In 1961, Whitby *et al.*[100] found that after the administration of 1 μC of NE-³H to a mouse, the label disappeared from the whole mouse rapidly at first and then more slowly. O-Methylation was proposed as the major catalytic process responsible for this rapid decline. Since large amounts of NE-³H were found in tissue long after its pharmacological effects had ceased, it was suggested that tissue binding played an important role in terminating the action of catecholamines. Similar results were obtained when radioactive EPI was administered.[101] In subsequent studies it was found that NE-³H disappeared from rat heart in a multiphasic fashion, leading to the conclusion that NE is stored within the sympathetic nerves in a multicompartment system where stores turned over at different rates.[102,103] Added support for this hypothesis was provided by the finding that tyramine could release only a limited amount of heart NE.[104] Moreover, if the heart was prelabeled with NE-^{3}H, tyramine released a greater quantity of NE-^{3}H than endogenous NE, again supporting multicompartmentation of the endogenous catecholamine store.[105,106] This view, corroborated by a number of other reports,[107–109] has received wide acceptance.

Several authors[109,110] have found that tyramine rapidly disappears from heart after injection. The tyramine-resistant pool of NE in heart seems to be an artifact resulting from the rapid metabolism of the drug, since frequent injections of tyramine elicit an exponential decline of heart, spleen, and uterus NE content to values about 10% of normal.[111,112] A recent report has shown that *in vitro* high concentrations of tyramine may block catecholamine synthesis.[113] It was inferred that high doses of tyramine would deplete heart NE completely because the releasing effect of the drug would summate with blockade of NE synthesis. This possibility needs verification, for in rats given tyramine, the concentrations of this drug at the cardiac sites where tyrosine hydroxylase occurs have not been estimated. Moreover, the decline of cardiac NE after tyramine is exponential but its fractional rate is an order of magnitude greater than that observed after blockade of NE synthesis.[3] In addition, pretreatment with a MAO inhibitor enhances tyramine tissue levels but fails to increase the rate constant of heart NE decline elicited by tyramine.[111] These considerations suggest that a bimodal compartmentation of heart NE is not the necessary corollary of this tyramine study.

The first concerted effort to measure actual rates of cardiac NE synthesis after injections of NE-^{3}H was presented in 1963.[98] After labeling the heart with NE-^{3}H, the specific activity of NE declined diphasically, suggesting that endogenous NE might be stored in an open, two-compartment system. From a mathematical evaluation of the decline, it was estimated that the size of the two compartments was about 40% and 60% respectively. The calculation of NE turnover rate assuming a two-compartment open system was reported by Beaven.[114]

Subsequent studies have shown that after administering small intravenous doses of radioactive NE (about 0.1 μg/kg of NE-^{3}H, as compared with 1.0 μg in earlier studies), a single exponential decline of NE-specific activity is produced.[3,11,23,27,66,95] These results, along with a report that tyramine given after a small dose of NE-^{3}H will release NE with the same specific activity as that present in heart at all times,[11] indicate that the multicompartment storage system based on early reports[105,106] is an artifact resulting from the injection of doses of NE-^{3}H greater than tracer amounts.

Kinetically, a single exponential decline of heart NE-specific activity may be consistent with a number of hypothetical models, including a two-compartment open system where the size of one of the cardiac compartments is about 15% of the other.[13] As an operational simplification, the models for calculating monoamine turnover rates considered in this chapter are based on a single open-compartment system.

The exponential decline of the specific activity of heart NE and other information summarized above are indications favoring a single open compartment where NE levels are maintained by a balance between synthesis and efflux. In such a system, the change of the specific activity of NE($\boxed{NE}$) with time is proportional to the $\boxed{NE}$ present at that time, or expressed in differential form:

$$\frac{d\boxed{NE}}{dt} = -k\boxed{NE} \tag{1}$$

where k is fractional rate constant.

Upon integration this equation becomes

$$\boxed{NE} = \boxed{NE}_o\, e^{-kt} \tag{2}$$

For convenience, this is usually expressed in common logarithms

$$\log \boxed{NE} = \log \boxed{NE}_o - \frac{kt}{2.303} \tag{3}$$

By plotting on semilogarithmic graph paper cardiac $\boxed{NE}$ versus time, a graph similar to that shown in Fig. 1 is obtained. The rate of synthesis, which must be equal to the rate of loss, is calculated by multiplying k by the steady-state NE levels. The fractional rate constant, k, can be obtained graphically or by statistical analysis of the individual values. Graphically, the best fit line for the value is estimated by eye. By dividing 0.693 (ln 2) by the half-life ($T\frac{1}{2}$ = time required for $\boxed{NE}$ to decline by $\frac{1}{2}$) one obtains k.

$$k = \frac{0.693}{T\frac{1}{2}} \tag{4}$$

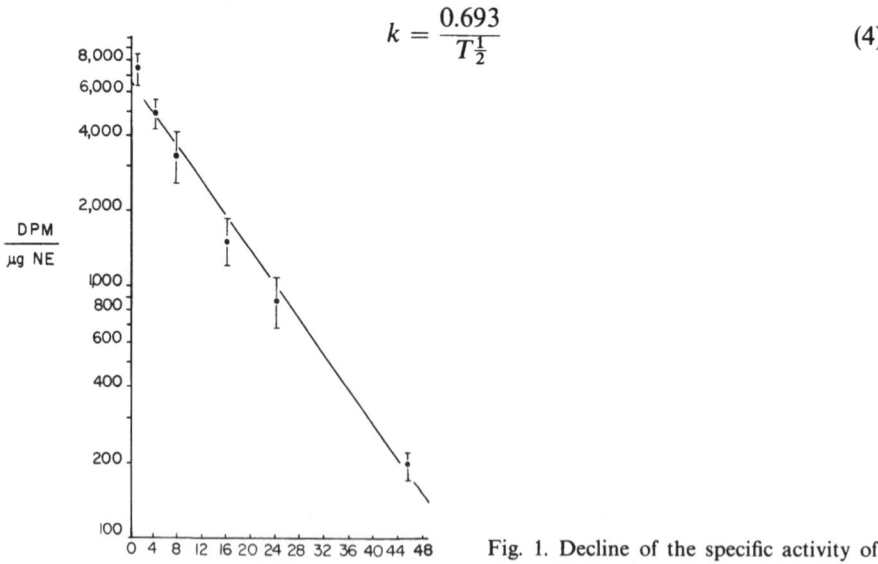

Fig. 1. Decline of the specific activity of heart NE after a small dose of NE-^{3}H.

The equation of this line is best calculated in the following form

$$\hat{Y} = \bar{Y} + b(X - \bar{X})$$

where $\hat{Y}$ is the expected value of Y (in this case $\boxed{\text{NE}}$) for any theoretical value of X, $\bar{Y}$ is the mean of the observed values of Y, $\bar{X}$ is the average value of X, and b is the regression coefficient or slope of the line computed as

$$b = \frac{\Sigma xy}{\Sigma x^2} \tag{6}$$

The quantities required for the calculation of the regression equation are determined from the following equations

$$\Sigma xy = \Sigma XY - \frac{(\Sigma X)(\Sigma Y)}{n} \tag{7}$$

$$\Sigma x^2 = \Sigma X^2 - \frac{(\Sigma X)^2}{n} \tag{8}$$

where X = time of sacrifice; Y = log of $\boxed{\text{NE}}$; n = number of observations. The fraction rate constant of the NE compartment under study is $2.303b$. To test whether the relationship is truly linear, an analysis of variance for linearity of regression should be performed.[115]

In a recent report,[13] it was suggested that the above method estimates the turnover rate of NE but not the rate of NE synthesis. This suggestion stems from the hypothetical case that: (a) cardiac NE is stored in an open two-compartment system; (b) the hypothetical size of one compartment is so small as to prove that the change in the slope of the $\boxed{\text{NE}}$ occurs at an early time when the precise estimation of k is impossible because of the racemic nature of the NE-^{3}H injected. Although this objection cannot be rejected, it should be stated that the single open-compartment model system of neuronal NE stores has resisted a number of challenges and is consistent with available data. In contrast, the open two-compartment model is not essential to explain results of kinetic studies of NE compartmentation. Specifically, the validity of the isotopic procedure, challenged by the recent report,[13] rests on the following simplifications: (a) at steady-state equilibrium, the total rate of NE synthesis equals efflux; (b) the neuron neither distinguishes between radioactive and nonradioactive molecules nor does it differentiate between old and newly formed NE molecules; (c) the label is incorporated into a population of biosynthetic units which behave as a reliable sample of the kinetic properties of the tissue under investigation. Since these assumptions are only partly supported by experimental data, it is impossible to rely exclusively upon this isotopic procedure to measure the turnover rate of endogenous NE stores. Computation based upon other methods should be employed to confirm the finding obtained with this isotopic procedure.

2. Labeling of Brain Stores with Radioactive Catecholamines

Iversen and Glowinski[116,117] have proposed that the rate of decline of radioactive NE (NE*) from brain after introducing NE-^{3}H into the lateral ventricle of the rat brain can be used to compute the rate of turnover of endogenous amine. The following problems remain to be resolved: (a) Fluorescent microscopic studies have shown that the uptake of intraventricularly injected catecholamines is significant in cells surrounding the ventricle, including ependymal cells which normally do not store monoamines.[118] (b) Monoamine oxidase of rate brain is l-stereospecific;[88] since DL NE-^{3}H is injected intraventricularly, it is hard to formulate a method that decides when the unphysiological stereoisomer is eliminated from brain. (c) The decline of radioactive NE is multiphasic, denoting multicompartmentation.[117] Since morphological studies support the view that this multiphasic decay depends on binding of the label to dopaminergic neurons and to ependymal cells, no substantial base is available to decide which portion of the multiphasic decline describes the turnover rate of endogeneous catecholamines.

Iversen and Glowinski[117] ignore this difficulty and calculate the turnover of NE from the portion of the decline which approximated the rate constant of the brain catecholamine efflux calculated by other methods. However, when appropriate corrections are made, the results of this calculation no longer coincide with the estimation of brain catcholamine turnover obtained with other methods.[3] These considerations suggest that it is premature to compute catecholamine turnover rates from the decline of NE* after the intraventricular injection of the radioactive amine. Furthermore, when this method is used to study drug effects on catecholamine turnover, the nonspecific uptake of labeled NE by brain cells adds an unnecessary source of error.†

3. Labeling of Catecholamine Stores with Precursor Amino Acids

Amino acid precursors of catecholamines, tyrosine-^{14}C (T*) and 3,4-dihydroxyphenylalanine-^{14}C(DOPA*)[8–10,13,34,41,51,54,55,67,68,113,114, 120–122] have been used to measure turnover of brain catecholamine *in vivo*

† In a recent report,[119] a new technique was proposed to measure the metabolism of norepinephrine in the CNS. This is a double isotopic technique where intracisternal injections of dl-NE-^{3}H and intravenous injections of dl-NE-O^{14}C are given to measure the turnover of endogenous brain NE from urinary analysis. Appropriate functions were derived from the quantities and rates of urinary appearance of labeled NE, normetanephrine, vanilmandelic acid, and 3-methoxy-4-hydroxyphenylglycol (MHPG). The rate of feed of NE-^{3}H to the body pool was obtained by a Laplace transformation of the appropriate functions, convolution, and subsequent inversion. This elaborate method partially corrects for leakage of intracisternally injected label into the circulation but neither corrects for nor takes into consideration the following: (a)The stereospecificity of MAO; and (b) the nonspecific uptake of NE in brain by cells which normally do not store NE. Both problems should be solved before this technique can produce reliable measurements of turnover rates of endogenous NE stores. This study indicates that about 22 to 27% of the MHPG in urine has its origin from the brain pool of NE-^{3}H. However, for the reasons given above, there is no proof that this metabolite originated in the endogenous pool of brain NE.

and *in vitro*. Since it is presently believed that tyrosine hydroxylation is rate-limiting for the biosynthesis of catecholamines *in vivo*, a comparison of the ^{14}C conversion into catecholamine when either T* or DOPA* is injected has been commonly used as a tool to detect whether a given experimental variable is acting as the rate-limiting event of catecholamine biosynthesis. The mechanism whereby catecholamine levels that are greater than control levels inhibit the rate of NE biosynthesis[23,66] was studied by comparing the rate of conversion of DOPA* or T* into NE.[68] These experiments showed that ^{14}C incorporation into NE in rats given a MAO inhibitor acutely, was inhibited when T* but not when DOPA* was the precursor injected. A similar approach was followed to study the nature of the increased rate of NE synthesis when sympathetic output was enhanced. This study revealed that nerve stimulation,[8,123] exposure to cold, and muscular exercise increase the rate of incorporation of T* into NE, but do not alter the rate of incorporation of DOPA* into NE. This difference has been related to the saturation of tyrosine hydroxylase by tissue tyrosine while the levels of DOPA are very low[124] and do not saturate the enzyme, aromatic amino acid decarboxylase; hence, injected DOPA is converted to catecholamine at a high rate.

    In vivo catecholamine turnover rates were estimated from the conversion of T* to labeled catecholamines when tissue compartments of tyrosine were labeled at steady-state[13,121] and in nonsteady-state conditions.[68] The latter has occurred as an unwanted complication of labeling with pulse injections of T*. In these experiments, a relatively large amount of T* must be injected because only a small part of injected T* is converted to catecholamines. Since the specific activity of the commercially available T* is rather low, it is impossible to rapidly inject a truly tracer dose of T*. Small amounts of T* can be injected by administering a slow constant infusion of T*. When the plasma levels of tyrosine are elevated by a rapid injection of T*, a large amount is metabolized to DOPA* in liver. In normal rats, aromatic amino acid decarboxylase and MAO promptly reduce the levels of DOPA* and radioactive DM in liver and other tissues. When MAO is inhibited, the rate of catabolism may be decreased. Hence, published results[68] show that immediately after injecting a MAO inhibitor in guinea pigs, when the catecholamine tissue levels are still equal to untreated animals, the labeling of the catecholamine stores by a pulse injection of T* is greater than in controls. Since such a difference does not occur after a slow rate of infusion of T*,[121] one should be aware that additional sources of error are created when turnover of NE is estimated by pulse injections of T*.

    When T* is infused intravenously at a constant rate, the accumulation of labeled catecholamines with time is dependent on the rate of NE synthesis.[125] Since it is impossible to measure the specific activity of tissue tyrosine ($\boxed{TT}$) in the adrenergic neuronal compartment involved in catecholamine synthesis, estimations of NE synthesis involve formulating a number of assumptions and simplifications. Some authors[13] have assumed that the area of the plot of brain $\boxed{TT}$ with time of infusion bears some relationship to the $\boxed{TT}$ in the neuronal compartment that synthesizes

NE. These experiments can be questioned for two reasons: (a) The measurements of $\boxed{\text{TT}}$ were incorrect because they included an amphoteric radioactive contaminant.[126] In brain, the percentage of radioactivity contributed by the amphoteric contaminant increases with time of tyrosine infusion; thus the mean $\boxed{\text{TT}}$ as calculated by Sedvall et al.[13] does not estimate mean brain $\boxed{\text{TT}}$. (b) Brain catecholamines are formed in a tyrosine compartment having a specific activity greater than that of brain $\boxed{\text{TT}}$ because the specific activity of brain DM is greater than that of total brain tyrosine purified from the amphoteric contaminant.[126]

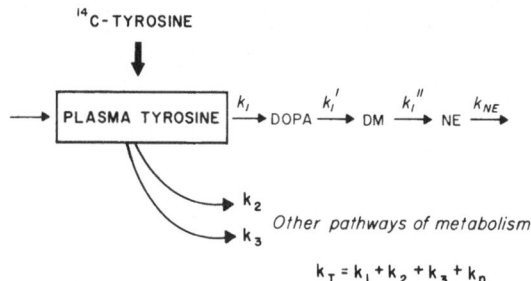

Fig. 2. Hypothetical model representing the formation of the norepinephrine from tyrosine.

These simplifications can be questioned because synthesis rates are measured from an average $\boxed{\text{TT}}$ during the time of infusion rather than correcting for the turnover rate of endogenous tyrosine. We suggest that the specific activity of catecholamines might bear a relationship to the specific activity of plasma tyrosine ($\boxed{\text{T}}$). This relationship might be described by the hypothetical model of Fig. 2.[125] In this model k_1 is the fractional rate constant for the limiting event of catecholamine formation, the hydroxylation of tyrosine. It is assumed that k_1 is smaller than K'_1, K''_2, K_{NE}. As presented in Fig. 2, $\boxed{\text{T}}$ would be expected to change with time as shown in the following differential equation.

$$\frac{d\boxed{\text{T}}}{dt} = K - k_T \boxed{\text{T}} \tag{9}$$

Where K is the rate of entrance of T* into the plasma compartment. On integration and imposing the condition that $\boxed{\text{T}} = 0$ at $t = 0$, Eq. (1) becomes

$$\boxed{\text{T}} = \frac{K}{k_T}(1 - e^{-k_T t}) \tag{10}$$

where t = duration of infusion.

Because catecholamine levels are maintained in a steady state by a balance between their rates of formation and efflux, presumably in an open single-compartment system, the change of NE* with time would be related to T* as follows

$$\frac{d\,NE^*}{dt} = k_1 T^* - k_{NE}NE^* \tag{11}$$

and dividing by tissue NE levels

$$\frac{d\,\boxed{NE}}{dt} = \frac{k_1 T^*}{NE} - k_{NE}\,\boxed{NE} \tag{12}$$

During steady-state conditions the following relationship must be true

$$k_1 T = k_{NE}NE$$

$$\tag{13}$$

or

$$NE = \frac{k_1 T}{k_{NE}}$$

and substituting for NE in Eq. (12)

$$\frac{d\,\boxed{NE}}{dt} = k_{NE}(\boxed{T} - \boxed{NE}) \tag{14}$$

Equation (10) can now be substituted in Eq. (14) for $\boxed{T}$.

$$\frac{d\,\boxed{NE}}{dt} = k_{NE}\left[\frac{K}{k_T}(1 - e^{-k_T t}) - \boxed{NE}\right] \tag{15}$$

Since $\boxed{NE} = 0$ at $t = 0$, Eq. (15) integrates into

$$\boxed{NE} = \frac{K}{k_T}\left\{1 + \frac{1}{k_{NE} - k_T}(k_T e^{-k_{NE}t} - k_{NE}e^{-k_T t})\right\} \tag{16}$$

By substituting various theoretical values of k_{NE} for 20, 40, and 60 min of infusion in Eq. (16), Fig. 3 can be constructed. The value of k_{NE} for any NE tissue store is found by determining the specific activity of NE in the tissue and finding the corresponding value for k_{NE} in Fig. 3. NE synthesis rate is calculated by multiplying k_{NE} by the NE level.

When T* (351–368 mC/mmole) is infused intravenously at the rate of 50 μC/hr, $\boxed{T}$ increases curvilinearly according to Eq. (10) (Fig. 4). The solution of this equation is

$$\boxed{T} = \frac{120\,dpm/m\mu mole/min}{0.047/min}(1 - e^{-(0.047/min)\,(t)})$$

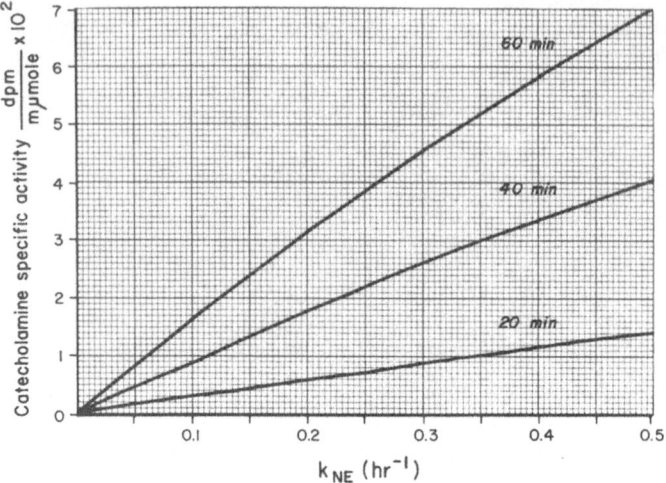

Fig. 3. Fractional rate constant for different values of catecholamine-specific activity calculated for 20, 40, 60 min of infusion tyrosine-[14]C.

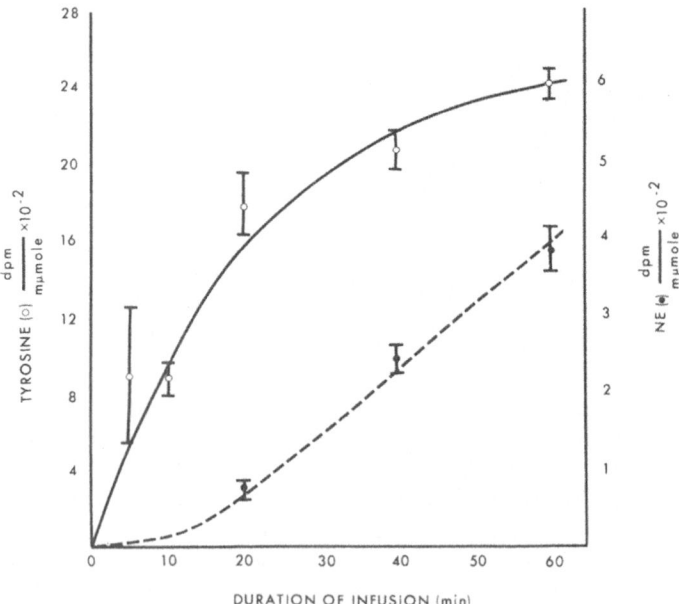

Fig. 4. Increase of the specific activity of plasma tyrosine and brain NE during a constant-rate infusion of tyrosine-[14]C (50 μC/hr; S.A. 350 mC/m mol).

Where 120 dpm/mμmole/min is the apparent rate of entry of radioactive tyrosine into the plasma compartment and 0.047/min equals k_T. Accordingly, the average lifetime ($1/k_T$) of a plasma tyrosine molecule is approximately 21 min. As predicted from Eq. (16), during the infusion of tyrosine $\boxed{\text{NE}}$ also increased (Fig. 4). Calculations from the data presented in this figure show that synthesis rates calculated from $\boxed{\text{NE}}$ after various infusion times of T* are equal.[125] The similarity of the results of these calculations suggests that the rate of conversion of T* into NE proceeds at an almost constant rate. This finding is consistent with the hypothetical model system of a single open compartment.

B. Nonisotopic Methods

1. Inhibition of Catecholamine Biosynthesis

When tyrosine hydroxylase is inhibited by appropriate intravenous doses of α-methyltyrosine (MT),[127] catecholamines are no longer formed and the amine levels decline at a rate that is proportional to the concentration of the remaining amine.[27] Mathematically, this relationship is described by Eq. (17).

$$\frac{d\,\text{NE}}{dt} = -k\text{NE}$$

Upon integration, Eq. (17) becomes:

$$\text{NE} = \text{NE}_o\, e^{-kt} \tag{18}$$

and converting to common logarithms

$$\log \text{NE} = \log \text{NE}_o - 0.434\,kt \tag{19}$$

Where NE_o is the initial steady-state amine concentration and NE is the concentration at any time t after the administration of MT. A plot of log NE versus time yields a straight line, the slope of which is 0.434 times the rate constant of NE efflux (k). Since MT inhibits tyrosine hydroxylase by competing with tyrosine,[39] an optimal tissue level of the inhibitor should be maintained throughout the experiment to assure maximal inhibition of the process that limits catecholamine biosynthesis. In rats, maximal *in vivo* inhibition of tyrosine hydroxylase can be obtained if tissue concentrations of MT are maintained at about 5×10^{-4} M.[127] This concentration persists for about 8 hr if a first dose of 200 mg/kg i.v. of MT is followed 2 hr later by a second dose of 100 mg/kg i.v. As shown in Table I, the rate constant of NE efflux after blockade of synthesis with MT is identical to the rate constant of the decline of $\boxed{\text{NE}}$ after the administration of tracer doses of NE*. This identity suggests that MT does not release NE, does not change the compartmentation of NE, and does not interfere with the enzymes that catabolize NE *in vivo*. Table I shows that k values are almost identical when the fractional rate constant of NE efflux is measured from the incorporation of T* into catecholamines or from the decline of endogenous NE after blockade of synthesis. Since the k values obtained with isotopic and nonisotopic methods are the same, turnover rate can be calculated by either method.

TABLE I

Comparison of the Results of Three Methods for Calculating the Turnover Rate of Tissue NE Stores

Tissue	Method	NE level (μg/g)	Rate constant of amine loss (hr^{-1})	Turnover rate (μg/g/hr)
Brain	T*[a]	0.50 ± 0.03	0.25 ± 0.02	0.12
	αMT[b]	0.56 ± 0.02	0.17 ± 0.04	0.095
Heart	T*[a]	1.09 ± 0.008	0.089 ± 0.01	0.095
	αMT[b]	0.92 ± 0.05	0.10 ± 0.01	0.092
	NE-³H[c]	1.16 ± 0.01	0.086 ± 0.02	0.10

[a] T* was infused intravenously at a constant rate (50 μC/hr, 350 mC/mmole).[125]
[b] NE synthesis was inhibited by administering α-methyltyrosine (αMT).[27]
[c] NE-³H was administered intravenously (5μC/kg).[27]

2. Comparison of Isotopic and Nonisotopic Methods

It was reported[13] that the minimal estimates of the rate of NE synthesis calculated by the ratio of $\boxed{T}$ to $\boxed{NE}$ in tissues were two- to threefold greater than previously published estimates of turnover rates obtained with other methods. These authors[13] conclude that it is unlikely that a discrepancy of this magnitude is due solely to a difference in experimental conditions. Sedvall et al.[13] theorized that if NE were stored in an open two-compartment system, then the size and k of the respective compartments could be tailored to prove that labeling with NE* or blocking the NE biosynthesis with MT measures the turnover rate of cardiac NE rather than estimates the NE synthesis rate. Since both methods were not sufficiently accurate to detect a small compartment of heart NE having a turnover time of about 40 min, the authors[13] concluded that the synthesis rate of heart NE is several times greater than its turnover rate. If this were the case, then measurements of NE synthesis from the incorporation of plasma T* into catecholamines[125] should change with the time of infusion. However, no differences were found when the infusion time of tyrosine was 20, 40, or 60 min,[125] suggesting that the hypothetical small compartment must have a turnover time of less than 40 min.

As a corollary to this discussion, in Table II we present a comparison of measurements of catecholamine turnover rates performed in rats of the

Sprague–Dawley strain from different commercial sources. The data show that the catecholamine turnover rate varies considerably.

TABLE II

Comparison of Amine Level, Rate Constant of Amine Loss, and Turnover Rate of Sprague–Dawley Rats from Different Suppliers[a]

Source of rats	Tissue	Amine	Amine level (μg/gm)	Rate constant of amine loss (hr^{-1})	Turnover rates (μg/g/hr)
Marland Farms[b]	Brain	NE	0.31 ± 0.001	0.12 ± 0.006	0.036
NIH[b]	Brain	NE	0.42 ± 0.03	0.17 ± 0.008	0.071
N.Y. Breeding Farms[c]	Brain	NE	0.56 ± 0.02	0.17 ± 0.04	0.095
Marland Farms[b]	Brain	DM	0.75 ± 0.02	0.28 ± 0.09	0.21
NIH[b]	Brain	DM	1.15 ± 0.03	0.37 ± 0.05	0.43
Marland Farms[b]	Heart	NE	0.64 ± 0.01	0.076 ± 0.005	0.049
N.Y. Breeding Farms[c]	Heart	NE	0.92 ± 0.05	0.10 ± 0.01	0.092

[a] Rate constant of amine loss and turnover rate were calculated by blocking catecholamine synthesis with α-methyltryrosine as described previously.[27]
[b] Data from Brodie et al. (1966).[27]
[c] Data from Neff et al. (1968).[125]

V. SEROTONIN TURNOVER

A. Synopsis of Serotonin Biosynthesis

The biosynthesis of 5-HT from L-tryptophan proceeds through two consecutive reactions: hydroxylation of L-tryptophan to form 5-hydroxy-tryptophan, followed by decarboxylation to 5-HT. Only recently has the hydroxylation of L-tryptophan been demonstrated in brain,[128] probably because the tryptophan hydroxylase activity in this tissue is relatively low. Of the two enzymatic events, hydroxylation appears to be rate limiting.[129] The *in vitro* assay of brain stem tryptophan hydroxylase requires a reduced pteridine, O_2, 2-mercaptoethanol (which appears to stabilize both enzyme and cofactors), and ferrous iron. Nakamura et al.[131] reported that $NADPH_2$ was also necessary for optimal enzymatic activity. However, Lovenberg et al.[42] postulated that $NADPH_2$ is not required when high concentrations of 2-mercaptoethanol are present in the incubation media, as this compound can maintain the pteridine cofactor in a reduced state. Brain tryptophan hydroxylase is found primarily in nerve ending (synaptosome) and mitochondrial fractions.[132]

Aromatic amino acid decarboxylase (see Section III, A) converts 5-hydroxytryptophan to 5-HT. 5-HT is bound within nerve terminals[59] in vesicles that are between 200 and 1200 Å.[133] Pellegrino de Iraldi et al.[16] reported that many of the nerve endings that store 5-HT are very small, below the resolving power of the light microscope. Nerve ending particles[134] can accumulate labeled 5-HT against a concentration gradient. Serotonin may be bound within storage vesicles by forming ternary complexes with ATP and a divalent cation.[135]

B. Synopsis of Serotonin Metabolism

More than 90% of the 5-HT formed in brain is eventually eliminated from brain as 5-hydroxyindoleacetic acid (5-HIAA).[136] Monoamine oxidase (see Section III, B) converts 5-HT to 5-hydroxyindoleacetaldehyde. The aldehyde can be converted to either 5-HIAA by an NAD or NADP-linked aldehyde dehydrogenase or to 5-hydroxytryptophol by an NADH- or NADPH-linked alcohol dehydrogenase.[137–139]

The acidic metabolites of the 5-HT are transported through the brain–blood barrier by a specialized mechanism (acid transport?) that can be blocked by probenecid[136,140–142] and 2,4-dinitrophenol, suggesting that an energy-requiring system participates in the removal of 5-HIAA.

VI. METHODS OF ESTIMATING SEROTONIN TURNOVER

A. Isotopic Methods

1. Labeling Brain Stores with Radioactive Serotonin

After radioactive 5-HT (5-HT*) is injected into the cerebral ventricle, the label can be found in brain areas adjacent to the ventricle.[143] However, there is a strong suggestion of localization at nerve terminals. The half-life of 5-HT* in brain after an intraventricular injection is about 4–5 hr.[143,144] In contrast, radioactive 5-HT disappears very rapidly from brain after an intracisternal (cisterna magna) injection.[145] Apparently the passage of 5-HT into brain tissue from CSF is by passive diffusion,[146] therefore specific 5-HT-containing neurons would not be preferentially labeled. Because of the nonspecific labeling of brain tissue and the great variation in the disappearance rate of radioactive 5-HT, intracisternal or intraventricular injections of 5-HT* probably will not give an accurate estimate of the turnover rate of endogenous brain 5-HT stores.

Bulat and Supek[147,148] have shown that 5-HT will pass directly from blood to brain if large doses of the amine are administered. The increase in brain 5-HT is linearly related to the amount of 5-HT injected intravenously from 1.25 to 20 mg/kg. Probably passage of 5-HT into brain is also by passive diffusion, as the accumulation of blood-borne 5-HT in brain neither reached saturation nor was blocked by ouabain treatment. Again, the nonspecificity of the procedure precludes its use as a method for estimating the turnover of endogenous 5-HT stores in brain.

2. Labeling Brain Stores with Radioactive Precursors

Udenfriend and Weissbach,[149] in one of the first attempts to measure the turnover of 5-HT in brain with a radioactive precursor, administered large amounts of relatively low specific activity DL-5-hydroxytryptophan (5-HTP) to rabbits and determined the specific activity of the 5-HT formed in brain. The half-life of 5-HT* was about 10 hr. The entrapment of blood-borne 5-HT* was considered to interfere significantly with the measurement of the turnover rate of brain 5-HT. This view was corroborated by estimations of the doubling time of brain 5-HT in rabbits given MAO inhibitors. After iproniazid the brain 5-HT level doubled in about 2 hr.[150]

Presently, it is accepted that hydroxylation of tryptophan is the rate-limiting event in brain 5-HT biosynthesis.[129] Therefore, if the turnover of the amine is measured from the rate of brain 5-HT increase after large amounts of 5-HTP or from the incorporation of radioactive 5-HTP into 5-HT, these estimated rates are related to the amount of decarboxylase present in brain and do not reflect normal synthesis. Furthermore, the ubiquitous distribution of the aromatic amino acid decarboxylase[151] and its lack of specificity may result in the formation of either 5-HT or 5-HT* at sites where the amine is normally absent. For example, Brodie *et al.*[152] showed that the intraperitoneal administration of large doses of 5-HTP to rabbits or rats resulted in depletion of brain NE. Johnson *et al.*[153] have suggested that the administration of large amounts of 5-HTP can inhibit the synthesis of NE in brain, thus providing biochemical evidence that quantities of exogenously administered 5-HTP or its metabolite 5-HT may enter noradrenergic neurons. Additional support for this hypothesis was provided by Aghajanian and Bloom[7b] by demonstrating that 5-HT can enter neurons that normally store NE.

In vivo estimates of the true rate of formation of endogenous neuronal constituents from their precursor must be performed without disturbing the normal steady-state equilibrium of the components involved in the reaction. One approach which appears promising in this regard is to administer high specific activity tryptophan-^{14}C (TP*) intravenously at a constant rate to rats and correlate the change of the specific activity of brain 5-HT ($\boxed{\text{5-HT}}$) and $\boxed{\text{TP}}$ with time.[154]

During this procedure, plasma tryptophan-specific activity ($\boxed{\text{TP}}$) increases in a curvilinear manner similar to plasma tyrosine-specific activity during an infusion of tyrosine-^{14}C. Therefore, equations similar to those describing the change of $\boxed{\text{T}}$ with time (Section IV, A, 3) have been developed.

In a typical experiment, L-tryptophan-3-^{14}C is infused intravenously into the tail vein of fasted restrained rats. The increase of the $\boxed{\text{TP}}$ can be described by the following equation:

$$\boxed{\text{TP}} = \frac{K}{k_{\text{TP}}}(1 - e^{-k_{\text{TP}}t}) \tag{20}$$

Where K = apparent rate of entry of TP* into the plasma compartment, k_{TP} = rate constant for the disappearance of tryptophan from plasma, and t = duration of infusion of TP*.

Assuming that 5-HT is stored in an open single compartment, the following equation can be derived for the change $\boxed{\text{5-HT}}$ with the duration of TP* infusion.

$$\boxed{\text{5-HT}} = \frac{K}{k_{TP}} \left\{ 1 + \frac{1}{k_{5\text{-HT}} - k_T} (k_{TP}e^{-k_{5\text{-HT}}t} - k_{5\text{-HT}}e^{-k_{TP}t}) \right\} \quad (21)$$

where $k_{5\text{-HT}}$ = fractional rate constant of the 5-HT compartment. The derivation of this equation is exactly like that for catecholamines (See Section IV, A, 3). A graph of $\boxed{\text{5-HT}}$ vs. $k_{5\text{-HT}}$ for various infusion periods is constructed and 5-HT synthesis calculated by multiplying $k_{5\text{-HT}}$ by the steady-state 5-HT concentrations.

B. Nonisotopic Methods

In 1957, Udenfriend et al.[150] administered the MAO inhibitor iproniazid to rabbits and rats and found that 5-HT levels in brain doubled in about 2–3 hr. This suggested to them that 5-HT was rapidly synthesized in brain. In subsequent studies using harmaline as MAO inhibitor,[149] they found that the doubling time was about 30 min. From these original studies, methods for estimating the synthesis rate of brain catecholamines[155] and serotonin[27] have been developed.

1. Accumulation of Serotonin After Inhibition of Monoamine Oxidase

After the intraperitoneal injection of pargyline hydrochloride (75 mg/kg), a nonreversible MAO inhibitor, brain levels of 5-HT increase linearly for about 2 hr and then plateau at about threefold their normal steady-state concentration. Synthesis is estimated from the initial linear portion of the curve.[156] The method of Bogdanski et al.[157] is sufficiently sensitive to measure synthesis rate of 5-HT in whole brain or in brain stem. For small tissue samples, more sensitive methods for assaying 5-HT should be used (see Snyder et al.[158] or Maickel and Miller[159]). When synthesis rates are measured by this procedure, it is assumed: (a) the MAO inhibitor acts almost immediately; (b) none of the 5-HT is lost from the brain by diffusion; and (c) the metabolism of 5-HT by other pathways is negligible. The synthesis rate of brain 5-HT as calculated from the initial rise of 5-HT levels after pargyline is presented in Table III.

2. Decline of 5-Hydroxyindoleacetic Acid Levels After Inhibition of Monoamine Oxidase

Assuming that brain 5-HT is metabolized solely to 5-HIAA by MAO, then the steady-state rate of 5-HIAA efflux from brain must be equal to the rate of formation of 5-HT. Following an intraperitoneal injection of pargyline (75 mg/kg) or tranylcypromine (10 mg/kg), brain levels of 5-HIAA

TABLE III

Rate of Synthesis of Serotonin and 5-Hydroxyidoleacetic Acid and Rate of Elimination of 5-Hydroxyidoleacetic Acid from Rat Brain

Method used to determine rate	Synthesis rate[a]		Elimination rate
	5-HT	5-HIAA	
Rise of 5-HT following pargyline	2.5[b] (0.44)	—	—
Decline of 5-HIAA following pargyline	2.3 (0.40)	2.3 (0.44)	2.3[b] (0.44)
Rise of 5-HIAA following probenecid	2.1 (0.37)	2.1[b] (0.40)	

[a] Rates are in $m\mu mole/g$ of brain/hr; those in parenthesis, $\mu g/g$ of brain/hr. From Neff and Tozer.[136]

[b] Obtained by direct measurement. All preceding reaction rates must be equal to or greater than this value.

decline exponentially. By using the same rationale employed for the calculation of NE biosynthesis from the exponential decline of the levels after blockade of synthesis (Section IV, B, 1), the equation for the decline of 5-HIAA can be derived as follows:

$$\log 5\text{-HIAA} = \log 5\text{-HIAA}_0 - 0.434\,kt \tag{22}$$

where 5-HIAA_0 is the steady-state or extrapolated zero time value for the acid, 5-HIAA is the level of acid in brain at any time t after inhibition of MAO, and the slope is 0.434 times k. The present belief that 5-HIAA is the main catabolite of brain 5-HT implies that the synthesis rate of 5-HIAA must be equal to the synthesis rate of 5-HT. Therefore, the latter is calculated by multiplying the zero time concentration of 5-HIAA (mole/g) by k, the fractional rate constant of decline of 5-HIAA.[156] Standard statistical methods are used to calculate linearity of regression and the standard error of the regression coefficient.[115] The synthesis rate of brain 5-HIAA and 5-HT as calculated by this procedure is presented in Table III.

The following assumptions are made when the turnover of 5-HIAA and 5-HT are measured by this procedure: (a) MAO is completely inhibited; (b) 5-HT is not lost from the brain at an appreciable rate; (c) oxidative deamination is the major catabolic pathway for 5-HT; and (d) the MAO inhibitor does not interfere with any aspect of 5-HT metabolism other than blocking MAO.

Not all MAO inhibitors satisfy assumption (a); consequently, not all of them can be used to study synthesis. Pargyline and tranylcypromine have been employed with success.[156] In contrast, the MAO inhibiting action of harmaline is too short-lasting to allow accurate synthesis estimates of brain 5-HT and the action of phenelzine is delayed. However, the slope of exponential decline of 5-HIAA after phenelzine treatment is the same as that after pargyline or tranylcypromine. As shown in Table III, about 2.3 mμmoles of 5-HIAA are formed and eliminated per gram of brain each hour. A transport system removes this acid from brain.[160]

3. Accumulation of 5-HIAA After Probenecid Administration

Data in the literature indicate that 5-HIAA is removed from brain by an acid transport.[160] After the administration of probenecid, 5-HIAA accumulates in spinal fluid[141,161] and in brain tissue[160] of many animal species. Synthesis of 5-HT or 5-HIAA in rat brain is studied by administering probenecid (200 mg/kg i.p.) and measuring the rate of accumulation of 5-HIAA with time (Table III). After this dose of probenecid, 5-HIAA levels in brain increase linearly for more than 90 min. When the formation of 5-HIAA is inhibited by pargyline, probenecid treatment will no longer induce a rise of 5-HIAA, but this drug prevents further decline of 5-HIAA levels after pargyline.[160]

The accumulation of 5-HIAA is related to the dose of probenecid administered. The amount of acid accumulated in 2 hr is maximum after 200 mg/kg of probenecid. Increasing the dose does not increase the rate of accumulation of the acid.[160] 5-HIAA levels in rat brain will increase to more than 6 times normal during an 8-hr period if probenecid is administered every 2 hr.[136] Probenecid is relatively nontoxic and can be administered to man and animals in large doses. In cat, the rise of 5-HIAA levels in spinal fluid after probenecid appears related to the synthesis of 5-HT in brain.[161] From these experiments it may be inferred that the rise of acidic compound in the CSF of man may be a useful index of catecholamines and 5-HT biosynthesis in brain. Several attempts have already been made to correlate CSF catechol and indole acid metabolite levels with human pathology. For example, both homovanillic acid and 5-HIAA levels are low in Parkinson's disease[162] and 5-HIAA levels are below normal in depressed patients.[163] By administering probenecid, an estimate of the synthesis rate of these compounds in human brain might be obtained from CSF.

VII. EFFECT OF DRUGS ON MONOAMINE TURNOVER

Drugs may interfere with the various mechanisms involved in the control of monoamine turnover. Figure 5 attempts to depict the possible sites and mechanisms where drugs can interfere with NE turnover.

A. Catecholamines

α-MT blocks tyrosine hydroxylase and reduces the tissue levels of catecholamines.[164] Several analogues of this compound, including 3-iodo-α-methyltyrosine, 3,5-diiodo-L-tyrosine, 3,5-diiodo-α-methyltyrosine, and α-methylphenylalanine, also inhibit tyrosine hydroxylase *in vitro*[165] and *in*

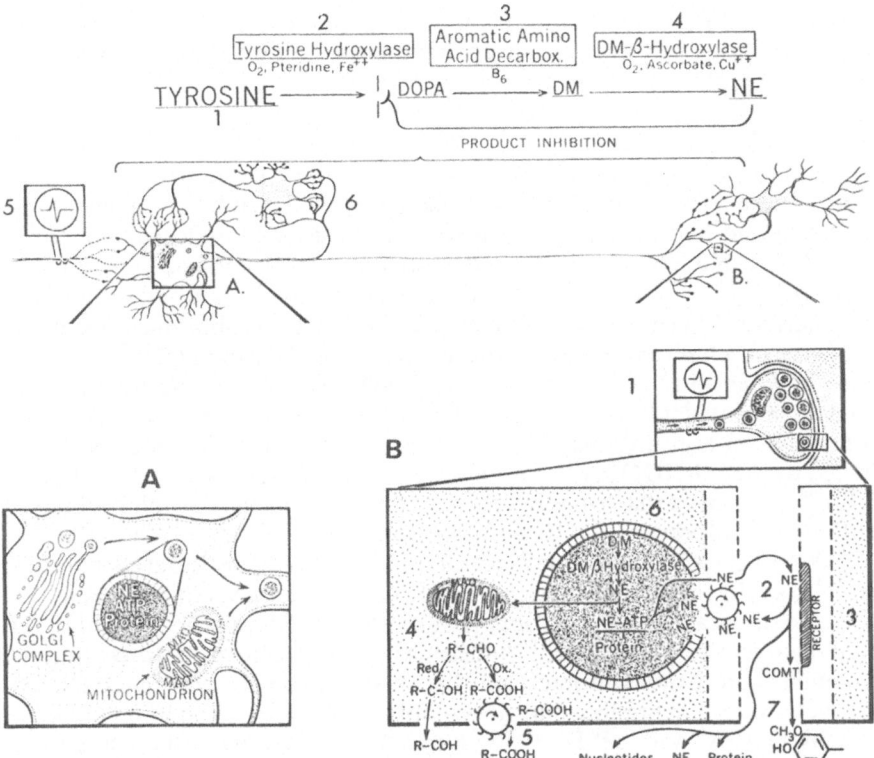

Fig. 5. Schematic representation of a noradrenergic neuron depicting where drugs might modify nerve function. Upper: Drugs acting on adrenergic neurons may: (1) Compete with tyrosine for tyrosine hydroxylase (α-methyltyrosine). (2) Compete with pteridine cofactor (product inhibition by catecholamines). (3) Inhibit DOPA decarboxylase (decaborane; α-methylDOPA). (4) Inhibit DM β-hydroxylase (disulfiram). (5) Inhibit driving input (ganglionic blocking agents). (6) Influence nerve activity by altering neuronal feedback loops. Lower: A. Drugs may suppress synthesis of enzymes and neuronal constituents needed for metabolism and binding of catecholamines. B. Drugs may: (1) Prevent nerve impulse from releasing NE (bretylium; guanethidine). (2) Inhibit NE retrieval (imipraminelike drugs). (3) Block receptors (adrenergic blocking agents). (4) Inhibit catecholamine deamination, consequently inhibiting tyrosine hydroxylase by product inhibition (pargyline). (5) Block transport of acidic catabolites (probenecid). (6) Inhibit catecholamine binding (reserpine). (7) Inhibit COMT (pyrogallol).

vivo.[166] Some iodinated derivatives are more potent than MT; however, their pharmacological value is rather limited because they disappear rapidly from tissues.[127] These compounds are of interest because, with the exception of 3-iodo-α-methyltyrosine, they are normal metabolites of thyroglobulin and accumulate in patients with a deficiency of tissue dehalogenase.[167] While the tyrosine analogues inhibit tyrosine hydroxylation *in vivo* by competing with tyrosine, several catechols, including 3,4-dihydroxyphenyl-propylacetamide[168] and 4-isopropyltropolone, inhibit tyrosine hydroxylation, apparently by interacting with the pteridine cofactor.[39] Many of these tyrosine hydroxylase inhibitors, given in appropriate doses, lower the tissue catecholamine levels.[127]

A number of compounds are known to inhibit the decarboxylation of DOPA: competitive inhibitors such as α-methyl DOPA;[170] a group of hydrazine and hydroxylamine derivatives such as N-(3-hydroxybenzyl)-N-methylhydrazine (NSD 1034) and m-hydroxy-p-bromobenzyl-oxyamine (NSD 1055),[171,172] and strong reducing agents such as decaborane.[173] The pharmacological effects of α-methylDOPA and decaborane are similar in that they both selectively lower the levels of brain NE and DM[174,175] and cause hypotension.[170,176] In contrast, NSD 1055 is a strong inhibitor of aromatic amino acid decarboxylase, but it does not produce a fall of tissue monoamine concentrations.[171] The rate of brain catecholamine depletion elicited by α-methylDOPA[174] is an order of magnitude greater than that of MT,[164] indicating that α-methyl DOPA may not act solely by inhibiting NE synthesis.[53] Moreover, the brain dopamine levels of rats given α-methyl-DOPA return to control values several hours earlier than the levels of brain NE,[174] again indicating that the depletion of NE may not be related to blockade of decarboxylase. Probably, α-methylDOPA impairs the ability of brain neurons to retain NE[174] because it is transformed to α-methyl-dopamine and α-methylnorepinephrine, which compete with endogenous catecholamines for binding sites in synaptic vesicles.[177] Experiments with α-methylDOPA and NSD 1055 support the inference that neuronal decarboxylase is in great excess and even its almost total inhibition may not cause a decrease of brain monoamine levels.

The reports on the mode of action of decaborane are ambivalent; the drug causes an accumulation of brain DOPA[173] and a prolonged depletion of tissue NE levels without impairing the uptake of NE-³H by heart.[177] Decaborane reduces the urinary excretion of NE in adrenalectomized rats exposed to 3°C.[178] Because of the strong reducing activity of decaborane, the drug might interact with pyridoxal phosphate, the cofactor of the decarboxylase system.[173] However, in vitro, decarborane accelerates the release of NE from synaptic vesicles,[177] inhibits the ATP-dependent NE uptake by these vesicles,[15] and releases NE from the guinea pig heart lung preparation.[178a]

Phenethylamine isosteres inhibit dopamine-β-hydroxylase in vitro[179] and in vivo. However, inhibition of the enzyme is not complete in the doses tested, which might explain why brain NE is only partially depleted in the living animal.[180] Various copper chelating agents are also potent inhibitors of this enzyme.[181a] Among these, disulfiram was found to be an effective inhibitor of dopamine-β-hydroxylase in vivo[182] as well as in vitro.[183] The assumption that copper is necessary for β-hydroxylation was substantiated by a report that copper-deficient animals show a reduced enzyme activity.[184] While dopamine-β-hydroxylase is not the rate-limiting reaction under physiological conditions, an effective inhibitor of this enzyme, such as disulfiram, will lower endogenous NE levels. The tissue concentration of NE adjusts to a lower steady-state level where destruction is equal to synthesis. Thus, in the presence of disulfiram, dopamine-β-hydroxylase may become rate limiting.[182] In support of this view, disulfiram inhibits the repletion of brain NE in rats given DOPA after reserpine.[185]

Glucocorticoids markedly influence the enzymatic synthesis of EPI in the rate adrenal medulla. Removal of the pituitary causes a decline in phenylethanolamine-N-methyltransferase activity and EPI content.[79] Moreover, the removal of the pituitary is associated with an increase of NE levels,[186] rough endoplasmic reticulum, and the Golgi complex in the medullary cells that store EPI.[187] These changes can be reversed by administering ACTH or glucocorticoids.[79,187] Since actinomycin-D prevents reversal by glucocorticoids, it is probable that these hormones exert their effect at the level of RNA transcription from DNA.

Blockade of NE catabolism by MAO inhibitors causes an increase of NE levels and a decrease of catecholamine synthesis rates.[188] This finding suggests that catecholamines could influence their own synthesis rate[188] by inhibiting tyrosine hydroxylase.[39,189] This possibility is corroborated by a report showing that the synthesis of cardiac NE is not changed immediately after the injection of a MAO inhibitor when the catecholamine levels are not yet increased even though MAO is almost completely blocked.[121]

Several years ago, a report appeared suggesting that chloropromazine may reduce the synthesis rate of brain monoamines.[190] Subsequently, it was shown that this effect was generated by the hypothermia elicited by this drug.[191] When hypothermia was prevented, chlorpromazine enhanced the rate of catabolism of labeled catecholamines formed in brain from injected DOPA*.[192]

In agreement with these results, Andén et al.[193] found an increase of endogenous phenylcarboxylic acids in the striatum of rabbits injected with chlorpromazine or haloperidol. An increase of homovanillic acid in the spinal fluid of psychiatric patients given haloperidol intravenously has been reported.[194] These findings have raised the question of whether neuroleptic and antidepressant drugs accelerate the synthesis of brain catecholamines either as a result of blockade of postjunctional receptors or as a result of alterations of the amine storage mechanism. An answer to this question was provided by measuring turnover rates of NE and DM in rats given nonhypothermic doses of chlorpromazine, desmethylimipramine or protryptiline.[52] These studies showed that chlorpromazine increases the turnover of brain dopamine, while the two cyclic antidepressants enhance the turnover rate of brain NE. An increase of brain dopamine turnover in animals given chlorpromazine was confirmed using radioactive[195,122,196] and histochemical methods.[197] Furthermore, the selectivity of the effects of chlorpromazine on brain dopamine biosynthesis was corroborated by reports showing that this tranquilizer does not increase the brain levels of 5-hydroxyindoleacetic acid.[198]

Hormones may affect turnover rates by a number of mechanisms. Although the NE synthesis rate in the heart of thyrotoxic mice is equal to that of controls,[99] thyroxine reduces the increase of catecholamine turnover rate elicited by removal of the hypophysis. Perhaps thyroxine acts either directly or through a metabolite on tyrosine hydroxylase.[199]

Hypophysectomy increases the turnover rate of heart NE but does not affect NE synthesis in brain, spleen, and salivary gland.[199] The restoration to normal of the cardiac NE turnover rate of hypophysectomized rats by ganglionic blockade[199] implicates a centrally mediated increase of sympathetic activity as a cause of the increased turnover rate of heart NE. A similar genesis has been invoked to explain the increased turnover of cardiac NE recorded in rats 2 weeks after adrenal demedullation.[200] Even in these experiments, brain catecholamine biosynthesis was not increased.

The dependence of the turnover rate of brain monoamines on nerve impulse flow was studied by biochemical and histochemical methods in the spinal cord after transection.[201] This surgical procedure reduces the rate of monoamine depletion elicited by drugs that inhibit monoamine biosynthesis, but does not affect the depletion elicited by drugs that affect amine storage (reserpine, tetrabenazine). These results mirror previous reports showing that 5-HIAA is lost from the rabbit spinal cord caudally to a transection before depletion of 5-HT.[202]

A corollary to the concept that amine synthesis varies with neuronal activity is the inference that synthesis is a continuous process which does not proceed at a constant rate. Pharmacological studies showing that chlorisondamine, pempidine, and bretyliumlike drugs reduce the turnover rate of heart NE[203] support these views. Additional supporting data come from reports showing that immobilization stress,[204] mild electric shocks to the feet,[205] and ethanol[206] increase the turnover of brain NE but not that of brain DM.

In closing this survey of the effect of drugs on catecholamine turnover rates, it is pertinent to mention that recent experiments have indicated that the rate of catecholamine synthesis may affect a drug's action on animal behavior. For instance, α-MT antagonizes the stimulatory effects elicited by amphetamine[207,208] and disrupts several types of behavior.[209–211] The time course of these effects is correlated with the depletion and recovery of brain catecholamines. Furthermore, reserpine increases the turnover rate of brain catecholamines[212,66] and potentiates the effects of amphetamine.[213]

Not all stimulant drugs affect the rate of brain NE synthesis. The behavioral stimulation elicited by a combination of desipramine and tetrabenzine depends on the availability of NE and is independent of its rate of synthesis.[214]

Despite the clear relationship between biosynthesis of monoamines and the CNS stimulatory effect of drugs, the available facts are not sufficient to exclude the possibility that the increase of NE synthesis may be the cause rather than the consequence of the increase of central noradrenergic function by drugs.

B. Serotonin

The 5-HT content of rabbit blood platelets and rat brain can be lowered by p-chlorinated aralkylamines, which act on the 5-HT stores by different mechanisms.[215] Among these compounds, 4-chloro-N-α-dimethyl phenethylamine depletes the 5-HT and 5-HIAA content of brain in a parallel fashion without changing brain NE levels.[216] This finding suggests that the compound may interfere with 5-HT biosynthesis. Although there is not definitive proof that these amines inhibit tryptophan hydroxylase, their parent amino acids inhibit liver phenylalanine and tryptophan hydroxylase[217,218] and brain tryptophan hydroxylase.[219] p-Cl-Phenylalanine (PCP) is a substrate for hydroxylase enzymes.[220] During the reaction 5 to 8 moles of NADPH per mole of reactant are consumed.[221] Since NADPH is required for the reduction of oxidized tetrahydropteridines, availability of tetrahydropteridines may become rate limiting in the synthesis of 5-HT. Support for this hypothesis is provided by reports that tetrahydropteridines can reverse the inhibition of hydroxylase induced by PCP.[218] The inhibition of hydroxylases elicited *in vivo* by PCP is of long duration.[42] Four hours after PCP injections, the inhibition of liver phenylalanine hydroxylase[222] and brain tryptophan hydroxylase[42] is partially reversed by dialysis, suggesting that at this time inhibition is due both to enzyme inactivation and to the presence of high levels of PCP in the tissue. At later times, the inhibition is nonreversible, indicating enzyme inactivation. This might be caused by inhibition of hydroxylase biosynthesis, incorporation of PCP into the enzyme during its biosynthesis, or irreversible interaction of PCP (or a metabolite) with the hydroxylase to yield inactive enzyme. It is important to note that p-Cl-α-methylphenylalanine is a good inhibitor of liver tryptophan hydroxylase, but is a weak inhibitor of brain tryptophan hydroxylase *in vivo* and *in vitro*.[218] This finding is of particular interest considering the brain 5-HT depletion elicited by its α-methylamine derivative.[216]

The inhibition of 5-HT biosynthesis by PCP has opened two promising avenues for future investigation of the biological role of brain 5-HT. Tenen[223] has recently reported that PCP antagonized the analgesic effects of morphine. Moreover, a preliminary report[224] shows that in mice rendered tolerant to morphine, the turnover rate of brain 5-HT is increased chronologically with the development of tolerance to morphine. Injection of PCP in tolerant mice seems to diminish withdrawal symptoms. The other line of endeavor supports the view that 5-HT is involved in the physiological mechanisms that control sleep.[225] In cats, ablation of the raphe medianus (a locus of large aggregates of neurons containing 5-HT[225] or injections of PCP[225] reduce the duration of slow-wave sleep; injection of small quantities of 5-HTP reverses the effect of PCP and induces slow-wave sleep.

Reserpine in doses that deplete brain 5-HT levels stimulates the synthesis of 5-HT.[156,160] Stimulated synthesis of 5-HT after reserpine treatment was demonstrated in rat brain by the accumulation of 5-HIAA after probenecid treatment and from the decline of 5-HIAA levels after pargyline.

It has been suggested that negative feedback control of indole synthesis by 5-HT might be responsible for this increase[136], but the data available do not support this suggestion.[45a]

Ether anesthesia causes a significant increase of rat brain 5-HIAA levels without affecting 5-HT levels.[43] In addition, 5-HT synthesis is accelerated by ether anesthesia. Stimulated synthesis, however, is not a general effect of anesthesia, as halothane and cyclopropane do not change indole metabolism.[43]

Imipramine reduces the turnover rate of brain 5-HT[226] but this effect is not elicited by desipramine.[227] This difference in the action of these two tricyclic antidepressants has been related to their action on the retrieval of 5-HT by CNS serotonergic neurons. Imipramine blocks uptake[228] while desipramine does not.[229]

A possible interaction between LSD and 5-HT at CNS neuronal receptors has been suggested following reports showing that LSD could antagonize[230] or facilitate[231] the effects of 5-HT on smooth-muscle preparations. Moreover, LSD induces a small but consistent increase of brain 5-HT concentrations[232] and a decrease of brain 5-HIAA[233,43] without inhibiting MAO. Another compound, which like LSD causes excitation and piloerection, β-tetrahydronaphthylamine, also decreases brain 5-HIAA without inhibiting MAO.[236] LSD reduces the turnover rate of brain 5-HT[43,234] and lowers brain NE and dopamine levels.[43] This effect may be associated with the hallucinogenic activity of LSD because its nonhallucinogenic congeners do not change brain 5-HT turnover.[235]

The decreased turnover of brain 5-HT elicited by LSD has been interpreted as feedback inhibition of synthesis because LSD mimics 5-HT at the postsynaptic receptor sites of serotonergic neurons.[234] This view is supported by a report that the spontaneous firing of serotonergic raphe medianus neurons is inhibited by LSD[237] injected parenterally.

VIII. CONCLUSIONS

Catecholamines and 5-HT are neuronal constituents synthesized and stored in synaptosomes. This localization and other findings summarized in this chapter suggest that their release may influence the probability of nerve impulse generation in adjacent neurons. Turnover rate estimations of monoamines may be a valid method for assessing the functional output of a biochemically defined category of neurons in complex neuronal populations. Our incomplete understanding of the metabolic and morphological processes participating in the physiological control of monoamine biosynthesis limits the overall implications of turnover rate measurements. A concept gradually taking shape in the pharmacology of sympathetic neurons is that drugs which decrease sympathetic output without lowering tissue stores of NE usually reduce the synthesis rate of NE. However, a recent report[238] limits the application of such generalized statements by showing that α-adrenergic receptor blocking agents increase the rate of conversion of tyrosine-^{14}C to NE in heart, brain, and adrenal medulla. Since these drugs do not reduce tissue NE and tyrosine levels, blockade of α-adrenergic receptors may be associated with increased biosynthesis of NE.

A final note of caution should be added to the current belief that 5-HT, NE, and DM are localized in distinct neurons. Recent experiments[239,240] show that sympathetic nerve endings in pineal gland contain both NE and 5-HT. The 5-HT is not synthesized in these nerves but is taken up and stored in sympathetic synaptic vesicles. At present, we cannot assess whether the pineal is an exception to the current belief that each nerve has one and only one transmitter. Theoretically, it can be surmised that when a monoaminergic nerve makes synaptic contacts with another monoaminergic nerve, this nerve ending could take up an amine that is not normally synthesized in that neuron. Should this possibility become substantiated by experiments, then we must be ready to reevaluate the validity of certain studies of 5-HT, NE, and DM turnover rates. For instance, it may well be the case that LSD prevents access of 5-HT to MAO because it acts by preventing the access of 5-HT to some noradrenergic or dopaminergic neurons situated postsynaptically to certain serotonergic terminals.

These considerations limit and caution any proposals that turnover studies of brain monoamines may become, in the near future, a satisfying neurobiological method to formulate a valid biochemical theory of behavior. The present understanding of the turnover rate of brain monoamine should merely invite further study of the basic mechanisms involved in the control of monoamine biosynthesis.

IX. REFERENCES

1. B. Falck and C. Owman, A detailed methodological description of the fluorescence method for the cellular demonstration of biogenic monoamines, *Acta Univ. Lund II* 7:1–23 (1965).
2. J. Axelrod, in *The Clinical Chemistry of Monoamines*, pp. 5–18, Elsevier, Amsterdam (1963).
3. E. Costa and N. H. Neff, in *Topics in Medicinal Chemistry*, Vol. 2, pp. 65–95, Interscience, New York (1968).
4. M. A. Grillo, Electron microscopy of sympathetic tissues, *Pharm. Rev.* 18:387–399 (1966).
5. V. P. Whittaker, I. A. Michaelson, and R. J. A. Kirkland, The separation of synaptic vesicles from nerve ending particles ("Synaptosome"), *Biochem. J.* 90:293–305 (1964).
6. A. Oliverio and L. Stjarne, Acceleration of norepinephrine turnover in mouse heart by cold exposure, *Life Sci.* 4:2339–2343 (1965).
7a. G. K. Aghajanian and F. E. Bloom, Electron-microscopic autoradiography of rat hypothalamus after intraventricular ^{3}H-norepinephrine, *Science* 153:308–310 (1966).
7b. G. K. Aghajanian and F. E. Bloom, Localization of tritiated serotonin in rat brain by electron-microscopic autoradiography, *J. Pharmacol. Exp. Therap.* 156:23–30 (1967).
8. N. Weiner and M. Rabadjya, The effect of nerve stimulation on the synthesis and metabolism of norepinephrine in the isolated guinea pig hypogastic nerve-vas deferens preparation, *J. Pharmacol. Exp. Therap.* 160:61–71 (1968).
9. G. C. Sedvall and I. J. Kopin, Influence of sympathetic denervation and nerve impulse activity on tyrosine hydroxylase in the rat submaxillary gland, *Biochem. Pharmacol.* 16:39–46 (1967).

10. R. Gordon, S. Spector, A. Sjoerdsma, and S. Udenfriend, Increased synthesis of norepinephrine and epinephrine in the intact rat during exercise and exposure to cold, *J. Pharmacol. Exp. Therap.* **153**:440–447 (1966).

11. N. H. Neff, T. N. Tozer, W. Hammer, E. Costa, and B. B. Brodie, Application of steady-state kinetics to the uptake and decline of H^3NE in the rat heart, *J. Pharmacol. Exp. Therap.* **160**:48–52 (1968).

12. H. Weil-Malherbe, L. G. Whitby, and J. Axelrod, The uptake of circulating [^{3}H] norepinephrine by the pituitary gland and various areas of the brain, *J. Neurochem.* **8**:55–64 (1961).

13. G. C. Sedvall, V. K. Weise, and I. J. Kopin, The rate of norepinephrine synthesis measured *in vivo* during short intervals, influence of adrenergic nerve impulse activity, *J. Pharmacol. Exp. Therap.* **159**:274–282 (1968).

14. A. Carlsson, Pharmacological depletion of catecholamine stores, *Pharmacol. Rev.* **18**:541–549 (1966).

15. V. P. Whittaker, Structure and function of animal cell membranes, *Brit. Med. Bull.* **24**:101–106 (1968).

16. A. Pellegrino de Iraldi, L. M. Zieher, and G. J. Etcheverry, Neuronal compartmentation of 5-hydroxytryptamine stores, *Adv. Pharmacol.* **6**:Part A, 257–270 (1968).

17. V. P. Whittaker, A comparison of the distribution of lysosome enzymes and 5-hydroxytryptamine with that of acetylcholine in subcellular fractions of guinea pig brain, *Biochem. Pharmacol.* **1**:351 (1958).

18. L. M. Zieher and E. DeRobertis, Subcellular localization of 5-hydroxytryptamine in rat brain, *Biochem. Pharmacol.* **12**:596–598 (1963).

19. S. Fahn, J. S. Rodman, and L. J. Côté, Association of Tyrosine Hydroxylase with Synaptic Vesicles in Bovine Caudate Nucleus, *J. Neurochem.* **16**:1293–1300 (1969).

20. P. Laduron and F. Belpaire, Evidence for an extragranular localization of tyrosine hydroxylase, *Nature* **217**:1155–1156 (1968).

21. L. J. Reed, *in The Neurosciences* (G. C. Quarton, T. Melnechukt, and F. O. Schmitt, eds.), pp. 79–90, The Rockefeller University Press, New York (1967).

22. A. Dahlström and J. Haggendal, Studies on the transport and life-span of amine storage granules in a peripheral adrenergic neuron system, *Acta Physiol. Scand.* **67**:278–288 (1966).

23. E. Costa and N. H. Neff, *in Biochemistry and Pharmacology of the Basal Ganglia* (E. Costa, L. J. Coté and M. D. Yahr, eds.), pp. 141–155, Raven Press, New York (1966).

24. A. Pellegrino de Iraldi and E. DeRobertis, The neurotubular system of the axon and the origin of granulated and non-granulated vesicles in regenerating nerves, *Z. für Zellforschung* **87**:330–344 (1968).

25. A. Dahlström, Transport of catecholamine storage granules, *Neuroscience Res. Prog. Bull.* **5**:317–322 (1967).

26. K. E. Eakins, E. Costa, R. L. Katz, and C. L. Reyes, Effect of pentobarbital anesthesia on the turnover of [^{3}H] noradrenaline in peripheral tissues of cats, *Life Sci.* **7**:71–76 (1968).

27. B. B. Brodie, E. Costa, A. Dlabac, N. H. Neff, and H. H. Smookler, Application of steady-state kinetics to the estimation of synthesis rates and turnover time of tissue catecholamines, *J. Pharmacol. Exp. Therap.* **154**:493–498 (1966).

28. K. C. Richardson, The fine structure of autonomic nerve endings in smooth muscle of the rat vas deferens, *J. Anat. Lond.* **96**:427–442 (1962).

29. G. Burnstock and M. E. Holman, Spontaneous potentials at sympathetic nerve endings in smooth muscle, *J. Physiol.* **160**:446–460 (1962).

30. W. W. Douglas, A. M. Poisner, and R. P. Rubin, Efflux of adenine nucleotides from perfused adrenal glands exposed to nicotine and other chromaffin cell stimulants, *J. Physiol.* **179**:130–137 (1965).

30a. N. Kirshner, H. J. Sage, W. J. Smith, and A. G. Kirshner, Release of catecholamines and specific protein from adrenal glands, *Science* **154**:529–531 (1966).

30b. P. Banks and H. Blaschko, Chromaffin tissue, *Pharmacol. Rev.* **18**:453–456 (1966).

31. E. DeRobertis, *in Hystophysiology of Synapses and Neurosecretion*, p. 185, Macmillan, New York (1964).

32. A. M. Poisner, T. M. Trifaro, and W. W. Douglas, The fate of the chromaffin granule during catecholamine release from the adrenal medula. II. Loss of protein and retention of lipid in subcellular fractions, *Biochem. Pharmacol.* **16**:2101–2108 (1967).

33. N. E. Andén and M. Henning, Adrenergic nerve function, noradrenaline level and noradrenaline uptake in cat nictitating membrane after reserpine treatment, *Acta Physiol. Scand.* **67**:498–504 (1966).

34. R. H. Roth, L. Stjarne, and U. S. von Euler, Acceleration of noradrenaline biosynthesis by nerve stimulation, *Life Sci.* **5**:1071–1075 (1966).

35. G. K. Aghajanian, J. A. Rosencrans, M. H. Sheard, Serotonin: release in the forebrain stimulation of midbrain raphe, *Science* **156**:402–403 (1967).

36. A. L. Lehninger, *in The Neurosciences* (G. C. Quarton, T. Melnechuk, and F. O. Schmitt, eds.), pp. 35–45, The Rockefeller University Press, New York (1967).

37. L. J. Reed and D. J. Cox, Macromolecular organization of enzyme system, *Ann. Rev. Biochem.* **35**:57–84 (1966).

38. D. E. Green, E. Murer, H. O. Hustin, S. H. Richardson, B. Salmon, G. P. Brierley, and H. Baum, Association of integrated metabolic pathways with membranes. I. Glycolytic enzymes of the red blood corpuscle and yeast, *Arch. Biochem. Biophys.* **112**:635–647 (1965).

39. S. Udenfriend, P. Zaltzman-Nirenberg, and T. Nagatsu, Inhibition of purified beef adrenal tyrosine hydroxylase, *Biochem. Pharmacol.* **14**:837–845 (1965).

40. T. Nagatsu, M. Levitt, and S. Udenfriend, Tyrosine hydroxylase: the initial step in norepinephrine biosynthesis, *J. Biol. Chem.* **239**:2910–2917 (1964).

41. R. H. Roth, L. Stjarne, and U. S. von Euler, Factors influencing the rate of norepinephrine biosynthesis in nerve tissue, *J. Pharmacol. Exp. Ther.* **158**:373–377 (1967).

42. W. Lovenberg, E. Jequier, and A. Sjoerdsma, Tryptophan hydroxylation in mammalian systems, *Adv. Pharmacol.* **6A**:21–36 (1968).

43. P. M. Diaz, S. H. Ngai, and E. Costa, Factors modulating brain serotonin turnover, *Adv. Pharmacol.* **6A**:75–92 (1968).

44. S. Eiduson, 5-Hydroxytryptamine in the developing chick brain: its normal and altered development and possible control by end-product repression, *J. Neurochem.* **13**:923–932 (1966).

45. S. F. Contractor and M. V. Jeacock, A possible feed-back mechanism controlling the biosynthesis of 5-hydroxytryptamine, *Biochem. Pharmacol.* **16**:1981–1987 (1967).

45a. R. C. Lin, N. H. Neff, S. H. Ngai, and E. Costa, Turnover rates of serotonin and norepinephrine in brain of normal and paragyline-treated rats. *Life Sci.* **18**:1077–1084 (1969).

46. B. Folkow, J. Haggendal, and B. Lesender, Extent of release and elimination of noradrenaline at peripheral adrenergic nerve terminals, *Acta Physiol. Scand. Suppl.* **307** (1967).

47. H. Corrodi, T. Malmfors, and C. Sachs, Differences in the uptake of secondary catecholamines by the adrenergic nerves, *Acta Physiol. Scand.* **67**:358–362 (1966).

48. L. L. Iversen, *The Uptake and Storage of Noradrenaline in Sympathetic Nerves*, Cambridge University Press, Cambridge, England (1967).

49. B. L. Brown, The release and fate of the transmitter liberated by adrenergic nerves, *Proc. Roy. Soc. Ser. B* **162**:1–19 (1965).

50. D. J. Boullin, E. Costa, and B. B. Brodie, Evidence that blockade of adrenergic receptors causes overflow of NE in cats colon after nerve stimulation, *J. Pharmacol. Exp. Therap.* **157**:125–134 (1967).

51. I. J. Kopin, G. R. Breese, K. R. Krauss, and V. K. Weise, Selective release of newly synthesized norepinephrine from the cat spleen during sympathetic nerve stimulation, *J. Pharmacol. Exp. Therap.* **161**:271–278 (1968).
52. N. H. Neff and E. Costa, in *Antidepressant Drugs*, Vol. 28, Excerpta Medical Foundation, International Congress Series No. 122, Amsterdam (1966).
53. H. Blaschko, The specific action of L-dopa decarboxylase, *J. Physiol. (Lond.)* **96**:50–51 (1939).
54. D. J. Demis, H. Blaschko, and A. D. Welch, The conversion of dihydroxyphenylanine-2-[14]C (Dopa) to norepinephrine by bovine adrenal medullary homogenates, *J. Pharmacol. Exp. Therap.* **113**:14–15 (1956).
55. S. Udenfriend and J. B. Wyngaarden, Precursors of adrenal epinephrine and norepinephrine in vivo, *Biochim. Biophys. Acta* **20**:48–52 (1956).
56. J. Pellerin and A. D'Jorio, Metabolism of DL-3,4-dihydroxyphenylalanine α-^{14}C in bovine adrenal homogenate, *Can. J. Biochem.* **35**:151–156 (1957).
57. P. Hagen, Biosynthesis of norepinephrine from 3,4-dihyroxyphenylethylamine (Dopamine), *J. Pharmacol. Exp. Therap.* **116**:26–27 (1956).
58. U. S. von Euler and A. Hillarp, Evidence for the presence of noradrenaline in submicroscopic structures of adrenergic axons, *Nature (London)* **177**:44–45 (1956).
59. I. A. Michaelson, V. P. Whittaker, R. Laverty, and D. F. Sharman, Localization of acetylcholine, 5-hydroxytryptamine, and noradrenaline within subcellular particles derived from guinea-pig subcortical brain tissue, *Biochem. Pharmacol.* **12**:1450–1453 (1963).
60. A. Carlsson, Detection and assay of dopamine, *Pharmacol. Rev.* **11**:300–304 (1959).
61. N. E. Andén, K. Fuxe, B. Hamberger, and T. Hokfelt, A quantitative study on the nigroneostriatal dopamine neuron system in the rat, *Acta Physiol. Scand.* **67**:306–312 (1966).
62. K. Fuxe, Evidence for the existence of monoamine neurons in the central nervous system. Distribution of monoamine nerve terminals in the central nervous system, *Acta Physiol. Scand.* **64**:Suppl. **247**:37–84 (1965).
63. H. Blaschko and A. D. Welch, Localization of adrenaline in cytoplasmic particles of the bovine adrenal medulla, *Arch. Exptl. Path. Pharmakol.* **219**:17–22 (1953).
64. S. P. Bagchi and P. L. McGeer, Some properties of tyrosine hydroxylase from the caudate nucleus, *Life Sci.* **3**:1195–1200 (1964).
65. L. Stjarne and F. Lishajko, Localization of different steps in noradrenaline synthesis to different fractions of a bovine splenic nerve homogenate, *Biochem. Pharmacol.* **16**:1719–1728 (1967).
66. N. H. Neff and E. Costa, Application of steady-state kinetics to the study of catecholamine turnover after monoamine oxidase inhibition or reserpine administration, *J. Pharmacol. Exp. Therap.* **160**:40–47 (1968).
67. A. Alousi and N. Weiner, The regulation of norepinephrine synthesis in sympathetic nerves: Effect of nerve stimulation, cocaine and catecholamine-releasing agents, *Proc. Nat. Acad. Sci. U.S.A.* **56**:1491–1904 (1966).
68. S. Spector, R. Gordon, A. Sjoerdsma, and S. Udenfriend, End-product inhibition of tyrosine hydroxylase as a possible mechanism for regulation of norepinephrine synthesis, *Mol. Pharmacol.* **3**:549–555 (1967).
69. P. Holtz, R. Heise, and K. Luotke, Fementativer abbau von L-dioxyphenylalanine (DOPA) durch niere, *Arch. Exp. Path. Pharmakol.* **191**:87–118 (1938).
70. W. Lovenberg, H. Weissbach, and S. Udenfriend, Aromatic L-aminoacid decarboxylase, *J. Biol. Chem.* **237**:89–93 (1962).
71. D. Lores Arnaiz Rodriguez and E. DeRobertis, 5-Hydroxytryptophan decarboxylase activity in nerve endings of the rat brain, *J. Neurochem.* **11**:213–219 (1964).
71a. E. Y. Levin, B. Levenberg, and S. Kaufman, The enzymatic conversion of 3,4-dihydroxyphenylethylamine to norepinephrine, *J. Biol. Chem.* **236**:2043–2049 (1961).

72. S. Friedman and S. Kaufman, 3,4-Dihydroxyphenylethylamine β-hydroxylase physical properties, copper content and role of copper in the catalytic activity, *J. Biol. Chem.* **240**:4763, 4773 (1965).

73. L. T. Potter and J. Axelrod, Studies of norepinephrine storage particles of the rat heart, *J. Pharmacol. Exp. Therap.* **142**:299–305 (1963).

74. L. Stjarne, Studies of noradrenaline biosynthesis in nerve tissue, *Acta Physiol. Scand.* **67**:441–454 (1966).

75. F. Belpaire and P. Laduron, Tissue fractionation and catecholamines. I. Latency and inactivation properties of dopamine-β-hydroxylase in adrenal medulla, *Biochem. Pharmacol.* **17**:421–441 (1968).

76. E. Bülbring, Methylation of noradrenaline by minced suprarenal tissue, *Brit. J. Pharmacol.* **4**:234–244 (1949).

77. J. Axelrod, Purification and properties of phenylethanolamine-N-methyl transferase, *J. Biol. Chem.* **237**:1898–1902 (1962).

78. J. Axelrod, Methylation reactions in the formation and metabolism of catecholamines and other biogenic amines, *Pharmacol. Rev.* **18**:95–113 (1966).

79. R. J. Wurtman and J. Axelrod, Control of enzymatic synthesis of adrenaline in the adrenal medulla by adrenal cortical steroids, *J. Biol. Chem.* **241**:2301–2305 (1966).

80. V. Z. Gorkin, Monoamine oxidases, *Pharmacol. Rev.* **18**:115–120 (1966).

81. C. Schnaitman, V. G. Erwin, and J. W. Greenwal, The submitochondrial localization of monoamine oxidase, *J. Cell Biol.* **32**:719–735 (1967).

82. K. F. Tipton, The sub-mitochondrial localization of monoamine oxidase in rat liver and brain, *Biochim. Biophys. Acta* **135**:910–920 (1967).

83. S. Nara and K. I. Yasunobu, Some recent advances in the purification of mitochondrial monoamine oxidase, *Pharmacol. Rev.* **18**:144 (1966).

84. K. T. Yasunobu, I. Igaue, and B. Gomes, The purification and properties of beef liver mitochondrial monoamine oxidase, *Adv. Pharm.* **6**:Part A, 43–59 (1968).

85. T. L. Sourkes, Properties of the monoamine oxidase of rat liver mitochondria, *Adv. Pharm.* **6**:Part A, 61–69 (1968).

86. V. Z. Gorkin, Partial separation of rat liver mitochondrial amine oxidase, *Nature* (*London*) **200**:77 (1963).

87. V. Z. Gorkin, On the selective inhibition of enzymatic deamination of biogenic monoamines by some monoamine oxidase inhibitors, *Abstracts VII*, International Congress Biochem. 1023 (1967).

88. A. Giachetti and P. A. Shore, Optical specificity of monoamine oxidase, *Life Sci.* **5**:1373–1378 (1966).

89. H. Blaschko, D. Richter, and H. Schlossmann, The inactivation of adrenaline, *J. Physiol.* **90**:1–17 (1937).

90. E. H. La Brosse, J. Axelrod, I. J. Kopin, and S. S. Kety, Metabolism of 7-³H-epinephrine-d-bitartrate in normal young man, *J. Clin. Invest.* **40**:253–260 (1961).

91. I. J. Kopin and E. K. Gordon, Metabolism of norepinephrine-³H released by tyramine and reserpine, *J. Pharm. Exp. Therap.* **138**:351–359 (1962).

92. C. W. Nash, E. Costa, and B. B. Brodie, The action of reserpine, guanethidine and metaraminol on cardiac catecholamine stores, *Life Sci.* **3**:441–449 (1964).

93. G. Hertting and J. Axelrod, Fate of tritiated noradrenaline at sympathetic nerve ending, *Nature* **192**:172–173 (1961).

94. D. J. Boullin, E. Costa, and B. B. Brodie, Apparent depletion of NE stores after repetitive stimulation of cat colon in presence of phenoxybenzamine, *Int. J. Neuropharmacol.* **5**:293–298 (1966).

95. E. Costa, D. J. Boullin, W. Hammer, W. Vogel, and B. B. Brodie, Interactions of drugs with adrenergic neurons, *Pharmacol. Rev.* **18**:577–597 (1966).

96. J. A. Axelrod and R. Tomchick, Enzymatic O-methylation of epinephrine and other catechols, *J. Biol. Chem.* **233**:702–705 (1958).

97. J. Axelrod, W. Albers, and C. D. Clemente, Distribution of catechol-O-methyltransferase in the nervous system and other tissues, *J. Neurochem.* **5**:68–72 (1959).

98. R. Montanari, E. Costa, M. A. Beaven, and B. B. Brodie, Turnover rate of norepinephrine in hearts of intact mice, rats, and guinea pigs using tritiated norepinephrine, *Life Sci.* **2**:232–240 (1963).

99. M. A. Beaven, E. Costa, and B. B. Brodie, The turnover of norepinephrine in thyrotoxic and non-thyrotoxic mice, *Life Sci.* **2**:241–246 (1963).

100. L. G. Whitby, J. Axelrod, and H. Weil-Malherbe, Fate of ^{3}H-norepinephrine in animals, *J. Pharmacol.* **132**:193–201 (1961).

101. J. Axelrod, H. Weil-Malherbe, and R. Tomchick, The physiological disposition of ^{3}H-epinephrine and its metabolite metanephrine, *J. Pharmacol. Exp. Therap.* **130**:367–369 (1959).

102. I. J. Kopin, G. Hertting, and E. K. Gordon, Fate of norepinephrine ^{3}H in the isolated perfused rat heart, *J. Pharmacol. Exp. Therap.* **138**:34–40 (1962).

103. E. Costa and B. B. Brodie, *in Progress in Brain Research* (H. E. Himwich and W. A. Himwich, eds.), Vol. 8, pp. 168–185, Elsevier, Amsterdam (1964).

104. J. Axelrod, E. Gordon, G. Hertting, I. J. Kopin, and L. T. Potter, On the mechanism of tacyphylaxis to tyramine in the isolated rat heart, *Brit. J. Pharmacol.* **14**:56–63 (1962).

105. L. T. Potter, J. Axelrod, and I. J. Kopin, Differential binding and release of norepinephrine and tachyphylaxis, *Biochem. Pharmacol.* **11**:254–256 (1962).

106. L. T. Potter and J. Axelrod, Studies on the storage of norepinephrine and the effect of drugs, *J. Pharmacol. Exp. Therap.* **140**:199–206 (1963).

107. L. L. Iversen and L. G. Whitby, The subcellular distribution of catecholamines in normal and tyramine depleted mouse hearts, *Biochem. Pharmacol.* **12**:582–584 (1963).

108. B. Bhagat and J. Gilliam, Effect of various procedures on the repletion of cardiac catecholamine stores after tyramine, *J. Pharmacol. Exp. Therap.* **150**:41–45 (1965).

109. R. Kuntzman and M. M. Jacobson, On the mechanism of heart NE depletion by tyramine, guanethidine and reserpine, *J. Pharmacol. Exp. Therap.* **144**:399–404 (1964).

110. B. B. Brodie, E. Costa, A. Groppetti, and C. Matsumoto, Interaction between desipramine, tyramine and amphetamine at adrenergic neurons, *Brit. J. Pharmacol.* **34**: 648–658 (1968).

111. N. H. Neff, T. N. Tozer, W. Hammer, and B. B. Brodie, Kinetics of release of norepinephrine by tyramine, *Life Sci.* **4**:1860–1875 (1965).

112. Y. Gutman and H. Weil-Malherbe, Kinetics of catecholamine release by tyramine in rat heart, spleen and uterus, *Life Sci.* **5**:1293–1298 (1966).

113. N. Weiner and I. Selvaratnam, The effect of tyramine on the synthesis of norepinephrine, *J. Pharmacol. Exp. Therap.* **161**:21–33 (1968).

114. M. A. Beaven, Use of tracers in the study of biogenic amine compartments, *Adv. in Tracer Methodology* **2**:243–252 (1965).

115. C. W. Snedecor, *Statistical Methods*, Iowa State University Press, Ames (1956).

116. L. L. Iversen and J. Glowinski, Regional differences in the rate of turnover of norepinephrine in the rat brain, *Nature* **210**:1006–1008 (1966).

117. L. L. Iversen and J. Glowinski, Regional studies of catecholamines in the rat brain. II. Rate of turnover of catecholamines in various brain regions, *J. Neurochem.* **13**:671–682 (1966).

118. K. Fuxe and U. Ungerstedt, Localization of catecholamine uptake in rat brain after intraventricular injection, *Life Sci.* **5**:1817–1824 (1966).

119. J. W. Maas and D. H. Landis, *In vivo* studies of the metabolism of norepinephrine in the central nervous system, *J. Pharmacol. Exp. Therap.* **163**:147–162 (1968).

119a. S. Udenfriend and P. Zaltzman-Nirenberg, NE and dopamine turnover in guinea pig brain *in vivo*, *Science* **142**:394–396 (1963).

120. L. Stjarne, F. Lishajko, and R. H. Roth, Regulation of noradrenaline biosynthesis in nerve tissue, *Nature* **215**:770–772 (1967).

121. S. H. Ngai, N. H. Neff, and E. Costa, Effect of pargyline on the rate of conversion of tyrosine ^{14}C to norepinephrine ^{14}C, *Life Sci.* **7**:847–855 (1968).

122. K. F. Gey and A. Pletscher, Acceleration of turnover of ^{14}C catecholamines in rat brain by chlorpromazine, *Experientia* **24**:335–336 (1968).

123. R. Gordon, J. V. O. Reid, A. Sjoerdsma, and S. Udenfriend, Increased synthesis of norepinephrine in the rat heart on electrical stimulation of the sellate ganglion, *Mol. Pharmacol.* **2**:606–613 (1966).

124. A. H. Anton and D. F. Sayre, The distribution of dopamine and DOPA in various animals and a method for their determination in diverse biological material, *J. Pharmacol. Exp. Therap.* **145**:326–336 (1964).

125. N. H. Neff, S. H. Ngai, C. T. Wang, and E. Costa, Calculation of the rate of catecholamine synthesis from the rate of conversion of ^{14}C tyrosine to catecholamines: effect of adrenal demedullation on synthesis rates, *Mol. Pharmacol.* **5**:90–99 (1969).

126. T. Lewander and J. Jonsson, Isolation and determination of free endogenous and radioactive tyrosine in studies of catecholamine synthesis in the rat brain, *Life Sci.* **7**:387–394 (1968).

127. S. Spector, Inhibition of endogenous catecholamine biosynthesis, *Pharmacol. Rev.* **18**:599–610 (1966).

128. E. M. Gal, M. Poczek, and F. D. Marshall, Hydroxylation of tryptophan to 5-hydroxytryptophan by brain tissue *in vivo*, *Biochem. Biophys. Res. Commun.* **12**:39ff (1963).

129. W. Lovenberg, E. Jequier, and A. Sjoerdsma, Tryptophan hydroxylation: measurements in pineal gland, brainstem, and carcinoid tumor, *Science* **155**:217–219 (1967).

130. S. Nakamura, A. Ichiyama, and O. Hayashi, Purification and properties of tryotophan hydroxylase in brain, *Fed. Proc.* **24**:604 (1965).

131. E. M. Gal, J. C. Armstrong, B. Ginsberg, The nature of *in vivo* hydroxylation of L-tryotophan by brain tissue, *J. Neurochem.* **13**:643–653 (1966).

132. A. Ichiyama, S. Nakamura, Y. Nishizuka, and O. Hayaishi, Tryptophan-5-hydroxylase in mammalian brain, *Adv. Pharmacol.* **6A**:5–17 (1968).

133. E. W. Maynert and K. Kuriyama, Some observations on nerve-ending particles and synaptic vesicles, *Life Sci.* **3**:1067–1087 (1964).

134. R. M. Marchbanks, Serotonin binding to nerve ending particles and other preparations from rat brain, *J. Neurochem.* **13**:1481–1493 (1966).

135. G. C. K. Roberts, The formation of complexes between 5-hydroxytryptamine, adenosine triphosphate and bivalent cations *in vitro*, *Biochem. J.* **100**:30P (1966).

136. N. H. Neff and T. N. Tozer, *In vivo* measurement of brain serotonin turnover, *Adv. Pharmacol.* **6A**:97–109 (1968).

137. H. Weissbach, B. G. Redfield, and S. Udenfriend, Soluble monoamine oxidase, its properties and actions on serotonin, *J. Biol. Chem.* **229**:953–963 (1957).

138. S. Kveder, S. Iskric, and D. Keglevic, 5-hydroxytryptophol: A metabolite of 5-hydroxytryptamine in rats, *Biochem. J.* **85**:447–449 (1962).

139. A. Feldstein and K. K.-K. Wong, Enzymatic conversion of serotonin 5-hydroxytryptophol, *Life Sci.* **4**:183–191 (1965).

140. B. Werdinius, Effect of probenecid on the level of homovanillic acid in the corpus striatum, *J. Pharm. Pharmacol.* **18**:546–547 (1966).

141. H. C. Guldberg, G. W. Ashcroft, and T. B. B. Crawford, Concentrations of 5-hydroxyindoleacetic acid and homovanillic acid in the cerebiospinal fluid of the dog before and during treatment with probenecid, *Life Sci.* **5**:1571–1575 (1966).

142. W. D. Reid, L. Volicer, and B. B. Brodie, Inhibition of 5-hydroxyindoleacetic acid transport from the brain by 2,4-dintrophenol, *Life Sci.* **7**:577–581 (1968).

143. G. K. Aghajanian, F. E. Bloom, R. A. Lovell, M. H. Sheard, and D. X. Freedman, The uptake of 5-hydroxytryptamine-^{3}H from the cerebral ventricles: autoradiographic localization, *Biochem. Pharmacol.* **15**:1401–1403 (1966).

144. E. M. Gal, M. Morgan, S. K. Chatterjee, and F. D. Marshall, Jr., Hydroxylation of tryptophan by brain tissue *in vivo* and related aspects of 5-hydroxytryptamine metabolism, *Biochem. Pharmacol.* 13:1639–1653 (1964).
145. M. Bulat and Z. Supek, Fate of intracisternally injected 5-hydroxytryptamine in rat brain, *Nature* 211:637–638 (1966).
146. M. Bulat and Z. Supek, Mechanism of 5-hydroxytryptamine penetration through the cerebrospinal fluid brain barrier, *Nature* 219:72–73 (1968).
147. M. Bulat and Z. Supek, The penetration of 5-hydroxytryptamine through the blood–brain barrier, *J. Neurochem.* 14:264–271 (1967).
148. M. Bulat and Z. Supek, Passage of 5-hydroxytryptamine through the blood–brain barrier, its metabolism in the brain and elimination of 5-hydroxyindoleacetic acid from the brain tissue, *J. Neurochem.* 15:383–389 (1968).
149. S. Udenfriend and H. Weissbach, Turnover of 5-hydroxytryptamine (serotonin) in tissues, *Proc. Soc. Exptl. Biol. Med.* 97:748–751 (1958).
150. S. Udenfriend, H. Weissbach, and D. F. Bogdanski, Effect of iproniazid on serotonin metabolism *in vivo*, *J. Pharmacol. Exp. Therap.* 120:255–260 (1957).
151. R. Kuntzman, P. A. Shore, D. Bogdanski, and B. B. Brodie, Microanalytical procedures for fluorometric assay of brain Dopa—5-HTP decarboxylase, norepinephrine and serotonin and a detailed mapping of decarboxylase activity in brain, *J. Neurochem.* 6:226–232 (1961).
152. B. B. Brodie, M. S. Comer, E. Costa, and A. Dlabac, The role of brain serotonin in the mechanism of the central action of reserpine, *J. Pharmacol. Exp. Therap.* 152:340–349 (1966).
153. G. A. Johnson, E. G. Kim, and S. J. Boukma, Mechanism of norepinephrine depletion by 5-hydroxytryptophan, *Proc. Soc. Exp. Ther.* 128:509–512 (1968).
154. R. C. Lin, E. Costa, N. H. Neff, C. T. Wang, and S. H. Ngai, In vivo Measurement of 5-Hydroxytryptamine Turnover rate in the rat brain from the conversion of ^{14}C-tryptophan to ^{14}C-5-hydroxytryptamine, *J. Pharmacol. Exp. Therap.* (in press).
155. S. Spector, C. W. Hirsch, and B. B. Brodie, Association of behavioral effects of pargyline, a non-hydrazide MAO inhibitor, with increase in brain norepinephrine, *J. Neuropharmacol.* 2:81–93 (1963).
156. T. N. Tozer, N. H. Neff, and B. B. Brodie, Application of steady-state kinetics to the synthesis rate and turnover time of serotonin in the brain of normal and reserpine-treated rats, *J. Pharmacol. Exp. Therap.* 153:177–182 (1966).
157. D. F. Bogdanski, A. Pletscher, B. B. Brodie, and S. Udenfriend, Identification and assay of serotonin in brain, *J. Pharmacol. Exp. Therap.* 117:82–88 (1956).
158. S. H. Snyder, J. Axelrod, and M. Zweig, A sensitive and specific fluorescence assay for tissue serotonin, *Biochem. Pharmacol.* 14:831–835 (1965).
159. R. P. Maickel and F. P. Miller, The fluorometric determination of indolealkylamines in brain and pineal gland, *Adv. Pharmacol.* 6A:71–77 (1968).
160. N. H. Neff, T. N. Tozer, and B. B. Brodie, Application of steady-state kinetics to studies of the transfer of 5-hydroxyindolacetic acid from brain to plasma, *J. Pharmacol. Exp. Therap.* 158:214–218 (1967).
161. M. B. Bowers, Jr. and F. Gerbode, CSF 5-HIAA: Effects of probenecid and parachlorophenylalanine, *Life Sci.* 7:773–776 (1968).
162. B. Johansson and B. E. Roos, 5-Hydroxyindoleacetic and homovanillic acid levels in the cerebrospinal fluid of healthy volunteers and patients with Parkinson's syndrome, *Life Sci.* 6:1449–1454 (1967).
163. G. W. Ashcroft, T. B. B. Crawford, D. Eccleston, D. F. Sharman, E. T. MacDougall, J. B. Stanton, and J. K. Binns, 5-Hydroxyindole compounds in the cerebrospinal fluid of patients with psychiatric or neurological diseases, *Lancet* 2:1049–1052 (1966).

164. S. Spector, A. Sjoerdsma, and S. Udenfriend, Blockade of endogenous NE synthesis by α-methyltyrosine, an inhibitor of tyrosine hydroxylase, *J. Pharmacol. Exp. Therap.* **147**: 85–95 (1965).

165. M. Goldstein and Z. Weiss, Inhibition of tyrosine hydroxylase by 3-iodo-L-tyrosine, *Life Sci.* **4**:261–264 (1965).

166. S. Spector, R. Ortega-Mata, A. Sjoerdsma, and S. Udenfriend, Biochemical and pharmacological effects of iodotyrosines: relation to tyrosine hydroxylase inhibition *in vivo*, *Life Sci.* **4**:1307–1311 (1965).

167. J. B. Stanbury, J. W. A. Meijer, and A. A. H. Kessenaer, The metabolism of iodotyrosine. The metabolism of mono and di-iodo tyrosine in certain patients with familial goiter, *J. Clin. Endocr. Metab.* **16**:848–868 (1956).

168. A. Carlsson, H. Corrodi, and B. Waldek, α-Substituierte Dopacetamide als Hemmer der Catechol-O-methyltransferase und der enzymatischen Hydroxylierung aromatischer Aminosäuren in den Catecholamin Metabolismus eingreifende Substanzen, 2, *Helv. Chim. Acta* **46**:2271–2285 (1963).

169. M. Goldstein, H. Gang, and B. Anagnoste, The inhibition of tyrosine hydroxylase by 4-isopropyltropolone, *Life Sci.* **6**:1457–1461 (1967).

170. J. A. Oates, L. Gillespie, S. Udenfriend, and A. Sjoerdsma, Decarboxylase inhibition and blood pressure reduction by α-methyl 3,4-dihydroxy-DL-phenylalanine, *Science* **131**:1890–1891 (1960).

171. B. B. Brodie, R. Kuntzman, C. W. Hirsch, and E. Costa, Effects of decarboxylase inhibition on the biosynthesis of brain monoamines, *Life Sci.* **1**:81–84 (1962).

172. D. J. Drain, M. Horlington, R. Lazare, and G. A. Poulter, The effect of α-methyl DOPA and some other decarboxylase inhibitors on brain 5-hydroxytryptamine, *Life Sci.* **1**:93–97 (1962).

173. J. H. Merritt and E. J. Schultz, The effect of decaborane on the biosynthesis and metabolism of norepinephrine in the rat brain, *Life Sci.* **5**:27–32 (1966).

174. S. M. Hess, R. H. Connamecher, M. Ozaki, and S. Udenfriend, The effects of α-methyl-meta-tyrosine on the metabolism of norepinephrine and serotonin *in vivo*, *J. Pharmacol. Exp. Therap.* **134**:129–138 (1961).

175. J. H. Merritt, E. J. Schultz, and A. A. Wykes, Effect of decaborane on the norepinephrine content of rat brain, *Biochem. Pharmacol.* **13**:1364ff (1964).

176. E. H. Krackow, Toxicity and health hazards of boron hydrides, *AMA Arch. Indus. Hug. and Occup. Med.* **8**:335–339 (1953).

177. A. Carlsson and M. Lindqvist, *In vivo* decarboxylation of α-methyl-*m*-tyrosine, *Acta Physiol. Scand.* **54**:87–94 (1962).

177a. U. S. von Euler and F. Lishajko, Catecholamine depletion and uptake in adrenergic nerve vesicles and in rabbit organs after decarborane, *Acta Physiol. Scand.* **65**:324–330 (1965).

178. D. G. Johnson, The effect of cold exposure on the catecholamine excretion of rats treated with decarborane, *Acta Physiol. Scand.* **68**:129–133 (1966).

178a. A. Oliverio, Release of cardiac noradrenaline by decaborane in the heart lung preparation of guinea pig, *Biochem. Pharmacol.* **14**:1689–1692 (1965).

179. C. R. Creveling, J. B. Van Der Schoot, and S. Udenfriend, Phenethylamine isosteres as inhibitor of dopamine β-oxidase, *Biochem. Biophys. Res. Commun.* **8**:215–219 (1962).

180. R. Kuntzman, E. Costa, C. Creveling, C. W. Hirsch, and B. B. Brodie, Inhibition of norepinephrine synthesis in mouse brain by blockade of dopamine-β-oxidase, *Life Sci.* **1**:85–92 (1962).

181. B. Nikadijevic, C. R. Creveling, and S. Udenfriend, Inhibition of dopamine-β-oxidase *in vivo* by benzyloxyamine and benzylhydrazine analogs, *J. Pharmacol. Exp. Therap.* **140**:224–228 (1963).

181a. M. Goldstein, E. Lauber, and M. R. McKeregan, Studies of the purification and characterization of 3,4-dihydroxyphenylethylamine β-hydroxylase, *J. Biol. Chem.* **240**:2066–2072 (1965).

182. J. M. Musacchio, M. Goldstein, B. Anagnoste, G. Poch, and I. J. Kopin, Inhibition of dopamine-β-hydroxylase by disulfiram *in vivo*, *J. Pharmacol. Exp. Therap.* **152**:56–61 (1966).

183. M. Goldstein, B. Anagnoste, E. Lauber, and M. R. McKeregan, Inhibition of dopamine-β-oxidase by disulfiram, *Life Sci.* **3**:763–767 (1964).

184. K. Missala, K. Lloyde, G. Gregoriads, and T. L. Sourkes, Conversion of ^{14}C-dopamine to cardiac ^{14}C-noradrenaline in the copper deficient rat, *Eur. J. Pharmacol.* **1**:6–10 (1967).

185. M. Goldstein and K. Nakajima, The effects of disulfiram on the repletion of brain catecholamine stores, *Life Sci.* **5**:1133–1138 (1966).

186. R. J. Wurtman, Control of epinephrine synthesis in the adrenal medulla by the adrenal cortex: hormonal specificity and dose-response characteristic, *Endocrinology* **79**:608–614 (1966).

187. L. Pohorecky and J. H. Rust, Studies on the cortical control of the adrenal medulla in the rat, *J. Pharmacol. Exp. Therap.* **162**:227–238 (1968).

188. N. H. Neff and E. Costa, The influence of monoamine oxidase inhibition on catecholamine synthesis, *Life Sci.* **5**:951–959 (1966).

189. S. Udenfriend, Tyrosine hydroxylase, *Pharmacol. Rev.* **18**:43–51 (1966).

190. K. F. Gey and A. Pletscher, Influence of chlorpromazine and chlorprothixene on the cerebral metabolism of 5-hydroxytryptamine, norepinephrine and dopamine, *J. Pharmacol. Exp. Therap.* **133**:18–24 (1961).

191. E. Costa, G. L. Gessa, and B. B. Brodie, Influence of hypothermia on chlorpromazine-induced changes in brain amine levels, *Life Sci.* **1**:315–319 (1962).

192. K. F. Gey and A. Pletscher, Effects of chlorpromazine on the metabolism of *dl*-2-C^{14} DOPA in the rat, *J. Pharmacol. Exp. Therap.* **145**:337–343 (1964).

193. N. E. Andén, B. E. Roos, and B. Werdinius, Effects of chlorpromazine, haloperidol and reserpine on the levels of phenolic acids in rabbit corpus striatum, *Life Sci.* **3**:149–158 (1964).

194. T. Person and B. E. Roos, Clinical and pharmacological effects of monoamine precursors or haloperidol in chronic schizophrenia, *Nature* **217**:854 (1968).

195. W. P. Burkard, K. F. Gey, and A. Pletscher, Activation of tyrosine hydroxylation in rat brain *in vivo* by chlorpormazine, *Nature (London)* **213**:732–733 (1967).

196. H. Nybäck, G. Sedvall, and I. J. Kopin, Accelerated synthesis of dopamine-C^{14} from tyrosine-C^{14} in rat brain after chlorpromazine, *Life Sci.* **6**:2307–2312 (1967).

197. H. Corrodi, K. Fuxe, and F. Hökfelt, The effect of neuroleptics on the activity of central catecholamine neurons, *Life Sci.* **6**:767–774 (1967).

198. M. Da Prada and A. Pletscher, Acceleration of the cerebral dopamine turnover by chlorpromazine, *Experientia* **15**:465–466 (1966).

199. L. Landsberg and J. Axelrod, Influence of pituitary, thyroid, and adrenal hormones on norepinephrine turnover and metabolism in the rat heart, *Circulation Res.* **22**:559–571 (1968).

200. E. Costa and N. H. Neff, The dynamic process of catecholamine storage as a site for drug action, pp. 757–764, *Proc. V Int. Cong. of the Collegium Int. Neuro-Psycho Pharmacol.* Excerpta Med. Int. Cong., Series No. 129, Amsterdam (1966).

201. N. E. Andén, K. Fuxe, and T. Hökfelt, Effect of some drugs on central monoamine nerve terminals lacking nerve impulse flow, *Eur. J. Pharmacol.* **1**:226–232 (1967).

202. N. E. Andén, T. Magnusson, B. E. Roos, and B. Werdinius, 5-hydroxy-indoleacetic acid of rabbit spinal cord normally and after transection, *Acta Phys. Scand.* **64**:193–196 (1965).

203. R. Volicer and W. D. Reid, Effect of drugs on turnover rate of heart norepinephrine, *Int. J. Neuropharmac.* **8**:1–7 (1968).

204. H. Corrodi, K. Fuxe, and T. Hökfelt, The effect of immobilization stress on the activity of central monoamine neurons, *Life Sci.* **7**:107–112 (1968).
205. A. M. Thierry, F. Javoy, J. Glowinski, and S. S. Kety, Effects of stress on the metabolism of NE, DM, and serotonin in the central nervous system of the cat. I. Modifications of norepinephrine turnover, *J. Pharmacol. Exp. Therap.* **163**:163–171 (1968).
206. H. Corrodi, K. Fuxe, and T. Hökfelt, The effect of ethanol on the activity of central catecholamine neurons in rat brain, *J. Pharm. Pharmacol.* **18**:821–822 (1966).
207. A. Weissman, B. K. Koe, and S. S. Tenen, Antiamphetamine effects following inhibition of tyrosine hydroxylase, *J. Pharmacol. Exp. Therap.* **151**:339–352 (1966).
208. A. Randrup and I. Munkvad, Role of catecholamines in the amphetamine excitatory response, *Nature (London)* **210**:540 (1966).
209. R. H. Rech, H. K. Borys, and K. E. Moore, Alterations in behavior and brain catecholamine levels in rats treated with α-methyltyrosine, *J. Pharmacol. Exp. Therap.* **153**:412–419 (1966).
210. K. E. Moore and R. H. Rech, Antagonism by monoamine oxidase inhibitors of α-methyltyrosine induced catecholamine depletion and behavioral depression, *J. Pharmacol. Exp. Therap.* **156**:70–75 (1967).
211. H. Corrodi and L. C. F. Hanson, Central effects of an inhibitor of tyrosine hydroxylase, *Psychopharmacologica* **10**:116–125 (1966).
212. N. A. Hillarp, K. Fuxe, and A. Dahlström, Adrenergic mechanisms in the nervous system, C. Demonstration and mapping of central neurons containing dopamine, noradrenaline and 5-hydroxytryptamine and their reaction to psychopharmaca, *Pharmacol. Rev.* **18**:727–741 (1966).
213. J. M. Stolk and R. H. Rech, Enhanced stimulant effects of *d*-amphetamine on the spontaneous locomotor activity of rats treated with reserpine, *J. Pharmacol. Exp. Therap.* **158**:140–149 (1967).
214. F. Sulser, M. L. Owens, M. R. Norvich, and J. V. Dingell, The relative role of storage and synthesis of brain norepinephrine in the psychomotor stimulation evoked by amphetamine or by desipramine and tetrabenazine, *Psychopharmacologia* **12**:322–332 (1968).
215. A. Pletscher, M. Da Prada, G. Bartholini, W. P. Burkard, and H. Bruderer, Two types of monoamine liberation by chlorinated aralkylamines, *Life Sci.* **4**:2301–2308 (1965).
216. A. Pletscher, M. Da Prada, W. P. Burkard, G. Bartholini, F. A. Steiner, H. Bruderer, and F. Bigler, Aralkylamines, with different effects on the metabolism of aromatic monoamines, *J. Pharmacol. Exp. Therap.* **154**:64–72 (1966).
217. K. B. Koe and A. Weissman, *p*-Chlorophenylalanine: A specific depletor of brain serotonin, *J. Pharmacol. Exp. Therap.* **154**:499–516 (1966).
218. B. K. Koe, Inhibiting action of *p*-chlorophenylalanine and α-methyl-*p*-chlorophenylalanine on liver phenylalanine and tryptophan hydroxylase, *Med. Pharmacol. Exp.* **17**:129–138 (1967).
219. E. Jequier, W. Lovenberg, and A. Sjoerdsma, Tryptophan hydroxylase inhibition: The mechanism by which *p*-chlorophenylalanine depletes rat brain serotonin, *Mol. Pharmacol.* **3**:274–278 (1967).
220. G. Guroff, K. Kondo, and J. Daly, The production of meta-chlorotyrosine from parachlorophenylalanine by phenylalanine hydroxylase, *Biochem. Bioph. Res. Commun.* **25**:622–628 (1966).
221. S. Kaufman, in *Oxygenases* (O. Hayaishi, ed.), pp. 129–180, Academic Press, New York (1962).
222. M. A. Lipton, R. Gordon, G. Guroff, and S. Udenfriend, *p*-Chlorophenylalanine-induced chemical manifestations of phenylketonuria in rats, *Science* **156**:248–250 (1967).
223. S. S. Tenen, Antagonism of the analgesic effect of morphine and other drugs by *p*-chlorophenylalanine, a 5-HT depletor, *Psychopharmacologia* **12**:278–285 (1968).
224. S. Fu-hsiung, H. H. Loh, and E. L. Way, Brain serotonin turnover and tolerance development to morphine, *Pharmacologist* **10**:322 (1968).

225. M. Jouvet, Insomnia and decrease of cerebral 5-hydroxytryptamine after destruction of the raphe system in the cat, *Adv. Pharmacol.* **6**:Part B:265–279 (1968).
226. H. Corrodi and K. Fuxe, The effect of imipramine on central monoamine neurons, *J. Pharm. Pharmacol.* **20**:230–231 (1968).
227. H. Corrodi, K. Fuxe, and T. Hökfelt, The effect of some psychoactive drugs on central monoamine neurons, *Eur. J. Pharmacol.* **1**:363–368 (1967).
228. A. Carlsson, K. Fuxe, and U. Ungerstedt, The effect of imipramine on 5-hydroxytryptamine neurons, *J. Pharm. Pharmacol.* **20**:150–151 (1969).
229. K. Fuxe and U. Ungerstedt, Localization of 5-hydroxytryptamine uptake in rat brain after intraventicular injection, *J. Pharm. Pharmacol.* **19**:335–337 (1967).
230. J. H. Gaddum and K. A. Hameed, Drugs which antagonize 5-hydroxytryptamine, *Brit. J. Pharm.* **9**:240–247 (1954).
231. E. Costa, Effects of hallucinogenic and tranquilizing drugs on serotonin evoked uterine contractions, *Proc. Soc. Exp. Biol. Med.* **91**:39–41 (1956).
232. D. X. Freedman and N. J. Giarman, LSD-25 and the status and level of brain 5-HT, *Ann. N.Y. Acad. Sci.* **96**:97–106 (1962).
233. J. A. Rosecrans, R. A. Lovell, and D. X. Freedman, Effects of lysergic acid diethylamide on the metabolism of brain 5-hydroxytryptamine, *Biochem. Pharmacol.* **16**:2011–2021 (1967).
234. N. E. Andén, H. Corrodi, K. Fuxe, and T. Hökfelt, Evidence for a central 5-hydroxy-tryptamine receptor stimulation by Lyseric acid diethylamide, *Brit. J. Pharmacol.* **34**:1–7 (1968).
235. P. Diaz, S. H. Ngai, and E. Costa, The effect of LSD on the metabolism of brain serotonin, *Pharmacologist* **9**:251 (1967).
236. D. F. Sharman, Changes in the metabolism of 3,4-dihydroxyphenlethylamine (dopamine) in the striatum of the mouse induced by drugs, *Brit. J. Pharmacol. Chemother.* **23**:153–163 (1966).
237. G. K. Aghajanian, W. E. Foote, and M. H. Sheard, Lysergic acid diethylamide: Sensitive neuronal units in the midbrain raphé, *Science* **161**:707–709 (1968).
238. W. Dairman, S. Gordon, S. Spector, S. Sjoerdsma, and S. Udenfriend, Effect of α-blockers on catecholamine biosynthesis, *Fed. Proc.* **27**:240 (1968).
239. C. Owman, Sympathetic nerves probably storing two types of monoamines in the rat pineal gland, *Int. J. Neuropharmacol.* **3**:105 (1964).
240. N. H. Neff, R. E. Barrett, and E. Costa, Compartmentation of serotonin (5-HT) in rat pineal gland, *Fed. Proc.* **27**:411 (1968).

Chapter 4

STORAGE AND RELEASE OF MONOAMINES IN THE CENTRAL NERVOUS SYSTEM

Jacques Glowinski

Section de Neuropharmacologie Biochimique du Laboratoire de Neurophysiologie Générale du College de France

I. INTRODUCTION

A. Historical Background

During the last few years, new tools have been used successfully by physiologists, pharmacologists, and biochemists to elucidate the metabolism and functions of some monoamines in the central nervous system. The development of histochemical fluorescence techniques, the utilization of labeled monoamines or of their precursors of high specific activity, and the discovery of new drugs have contributed particularly to the accumulation of much information related to the central metabolism and disposition of norepinephrine (NE), dopamine (DA), and serotonin (5-HT).

The occurrence of NE in the mammalian brain was first described by Von Euler in 1946 and its uneven distribution by Vogt in 1954.[1] The same year, Amin *et al.* found 5-HT in brain tissues.[2] More recently, Montagu, Weil-Malherbe, and Bone found DA in the central nervous system[3] and in 1959, Carlsson *et al.* found high concentrations of this amine particularly in the striatum.[3] Strong experimental evidence suggests that NE, DA, and · 5-HT act, like acetylcholine, as neurotransmitters at some central synapses. These monoamines have already been shown to meet many of the requirements for the establishment of a transmitter role of a given substance. NE, DA, and 5-HT have all been shown to be located intraneuronally and to be highly concentrated in the axon terminals of special neuron systems which have been extensively described, particularly in the rat and more recently in other species.[4] These amines are synthesized intraneuronally and partly bound in intraneuronal amine granules or vesicles in an inactive form and thus are protected from inactivation by enzymes*. Local extraneural or intraneural inactivating systems including inactivating enzymes and re-uptake processes have been described. Variations in the utilization of these biogenic amines under different physiological conditions have been recently observed. Some experiments have shown a release of these amines by electrical stimulation of specific central nuclei.

The various aspects of the metabolism and functions of central NE,[1] DA,[3] and 5-HT[2] have been reviewed extensively. In this chapter, we will focus our attention mainly on the storage and release processes of these aminergic neurons. However, it is first necessary to briefly describe the localization or mapping of these neuron systems in the central nervous system.

B. Localization of Central Amine-Containing Neurons

As demonstrated by Hillarp's group,[4] the different systems of aminergic neurons can be completely visualized by a histochemical fluorescence method† Several ascending and bulbospinal NE neurons have been de-scribed. Their cell bodies are mainly located in the lower ventral brain stem; however, small groups of cells have been found as well in the hypothalamus. Their axons run together in bundles and terminate in fine arborizations which form dense plexuses and contain many "varicosities." They are mainly found in the vegetative hypothalamus, various parts of the limbic system, in many of the visceral nuclei of the cranial nerves, but are also seen in the ventral horn, the sympathetic lateral column, and the substantia gelatinosa of the spinal cord. A large nigro-striatal system of DA neurons has been de-scribed. The cell bodies are located in the pars compacta of the substantia nigra and the terminals are mainly found in the neostriatum. Furthermore, a tubero-infundibular DA neuron system has been demonstrated. As in the case of NE neuron systems, there are several ascending and bulbospinal systems of 5-HT neurons (Fig. 1). Their cell bodies are mainly located in the raphe nuclei and in the ventral and medial part of the reticular formation. In the brain, 5-HT terminals have been detected mainly in the lower brain stem, particularly in visceral nuclei (only certain nuclei are rich in 5-HT terminals in the diencephalon and telencephalon). In the spinal cord 5-HT terminals are seen in the sympathetic lateral column and the ventral horn.

Central NE, DA, and 5-HT neurons have been observed in the mouse, rat, guinea pig, dog, cat, and monkey, and also in birds. Adrenergic neurons are present in the frog brain which contains large amounts of adrenaline. This catecholamine is detected only in small amounts in the mammalian brain.[4,5]

Biochemical determinations of the endogenous content of brain amine have been made after specific lesions in an area containing cell bodies or non-terminal axons of aminergic neurons and the disappearance of neuronal

* See Chapter 9 on Metabolism of Catecholamines and Chapter 11 on Serotonin.

† References on the details of the histochemical fluorescence method are given by Hillarp et al.[4]

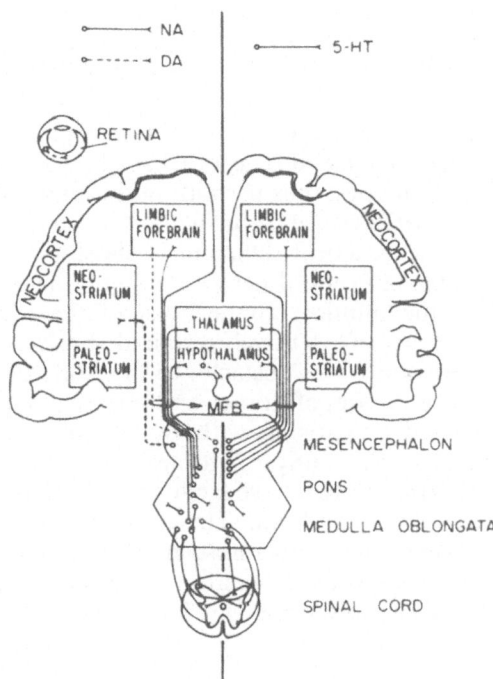

NA
DA
5-HT
RETINA
NEOCORTEX
LIMBIC FOREBRAIN
LIMBIC FOREBRAIN
NEOCORTEX
NEO-STRIATUM
NEO-STRIATUM
THALAMUS
PALEO-STRIATUM
HYPOTHALAMUS
PALEO-STRIATUM
MFB
MESENCEPHALON
PONS
MEDULLA OBLONGATA
SPINAL CORD

Fig. 1. Schematic drawing showing, in highly simplified form the main monoamine neuron systems in the central nervous system. (From Andén et al.[12])

amines has been related to the effect of the lesions. Thus tegmental lesions in monkey and rats produced a decrease in dopamine in the striatum.[6,7,8] Lesions in the raphe nuclei were followed by a disappearance of brain serotonin in the cat.[9] Lesions in the middle forebrain bundle of the lateral hypothalamus and in the dorsomedial brain stem tegmentum significantly decreased both NE and 5-HT in the whole rat brain.[10] A thoracic cord section led to a decrease in 5-HT and NE in the spinal cord.[11]

C. Definition of Storage and Release Processes

The central disappearance or utilization of NE, DA, and 5-HT is the result of two different processes: the extraneuronal release of the amine in a physiologically active form and the intraneuronal release from binding sites. In the latter case, a great part of the amine is immediately enzymatically inactivated in the neuron and is depleted from the nerve in an inactive form. It should be mentioned also that the extraneuronal release can result from two processes: a spontaneous leakage or diffusion of the amine through the axon membrane and an active release by nerve impulses. In order to compensate for the used amines, new amines are synthesized and stored mainly in nerve endings. However, an important part of the extraneuronally released amines are reabsorbed by the neuron by two independent powerful mechanisms which have been mainly studied in the case of catecholaminergic neurons. On the one hand, there is an "uptake" or active transport of the amines across the cell membrane in the intraneuronal space and, on the other hand, there is an intraneuronal redistribution which includes an uptake and a retention of the amines mainly in specific intracellular storage sites: the granules or synaptic vesicles. The important reuptake process occurring at the neuronal membrane level has often been used to take up and to store exogenous amines, thus permitting the study of their subsequent disposition in *in vitro* or *in vivo* experiments.

The practical difficulties inherent in *in vivo* experimentation on the mammalian central nervous system, the complexity of the brain organization, the multiple connections of the monoaminergic neurons, and the occurrence of extensively terminal aborizations are some of the factors which represent obvious obstacles to the study of the mechanisms and regulations of the processes of storage and release of monoamines in the central nervous system. Comparisons with and sometimes extrapolations from results obtained in the study of storage and release mechanisms in the peripheral adrenergic system have facilitated the approach and understanding of these questions in the brain. These comparisons are possible because peripheral and central catecholaminergic (CA) neurons show similar basic properties such as biosynthesis, catabolism, or uptake processes, and also similar sensitivity to various drugs.

II. STORAGE OF MONOAMINES

A. The Intraneuronal Gross Distribution of Catecholamines and Serotonin

As observed with histochemical fluorescence methods, cell bodies and nonterminal axons of NE, DA, and 5-HT neurons contain only very low concentrations of monoamine; almost all the monoamines of the central nervous system are stored in the numerous "varicosities" of the terminal arborizations of the monoaminergic neurons. These varicosities, which are most likely presynaptic structures, vary in size from one structure to another. The amine concentration in cell bodies, in nonterminal axons, and in varicosities has been estimated and is probably on the order of 10–100, 150–500, and 10,000 μg/g by weight for the three locations respectively.[4] The whole terminal system of one neuron of the nigro-neostriatal dopamine neuron system in the rat has been estimated to be about 60 cm long and to contain about 500,000 varicosities.[12] However, there are some anatomical and size differences between various groups of monoaminergic central neurons. Variations of fluorescence intensity have been observed between different groups of catecholaminergic neurons, which may result from differences in the total volume of terminals belonging to a given neuron.

Inhibitors of monoamine oxidase (MAO) have been very useful for the mapping of 5-HT neurons; after MAO inhibition an intense increase in yellow fluorescence appears in all parts of 5-HT neurons. The MAO inhibition alone does not produce such an increase in the green fluorescence of catecholaminergic neurons, but a combined administration of both with MAO inhibitors and the precursor DOPA results in an increased fluorescence in all parts of these neurons. Moreover, these data show that the inactivating enzyme MAO is unevenly distributed in the entire length of aminergic neurons and suggest that it may play an important role in regulating the accumulation and overflow of unbound amines.[4] It is very likely that the most important storage sites of monoamines, the granules or vesicles, are manufactured in cell bodies and rapidly transported down to the varicosities of the terminals, where they accumulate. This is indicated by the picture of the amine fluorescence recovery following reserpine or α-methyl-p-tyrosine treatment, by effects of axotomy, and, more strikingly, by histochemical and biochemical data obtained after constriction of peripheral adrenergic neurons.[4,13]

B. Storage of Exogenous Monoamines in Central Aminergic Neurons

Further evidence of the ability of amines to accumulate in different parts of central neurons has been obtained by studies of uptake and accumulation of exogenous amines. When exogenous catecholamines are administered in the peripheral circulatory system, there is only a limited accumulation of these catecholamines in the brain: it is restricted to certain areas, such as the area postrema and some parts of the hypothalamus where the blood–brain barrier for monoamines is not effective.[1,14] Histochemical studies have shown that this local accumulation occurs in terminals and cell bodies of aminergic neurons.[4,15] The blood–brain barrier for monoamines can be circumvented by introducing catecholamines[16] or 5-HT[17] directly into the cerebrospinal fluid. Biochemical and pharmacological studies have demonstrated that labeled DA or NE is taken and retained in catecholamine-containing neurons.[1,16] Furthermore, it has been observed by light radioautography[18] and more precisely by electron microscopic radioautography[19–20]* that after the intraventricular administration of labeled NE, the radioactivity is localized in different parts of unmyelinated neurons containing granular vesicles. That these neurons are certainly catecholamine-containing neurons is evident by their distribution and their sensitivity to drugs,[16] and by the results of lesion studies.[8] The specific accumulation of exogenous CA in catecholaminergic neurons has been confirmed by *in vivo* and *in vitro* histochemical studies. In the latter case, brain slices, whose amine content had been previously depleted by reserpine, were incubated with α-methyl norepinephrine or with catecholamines after MAO inhibition.[21] It can be concluded from the various data obtained that as in peripheral NE neurons, the uptake process operating at the neuronal membrane level is acting on the entire length of the central catecholaminergic neurons.

* See pictures in the chapter of B. Droz on Metabolic Information derived from Radioautography.

However, limitations of the resolutions of these experimental approaches excluded any definitive conclusions on the nature of the intraneuronal anatomical structure on or in which the exogenous amines are retained and stored. Furthermore, studies on accumulation and storage of exogenous amines raised some questions about the specificity of uptake and storage processes in central aminergic neurons. Exogenous NE accumulates in central DA as well as in NE-containing neurons. Similarly, DA can utilize the uptake process of noradrenergic neurons (Fig. 2). Nevertheless, the two systems of uptake are different; desmethylimipramine (DMI) inhibits the uptake of NE but has no effect on the uptake of dopamine in dopaminergic neurons.[1]

Exogenous 5-HT can also accumulate in brain tissues as indicated by *in vivo* and *in vitro* studies.[17,22] In recent electron radioautographic studies performed after intraventricular administration of the labeled amine, it was observed that the localization of 5-HT-^{3}H is quite different from that of NE-[3]H.[17] These results suggest that 5-HT is taken and retained in nerve endings and axons of 5-HT neurons. Recent investigations suggest that the stored exogenous amine may behave similarly to the endogenous 5-HT. In a stress situation in which central 5-HT neurons of rats are activated, the rates

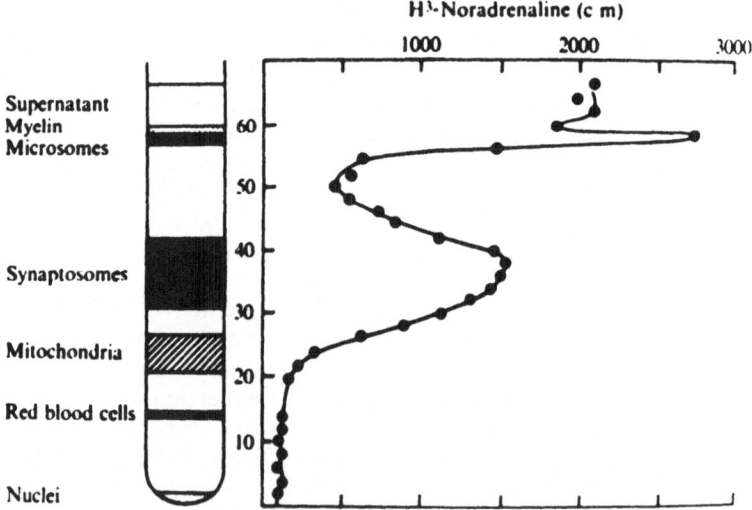

Fig. 2. Accumulation of exogenous norepinephrine-^{3}H in "synaptosomes" of rat striatum. Distribution of NE-^{3}H on a sucrose density gradient after centrifugation of a sample of a striatum homogenate obtained 4 hr after the intraventricular injection of DL-NE-[3]H (20 μC). Appearance of the gradient is illustrated diagrammatically on the left. Five drops of samples were collected after puncturing the bottom of the tube, alternate fractions were assayed for NE-[3]H. Most of the exogenous NE accumulates in nerve endings of dopamine-containing neurons. (From Glowinski and Iversen.[28])

Fig. 3. Comparison of exogenous and endogenous serotonin utilization in the central nervous system. Groups of rats received 5-HT-^{3}H (3, 4 μC) or tryptophan-^{3}H (^{3}H TRY) (30 μC) intracisternally, 1 hr later, they were submitted to a stress produced by mild electrical shocks applied intermittently to the feet for 3 hr. Control and stressed animals were killed 4 hr after the injection of the labeled compounds, and the 5-HT-^{3}H content was estimated in the brain stem + mesencephalon. The stress induced an increased utilization of both 5-HT-3H formed endogenously from its precursor tryptophan-^{3}H and exogenous 5-HT. Results are expressed in percent of the respective control values. (From Thierry, Fekete and Glowinski, unpublished observations.)

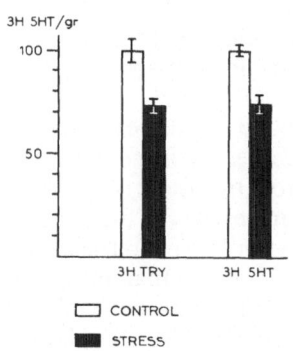

of utilization of the exogenous amine previously stored in the brain and of 5-HT-^{3}H endogenously synthesized from its precursors are parallelly increased (Fig. 3). However, further studies are required to ascertain and clarify the question of uptake and storage processes of exogenous 5-HT in central 5-HT neurons.

C. The Subcellular Localization of Monoamines

Ultracentifugation and electron microscopic techniques have been used to study the intraneuronal structures which store monoamines. In peripheral adrenergic nerves, high concentrations of NE are found in subcellular particles, and relatively pure NE-containing vesicles have been prepared. After the homogenization of brain tissues in isotonic sucrose and their subsequent centrifugation, an important part of the tissue content of NE, 5-HT, and DA is recovered in the particulate fraction. A fraction rich in bound acetylcholine and other monoamines can be separated from mitochondria, myelin, and other subcellular particles by differential or density-gradient centrifugation methods. This purified fraction consists mainly of pinched-off nerve endings, which are often accompanied by postsynaptic membrane elements. During the process of homogenization, nerve endings seal themselves and thus retain a part of their contents, particularly the vesicles and small mitochondria. These particles have recently been given the name of synaptosomes[23] and seem to be particularly rich in acetylcholine,[24,25] 5-HT,[2] NE,[1] and DA.[3] The varicosities seen by histochemical techniques probably correspond to the synaptosomes which contain catecholamines or 5-HT. De Robertis et al.[25] were able to separate these synaptosomes from those rich in acetylcholine, but Michaelson and Whittaker observed only a slight fractionation between acetylcholine and 5-HT synaptosomes.[23]

Synaptosomes can be disrupted by ultrasound or by hypotonic shock, thus liberating particles having sedimentation properties closely similar to those of microsomes. These particles contain relatively high proportions of monoamines and are probably synaptic vesicles, as indicated by their ultramicroscopic picture.[23,25-26] Different hypotheses have been made concerning the origin of the remaining part of the amine brain content recovered in the high-speed supernatant fraction after centrifugation of sucrose brain homogenates. The free amines found in the cell sap diluted with sucrose may have a complex origin. They can be derived either from disrupted synaptosomes, or from disrupted axons and cell bodies. They can correspond either to a free form of neuronal amines or to loosely bound forms of amines in or on vesicles or other neuronal structures, or even to a fraction of the strongly bound amine released during the process of homogenization.

More than 60% of the rat brain content of acetylcholine,[24] NE,[1] and 5-HT[2] is recovered in the particulate fraction rich in synaptosomes and mitochondria. Similar results were obtained with other species. However, the subcellular distribution of DA in the striatum appears to be different from that of other amines. Less than 40% of the striatal DA has been recovered in the brain particulate fraction of various species.[1,3] The following hypotheses were made to explain such discrepancies with the other known monoamines: (1) that DA may occur in a free form throughout the cytoplasm of dopaminergic terminals, or (2) that, due to the unusual fineness of the DA-containing terminals they might be more fragile and thus less resistant to the process of homogenization. (3) that DA has a lower affinity than NE for binding sites on synaptic vesicles. This last hypothesis is supported by the results of Musacchio et al.[27] on peripheral sympathetic nerves and by a study of subcellular distribution pattern of ^3H catecholamines in the rat hypothalamus and striatum after intraventricular administration of DOPA-^3H, in which it was demonstrated that the proportion of the labeled NE in the microsomal fraction rich in synaptic vesicles was higher than that of labeled dopamine.[28] The subcellular distribution of a monoamine may differ from one brain structure to another. For example, regions of the rat brain with high concentration of endogenous NE, such as hypothalamus and medulla oblongata, have relatively higher proportions of this amine localized in particulate fractions than structures with low concentration of endogenous NE, such as cerebellum and cerebral cortex.[28]

As already mentioned, exogenous amines can accumulate in central aminergic neurons. This has been further confirmed by subcellular *in vivo* and *in vitro* studies. After intraventricular administration of labeled catecholamines, a great proportion of the amines are found in the synaptosomal fraction (Fig. 2).[28] It has been shown that a temperature-dependent uptake of exogenous NE and 5-HT may occur in isolated vesicles prepared by hypotonic disruption of bovine hypothalamus synaptosomes.[26] The accumulation of exogenous amines in this vesicular preparation could be inhibited by reserpine, which also caused a depletion of the stored amine.[26] The incorporation of exogenous amines in central vesicles may require the presence of ATP and Mg ions as for adrenal medullary granules and splenic nerve granules; however, the importance of ATP and Mg^{2+} in the uptake storage processes of monoamines at the vesicular level in the central nervous system has not been well established.[1]

D. Characteristics of Storage Vesicles of Monoaminergic Neurons

Electron microscopical studies have revealed the occurrence of large (about 1000 Å) and small (about 500 Å) granular or dense core vesicles in peripheral nerves. Strong evidence has been provided by different experimental approaches that the small granular vesicles are the main storage sites of NE.[19-20,29-30] A decreased proportion of granular vesicles is observed when drugs affecting storage, such as reserpine and metaraminol, are given. In contrast, an increased proportion of these vesicles is seen after MAO inhibition or norepinephrine administration. Furthermore, electrical stimulation of peripheral NE nerves induced a disappearance of the dense core of small granular vesicles after NE synthesis inhibition.[30]

In the central nervous system, the early studies have centered around large dense core vesicles (about 1000 Å) which are found in small number in discrete areas of the brain rich in endogenous NE and 5-HT and in aminergic varicosities[19-20,30-31] (Fig. 4).

A relatively good correlation was found between the number of large granular vesicles and the content of NE and 5-HT in several areas of the

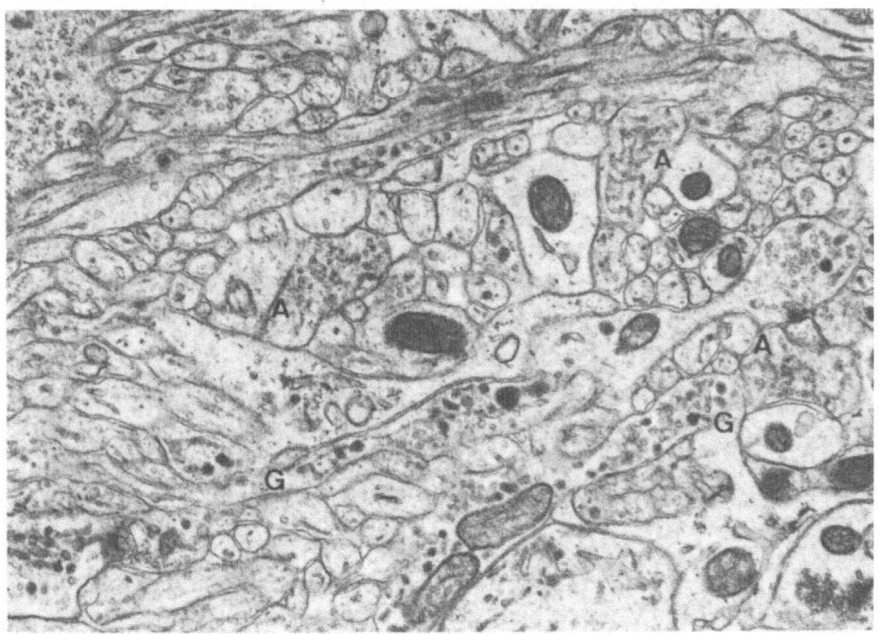

Fig. 4. Low-power electron micrograph of subependymal portion of rat hypothalamus showing the frequent occurrence of neural processes containing large granular vesicles (G). × 36,000. Nerve endings containing only agranular vesicles can also be seen (A). (From Bloom and Aghajanian.[31])

brain of the rat.[31] The effects of various drugs on the number of these large vesicles reported by various authors have been discussed extensively by Bloom and Aghajanian.[31] In their own study, these authors did not find changes in vesicle electron opacity which agreed in magnitude with the expected change in amine levels induced by several drugs.

However, the large granular vesicles appear not to be the main storage sites for NE; as shown by Hökfelt,[30] the much more numerous small vesicles (about 500 Å in diameter) which appear agranular can be visualized when potassium permanganate is used for fixation, e.g., in nerve terminals of the rat locus coeruleus, but also in cell bodies of this area known to contain numerous NE cell bodies. Thus, it has been suggested that the small granular vesicles observed after permanganate fixation are probably storage sites of the amine in NE central neurons and may correspond to those visualized in the peripheral NE nerves.

Electron microscopic radioautographic studies have shown that labeled NE administered intraventricularly accumulated mainly in areas rich in CA and that it is mainly localized in axon and nerve endings containing a small number of large granular vesicles but also numerous agranular vesicles.[19–20] It should be noted that permanganate fixation was not used in these studies. Furthermore, the occurrence of large dense core vesicles does not appear to be by itself a good index for storage site of NE: in fact, exogenous 5-HT-^{3}H, which accumulates in structures rich in endogenous 5-HT and areas different from those storing NE, appears to be localized in endings containing also a small number of large granular vesicles (about 800 and 1000 Å).[19,31] New electron microscopic histochemical methods will perhaps give more direct evidence of the storage sites of amines in neurons. Recently, Wood[32] by means of a histochemical reaction was able to detect central vesicles containing NE and/or 5-HT but was unable to distinguish between them. A new development of this technique has been made recently and vesicular NE and 5-HT could be distinguished in peripheral neurons.[33]

E. Different Forms of Storage

Kinetic studies of uptake and release of catecholamine in peripheral tissues or in granular preparations and effects on NE disposition induced by drugs, particularly reserpine and tyramine, have provided considerable evidence that norepinephrine in sympathetic nerves may be stored in various compartments or pools having different physiological significance.[29] But the structural bases of these different forms of storage are still unknown. These various compartments may correspond to different binding sites within a single vesicle, to different populations of vesicles in nerve endings, to different localization of the vesicles in the neurons, and as discussed before, to intra- versus extravesicular localization of the amine. In the central nervous system, the question is even more complex and results suggesting different forms of storage are difficult to interpret, especially because one has to take into consideration that brain structures analyzed may contain heterogeneous systems of neurons containing the same amine but showing different characteristics of metabolism and different sensitivity to various drugs.

Nevertheless, many results suggest the existence of different intracellular compartments. As indicated previously, subcellular studies have demonstrated the occurrence of at least two forms of NE and 5-HT generally called bound and free amines. There is little doubt that part of the amines recovered in the supernatant fraction correspond to a different form of storage. This is further confirmed by pharmacological studies which indicate that reserpine initially depleted mostly NE or 5-HT found in the supernatant fraction after homogenization.[1,2] Amphetamine, on the other hand, has the opposite action on NE neurons acting on the NE recovered in the particulate fraction.

As in peripheral tissues, a reserpine-resistant form of NE has been demonstrated, by both biochemical and histochemical methods.[1] It should also be pointed out that the behavioral recovery observed after reserpine treatment paralleled the appearance of a small store of brain amines, at which time the brain tissue recovered its ability to accumulate small amounts of exogenous NE; this suggests that only a small store of the amine may be physiologically important.[1] Furthermore, the multiphasic disappearance of labeled exogenous or endogenously formed NE is another argument in favor of the existence of more than one compartment of storage for the amine.[34] Recently, Sedvall et al.[35] have also suggested the existence of a multicompartmental system in peripheral and central NE neurons to explain the fact that their quantitative estimation of true synthesis of the amine largely exceeded estimations of turnover, made by experimental approaches based on the measure of the amine utilization.

F. Effects of Drugs on the Storage of Monoamine

The effects of various psychotropic drugs on the metabolism of monoaminergic neurons have been extensively studied; moreover, drugs with specific effects, particularly on various biosynthesis steps, have been synthesized. Their mechanisms of action have been described and discussed at length in recent reviews or articles,[1-4,29,36-37] and we shall thus only briefly summarize some of the main known facts.

Various drugs can be arbitrarily classified from different studies on their effects on monamine metabolism in the central nervous system, and can be divided into two large groups:

Drugs which increase the extraneuronal level of monoamines, and thus lead to higher concentrations of transmitter in synaptic clefts.

Drugs which have the opposite effect, mainly by affecting storage or synthesis processes.

Among drugs of the first group, we can include:

a. The various precursors of synthesis and the catabolism inhibitors. The latter drugs are mostly represented by the MAO inhibitors which

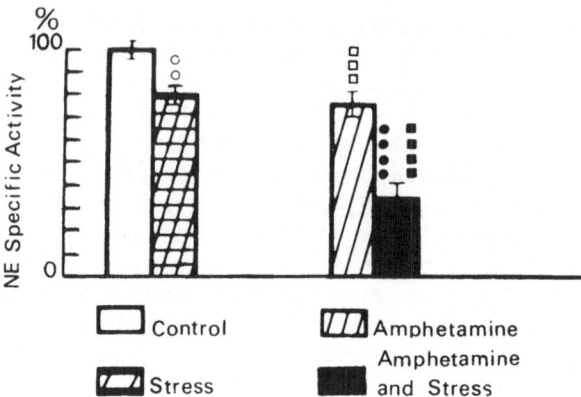

Fig. 5. Effect of amphetamine on the norepinephrine turnover in
the brain stem-mesencephalon of the rat. Groups of rats received
5 mg/kg (i.p.) of D amphetamine sulfate 165 min after the intracisternal injection of NE-^{3}H (7 μC). The animals were killed
6 hr after the administration of NE-^{3}H. Three hours after the
injection of the labeled amine, two groups of rats were submitted to electrical shocks applied to the feet. The shocks were
given for 10 min and alternated with 20 min of rest for an overall
period of 3 hr. Changes in NE turnover in the various situations
were estimated by the changes of NE-specific activity. Amphetamine alone and stress increased NE turnover. Amphetamine
potentiated the effect of stress. The white round and square
signs represent respectively the P values for the stressed and
amphetamine-treated groups as compared with the normal
group. The black round and square signs represent respectively
the P values for the stressed amphetamine pretreated group as
compared to the amphetamine-treated group and to the stressed
group. °°P less than 0.02; °°°P less than 0.01; °°°°P less than
0.01. (From Javoy, Thierry, Kety, and Glowinski.[54])

markedly increase amine levels. They also may have other effects, such as
enhancing the release of monoamines.[37]

    b. Drugs which act on the extraneuronal release process or increase
amine turnover. For example, amphetamine acts in this way on NE and DA
central neurons (Fig. 5). The neuroleptics enhanced particularly the DA
turnover in DA neurons.

    c. Drugs which interfere with the uptake process at the neuronal level.
They are mainly represented by compounds of the imipramine group, which
inhibit particularly the uptake of NE.

    A decrease of the extraneuronal levels of monoamines can be obtained
by various compounds such as:

    a. Drugs which diminish the turnover of central amines. The anes-
thetics affect particularly NE and 5-HT neurons by this mechanism.

b. Drugs which decrease the global concentrations of amines stored by (1) inhibition of their synthesis, (2) alteration of their retention, (3) alterations of the storage sites.

1. Inhibition of Synthesis

The inhibition of synthesis in catecholaminergic and 5-HT neurons is produced by α-methyl p-tyrosine and p-chlorophenylalanine respectively and their methyl ester derivatives. They affect the first step of biosynthesis. α-MethylDOPA acts on DOPA and 5-hydroxytryptophan decarboxylase. Disulfuram and diethyldithiocarbamate, which inhibit dopamine β-hydroxylase, decrease the NE stores.

2. Alteration of Retention

Methylated analogues of catecholamine precursors, such as α-methyl-DOPA and α-methyl-m-tyrosine, do not act only as inhibitors of synthetic enzymes. The generally accepted mechanism for catecholamine depletion by these drugs is displacement of the amines from neuronal storage sites by the amine analogues formed endogenously. Thus in NE neurons, α-methyl-norepinephrine and metaraminol are formed by decarboxylation and β-hydroxylation of α-methylDOPA and α-methyl-m-tyrosine respectively; they can replace part of the NE in granule storage sites and can be released instead of NE, acting as "false" neurochemical transmitters. They thus compete with the physiological monoamines.

3. Alteration of Storage Sites

The most familiar drug of this group is reserpine, which induces long-lasting diminution of brain 5-HT, DA, and NE. It is well established that the reserpine depleting effect is not due to an inhibition of the uptake process at a neuronal membrane level. Furthermore, it is unlikely that reserpine depletes biogenic amines by directly interfering with the enzymes involved in mono-amine biosynthesis. Reserpine has been shown to have the opposite effect in 5-HT neurons in which it activates the amine turnover.* The generally accepted mechanism of action of this drug is to block the uptake storage process of monoamines in granules, as demonstrated by numerous in vivo and in vitro studies. This alteration of the main storage sites led to the intra-cellular enzymatic inactivation by MAO of the unbound amines as shown by the appearance of an increased proportion of deaminated metabolites. Thus NE, DA, or 5-HT are almost completely destroyed before they escape from the neurons.

Other drugs, such as tetrabenazine or prenylamine, which have different chemical structures, seem to affect similarly the monoamine storage sites in central aminergic neurons, but they are less potent and have a shorter action. However, they can compete with reserpine and may prevent its potent effect when they have been previously injected.

* See Chapter 3 on Turnover of Chemical Transmitters.

III. RELEASE OF CENTRAL MONOAMINES

A direct demonstration of the *in vivo* release of transmitters in the central nervous system has been made in only a few studies. Whereas convincing data have been obtained in the case of acetylcholine, the evidence for extraneuronal release of monoamines which are thought to act as neurotransmitters is mostly indirect. *In vivo* studies have nevertheless given appreciable information on the occurrence of this process in central aminergic neurons. It should be mentioned that quantitative estimations of NE release in peripheral tissues[38] have confirmed the idea, based on pharmacological evidence, that the major part of NE content of adrenergic nerve fibers forms "a storage pool," while only a small percentage of the whole NE content is immediately available for extraneuronal release. Moreover, 70 to 80 % of the released amine may be rapidly inactivated by the reuptake process. The difficulty in demonstrating the *in vivo* release in central NE from NE terminals is obvious if similar conclusions are drawn for central NE-containing neurons. There are probably similar technical difficulties for undertaking the study of the release of central 5-HT or DA.

A. Direct Evidence of Release

Spontaneous and evoked release of acetylcholine in the central nervous system was directly demonstrated *in vivo* in elegant experimental approaches by collecting the transmitter in perfusion solutions. Similar approaches were used particularly to study the release of DA from the caudate nucleus. An enhanced monoamine output was also observed in *in vitro* studies after electrical stimulation of brain slices or spinal cord.

Studies on acetylcholine release in the central nervous system have already and will again in the future serve as experimental models to investigate release processes in DA, NE, and 5-HT neurons. Therefore, I will first comment on some of the results obtained in the case of acetylcholine.

1. In Vivo *Studies*

a. Acetylcholine. Thanks to the fact that released acetylcholine is mainly inactivated enzymatically, it has been possible, with inhibitors of cholinesterase, to obtain measurable amounts of this amine in collecting solutions. A collecting cup technique was first used to study release of acetylcholine from the surface of the brain.[39] A push-pull cannula introduced originally by Gaddum in 1961[39] was used to examine the release of acetylcholine in localized areas of the cortex or in deeper structures, particularly the striatum. This technique allows the perfusion of volumes in the order of 1 mm^3. An output of acetylcholine from the caudate nucleus was also observed in ventricular perfusion experiments developed by Felberg *et al.*[39] Finally, in a few experiments, Mitchell *et al.*,[39] using double concentric glass micropipettes, have detected acetylcholine in small volumes of fluid collected over long periods of time from spontaneously firing cortical cells.

With these various technical approaches, release of acetylcholine could be observed in various areas of the cortex of different animals, and particularly in the post cruciate gyrus, the parietal, temporal, and occipital lobes. A minor release of acetylcholine was observed also from the dorsal surface of the cerebellum, and a more important spontaneous release was observed from the caudate nucleus, which is known to contain very high levels of acetylcholine. The output of spontaneous release varies with the different structures and is also dependent on the experimental conditions. For instance, with the push-pull cannula technique, figures on the order of 1 to 5 ng/min of released acetylcholine have been obtained in the caudate nucleus. Generally, the rate of acetylcholine release is reduced by increased anesthesia. Chloralose appears particularly efficient in the inhibition of the spontaneous release of this amine in the cortex. A marked increase in the acetylcholine resting output in specific brain areas[39] has been produced by either electrical stimulation of some peripheral nerves or of some cortical areas, or by physiological stimulation such as retinal stimulation by light. These studies should provide much information about the organization of cholinergic pathways in the central nervous system.

b. Dopamine. MacLennan[40] with the push-pull cannula technique has first observed a resting release of dopamine of the order of 1 ng/min from the caudate nucleus. The output of dopamine in this structure could be

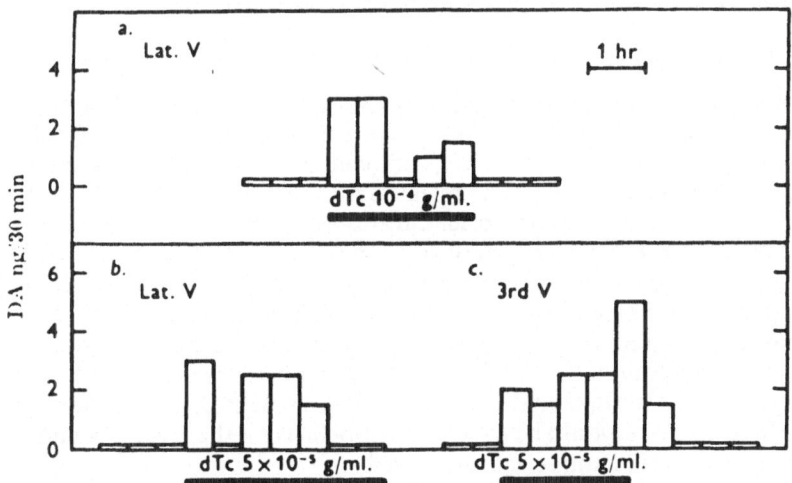

Fig. 6. Release of dopamine by prolonged cerebroventricular perfusions with tubocurarine (dTc). Dopamine (DA) determined in 30 min samples of artificial cerebrospinal fluid; dTc perfused through the anterior horn of a lateral ventricle (a and b), or through the third ventricle (c) of a cat. The columns represent concentrations in consecutive samples, expressed as ng/30 min, and not corrected for recovery. Heavy bars: perfusion of ventricle with tubocurarine 10^{-4} g/ml (a), or 5×10^{-5} g/ml (b and c). (From Portig, Scharman, and Vogt.[42])

increased by low-frequency stimulation of the nucleus centro-medianus of the thalamus. This stimulating effect is probably indirect and mediated by polysynaptic pathways. In this study the stimulation of the substantia nigra was not followed by an increased output of the amine in the caudate nucleus. This is surprising because histochemical and biochemical data suggest a monosynaptic nigro-striatal dopaminergic pathway. Nevertheless, stimulation of the pars compacta of the substantia nigra markedly increased the dopamine output in the putamen.[41] More recently, Portig et al.[42] have found in perfusates of the anterior horn of the lateral ventricle of the cat that measurable amounts of DA and its metabolite, homovanillic acid, were frequently released from the caudate nucleus, after an electrical stimulation of the substantia nigra. Moreover, these authors have detected a release of dopamine of the order of 0.1 ng/min after the addition of tubocurarine in the perfusion fluid. The release of the amine was only observed when the intraventricular perfusions of tubocurarine induced clinical signs such as shivering and tremor (Fig. 6). Development of microtechniques for the detection of DA and utilization of powerful inhibitors of the reuptake process would probably allow a more precise exploration of the mechanisms of DA release from the terminals of the nigro-striatal DA neuron system.

Local perfusion of the cerebral ventricles has permitted the detection, in the perfusion fluid, not only of acetylcholine and dopamine, probably released from the caudate nucleus, but also of 5-HT.[43] In the latter case the origin and the identity of the amine were not precisely established.

2. In Vitro *Studies*

Prolonged electrical stimulation of the isolated spinal cord of mice and frogs[44,4] has been shown to cause an increased release of NE and 5-HT in the bathing solution. More recently, radioactive NE initially accumulated in brain slices has been shown to be released into the medium by electrical field stimulation of low voltage and brief duration.[45] Similar findings were made with labeled 5-HT.[22] The release of both labeled NE and 5-HT could be enhanced by potassium. The release of NE was diminished by chlorpromazine of pentobarbital and inhibited when calcium was omitted in the incubating medium. The release of 5-HT was reduced by LSD. Moreover, differences in the amount of 5-HT released were observed between the various brain structures; the maximal effect was seen in structures rich in endogenous 5-HT. These results would be even more conclusive if the release of endogenously formed labeled amine was measured.

B. Indirect Demonstration of Release

The various data obtained can be divided into two groups. First, a decreased monoamine content has been observed in brain tissues of living animals after electrical stimulation of localized areas. Second, various physiological and pharmacological conditions have been shown to provoke changes in amine turnover of central neurons. When changes are observed in both types of studies, they represent changes in global utilization: i.e., both intra and extraneuronal utilization. It has been generally assumed, however, that the two kinds of utilization, though quantitatively different, change in parallel.

1. Modifications of Endogenous Content of Monoamines

In 1963, Gune and Reis reported that an important decrease in brain NE levels appeared in unrestrained cats which were submitted to intermittent stimulation of the amygdaloid nuclei for long time periods. Following such stimulation, NE fluorescence intensity of varicosities decreased in various areas of the brain and this decrease was even more marked when NE synthesis was inhibited.[46] Similarly, long-lasting electrical stimulation of the medulla oblongata of anesthetized rats caused a marked reduction in the NE and 5-HT stores in varicose nerve terminals of the spinal cord.[47] Aghajanian et al. have also demonstrated that an increased neuronal activity of central 5-HT-containing neurons in rat induced by prolonged electrical stimulation of the midbrain raphe, led to an enhanced utilization of the amine as shown by the decreased 5-HT and the elevation of 5-HIAA concentrations in the forebrain.[48] Normal impulse flow appears necessary for the extraneuronal release of monoamines; thus Andén et al.[49] have shown that inhibition of tyrosine hydroxylase or tryptophan hydroxylase did not affect the levels of NE and 5-HT stores in the caudal part of rat's spinal cord which was previously transected. To the contrary, reserpine and tetrabenazine, which block the uptake-storage mechanism in amine granules, were much less dependent on nerve impulse flow, and depletion of amine levels was observed.

2. Modifications of Monoamine Turnover

An enhanced utilization of central monoamine leads to an increased biosynthesis, probably by a positive feedback mechanism which very often masks the effect on endogenous monoamine levels of a moderate activation of central neurons. Therefore, the estimation of catecholamine or 5-HT turnover is a more reliable and precise procedure for detecting changes in neuronal activities. The different experimental approaches used to estimate turnover are reported and discussed in another chapter,* so I shall thus comment only briefly on the main change in turnover observed under various physiological situations.

Various severe stress situations can induce changes in amine levels; NE content particularly is decreased in such conditions.[50] Under intense muscular activity,[51] exposure to high temperature,[52] or mild repeated electrical foot shocks,[53] stress which did not appreciably affect NE levels, a marked increased turnover of central NE has been demonstrated (Fig. 5). In some cases[52,53] changes in NE turnover were associated with an increased

* See Chapter 3 on Turnover of Chemical Transmitters.

content of brain normethoxyadrenaline. This is compatible with the idea of a higher proportion of the amine reaching the extraneuronal space. The increased neuronal activity may persist for a long period of time and sustained effect on NE metabolism can be observed. Thus Kety et al.[54] have shown that rats submitted to chronic electroshock sessions present a marked increase in their brain NE turnover in the various structures of the central nervous system up to more than 24 hr after the last electroconvulsive shock.

Generally, it has not been observed that striatal dopaminergic neurons are activated by experimental physiological situations which affect NE neurons.[52,53] But changes in neuronal activities of 5-HT neurons have been detected. A stress induced by electrical foot shock increases the rate of utilization of brain serotonin (Fig. 3). Isolated animals develop an aggressive behavior which is associated with a decrease of brain 5-HT turnover.[55]

These modifications of activity of monoaminergic neurons probably lead to changes in the synaptic concentration of transmitters which may be related to the physiological or comportmental effects observed.

In this respect, the studies of the effects of various psychotropic drugs on neuronal activities are fruitful and have already given some insight into their mechanism of action and provoked hypotheses concerning the function of the released amines. However, knowledge about mechanisms of interrelations and regulations between these various aminergic systems of neurons is not yet very broad. These interrelations may be direct or indirect; histochemical studies have revealed, in some cases, close contact between 5-HT terminals and NE cell bodies.[4] The discovery of powerful inhibitors of catecholamine and 5-HT synthesis which indirectly block the physiological utilization of monoamines, should facilitate the understanding of these mechanisms.

C. Mechanisms of Release Processes

The mechanisms by which stored central monoamines are released at synapses are not yet well established, and most of the hypotheses which can be made are based on studies performed on peripheral sympathetic terminals. Three main questions can be raised. First, does the release process occur only at presynaptic structures, that is, from varicosities observed with histochemical fluorescence methods? Second, does the amine released originate from an extravesicular site of the neuron or from the vesicles themselves? Third, if they originate from the vesicles, is the whole vesicle content released, and how?

The release process which probably occurs mainly at the terminals could also operate, under particular conditions, at other sites of central neurons. In one histochemical study, Carlsson et al.[52] noticed that a diffuse NE fluorescence appeared around cell bodies of NE central neurons of stressed animals previously treated with a MAO inhibitor. It should be remembered that the uptake process in NE, DA, and 5-HT neurons has been demonstrated to occur over the entire length of these neurons.

A small part of the neuron amine content is localized outside the vesicles. It has been suggested that this extravesicular form of amine was functionally more important than the intravesicular bound amine, which could serve as a reservoir for the extravesicular compartment. However, Malmfors,[56] has shown that in reserpinized peripheral tissues, extravesicular catecholamines accumulated experimentally could not be released by nerve stimulation. And more generally, many data suggest a vesicular origin of the released amine.

Electrophysiological studies have shown that transmitter liberation from sympathetic nerve endings occurs in quantal units;[57] this is compatible with the hypothesis that liberation of NE induced by nerve stimulation occurs by negative pinocytosis of vesicles (that is, by expulsion of the content of synaptic vesicles). Douglas and Poisner[58] observed that catecholamines and nucleotides are liberated from chromaffin cells of the isolated adrenal gland in the same proportion as those in which they are stored in the granular vesicles. A specific protein originating from vesicles has also been shown to be released with catecholamines and nucleotides.[59] Thoenen et al. have demonstrated that DA and NE could be released from the spleen by nerve stimulation in the same proportion in which they were stored.[60] This is in agreement with the findings of Musacchio et al.[61] who have shown that DA and α-methyl dopamine which were stored in vesicles could be liberated by nerve stimulation.

However, the idea that one NE quantal packet corresponds to the entire NE content of one granule is not generally accepted. Folkow et al.[38] have extensively discussed this question, and from calculations based on their experimental results, they have estimated that the quantal packet of NE released contains about 400 molecules, which originate from a fusion of groups of still smaller amount of molecules expelled from several granules. They concluded that only a small part of the NE granule content is expelled during the release process. This suggests that the amine is stored in at least two different pools within the granule or that the varicose membrane contains a limited number of release points, where some granules are attached and from which a small amount of NE granule content is then "fired" out into the junction gap.

Fundamental information about the functions of ions in the mechanism of release of catecholamines has already been obtained in peripheral tissues.[58] Entry of calcium into the cell appears necessary to the onset of the exclusion of the amine from the nerve terminal. In in vitro studies made with hypothalamic vesicles, it has been demonstrated that calcium causes a dose-dependent release of the stored NE.[62] However, the full explanation of amine release awaits the elucidation of many still poorly understood steps in the process, and at our stage of knowledge it appears that relatively simply experimental models are required to integrate and study the various factors of this complex mechanism at the neuronal level.

IV. CONCLUSION

Processes of storage and release in central monoaminergic neurons are closely related, but it appears in some cases that change in the amount of stored amines does not interfere with the normal extraneuronal release induced by the neuronal impulse flow since only a small amount of the stored amine is immediately available for release. Slight modifications of release are often associated with changes in synthesis at different intraneuronal sites regulated by fine mechanisms which are not yet completely understood. Some drugs indirectly affect the release process either by acting on synthesis mechanisms or by interfering with storage processes. When they are acting specifically on one type of aminergic neurons, they are particularly useful for elucidating some of the functions of monoamines which are believed to act as neurotransmitters. Inhibitors of 5-HT and catecholamine synthesis on one hand, and compounds such as DOPA analogues, which lead to the formation of false transmitters, have already been widely used as pharmacological tools for modifying the amount of amine normally released.

Striking results on the functions of NE and 5-HT neurons in sleep mechanisms have been recently reported by Jouvet and his co-workers.[9,63] Inhibition of central serotoninergic activity induced by lesion of the raphe nuclei or by inhibition of 5-HT synthesis suppresses both slow and paradoxical states of sleep. In the latter case, new formation of 5-HT obtained by administration of 5-hydroxytryptophan rapidly suppressed the effect of the inhibitor. Sleep is recovered even when small amounts of amine are stored in

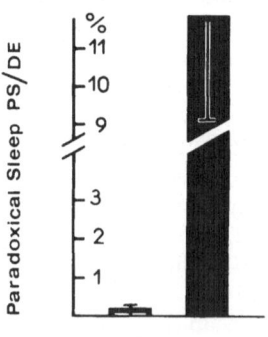

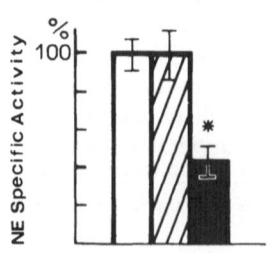

Fig. 7. Changes in norepinephrine turnover during paradoxical sleep rebound in the brain stem + mesencephalon of the rat. Three groups of rats received 7 μC of NE-^{3}H intracisternally and were killed 5 hr later. Rats of two groups were previously deprived of paradoxical sleep (PS) for 91 hr by putting them on small supports surrounded by water. In one group the animals were put for 5 hr in isolated boxes for sleep recuperation after the injection of NE-^{3}H (black bars); animals of the other group were again deprived of PS for 5 hr after the injection of NE-^{3}H (hatched bars). The third group of rats served as controls (white bars). All the animals were recorded during this 5-hr period and percentage of paradoxical sleep during the 5 hr were estimated (PS/DE). Five hours after the injection of the labeled amine, the NE-specific activity was only decreased in the group of animals which exhibit a PS rebound, thus indicating an enhanced turnover of this amine in the brain stem + mesencephalon.

central neurons. Furthermore, selective suppression of paradoxical sleep can be obtained by destruction of the NE neurons of the locus coeruleus or by administration of α-methylDOPA or *m,m*-tyrosine, which, as already mentioned, indirectly compete with NE release. Moreover, an important increase of central NE turnover has been noticed during paradoxical sleep rebound which was produced by a previous selective deprivation of this state of sleep (Fig. 7).[64] There is thus little doubt that changes in regulation of both types of neurons occur and are involved in sleep mechanisms.

Blockade of NE release can abolish the normal function of the complex "rewarding system," as observed by the abolition of the process of self-stimulation, and may suppress responses to conditioned stimuli. Monoamines also appear to be involved in hormonal regulation. Thus, blockade of central serotoninergic neuron activity suppresses the normal inhibiting effect of IMAO on estrous behavior[65] and the ovulation process.[66] Many facts also suggest that the system of nigro-striatal dopaminergic neurons is particularly involved in extrapyramidal functions.[3]

As shown by these examples (and other information obtained with psychotropic drugs), most of the results on the functional roles of monoaminergic neurons have been obtained by the use of lesions and drugs. It is likely that in the near future new approaches based on other methods, such as multicompartmental analysis or direct estimation of release in localized areas of the brain, will contribute to the understanding of the interrelated functions of NE, DA, and 5-HT neurons at different levels of the central nervous system organization.*

V. REFERENCES

1. J. Glowinski and R. J. Baldessarini, Metabolism of norepinephrine in the central nervous system, *Pharmacol. Rev.* **18**:1201–1238 (1966).
2. S. Garattini and L. Valzelli, *Serotonin*, Elsevier, New York (1965).
3. O. Hornykiewicz, Dopamine (3-hydroxy-tyramine) and brain functions, *Pharmacol. Rev.* **18**:925–964 (1966).
4. N. A. Hillarp, K. Fuxe, and A. Dahlström, in *Mechanisms of Release of Biogenic Amines* (U. S. Von Euler, S. Rosell, and B. Uvnäs, eds.), pp. 31–57, Pergamon Press, New York (1966).
5. L. M. Gunne, Relative adrenaline content in brain tissue, *Acta Physiol. Scand.* **56**:324–333 (1962).
6. L. J. Poirier and T. L. Sourkes, Influence of the substantia nigra on the catecholamine content of the striatum, *Brain* **88**:181–192 (1965).
7. N. E. Andén, A. Carlsson, A. Dahlström, K. Fuxe, N. A. Hillarp, and K. Larsson, Demonstration and mapping out of nigro-neostriatal dopamine neurons, *Life Sci.* **3**:523–530 (1964).
8. M. Goldstein, B. Anagnoste, W. S. Owen, and A. F. Battista, The effects of ventro medial tegmental lesions on the disposition of dopamine in the caudate nucleus of the monkey, *Brain Res.* **4**:298–300 (1967).
9. M. Jouvet, P. Bobillier, J. F. Pujol, and J. Renault, Suppression du sommeil et diminution de la sérotonine cérébrale par lésions du système du raphé, *Compt. Rend. Acad. Sci. Paris* **264**: 360–362 (1967).

* Submitted for publication in 1968.

10. A. Heller and R. Y. Moore, Effect of central nervous system lesions on brain monoamines in the rat, *J. Pharmacol.* **150**:1–9 (1965).
11. T. Magnusson and E. Rosengren, Catecholamines of the spinal cord normally and after transection, *Experientia* **19**:229 (1963).
12. N. E. Andén, K. Fuxe, B. Hamberger, and T. Hökfelt, A quantitative study on the nigro neostriatal dopamine neuron system, *Acta Physiol. Scand.* **67**:306–312 (1966).
13. A. Dahlström and J. Haggendal, Studies on the transport and lifespan of amine storage granules in a peripheral adrenergic neuron system, *Acta Physiol. Scand.* **67**:278–288 (1966).
14. H. Weil-Malherbe, L. G. Whitby, and J. Axelrod, in *Regional Neurochemistry* (S. S. Kety and J. Elkes, eds.), pp. 284–292, Pergamon Press, Oxford (1961).
15. W. Lichtensteiger and H. Langemann, Uptake of exogenous catecholamines by monoamine containing neurons of the central nervous system: Uptake of catecholamines by arcuato-infundibular neurons, *J. Pharmacol.* **151**:400–408 (1966).
16. J. Glowinski and J. Axelrod, Effects of drugs on the disposition of H^3 norepinephrine in the rat brain, *Pharmacol. Rev.* **18**:775–785 (1966).
17. G. K. Aghajanian and F. E. Bloom, Localization of tritiated serotonin in rat brain by electron-microscopic autoradiography, *J. Pharmacol.* **156**:23–30 (1967).
18. M. Reivich and J. Glowinski, An autoradiographic study of the distribution of C^{14} norepinephrine in the brain of the rat, *Brain* **90**:633–646, Part III (1967).
19. G. K. Aghajanian and F. E. Bloom, Electron-microscopic localization of tritiated norepinephrine in rat brain: Effects of drugs, *J. Pharmacol.* **156**:407–416 (1967).
20. N. J. Lenn, Localization of uptake of tritiated norepinephrine by rat brain *in vivo* and *in vitro* using electron microscopic autoradiography, *Am. J. Anat.* **120**:337–390 (1967).
21. B. Hamberger and D. Masuoka, Localisation of catecholamines uptake in rat brain slices, *Acta Pharmacol. Tox. KSH* **22**:363–368 (1965).
22. T. N. Chase, G. R. Breese, and J. J. Kopin, Serotonin release from brain slices by electrical stimulation: Regional differences and effect of LSD, *Science* **157**:1461–1463 (1967).
23. V. P. Whittaker, in *Progress in Brain Research* (H. E. and W. A. Himwich, eds.), Vol. 8, pp. 90–117, Elsevier, Amsterdam (1964).
24. V. P. Whittaker, in *Mechanisms of Release of Biogenic Amines* (U. S. Von Euler, S. Rosell, and B. Uvnäs, eds.), Vol. 5, pp. 147–163, Wenner-Gren Center International Symposium Series, Pergamon Press, New York (1966).
25. E. DeRobertis, in *Progress in Brain Research* (H. E. and W. A. Himwich, eds.), Vol. 8, pp. 118–136, Elsevier, Amsterdam (1964).
26. E. W. Maynert and K. Kuriyama, Some observations on nerve endings particles and synaptic vesicles, *Life Sci.* **3**:1067–1087 (1964).
27. J. M. Musacchio, I. J. Kopin, and V. K. Weise, Subcellular distribution of some sympathomimetic amines and their β-hydroxylated derivatives in the rat heart, *J. Pharmacol.* **148**:22–28 (1965).
28. J. Glowinski and L. L. Iversen, Regional studies of catecholamines in the rat brain. III. Subcellular distribution of endogenous and exogenous catecholamines in various brain regions, *Biochem. Pharmacol.* **15**:977–987 (1966).
29. L. L. Iversen, *The Uptake and Storage of Noradrenaline in Sympathetic Nerves*, Cambridge University Press, Cambridge (1967).
30. T. Hökfelt, On the ultrastructural localization of noradrenaline in the central nervous system of the rat, *Z. Zellforschung* **79**:110–117 (1967).
31. F. E. Bloom and G. K. Aghajanian, An electron microscopic analysis of large granular synaptic vesicles of the brain in relation to monoamine content, *J. Pharmacol.* **159**:261–273 (1968).
32. J. G. Woad, Electron microscopic localization of amines in central nervous tissue, *Nature (London)* **209**:1131–1133 (1966).

33. J. G. Etcheverry and L. M. Zieher, Cytochemistry of 5-hydroxytryptamine at the electron microscope level. II. Localizing 5-HT in pineal gland by electron microscopy cytochemistry. Personal communication. (In press.)

34. L. L. Iversen and J. Glowinski, Regional studies of catecholamines in the rat brain. II. Rate of turnover of catecholamines in various brain regions, *J. Neurochem.* **13**:671–682 (1966).

35. G. C. Sedvall, V. K. Weise, and I. J. Kopin, The rate of norepinephrine synthesis measured *in vivo* during short intervals; influence of adrenergic nerve impulse activity, *J. Pharmacol.* **159**:274–282 (1968).

36. A. Carlsson, in *Mechanisms of Release of Biogenic Amines* (U. S. Von Euler, S. Rosell, and B. Uvnäs, eds.), Vol. 5, Wenner Gren Center International Symposium Series, Pergamon Press, New York (1966).

37. A. Carlsson, in *Handbuch der experimentellen Pharmakologie*, pp. 529–592, Springer-Verlag, Berlin (1966).

38. B. Folkow, J. Häggendal, and B. Lisander, Extent of release and elimination of noradrenaline at peripheral adrenergic nerve terminals, *Acta Physiol. Scand.* **307**:1–38 (1967).

39. J. F. Mitchell, in *Mechanisms of Release of Biogenic Amines* (U. S. Von Euler, S. Rosell, and B. Uvnäs, eds.), Vol. 5, pp. 425–438, Wenner Gren Center International Symposium Series, Pergamon Press, New York (1966).

40. H. Mclennan, The release of acetyl-choline and of 3-hydroxytyramine from the caudate nucleus, *J. Physiol. (Lond.)* **174**:152–156 (1964).

41. H. Mclennan, The release of dopamine from the putamen, *Experientia (Basel)* **21**:725 (1965).

42. P. J. Portig, D. F. Sharman, and M. Vogt, Release by tubocurarine of dopamine and homovanillic acid from the superfused caudate nucleus, *J. Physiol.* **194**:565–572 (1968).

43. W. Feldberg and R. D. Myers, Appearance of 5-hydroxytryptamine and an unidentified pharmacologically active lipid acid in effluent from perfused cerebral ventricles, *J. Physiol.* **184**:837–855 (1966).

44. N. E. Andén, A. Carlsson, N. A. Hillarp, and T. Magnusson, Noradrenaline release by nerve stimulation of the spinal cord, *Life Sci.* **4**:129–132 (1965).

45. R. J. Baldessarini and I. J. Kopin, Tritiated norepinephrine: Release of brain slices by electrical stimulation, *Science* **152**:1630–1631 (1966).

46. K. Fuxe and L. M. Gunne, Depletion of the amine stores in brain catecholamine terminals on amygdaloid stimulation, *Acta Physiol. Scand.* **62**:493–494 (1964).

47. A. Dahlström, K. Fuxe, D. Kernell, and G. Sedvall, Reduction in the monoamines stores in the terminals of bulbospinal neurons following stimulation in the medulla oblongata, *Life Sci.* **4**:1207–1212 (1965).

48. G. K. Aghajanian, J. A. Rosecrans, and M. H. Sheard, Serotonin, release in the forebrain by stimulation of midbrain raphe, *Science* **156**:402–403 (1967).

49. N. E. Andén, K. Fuxe, and T. Hökfelt, Effect of some drugs on central monoamine nerve terminals lacking nerve impulse flow, *European J. Pharmacol.* **1**:226–232 (1967).

50. E. W. Maynert and R. Levi, Stress-induced release of brain norepinephrine and its inhibition by drugs, *J. Pharmacol.* **143**:90–95 (1964).

51. R. Gordon, S. Spector, A. Sjoerdsma, and S. Udenfriend, Increased synthesis of norepinephrine and epinephrine in the intact rat during exercise and exposure to cold, *J. Pharmacol.* **153**:440–447 (1966).

52. A. Carlsson, A. Dahlström, K. Fuxe, and M. Lindqvist, Histochemical and biochemical detection of monoamine release from brain neurons, *Life Sci.* **4**:809–816 (1965).

53. A. M. Thierry, F. Javoy, J. Glowinski, and S. S. Kety, Effects of stress on the metabolism of norepinephrine, dopamine, and serotonin in the central nervous system of the rat, *J. Pharmacol.* **163**:163–171 (1968).

54. S. S. Kety, F. Javoy, A. M. Thierry, L. Julou, and J. Glowinski, A sustained effect of electro-convulsive shocks on the turnover of norepinephrine in the central nervous system of the rat, *Proc. Nat. Acad. Sci.* **53**:1249–1254 (1967).

55. S. Garattini and L. Valzelli, *in Symposium on Biological Role of Indolealkylamine Derivatives* May 1967 (in press).

56. T. Malmfors, Studies on adrenergic nerves. The use of rat and mouse iris for direct observations on their physiology and pharmacology at cellular and subcellular levels, *Acta Physiol. Scand.* **64**:248 (1965).

57. G. Burnstock and M. E. Holman, Junction potentials at adrenergic synapses, *Pharmacol. Rev.* **18**:481–493 (1966).

58. W. W. Douglas, *in Mechanisms of Release of Biogenic Amines* (U. S. Von Euler, S. Rosell, and B. Uvnäs, eds.), pp. 267–289, Wenner Gren Center International Symposium Series (1966).

59. P. Banks and K. Helle, The release of protein from the stimulated adrenal medulla, *Biochem. J.* **97**:40–41 C (1965).

60. H. Thoenen, W. Haefely, K. F. Gey, and A. Huerlimann, Quantitative aspects of the replacement of norepinephrine by dopamine as a sympathetic transmitter after inhibition of dopamine β-hydroxylase by disulfuram, *J. Pharmacol.* **156**:246–251 (1967).

61. J. M. Musacchio, J. E. Fisher, and I. J. Kopin, Subcellular distribution and release by sympathetic nerve stimulation of dopamine and α-methyl-dopamine, *J. Pharmacol. Exp. Ther.* **152**:51–61 (1966).

62. A. Phillipu and H. Przuntek, Noradrenalin Speicherung im Hypothalamus und Wirkung von Pharmaka auf die isolierten Hypothalamus Vesikel, *Naunyn, Schmiedebergs Arch. Pharmakol. U. Exp. Path.* **258**:238–250 (1967).

63. M. Jouvet, *Sleep and Altered States of Consciousness* (S. S. Kety and E. Evarts, eds.), pp. 86–126, Williams and Wilkins, Baltimore (1967).

64. J. F. Pujol, J. Mouret, M. Jouvet, and J. Glowinski, Increased turnover of cerebral norepinephrine during rebound of paradoxical sleep in the rat, *Science* **159**:112–114 (1968).

65. B. J. Meyerson, Central nervous monoamines and hormone-induced estrus behaviour in the spayed rat, *Acta Physiol. Scand.* **64**:248 (1965).

66. C. Kordon, F. Javoy, G. Vassent, and J. Glowinski, Blockade of superovulation in the immature rat by increased brain serotonin. *Eur. J. Pharmacol.* **4**:169–174 (1968).

Chapter 5

AMINO ACID TRANSMITTERS

David R. Curtis and Graham A. R. Johnston

Department of Physiology
Australian National University, Canberra

I. INTRODUCTION

Appreciable amounts of amino acids occur free in both vertebrate and invertebrate nervous systems. It is generally accepted that these are predominantly concerned with general metabolic processes, and with the maintenance of water and ion distributions across cellular membranes: however, certain amino acids may also function as synaptic transmitters. Recent reviews catalogue the many sites at which excitant or depressant actions of a variety of amino acids have been demonstrated,[1-3] and it is intended to restrict this present analysis to synapses where a transmitter or "mediator" function of amino acids is considered to be very likely. Prior to discussing the function of specific amino acids at particular synapses, the processes involved in synaptic transmission are outlined, together with general problems relating to transmitter function and the techniques available for their investigation. It is not intended to discuss "modulator" substances which have been postulated to function as regulators of the activity of neurons or effector cells, but which are not essential for synaptic transmission;[2,4] amino acids may be well suited to this function.

II. INVESTIGATION OF AMINO ACID TRANSMITTER FUNCTION

The concept of chemical synaptic transmission includes the synthesis and storage of the transmitter within the presynaptic nerve cell, its release from the presynaptic terminal, the interaction of transmitter with receptors on the postsynaptic neuron or effector cell, and the consequent transient alteration in the permeability of this subsynaptic membrane toward ions in the immediate environment which redistribute along electrochemical gradients.[5,6] In the case of excitatory transmitters, the membrane becomes more permeable to sodium and at least one other ion species, the resultant ionic currents depolarizing the postsynaptic membrane. In contrast, at inhibitory synapses there is an increase in membrane permeability to potassium or chloride ions, or both, which may result in hyperpolarization of the postsynaptic membrane. Both the increase in membrane conductance and the hyperpolarization counteract the effectiveness of depolarizing ionic currents set up at excitatory synapses, and thus lead to inhibition of the activity of the postsynaptic cell. Transmitter action is terminated either by desensitization of receptor sites in the continued presence of transmitter, or by a reduction in transmitter concentration in the vicinity of synapses, the transmitter being removed from the synaptic area by diffusion, enzymatic or chemical inactivation, or by uptake into cellular elements including the presynaptic terminal.

It follows from this outline of chemical transmission that certain criteria can be defined in relation to transmitter identification, compliance with which is desirable before a substance can be identified as the transmitter at a particular synapse. These criteria are based on the evidence obtained for the transmitter function of acetylcholine at vertebrate neuromuscular and ganglionic synapses,[7] and their rigid application to transmitter identification, within the CNS in particular, has been rightly questioned.[8-10] While it is relatively easy to redefine criteria and to argue about their interpretation, experimental difficulties remain and compromises seem desirable. It is clearly essential that the substance proposed as a transmitter has a postsynaptic action identical with that of the synaptically released transmitter; identity of action includes antagonism by compounds acting in a specific fashion at postsynaptic receptor sites. Information regarding the synthesis, storage, release, and inactivation of the transmitter would provide further evidence as to its nature. However, the stage at which identification becomes acceptable will undoubtedly reflect personal experiences and prejudices.

A. Synthesis and Storage Within the Presynaptic Neuron

It is necessary that the transmitter be synthesized in or accumulated by the presynaptic neuron, particularly at the terminals but perhaps also in the cell body. However, a substantial presynaptic store of transmitter may not be required if there is an adequate supply of precursors and its synthesis on demand is rapid and efficient. If in addition to being a transmitter, the substance has other intracellular functions, as do most amino acids, a localized presynaptic store may not be apparent. Thus, although it may be possible to isolate nerve endings (synaptosomes) from brain tissue,[11,12] the concentration of a transmitter may not necessarily be higher in these than elsewhere in the neuron. These studies of the subcellular distribution of possible transmitters, precursors, and synthesizing enzymes may be further complicated by the differing mechanical properties of neurons subjected to the fractionation technique, and the success achieved in separating cell bodies from endings. Subcellular fractionation techniques can be extended to the isolation of synaptic "vesicles," although the necessary association of these structures with *in vivo* transmitter storage may be open to question.

In suitable cases it has been possible to separate neurons and nerve fibers of different function in peripheral tissues and simple nervous systems, and to subject these to chemical and enzyme analyses. Such a highly desirable method is not readily applicable to the mammalian CNS, although it may be possible to demonstrate that the distribution of a given compound or enzyme corresponds with that of a particular type of neuron, and that this distribution changes after destruction of the neurons or interruption of their axons. Evidence for the synthesis, and storage, of a particular compound at synapses may be provided by the use of substances which hinder transmitter synthesis in a specific fashion, either by limiting the access of a precursor to the site of synthesis, or by inhibiting an essential enzyme.

B. Release from Presynaptic Terminals

The investigation of transmitter release raises considerable problems. In the first place, conditions must be such that all of the released transmitter is not destroyed by any inactivating process; for example, if inactivation is predominantly enzymatic then a specific inhibitor of the enzyme may be required. Second, difficulties arise in choosing a method for collecting the transmitter. It may be possible to perfuse peripheral tissues or nervous systems having relatively unsophisticated metabolic requirements. However, for the mammalian central nervous system, collection can only be remote from the site of release, either from the surface of the tissue, from the venous effluent, or from an artificial tissue space created by a "push-pull" type of cannula.[13] Third, with all tissues it may be difficult to determine whether the detected substance originates from nerve terminals or is merely associated with metabolic activity of pre- or postsynaptic elements.

The amount of transmitter collected should depend on the frequency and duration of stimulation of a particular nervous pathway, and should be affected by specific agents or conditions which alter synaptic transmission by influencing the amount of transmitter released per impulse. It may be hard to evaluate whether the amounts collected are "adequate" for postsynaptic action, particularly in the absence of precise knowledge of the proportion recovered of the total amount released, of the number of nerve endings activated by a given stimulus, and of the concentrations required at cell receptors for a particular physiological effect.

C. Postsynaptic Action

The essential requirement that a suspected transmitter has the same postsynaptic action as the synaptically released substance raises problems with regard to determining both its site and mode of action. As a consequence, methods of administration and observation become highly important, and for a variety of reasons most studies are best carried out with a combination of localized drug administration, usually by microelectrophoretic ejection, in combination with intracellular recording of the postsynaptic response. Even with ideal conditions, care must be taken to exclude the possibility that the observed effect is indirect, and mediated either by excitation or depression of synaptically connected neurons or by drug-induced alterations in the release of transmitter from presynaptic terminals.

It would be reasonable to assume that the postsynaptic membrane would be most sensitive to transmitters in the close vicinity of the appropriate synapses, but at least with neurons in the CNS it is clearly impossible to restrict the administered substance to the immediate environment of particular synapses. Indeed, it is impossible to ensure that the drug affects the same area of membrane as is involved in synaptic processes associated with impulses in a selected nervous pathway: close similarity of these regions is essential for comparing equilibrium potentials of ionic events associated with alterations of membrane potential.

Administration of the suspected transmitter in concentrations approximating those attained during the synaptic process must produce the same change in membrane properties as the actual transmitter: an alteration in membrane potential in the same direction and an increase in membrane conductance. Two major methods are available for comparing the ionic permeability changes induced by transmitters and artificially administered compounds.

The first involves the measurement of equilibrium potentials, the membrane potential at which the net ionic current flow at activated synapses is zero. At levels further from the resting value than this equilibrium potential, synaptic potentials are reversed in direction. In view of the very limited number of ions involved (Na^+, K^+, Ca^{2+}, Cl^-), the identity of equilibrium potentials for two membrane processes would provide very strong evidence that these were associated with permeability increases toward the same ions. However, in such investigations, membrane polarizing currents are generally passed through an intracellular microelectrode, and the ability to change the potential at a given synapse will depend on the geometry of the postsynaptic structure and the distance of that synapse from the polarizing electrode. Furthermore, the recording electrode may be identical with or immediately adjacent to this polarizing electrode, and hence measured reversal potentials generally provide an incorrect estimate of the equilibrium potential of ionic events at synapses unless all of the membrane involved in the synaptic process can be uniformly polarized. Thus difficulties exist, particularly in the CNS, when comparing equilibrium potentials of ionic events which may involve different or noncorresponding areas of membrane.

The second method available for assessing the nature of the permeability increases at activated synapses involves an analysis of the ion fluxes, either by altering extra- or intracellular concentrations of a particular ion or ions, or by studying the actual movement of labeled ions. Reversal or accentuation of the electrochemical gradient for a particular ion species can lead to measurable alterations in synaptic potentials, and equilibrium potentials. These changes may be transient following the electrophoretic administration of ions into the intracellular environment, the only method available for such studies on neurons in the intact mammalian CNS, and the use of a series of ions of different hydrated size may provide some measure of the diameter of membrane "pores" associated with ion redistribution at synapses.[5,6] Again, as with equilibrium potentials, identical alterations produced in synaptic and chemically induced membrane potentials, as a consequence of changes in the ionic environment, strongly support the assumption that the same ionic fluxes are involved in the synaptic and induced membrane processes.

It has been frequently assumed that rapidity of action, ready reversibility, and high potency of an electrophoretically administered agent characterize transmitterlike properties. However, it may be queried whether the time course of action of a substance administered rather diffusely near a neuron is strictly relevant to that of a transmitter released very close to the receptors. Furthermore, although the duration of transmitter action may be required to be brief at certain synapses, the prolonged synaptic potentials recorded from some central neurons indicate that this is not always the case. Statements relating the apparent sensitivity of neurons to administered substances are also rather meaningless in the absence of accurate knowledge of the concentrations attained in the vicinity of receptors, both after artificial administration and synaptic release of the transmitter. It might be expected that close structural analogues of the transmitter could all have the same postsynaptic action, with relative potencies and time courses depending upon steric factors and local inactivating mechanisms.

D. Postsynaptic Antagonists

Compounds which block synaptic transmission by a specific postsynaptic effect must of necessity also prevent the action of a substance suspected of being the transmitter. Such competitive antagonism may also extend to close structural relatives of the transmitter, and, in addition, the same transmitter may have different antagonists at different synapses. The structure of the antagonist rarely provides clear evidence as to the structure of the transmitter, and possible antagonists may not be readily apparent from the structure of a synaptically active substance suspected on other grounds of being a transmitter. Furthermore. in the absence of a specific antagonist, it is difficult to demonstrate that an antagonist of synaptic transmission has a postsynaptic action, and conclusions interrelating antagonists and transmitters are often mutually dependent.

E. Inactivation and Removal

There are two main problems associated with inactivation and removal of synaptically released transmitter: the need to limit the immediate action within the subsynaptic region, and the prevention of accumulation in surrounding tissues. The requirement for an enzyme located extracellularly and close to the postsynaptic receptors is probably the most contentious of the transmitter criteria: receptor desensitization would be an effective means of limiting immediate action and, at certain synapses, even simple diffusion may account for the brief time course of transmitter action.[14] However, in the absence of local inactivation, accumulation in the surrounding extracellular space may lead to inappropriate alterations in the activity of neighboring neurons. This difficulty could also arise if inactivation takes place at sites quite remote from synapses, during or after transport in extracellular fluid or bloodstream. Whether local or remote, inactivation implies an alteration of the structure of the transmitter (not necessarily degradation) which prevents interaction with appropriate receptors. This alteration may result from enzyme action or from inherent chemical instability. If local inactivation is in part enzymatic, it may be possible to demonstrate enhancement and prolongation of the action of both synaptically released and artificially administered substances as a result of selective local enzyme inhibition.

Since intracellular injections of amino acids do not have transmitterlike actions on nerve or muscle cells, cellular uptake could be a very effective means of terminating transmitter action, and could preclude the need for enzymatic or chemical inactivation. Furthermore, uptake by the presynaptic terminals would provide a supply of transmitter for re-use. If uptake occurs close to the synapse it may be possible to block or facilitate local uptake systems in a specific fashion, enabling a comparison to be made between synaptically released and suspected transmitters. Uptake systems are well known in nervous tissue. In brain slices, amino acid levels are maintained by active transport systems[15] and structurally specific systems exist for GABA, small neutral (e.g., glycine), large neutral (leucine), small basic (diamino-butyrate), large basic (lysine), and acidic (glutamate) amino acids.[16]

In vitro studies of enzymes and uptake systems may not be directly applicable to processes occurring *in vivo*, particularly with central nervous tissues having complex metabolic, ionic, and organizational requirements. Amino acid transmitters are likely to have other important functions, and alterations in tissue levels, produced by interference with a particular enzyme or uptake system, may not necessarily provide significant information regarding that fraction more directly concerned with synaptic transmission.

III. γ-AMINOBUTYRIC ACID AND RELATED AMINO ACIDS

Interest in γ-aminobutyric acid (4-aminobutyric acid, GABA) originated primarily from its unique occurrence in the mammalian CNS, and its readily demonstrable effects on crustacean stretch receptors and muscle.

A. Invertebrate

GABA is generally accepted as an inhibitory transmitter at a variety of crustacean synapses, and particularly intensive studies have been carried out on the crayfish *Cambarus clarkii* and the lobster, *Homarus americanus*.[2,3,17]

GABA is found in both the central and peripheral nervous system of the lobsters, the content of inhibitory fibers and neurons being higher than that of excitatory fibers and neurons.[18] Excitatory fibers contain less than 1 % of the GABA content of inhibitory fibers, the concentration in the latter being of the order of 0.1 M, assuming uniform distribution. In contrast, the glutamic acid levels of inhibitory and excitatory fibers and neurons are approximately equal. The activity of glutamic acid decarboxylase is approximately eleven times higher in inhibitory than in excitatory fibers.[19]

GABA increases the membrane conductance of lobster muscle fibers and in this, and other crustaceans, the amino acid has a postsynaptic action similar to that of the inhibitory transmitter.[3] In the crayfish, the amino acid is inactive when administered intracellularly, the GABA-sensitive portions of the external surface of the muscle fibers correspond to synaptic regions, and the potential induced by GABA has the same reversal potential as that of the inhibitory junctional potential,[20] the activated inhibitory membrane becoming permeable predominantly to chloride ions.[21]

There is also evidence that GABA affects the presynaptic terminals in crustacean muscles in a manner indistinguishable from that of neural inhibition,[3,22] and is thus most probably the transmitter responsible for presynaptic inhibition in crustacea, a process which reduces the release of excitatory transmitter. GABA also mimics the inhibition of crustacean stretch receptors and of the giant motor fibers of the crayfish *Astacus fluviatilis*; in each case the equilibrium potential for GABA and inhibitory effects are very probably identical.[3] Furthermore, in all of these invertebrate preparations, GABA is the most potent of the naturally occurring amino acids which have transmitterlike postsynaptic actions.[23]

Good evidence for the similarity between GABA and the crustacean inhibitory transmitter is also provided by the action of picrotoxin, which blocks both synaptic inhibition and the effects of GABA upon crustacean muscle and stretch receptors.[3]

In experiments performed at low temperatures to diminish enzymatic inactivation and cellular uptake mechanisms, stimulation of inhibitory but not of excitatory nerves results in the release of GABA at neuromuscular junctions in the lobster.[24] This release, dependent on the number of impulses, is abolished by reducing the amount of calcium in the perfusing medium, a condition which at other synapses blocks transmitter release.

The levels of GABA-glutamate transaminase in inhibitory and excitatory fibers of the lobster are approximately equal.[19] The metabolic requirement for α-oxoglutarate suggests that this enzyme is unlikely to be located extracellularly, as would be necessary if it was important as the initial process in the removal of synaptically released GABA in the region of inhibitory synapses. Furthermore, neither amino-oxyacetic acid nor hydroxylamine, inhibitors of the transaminase, prolong the duration of the inhibitory process.[25] It is thus likely that the major means of inactivation of GABA at crustacean synapses is by a specific uptake mechanism which has been demonstrated for both lobster muscle[26] and crayfish stretch receptors.[27] This uptake process is sodium dependent,[27] and as the conductance increase induced in crustacean muscle by GABA is unaffected by alteration of the extracellular sodium ion concentration,[21] the mechanisms of uptake and transmitter action are clearly quite different.

Evidence for the involvement of GABA as an inhibitory transmitter in other invertebrates depends to a large extent on the demonstration that the amino acid produces a membrane permeability increase similar to that accompanying neural inhibition, and that the actions of both GABA and the transmitter are blocked by picrotoxin, as for example at the peripheral inhibitory synapses of the grasshopper and locust.[28]

B. Vertebrate

In mammals the unique occurrence of GABA in the CNS, together with enzymes associated with its metabolism, has been the subject of numerous reviews, and the function of this amino acid as an invertebrate inhibitory transmitter has led to considerable speculation and investigation regarding a similar role in the vertebrate nervous system.

Difficulty has been experienced in relating the distribution of GABA to that of inhibitory neurons. Subcellular distribution studies indicate that GABA is apparently a cytoplasmic constituent of neurons without selective localization within nerve endings,[29,30] although glutamic acid decarboxylase (GAD) has been localized in the endings.[29,31,32]

The layered arrangement of neurons in the cerebellar cortex, and the demonstration that Purkinje cells are inhibitory in function,[33] provide favorable circumstances for comparing the distribution of GABA with the presence of inhibitory neurons, particularly as GABA hyperpolarizes Deiters neurons,[34] cells with which Purkinje cells synapse. The GABA content and GAD activity of the layer containing Purkinje cell bodies is higher than that of the whole cerebellum or any other cerebellar region;[35,36] the GABA content of isolated Purkinje bodies, on a weight per volume basis, exceeds that of excitatory neurons, such as spinal motoneurons and ganglion cells.[37] Although GAD may be confined to synaptic terminals, these elegant studies do not establish whether GABA is a cytoplasmic constituent of Purkinje cells or is located within the numerous inhibitory terminals from basket cells which cover the outer surface of the bodies.[33] The GABA-transaminase-succinic semialdehyde dehydrogenase system (GABA-TS) is found particularly within neurons which receive inhibitory inputs, such as Purkinje cells, Golgi cells, and mossy fiber endings.[36,38] Apparently, in the cerebellum as elsewhere in the CNS,[39] the distribution of this enzyme system is not restricted to the cytoplasm of neurons of inhibitory function or to the extraneuronal environment of inhibitory synapses. High levels of GABA-TS are found in brain stem and spinal motoneurons, although the activity which surrounds hippocampal pyramidal cells may be associated with the inhibitory terminals of hippocampal basket cells.[39]

Within the spinal cord, GABA concentrations are higher in dorsal than in ventral gray segments, less but equal in dorsal and ventral white segments, very low in dorsal and ventral roots,[40] and are not altered significantly after hypoxic destruction of spinal interneurons.[41] However, as such interneuronal loss is most severe in the central portion of the cord, the distribution of GABA may not be incompatible with that of inhibitory interneurons located predominantly in the dorsal horn.

GABA levels are highest in the outer layers of the cerebral cortex,[35] but a detailed study has yet to be made of the distribution of GABA-TS. GABA has been detected in a perfusate of the cerebral cortex after perforation of the pia arachnoid,[42] and output is increased in association with electroencephalographic evidence of "sleep," a state in which cortical inhibitory processes perhaps predominate over excitory processes. However, other amino acids found in cortical perfusates evidently originate mainly from the plasma,[43,44] and a correlation between the release of GABA and the activity of a specific inhibitory pathway may provide more significant evidence regarding the origin of GABA from synaptic terminals. An increased release of GABA from the perfused fourth ventricle follows repetitive stimulation of the cerebellar cortex,[37] perhaps indicative of liberation from Purkinje cell terminals in Deiter's and cerebellar nuclei lying close to the ventricle.

When administered electrophoretically, GABA depresses the firing of neurons throughout the mammalian CNS, providing an explanation of observations made with other methods of administration.[3] GABA hyperpolarizes cortical neurons,[45] Deiters neurons,[34] and spinal motoneurons.[46] The earlier failure to detect the hyperpolarization of motoneurons and cortical neurons[3] can be ascribed to technical difficulties.[45,46]

Postsynaptic inhibition of mammalian motoneurons is associated with a membrane hyperpolarization as a result of the transient increase in the permeability to potassium and chloride ions,[5] and it is probable that a similar ionic mechanism accounts for the hyperpolarizing inhibitory postsynaptic potentials (IPSP's) of other spinal, cortical, and cerebellar neurons. Extracellularly administered GABA increases the membrane conductance of neurons, depressing action and synaptic potentials. The reversal potentials for IPSP's and GABA hyperpolarizations of cortical[45] and motoneurons[46] are at a more hyperpolarized level than the resting potential, the reversal potentials being equal for some motoneurons.[46] Hence it is probable that the equilibrium potentials for the associated conductance increases are similar or identical. Furthermore, intracellular injection of a series of anions and cations has the same effects upon inhibitory and GABA potentials. In particular, elevation of the intracellular chloride ion concentration converts IPSP's and GABA hyperpolarizations of cortical and spinal neurons into depolarizations, and experiments carried out with a series of anions of different hydrated diameter indicate that the membrane permeability change of the motoneuron membrane induced by GABA is almost certainly identical to that occurring during inhibition.[46] Neuroglial cells[47] and spinal axons are not affected by GABA, and it is possible that the hyperpolarizing action of the amino acid is restricted to inhibitory synapses. The reduction in the excitability of spinal presynaptic fibers by GABA[48] suggests the presence of hyperpolarizing inhibitory synapses on the terminals, and as presynaptic "inhibition" in the spinal cord is associated with a depolarization of afferent terminals,[5] GABA is unlikely to be the transmitter for this process.

No specific antagonist has been found for the depressant action of GABA on mammalian central neurons: a variety of amino acid analogues and convulsants have been tested, including picrotoxin and strychnine. Neither of these convulsants blocks the synaptic inhibition of cortical neurons,[49,50] the effect of GABA on cortical neurons,[49] or the inhibition of Deiter's neurons exerted by Purkinje cell axons.[34] Strychnine blocks the postsynaptic inhibition of spinal motoneurons by impulses in a variety of spinal and supraspinal pathways[51] without affecting the action of GABA.[46,52] However, it is by no means certain that all spinal postsynaptic inhibition is of the strychnine-sensitive type.

The hyperpolarizing action of GABA on mammalian neurons is rapidly reversible, and the effect on spinal interneurons is neither prolonged nor accentuated by the electrophoretic administration of inhibitors of the GABA-TS system, hydroxylamine and amino-oxyacetic acid, or of substances such as thiosemicarbazide, hydroxyethyl-hydrazine and p-mercuribenzoate.[52] This may not be surprising in view of the apparently predominantly intracellular location of this enzyme system. Electrophoretically administered amino-oxyacetic acid does not influence the Purkinje cell inhibition of Deiter's neurons, although such inhibitory hyperpolarizations are sometimes enhanced but not prolonged by hydroxylamine.[35]

This evidence suggests that cellular uptake mechanisms which have been demonstrated for cortical slices may be important in terminating the action of GABA. Certain GABA analogues, known to reduce this uptake[53] neither prolong nor accentuate GABA effects on spinal neurons.[51] However, the concentrations attained electrophoretically may have been inadequate to appreciably inhibit what may be a very efficient process for removing GABA from the extraneuronal space.

Of the structural analogues of GABA present in the CNS, only γ-amino-β-hydroxybutyric acid (GABOB) warrants consideration as a possible transmitter.[54] GABOB is only slightly less potent than GABA as a depressant of spinal and cortical neurons,[3] and its action is not blocked by strychnine.[52]

Thus there is good evidence that GABA mimics the action of inhibitory transmitters in the feline CNS, but at the present time there is no convincing positive evidence linking this amino acid with the operation of any particular inhibitory pathway. It is possible that GABA is an inhibitory transmitter in the cerebral cortex, at inhibitory synapses made by Purkinje cell axons, and in the spinal cord at sites of strychnine-resistant inhibition, particularly in the dorsal horn. However, confirmation of this function requires a specific GABA antagonist.

IV. GLYCINE AND RELATED AMINO ACIDS

The realization that glycine and not GABA is probably a transmitter at strychnine-sensitive inhibitory synapses in the feline CNS is relatively recent.

At present there is no evidence to indicate that glycine and related amino acids may have transmitter functions at invertebrate synapses, although large amounts of glycine, L-α-alanine, and taurine occur in certain invertebrate nerves.[55]

Moderate amounts of glycine are known to occur in vertebrate nervous tissue,[55] and early pharmacological investigations showed that glycine, applied topically or electrophoretically, depressed the activity of feline central neurons, as did many amino acids structurally related to GABA.[3] However, the distribution of glycine within the feline spinal cord indicated that it might be a transmitter in its own right,[56] and subsequent investigations have provided very strong evidence for this.

In normal rat brain, the glycine concentration in the pons-medulla is twice that in the midbrain, and more than three times that in the cerebral hemispheres and cerebellum; α-alanine and taurine levels are higher in the hemispheres and cerebellum than in the midbrain and pons-medulla, while serine is evenly distributed and cystathionine is more concentrated in the cerebellum than in the other brain regions.[57] Glycine, α-alanine, and serine are the major amino acids present in superfusates of the somatosensory cortex of cats anesthetized with pentobarbitone.[44] Glycine, α-alanine, and serine occur in subcellular fractions of guinea pig cortex containing nerve ending particles, but are not selectively localized in such fractions.[30] Subcellular distribution studies in spinal cord have not been published. In the feline spinal cord glycine is more concentrated in the ventral gray matter than in other regions,[40] the spinal levels being approximately five times those of cat brain.[55] α-Alanine and serine levels are higher in spinal gray than white matter, whereas cystathionine appears to be evenly distributed.[58] Traces of taurine also occur in spinal gray matter.[58]

Afferent fibers entering the spinal cord in dorsal roots are excitory in nature, and, in general, spinal inhibitory interneurons have short axons. Hence a spinal inhibitory transmitter should be confined predominantly to spinal gray matter, with a fraction perhaps occurring in propriospinal fibers within the white matter. Within the gray, the distribution will depend on the location of the bodies of the neurons and their terminals; a transmitter acting upon motoneurons may be more concentrated in the ventral than dorsal gray. The intraspinal distribution of glycine is thus not incompatible with an inhibitory transmitter function, particularly as spinal glycine levels fall in conjunction with the loss of small neurons, including inhibitory interneurons, which follows aortic occlusion.[41] It is not known whether the levels of α-alanine, cystathionine, serine, and taurine are similarly affected. Following intraspinal injection of radioactive glycine into cats, the distribution of the incorporated amino acid in the spinal cord is remarkably similar to that of free glycine, neurons in the ventral horn being heavily labeled, and presumably reveals sites of rapid nucleic acid and protein turnover.[59]

In general, glycine is a more potent depressant of spinal motoneurons and interneurons than GABA,[52,60] the reverse is true for cortical neurons[52,61] whereas the two amino acids are approximately equally effective upon

Renshaw cells[52] and cuneate neurons.[62] Such differences may reflect the relative number of synapses at which glycine and GABA function as inhibitory transmitters. Both cortical neurons and spinal motoneurons are hyperpolarized by glycine, thus providing an explanation of the depressant effects, but intensive investigation has only been possible with the latter type of cell.[46,63] The hyperpolarization of motoneurons is rapid in onset, readily reversible, and is accompanied by an increase in membrane conductance. The hyperpolarization tends to fade in the continued presence of the amino acid, the conductance change usually persists, and a late depolarizing phase, rather than receptor densitization, appears to be responsible for the reduction in potential.[46]

Inhibitory postsynaptic potentials of spinal motoneurons are readily altered in magnitude by passing current through an intracellular microelectrode,[5] the majority of synapses evidently being located on the soma and large dendrites.[64] Hence the finding that the reversal potential for hyperpolarization by electrophoretically administered glycine is similar to, or at a slightly less hyperpolarized level than, that of the IPSP indicates that equilibrium potentials for the causative permeability increases are probably identical.[46,63] Further evidence that alterations in the permeability toward the same ions occur is provided by experiments in which intracellular ion concentrations were altered. Changes in the levels of potassium and chloride concentrations, produced by intracellular injection of potassium, sodium or chloride ions, result in identical alterations of IPSP's and glycine hyperpolarization. For example, injection of potassium or chloride ions converts both types of potential into depolarizations, recoveries are of similar extent and time course, and are more rapid than the identical effects which follow sodium ion injection.[46]

The ionic mechanism of spinal inhibition has been analyzed by injecting anions of differing hydrated diameter into motoneurons, inversion of IPSP's indicating that the ions pass through the activated inhibitory membrane.[5] The critical ion diameter exceeds that of the hydrated potassium ion by a factor of 1.14–1.24, formate ions being anomalous in that IPSP's are inverted, yet the diameter apparently exceeds this range.[5] Although this method of analyzing ionic events at synapses may be open to question,[6,21] it is remarkable that intracellular injections of bromide, iodide, nitrite, thiocyanate, chlorate, and formate ions convert IPSP's and glycine hyperpolarizations into depolarizations, recovery for these two types of potential occurring with similar time courses after injection of any one ion, whereas intracellular injections of bromate, sulfate, or citrate ions affect neither type of hyperpolarization.[46] It is thus reasonable to conclude that the ionic fluxes giving rise to inhibitory and glycine hyperpolarizations are probably identical. Furthermore, the same mechanism is responsible for the hyperpolarization of these cells by GABA.[46]

The postsynaptic inhibition of motoneurons by impulses in a number of spinal and descending pathways is reversibly blocked by strychnine,[51] and the finding that this alkaloid also reversibly suppresses the hyperpolarization of these neurons by glycine, without affecting that produced by GABA,[46] is of vital importance in the identification of glycine as a major spinal inhibitory transmitter. Strychnine has a similar effect upon the depression of Renshaw cell activity by glycine, and the inhibition of these neurons by both spinal and descending pathways.[52] Bruceine, thebaine, and a series of synthetic compounds of diverse structure, which all diminish spinal inhibition,[51] are also glycine antagonists, but are less effective than strychnine.[52] Strychnine also blocks the depressant effect of glycine on cortical neurons,[52] but a strychnine-sensitive cortical inhibition has yet to be demonstrated.[49,50]

These results indicate that strychnine apparently does not block spinal synaptic inhibition by influencing transmitter release, but interacts with membrane sites at or near postsynaptic receptors, so hindering access of the amino acid/transmitter. On the basis of strychnine antagonism, amino acids with depressant effects upon spinal neurons can be classified as "glycine-like" or "GABA-like,"[52] and of the former, α-alanine, cystathionine, serine, and taurine, because of their presence in spinal tissue, also require consideration as inhibitory transmitters. However, glycine, alone having a distribution expected of such a transmitter and being the most abundant and most potent depressant of this series of naturally occurring amino acids in the spinal cord, is thus presumably of major significance.

Apart from strychnine and strychnine-like substances, no other specific glycine antagonist has been found in a study which included the convulsants picrotoxin and pentamethylene tetrazole, and a series of glycine analogues.[52]

As with other amino acids, cellular uptake may be important in terminating the action of glycine, but glycine analogues and substances which reduce amino acid uptake by brain slices and other tissues have no effect on the depressant action of glycine on spinal neurons.[52] Furthermore, the failure of strychnine to reduce the uptake of glycine by cortical and spinal slices[46] suggests that the uptake process differs from the membrane events producing hyperpolarization. The depressant effect of glycine on interneurons is neither prolonged nor accentuated by electrophoretically administered hydroxylamine, amino-oxyacetic acid, hydroxy-ethylhydrazine, kojic acid, p-aminobenzoic acid, p-aminohippuric acid or aminopterin.[52] However, considerable potentiation does follow administration of p-mercuribenzoate, but further investigation is required of the specificity of this substance toward the inactivation of "glycine-like" amino acids, of its influence on synaptic inhibitory processes, and of its inhibition of enzymes and transport systems associated with amino acid metabolism.

In conclusion, glycine has the same action upon spinal neurons as does the inhibitory transmitter: the alteration in membrane permeability is apparently identical and strychnine is an antagonist. Hence the evidence that glycine is the transmitter at strychnine-sensitive spinal inhibitory synapses is particularly convincing.

V. L-GLUTAMATE AND RELATED DERIVATIVES

Acidic amino acids generally have an excitant action on susceptible tissues,[3] and several merit consideration as possible excitatory transmitters, particularly L-glutamate. The metabolic importance of this amino acid is such that distribution studies have not been revealing with regard to transmitter function. This difficulty, and the lack of specific antagonists, has meant that the evaluation of L-glutamate as a transmitter depends heavily on the criteria of release and identity of postsynaptic action.

A. Invertebrate

L-Glutamate and related acidic amino acids have excitant actions on many central and peripheral invertebrate neurons,[2,3] but only at crustacean and insect neuromuscular junctions is there sufficient evidence to regard L-glutamate as a probable excitory transmitter.

In marked contrast to differences observed with GABA, excitatory and inhibitory axons and nerve cells of the lobster *Homarus americanus* contain similar amounts of L-glutamate.[18] L-Glutamate has been detected in perfusates of the leg of the crab *Carcinus maenus* in amounts proportional to the frequency of stimulation of the nerve trunk.[65]

Low concentrations of L-glutamate cause contraction of crustacean muscle, while D-glutamate and L-aspartate are much less effective.[3] L-Glutamate has a rapid and readily reversible depolarizing action when administered electrophoretically to the exterior of crayfish (*Cambarus clarkii*) muscle fibers at local concentrations of approximately 4×10^{-5} M.[66] The areas of highest sensitivity correspond to synaptic junctions, and the persistence of a depolarizing effect after denervation suggests that the release of excitatory transmitter by glutamate is an unlikely explanation of the depolarization. The reversal potential for this depolarization is near zero level: the equilibrium potential of the excitatory junctional potential is unknown. Desensitization occurs to both L-glutamate and the excitatory transmitter on prolonged administration of the amino acid, without affecting the depressant action of GABA.[20] Hence L-glutamate and the transmitter presumably interact with a receptor site which differs from the GABA receptor. There is no evidence for a specific antagonist for L-glutamate, or for a terminating mechanism for transmitter action, apart from desensitization and diffusion.

L-Glutamate has been identified in the perfusate from the cockroach (*Periplaneta americana*) leg during stimulation of the limb nerves[65] and the muscles are depolarized by this amino acid.[67] The metathoracic retractor unguis muscle of the locust is depolarized by concentrations of L-glutamate of the order of 6×10^{-7} M; the effect is diminished in the presence of glutamic acid decarboxylase and enhanced by an inhibitor of this enzyme, phenylhydrazine.[68]

B. Vertebrate

L-Glutamate has no effects at vertebrate neuromuscular junctions, and its excitant effects are apparently confined to the CNS.[3]

The concentration of L-glutamate, the most abundant amino acid in the brain, is higher in the central nervous system than in other vertebrate tissues.[55] Appreciable amounts of L-aspartate also occur, and there are regional differences in the levels of these amino acids. In normal rat brain, L-glutamate levels are considerably higher in the cerebral hemispheres and cerebellum than in the midbrain and pons-medulla, while L-aspartate is more evenly distributed.[57] Higher levels of L-glutamate are found in the dorsal gray quadrant of the feline spinal cord, the site of termination of the majority of primary afferent fibers, than in other regions, while L-aspartate is somewhat more concentrated ventrally in the gray matter.[40] The spinal levels of both amino acids fall significantly following the loss of interneurons as a consequence of aortic occlusion, but there is a correlation only between the count of small neurons and L-aspartate levels, suggesting that L-aspartate is perhaps associated with excitatory interneurons.[41] On the other hand, the high level of L-glutamate in dorsal roots, dorsal gray, and dorsal white matter may indicate the involvement of this amino acid in excitation by primary afferent fibers. Subcellular fractionation studies on rat and guinea pig brain homogenates do not indicate a selective localization of L-glutamate or L-aspartate with nerve ending particles.[30] L-Glutamate decarboxylase and L-aspartate aminotransferase are associated with nerve ending particles, whereas L-glutamate dehydrogenase and γ-aminobutyrate aminotransferase are associated with the mitochondria.[31,32]

The rate of release of L-glutamate from the perforated pial surface of the cat cerebral cortex is lower in "sleeping" animals than in "aroused" animals, the opposite of that observed for GABA, while there is no difference in the rate of release of L-aspartate.[42] In similar preparations the rate of release of L-glutamate during electrical stimulation of the reticular activating system is six times that occurring during "sleep"; under the same conditions the rate of release of acetylcholine is increased threefold.[69] Further studies are required, however, of amino acid release associated with stimulation of more specific excitatory pathways to the cortex. The release of L-glutamate from stimulated frog sciatic nerves[70] may be related to the function of this amino acid as the excitatory transmitter of primary afferent fibers, although the efflux may merely be that associated with the operation of the sodium pump.[71]

L-Glutamate reversibly depolarizes feline spinal motoneurons, providing an explanation of the enhanced excitability or actual firing of all types of central neuron so far investigated by this and structurally related acidic amino acids.[3] L-Aspartate and L-glutamate are excitants of moderate and similar strength compared with D-homocysteate, N-methyl-D-aspartate, and β-N-oxalyl-L-α,β-diaminopropionate. Both the onset and offset of L-glutamate and L-aspartate excitation are rapid, and the threshold extracellular concentration for the *firing* of neurons in the cortex and spinal cord is approximately 10^{-4} M; the threshold for depolarization has not been measured.

Although an increase in membrane sodium ion permeability must clearly be involved in a depolarizing process, an analysis of the complete ionic mechanism of the amino acid depolarization of neurons, or indeed of synaptic excitation, raises considerable difficulties. The widespread distribution of synapses over dendrites and soma may preclude both the measurement of the equilibrium potential of synaptic excitation by passing current through an intracellular microelectrode,[64] generally assumed to be located in the soma, and the comparison of this equilibrium potential with that of the depolarization evoked by an amino acid, the action of which is limited presumably to a restricted area of the membrane. Thus the finding that the reversal potential for the depolarization of spinal motoneurons by L-glutamate, administered from the outer barrel of a coaxial pair of micropipettes, is at a less depolarized level than that of monosynaptic excitation[72] is not incompatible with both types of depolarization having identical ionic mechanisms. Further study seems warranted, particularly of the effects of alterations of intracellular ion concentration on the depolarizing action of L-glutamate. Preliminary investigations suggest that intracellular injections of potassium or chloride ions sufficient to markedly alter inhibitory and glycine hyperpolarization are without effect on the depolarizing action of DL-homocysteate on spinal motoneurons. The ionic mechanism of this depolarization thus may well be similar to that of synaptic excitation.

It has not proved possible to either antagonize the excitant action of acidic amino acids in a specific fashion, or to prolong it by local administration of enzyme inhibitors;[3] indeed enzymatic inactivation seems improbable in view of the close similarities in the time courses of action of L- and D-glutamate. However, intravenously but not electrophoretically administered thiosemicarbazide has been reported to prolong and enhance the excitant action of L-glutamate, an effect reduced by prior local administration of pyridoxal-5'-phosphate.[73] Although this evidence apparently implicates glutamic acid decarboxylase in the inactivation of L-glutamate, such a conclusion would be more convincing if the inhibitor was active after local administration. Desensitization does not appear to be involved in the termination of the action of electrophoretically administered excitant amino acids.[3] Intracellular uptake may be important for removing synaptically released or artificially administered acidic amino acids from the extraneuronal space.

Thus, although much remains to be done before a central excitatory transmitter function can be unequivocally assigned to L-glutamate, the available evidence is strongly in favor of such a function. A similar case exists for the other excitant amino acids present in the CNS, and a successful search for specific antagonists would be most rewarding in terms of classifying the role of these amino acids in synaptic transmission.

The involvement of closely metabolically related neutral and acidic amino acids as transmitters of opposing function raises the fascinating possibility that slight alteration of intracellular metabolism could convert an inhibitory to an excitatory transmitter-producing neuron, and vice versa. Assuming potentially for the appropriate subsynaptic receptors, such inter-conversion, particularly between GABA and L-glutamate, may provide a basis for the functional plasticity of the CNS associated with learning, memory, and forgetting.

VI. CONCLUSION

The evidence has been discussed for the participation of amino acids as synaptic transmitters in the invertebrate and vertebrate nervous systems. In the former there are good reasons for considering GABA and L-glutamate as inhibitory and excitatory transmitters respectively. In the vertebrate CNS both glycine and GABA have many properties expected of inhibitory trans-mitters, and glycine is most probably a transmitter at strychnine-sensitive inhibitory synapses. L-Glutamate is a particularly strong candidate as an excitatory transmitter operating at many synapses throughout the mam-malian CNS. However, specific antagonists of GABA and L-glutamate action are needed to identify the synapses at which these amino acids function as transmitters.*

VII. REFERENCES

1. E. Florey, in *Major Problems in Neuroendocrinology* (E. Bajusz and G. Jasmin, eds.), pp. 17–41, S. Karger, Basel (1964).
2. E. Florey, Neurotransmitters and modulators in the animal kingdom, *Fed. Proc.* **26**:1164–1178 (1967).
3. D. R. Curtis and J. C. Watkins, The pharmacology of amino acids related to gamma-aminobutyric acid, *Pharmacol. Rev.* **17**:347–392 (1965).
4. J. E. P. Toman, Some aspects of central nervous pharmacology, *Ann. Rev. Pharmacol.* **3**:153–184 (1963).
5. J. C. Eccles, *Physiology of Synapses*, Springer-Verlag, Berlin-Heidelberg (1964).
6. B. L. Ginsborg, Ion movements in junctional transmission, *Pharmacol. Rev.* **19**:289–316 (1967).
7. W. D. M. Paton, Central and synaptic transmission in the nervous system (Pharmacological aspects), *Ann. Rev. Physiol.* **20**:431–470 (1958).
8. K. Krnjević, Transmitters in the cerebral cortex, *Proc. XXIII Int. Physiol. Congr.* 435–443 (1965).
9. R. Werman, Criteria for identification of a central nervous system transmitter, *Comp. Biochem. Physiol.* **18**:745–766 (1966).
10. D. R. Curtis, in *Structure and Function of Inhibitory Neuronal Mechanisms* (C. von Euler, S. Skoglund, and U. Söderberg, eds.), pp. 429–456, Pergamon Press, Oxford (1968).
11. V. P. Whittaker, The application of subcellular fractionation techniques to the study of brain function, *Prog. Biophys. Mol. Biol.* **15**:39–96 (1965).
12. E. DeRobertis, Ultrastructure and cytochemistry of the synaptic region, *Science* **156**:907–914 (1967).
13. J. C. Szerb, Model experiments with Gaddum's push-pull cannulas, *Can. J. Physiol. Pharmacol.* **45**:613–620 (1967).

* Submitted for publication in 1968.

14. J. C. Eccles and J. C. Jaeger, The relationship between the mode of operation and the dimensions of the junctional regions at synapses and motor end-organs, *Proc. Roy. Soc. Ser. B*, **148**:38–56 (1958).

15. G. Levi, J. Kandera, and A. Lajtha, Control of cerebral metabolite levels. 1. Amino acid uptake and levels in various species, *Arch. Biochem. Biophys.* **119**:303–311 (1967).

16. R. Blasberg and A. Lajtha, Heterogeneity of the mediated transport systems of amino acid uptake in brain, *Brain Res.* **1**:86–104 (1966).

17. S. W. Kuffler, Excitation and inhibition in single nerve cells, Harvey Lectures 1958–59, pp. 176–218, Academic Press, New York (1960).

18. M. Otsuka, E. A. Kravitz, and D. D. Potter, Physiological and chemical architecture of a lobster ganglion with particular reference to gamma-aminobutyrate and glutamate, *J. Neurophysiol.* **30**:725–752 (1967).

19. E. A. Kravitz, P. B. Molinoff, and Z. W. Hall, A comparison of the enzymes and substrates of gamma-aminobutyric acid metabolism in lobster excitatory and inhibitory axons, *Proc. Nat. Acad. Aci. U.S.* **54**:778–782 (1965).

20. A. Takeuchi and N. Takeuchi, Localized action of gamma-aminobutyric acid on the crayfish muscle, *J. Physiol. (London)* **177**:225–238 (1965).

21. A. Takeuchi and N. Takeuchi, Anion permeability of the inhibitory postsynaptic membrane of the crayfish neuromuscular junction, *J. Physiol. (London)* **191**:575–590 (1967).

22. A. Takeuchi and N. Takeuchi, On the permeability of the presynaptic terminal of the crayfish neuromuscular junction during synaptic inhibition and the action of γ-aminobutyric acid, *J. Physiol. (London)* **183**:433–449 (1966).

23. E. A. Kravitz, S. W. Kuffler, D. D. Potter, and N. M. van Gelder, Gamma-aminobutyric acid and other blocking agents in crustacea. II. Peripheral nervous system, *J. Neurophysiol.* **26**:729–738 (1963).

24. M. Otsuka, L. L. Iversen, Z. W. Hall, and E. A. Kravitz, Release of gamma-aminobutyric acid from inhibitory nerves of lobster, *Proc. Nat. Acad. Sci. U.S.* **56**:1110–1115 (1966).

25. Z. W. Hall and E. A. Kravitz, The metabolism of γ-aminobutyric acid (GABA) in the lobster nervous system. I. GABA-Glutamate transaminase, *J. Neurochem.* **14**:45–54 (1967).

26. L. L. Iversen and E. A. Kravitz, Uptake of γ-aminobutyric acid (GABA) in lobster nerve muscle preparation, *Fed. Proc.* **25**:714 (1966).

27. B. Sisken and E. Roberts, Radioautographic studies of binding of γ-aminobutyric acid to the abdominal stretch receptors of the crayfish, *Biochem. Pharmacol.* **13**:95–103 (1964).

28. P. N. R. Usherwood and H. Grundfest, Peripheral inhibition in skeletal muscle of insects, *J. Neurophysiol.* **28**:497–518 (1965).

29. H. Weinstein, E. Roberts, and T. Kakefuda, Studies of subcellular distribution of γ-aminobutyric acid and glutamic decarboxylase in mouse brain, *Biochem. Pharmacol.* **12**:503–509 (1963).

30. J. L. Mangan and V. P. Whittaker, The distribution of free amino acids in subcellular fractions of guinea-pig brain, *Biochem. J.* **98**:128–138 (1966).

31. L. Salganicoff and E. DeRobertis, Subcellular distribution of the enzymes of the glutamic acid, glutamine and γ-aminobutyric acid cycles in rat brain, *J. Neurochem.* **12**:287–309 (1965).

32. R. Balazs, D. Dahl, and J. R. Harwood, Subcellular distribution of enzymes of glutamate metabolism in rat brain, *J. Neurochem.* **13**:897–905 (1966).

33. J. C. Eccles, M. Ito, and J. Szentagothai, *The Cerebellum as a Neuronal Machine*, Springer-Verlag, New York (1967).

34. K. Obata, M. Ito, R. Ochi, and N. Sato, Pharmacological properties of the postsynaptic inhibition by Purkinje cell axons and the action of γ-aminobutyric acid on Deiters neurones, *Exp. Brain Res.* **4**:43–57 (1967).

35. H. E. Hirsch and E. Robins, Distribution of γ-aminobutyric acid in the layers of the cerebral and cerebellar cortex. Implications for its physiological role, *J. Neurochem.* **9**:63–70 (1962).

36. K. Kuriyama, B. Haber, B. Sisken, and E. Roberts, The γ-aminobutyric acid system in rabbit cerebellum, *Proc. Nat. Acad. Sci. U.S.* **55**:846–852 (1966).

37. K. Obata, Personal communication (1967).

38. N. M. van Gelder, The histochemical demonstration of γ-aminobutyric acid metabolism by reduction of a tetrazolium salt, *J. Neurochem.* **12**:231–237 (1965).

39. N. M. van Gelder, A comparison of γ-aminobutyric acid metabolism in rabbit and mouse nervous tissue, *J. Neurochem.* **12**:239–244 (1965).

40. L. T. Graham, Jr., R. P. Shank, R. Werman, and M. H. Aprison, Distribution of some synaptic transmitter suspects in cat spinal cord: glutamic acid, aspartic acid, γ-aminobutyric acid, glycine and glutamine, *J. Neurochem.* **14**:465–472 (1967).

41. R. A. Davidoff, L. T. Graham, Jr., R. P. Shank, R. Werman, and M. H. Aprison, Changes in amino acid concentrations associated with loss of spinal interneurons, *J. Neurochem.* **14**:1025–1031 (1967).

42. H. H. Jasper, R. T. Khan, and K. A. C. Elliott, Amino acids released from the cerebral cortex in relation to its state of activation, *Science* **147**:1448–1449 (1965).

43. A. van Harreveld and M. Kooiman, Amino acid release from the cerebral cortex during spreading depression and asphyxiation, *J. Neurochem.* **12**:431–439 (1965).

44. K. Crowshaw, S. J. Jessup, and P. W. Ramwell, Thin-layer chromatography of 1-dimethyl-amino-naphthalen-5-sulphonyl derivatives of amino acids present in superfusates of cat cerebral cortex, *Biochem. J.* **103**:79–85 (1967).

45. K. Krnjević and S. Schwartz, The action of γ-aminobutyric acid on cortical neurones, *Exp. Brain Res.* **3**:320–336 (1967).

46. D. R. Curtis, L. Hösli, G. A. R. Johnston, and I. H. Johnston, The hyperpolarization of spinal motoneurones by glycine and related amino acids, *Exp. Brain Res.* **5**:238–262 (1968).

47. K. Krnjević and S. Schwartz, Some properties of unresponsive cells in the cerebral cortex, *Exp. Brain Res.* **3**:306–319 (1967).

48. D. R. Curtis and R. W. Ryall, Pharmacological studies upon spinal presynaptic fibres, *Exp. Brain Res.* **1**:195–204 (1966).

49. K. Krnjević, M. Randic, and D. W. Straughan, Pharmacology of cortical inhibition, *J. Physiol. (London)* **184**:78–105 (1966).

50. T. H. Biscoe and D. R. Curtis, Strychnine and cortical inhibition, *Nature (London)* **214**:914–915 (1967).

51. D. R. Curtis, The pharmacology of central and peripheral inhibition, *Pharmacol. Rev.* **15**:333–364 (1963).

52. D. R. Curtis, L. Hösli, and G. A. R. Johnston, A pharmacological study of the depression of spinal neurones by glycine and related amino acids, *Exp. Brain Res.* **6**:1–18 (1968).

53. K. Sano and E. Roberts, Binding of γ-aminobutyric acid by mouse brain preparations, *Biochem. Pharmacol.* **12**:489–502 (1963).

54. T. Hayashi, in *Enzymes in Mental Health* (C. J. Martin and B. Kisch, eds.), Lippincott, Philadelphia (1966).

55. H. H. Tallan, in *Amino Acid pools* (J. T. Holden, ed.), pp. 471–485, Elsevier, Amsterdam (1962).

56. M. H. Aprison and R. Werman, The distribution of glycine in cat spinal cord and roots, *Life Sci. (Oxford)* **4**:2075–2083 (1965).

57. R. K. Shaw and J. D. Heine, Ninhydrin positive substances present in different areas of normal rat brain, *J. Neurochem.* **12**:151–155 (1965).

58. G. A. R. Johnston, Intraspinal distribution of some depressant amino acids, *J. Neurochem.* **15**:1013–1017 (1968).

59. H. Koenig, An autographic study of nucleic acid and protein turnover in the mammalian neuraxis, *J. Biophys. Biochem. Cytol.* **4**:785–792 (1958).

60. R. Werman, R. A. Davidoff, and M. H. Aprison, The inhibitory action of cystathionine, *Life Sci. (Oxford)* **5**:1431–1440 (1966).
61. K. Krnjević and J. W. Phillis, Iontophoretic studies of neurones in the mammalian cerebral cortex, *J. Physiol. (London)* **165**:274–304 (1963).
62. A. Galindo, K. Krnjević, and S. Schwarz, Micro-iontophoretic studies on neurones in the cuneate nucleus, *J. Physiol. (London)* **192**:359–377 (1967).
63. R. Werman, R. A. Davidoff, and M. H. Aprison, Inhibition of motoneurones by iontophoresis of glycine, *Nature (London)* **214**:681–683 (1967).
64. T. G. Smith, R. B. Wuerker, and K. Frank, Membrane impedence changes during synaptic transmission in cat spinal motoneurones, *J. Neurophysiol.* **30**:1072–1096 (1967).
65. G. A. Kerkut, L. D. Leake, A. Shapiro, S. Cowan, and R. J. Walker, The presence of glutamate in nerve-muscle perfusates of *Helix, Carcinus* and *Periplaneta, Comp. Biochem. Physiol.* **15**:485–502 (1965).
66. A. Takeuchi and N. Takeuchi, The effect on crayfish muscle of iontophoretically applied glutamate, *J. Physiol. (London)* **170**:296–317 (1964).
67. G. A. Kerkut and R. J. Walker, The effect of iontophoretic injection of glutamic acid and γ-amino-*n*-butyric acid on the miniature end-plate potentials and contractures of the coxal muscles of the cockroach *Periplaneta americana L, Comp. Biochem. Physiol.* **20**:999–1003 (1967).
68. P. N. R. Usherwood and P. Machili, Chemical transmission at the insect excitatory neuro-muscular synapses, *Nature (London)* **210**:634–636 (1966).
69. H. Jasper and I. Koyama, Rate of release of acetylcholine and glutamic acid from the cerebral cortex during reticular activation, *Fed. Proc.* **26**:373 (1967).
70. D. D. Wheeler, L. L. Boyarsky, and W. H. Brooks, Release of amino acids from nerve during stimulation, *J. Cell. Physiol.* **67**:141–148 (1966).
71. P. F. Baker, An efflux of ninhydrin-positive material associated with the operation of the Na⁺ pump in intact crab nerve immersed in Na⁺-free solutions, *Biochem. Biophys. Acta* **88**:458–460 (1964).
72. D. R. Curtis, *in Studies in Physiology, Presented to J. C. Eccles* (D. R. Curtis and A. K. McIntyre, eds.), pp. 34–42, Springer-Verlag, Heidelberg (1965).
73. F. A. Steiner and K. Ruf, Excitatory effects of L-glutamic acid upon single unit activity in rat brain and their modification by thiosemicarbazide and pyridoxal-5′-phosphate, *Helv. Physiol. Acta* **24**:181–192 (1966).

Chapter 6

BIOLOGICALLY ACTIVE PEPTIDES (SUBSTANCE P)

G. Zetler

Department of Pharmacology
Medizinische Akademie Lübeck
Lübeck, Germany

I. INTRODUCTION

Nervous tissue is capable of synthesizing pharmacologically active peptides. The hypothalamic releasing factors and the pituitary hormones constitute two groups of peptides, which are described and discussed in another chapter of this volume. This review concerns those peptides found in tissue extracts and known collectively as "Substance P" (SP), which were reported over 35 years ago. Euler and Gaddum[1] demonstrated in 1931 that alcoholic extracts of brain and intestines cause isolated gut to contract, and their intravenous injection elicited a fall in blood pressure.

The effects of SP cannot be attributed to acetycholine, since they are unaffected by atropine; similarly, the effects of SP did not correspond with histamine, adenylic acid, or adenosine.

Many previous investigations on the distribution and purification of SP assumed that the activity was associated with a single component, which was peptide in nature. This aspect of the work has been previously extensively reviewed.[2–5]

Because of many problems that will be discussed, the chemical structure and the physiological role of SP are not fully elucidated. Chromatographic fractionation of crude SP revealed that this material contained more than one active peptide.[6,7] The reviewer feels that the present body of evidence points to a family of unknown peptides that are chemically as well as historically related to SP. References will be made mainly to current advances. The term peptide will be used for all active components of tissue extracts or SP preparations that are inactivated by chymotrypsin.

II. THE OLD CONCEPT OF SUBSTANCE P

A. Extraction, Estimation, Purification

SP is extracted by mincing intestinal or nervous tissue with 2 vol distilled water and boiling at pH 4–5. Inactive proteins are then removed from the concentrated extract by addition of 4 vol ethanol at pH 8. After evaporation of ethanol and acidification to pH 4, active material is precipitated by addition of 2 vol saturated ammonium sulfate solution. Crude SP is obtained by redissolving this precipitate and salting it out once more. One gram (wet weight) of whole cattle brain without cerebellum yields about 0.4 mg dry SP powder (cf. Table I). Preparations made in this way were used by many workers as SP standards and for pharmacological experimentation. Inactive impurities were removed from aqueous solutions of this material by the addition of methanol up to 70 % in the experiments summarized in Table I. The tissue can also be extracted at room temperature by mincing in glacial acetic acid and adding ethanol. During a further extraction procedure the homogenized tissue is treated with 10 vol chloroform–methanol (2 + 1 vol) and SP is extracted from the dried organic phase by boiling with 0.1 N HCl. If the SP content of small pieces of tissue is to be determined, the tissue is first treated with acetone to remove 5-HT, and the boiled extract is cleared of

TABLE I

Amounts of Substance P Powder and Biological Activities Achieved from Human Brain Areas[14,15]

	Amount of tissue extracted (g wet wt.)	Yield	
Area of human brain		SP powder (mg/g wet wt.)	SP activity (u/g wet wt.)
Frontal cortex	117	3.8	4
Precentral gyrus	150	1.3	3
Occipital cortex	165	2.7	6
Cingulate gyrus	204	2.3	8
Amygdaloid nucleus	174	2.4	18
Caudate nucleus	183	3.2	24
Putamen	203	2.9	36
Globus pallidus	67	1.7	28
Substantia nigra	24.5	3.6	145
Red nucleus	60	1.2	8
Thalamus	195	1.8	6
Hypothalamus	88	5.8	12
Medulla oblongata	115	4.9	5

interfering adenosine derivatives by adenylic acid deaminase or by an anion-exchange resin.[8]

The SP activity of tissue extracts or SP preparations is estimated by means of biological assays and is expressed as units per milliliter or milligram. One SP-unit (u) corresponds to the gut contracting activity of 2–4 threshold doses for the isolated rabbit jejunum in 30 ml Tyrode solution, or to the activity of 7–10 threshold doses for the isolated guinea pig ileum in 3 ml bath fluid. Crude SP powder from cattle brain has 6–12 u/mg; SP from horse intestines has 15–35 u/mg. For estimation the isolated guinea pig ileum is the most useful preparation, but the estrus uterus of the rat and the rectal cecum of the hen, or measurement of the blood pressure of the atropinized rabbit can also be used. The activity of the unknown sample is determined by matching against a standard preparation according to the four-point assay scheme. The isolated organs should be pretreated with antagonists of histamine, 5-HT and ACh, which could interfere with the action of SP in impure tissue extracts.

Purification that enhances activity up to five- to tenfold can be achieved by extracting the crude SP powder with ethanol or glacial acetic acid, and then precipitating the active material with acetone or ethyl ether. Another possibility is to extract the crude powder with 70 % (v/v) solutions of ethanol or methanol, and to pass the extract through a column of aluminum oxide. SP activity is adsorbed and eluted by 50 % alcohol or distilled water.[2] The purest SP preparations, with activities of > 100,000 u/mg, were made from intestines[9,10] and brains[11,12] by repeated column chromatography on various materials (Al_2O_3, Amberlite IRC-50 and IR-45, CM- and Oxy-cellulose, Sephadex G-25, CM-Sephadex G-25), countercurrent distribution, preparative high-voltage electrophoresis, and preparative paper chromatography. This purification work was hampered by unexpected difficulties and is characterized by extremely low yields. Probably only 1 mg of pure SP peptide can be extracted from 100 kg of bovine brain.[12]

B. Chemical Properties

Tissue extracts and all SP preparations are inactivated by proteases such as chymotrypsin, trypsin, pepsin, papain, tissue cathepsin, and other proteolytic enzymes present in nerve tissue and in bacteria.

It must be concluded that SP activity is associated with a polypeptide structure since hydrolytic destruction was accompanied by release of amino acids (Table II). With highly purified preparations the amino acid composition varied between the different organs and species. The absence of Glu and Gly in SP from hog brain is especially impressive since these two amino acids were found in SP from goat brain and equine intestines. Large differences between two highly purified SP preparations from different sources were also found by means of carboxypeptidase B, which completely destroyed the activity of one material but left that of the other intact.[11] The molecular weight of bovine intestinal SP, determined by high-speed centrifugation, is

TABLE II

Amino Acid Composition of Highly Purified SP Preparations and a Partially Purified Preparation of the SP Peptide Fa

Amino acids	SP from hog brain[11]	SP from goat brain[11]	SP from bovine brain[12]	Fa[a] from bovine brain[28]
Histidine	0.32	0.87		
Serine	1.73	1.01	2	2–3
Lysine	1.58	1.84	1–2	3
Arginine	0.64	2.21	1–2	1–2
Aspartic acid	2.12	1.31	1	8–9
Glutamic acid	—	2.95	2	10–11
Proline	1.6	3.35	2	3–4
Glycine	—	2.68	2–3	6–7
Alanine	1.0	1.0	2–3	5
Valine	1.12	1.02	1	2–3
Leucine	2.55	2.62	2–3	2
Isoleucine	0.76	0.49	1	1
Phenylalanine			1–2	1
Threonine	0.65	0.83	1	1–2
Tyrosine	0.0	0.0	1	
Methionine + MetSO	?	0.5–1	1	

[a] Partially purified, fifty times more active than crude Fa.

1,650 ± 250;[10] this agrees well with the molecular weight of hog brain SP, which was found by means of gel filtration and thin film dialysis[11] to be a little higher than that of bacitracin (mol. wt. 1420). Thus, the SP molecule is larger than the other pharmacologically active polypeptides, e.g., the nonapeptide bradykinin (mol. wt. 1060) and the dekapeptide kallidin (mol. wt. 1188). SP is a basic peptide migrating at $pH < 7$ to the cathode.

The structure of the SP molecule is not known. According to degradation studies with proteolytic enzymes, the sequence Phe-Leu-Tyr is at the carboxyl end of bovine brain SP.[12] Like bradykinin, SP from equine intestines has the sequence Arg-Pro-Pro at the amino terminal position,[9] the rest of the molecule is probably different. A chemical relationship between SP and the linear endekapeptides physalemin (mol. wt. 1264) and eledoisin (mol. wt. 1206) has been supposed because of the very high similarity of pharmacological activities.[13] Elucidation of the SP structure was hindered not only by the scarcity of pure material but also by great instability of highly active solutions, which could not be overcome by antioxidants, low pH, or storage at low temperatures or under nitrogen. Adsorption on glass surfaces is probably the cause of the rapid losses of activity. Solutions of impure SP, however, withstand boiling at pH 1–7, but are rapidly inactivated above pH 8. SP in tissues is remarkably resistant to postmortem autolysis.

C. Distribution

SP is present in intestinal and nervous tissue of man and all vertebrates so far investigated, including fishes, amphibians, and birds. Within both organs, the distribution is uneven. However, not too much weight should be laid on differences between tissues or parts of organs not exceeding about 70%, since most investigations were made with impure extracts and were therefore complicated by the difficulties of extraction and estimation in the presence of interfering substances.[3,8]

Of the intestinal tract, duodenum and jejunum have the largest amounts of SP, esophagus and stomach the smallest. The SP content of small intestine tissue is about 50 u/g in monkeys and is lower in other species, the rabbit having only about 3 u/g. All layers of the intestinal wall contain SP, but the highest concentrations were found in the muscularis mucosae. This points to a relation of SP to Meissner's plexus and is in agreement with the finding that in Hirschsprung's disease the aganglionic, inactive part of the rectosigmoid contains much less and the proximal hyperactive part much more SP than normal tissue. However, high doses of neostigmine, atropine, and morphine did not influence the SP concentration in the small intestine, whereas vagal stimulation and filling the gut increased it.

In the central nervous system, gray matter usually contains much more SP (about 50 u/g) than white matter (2–5 u/g), but there are exceptions to this rule. For instance, the concentration in visual cortex is not higher than that in white matter, whereas the anterior cingulate gyrus has twice as much SP as the precentral or postcentral gyrus. Another exception concerns the white matter of dorsal roots and dorsal columns of the spinal cord. Here, the concentrations are about 5–10 times higher than in ventral roots or in the pyramidal tract. Only negligible amounts of SP are present in the cerebellum. SP concentrations vary greatly in subcortical areas and are highest in substantia nigra, for which after elimination of interfering nucleotides a value of 1550–1580 u/g was found (ox brain). Clearly less SP, but still at least five times more than that in cortex is present in area postrema and ala cinerea. In man, the posterior colliculi and the medial geniculate bodies (auditory system) contain considerably more SP than that of the anterior colliculi and the lateral geniculate bodies (visual system), and within each system the concentrations are higher in the colliculi than in the geniculate bodies. As extensively reviewed,[4,5] there is in many brain parts or nervous structures an inverse ratio of SP concentration to choline acetylase activity. This is especially clear for anterior and posterior roots and could mean that SP is located in noncholinergic neurons. The presence of SP in nerve and not in glial cells can be concluded from its absence in glial tumors, its presence in the retina, which is almost without glial tissue, and from its disappearance during nerve degeneration. SP is also present in the peripheral nervous system, but in lower concentration. The difference in SP content between the various kinds of nerves is not as large as that between anterior and posterior roots.

Practically all of this knowledge on the distribution of SP activity in brain and nerves is based on the estimation of very crude tissue extracts. Table I gives the activities recently obtained in the reviewer's laboratory[14,15] when SP powder was made from human brain areas according to the standard procedure described on page 136. It is obvious that cortical tissues yield less SP than subcortical structures do, and that there are great differences even between closely neighboring areas, e.g., substantia nigra and red nucleus.

Subcellularly, SP is located in brain tissue in the same fraction of synaptic vesicles and nerve endings as acetylcholine,[16,17] but in peripheral nerves it is in the microsome fraction.[18,19] Furthermore, there is probably a bound and a free form of SP.[16,18] About 70 % of total SP activity in mammalian brain is particle-bound and can be released by hypotonic shock, and exposure to acid or high temperatures.[19] A protease which destroys SP was found in a different compartment of brain tissue from that of SP itself, namely in the microsomes.[16]

The important problem whether SP concentration in brain changes with function of this organ cannot be clearly answered in spite of numerous efforts.[3-5] The application of various physiological and pharmacological tools yielded contradictory results; this applies even to the question whether darkness and illumination influence retinal SP concentration. The reasons for this unsatisfactory state are very probably the difficulties of extraction and estimation already mentioned.

D. Pharmacological Actions

SP contracts smooth muscles of the intestines and the bronchial tree, briefly lowers the blood pressure, and enhances the permeability of capillaries. Most of these SP actions were observed not only in isolated organs but also in whole animals and in man. Highly purified SP is a spasmogenic and a vasodilator of very strong activity acting in doses or concentrations of a few nanograms per kilogram or nanograms per milliliter. In this respect and on a molar basis, pure SP can be considered the most potent compound so far known. No antagonist is known that acts specifically against SP as atropine acts against acetylcholine, for example. The mechanism of action on smooth muscle is direct, but the receptors for SP are different from those for the peptides bradykinin and kallidin. After intraarterial injection, SP stimulates receptors in the vascular wall and thus initiates various reflexes. On contact with free sensory nerve endings, e.g., on a blister base, it produces pain similarly to bradykinin.

Great efforts to find central effects of SP were made by many workers.[3-5] Crude SP preparations given parenterally clearly had sedative and tranquilizing activity; they antagonized central stimulants and convulsants (e.g., strychnine), prevented or interrupted morphine analgesia, enhanced central depressants, and altered the bioelectrical activity of cortical and subcortical structures. These potencies disappeared when the preparations were incubated with proteases. However, several workers found that purified SP is devoid of central activity. Only recently it was shown that 40 pmoles/ml

(9 u/ml) of a practically pure SP peptide isolated from goat hypothalamus and having a molecular weight of about 3000[11] stimulated the electrical activity of goat hypothalamus when applied by perfusing the brain ventricles.[20] The centrally active SP component with antistrychnine potency was separated by chromatographic means and shown not to be gut-contracting.[21-23] Therefore crude SP preparations contain an unknown peptide or peptides which inhibit central nervous activity but do not influence smooth muscles.

III. CRUDE SUBSTANCE P AS A MIXTURE OF PEPTIDES

A. Chromatographic Basis

Contrary to previous expectations,[2] SP preparations were shown to consist of a family of related peptides.[6,7,19,23-25] Different fractions characterized by the gut-contracting ability were obtained by passage through Al_2O_3 columns (Fa), or by elution with distilled water (Fb), or by 0.1 N NaOH (Fc). Gel filtration yielded also Fa, Fb, and Fc.[25] Rechromatography of Fb on Al_2O_3 columns revealed that Fb yielded the other two components. We have recently studied the interrelationships between these three components to determine whether they arise by a process of denaturation during extraction and purification. These studies show that Fc is a normal constituent of cortical tissue and does not arise from a precursor protein.[26] Fb appears to contain Fc as an impurity which can be separated chromatographically; Fa is a component of the Fb complex that is split off during purification. The question whether a separate Fa exists in the brain in a free or bound form has not been resolved.

Since peaks on elution from Al_2O_3 (Fig. 1) tend to overlap, Fc can be characterized only by its biochemical and pharmacological properties (see also Section III, C). Even pure SP preparations are known to contain other peptides capable of contracting the gut. It was concluded on the basis of several chromatographic techniques that the materials are closely related in structure to SP.[12] Additional groups of such peptides have been reported as present in the goat hypothalamus, calf, frog, and pig brains.[11,27] Although chloroform–methanol extracted all mammalian SP, it extracted only 20% from the frog; the remaining 80% was extracted by boiling and gave an Fb-like material which did not split into Fa and Fc on Al_2O_3 columns. This material, however, gave three additional spots on paper that contracted the gut but were distinct from SP.[27] In bullfrog extracts an additional "Fa" was identified with a different electrophoretic mobility toward the cathode.[28]

B. Biochemical and Pharmacological Characteristics

Nerve tissue extracts with pharmacologically active peptides or with SP activity lose activity after incubation with chymotrypsin. Some active peptides, for example Fc, are resistant to trypsin and can be separated on Al_2O_3[6,7,14,19,23-26] or by countercurrent distribution.[11] It is not known whether the material separated by the different methods is identical. Unlike most biogenic peptides except gastrin, Fc is acidic in nature. Other properties that differentiate Fa and Fc are summarized in Table III. Fb can be differentiated by its faster migration toward the anode; Fc peptides from the human amygdala, red nucleus, and hypothalamus move toward the anode at pH 4.95.[15] The amino acid composition for Fc is unknown, that for Fa is given in Table II. Fa may be identical to the "SP-peptide" purified and isolated by others. Since it is inactivated by trypsin, it probably contains Lys and Arg residues. The report that Fa is trypsin resistant is not in accord with current observations.[19]

TABLE III

Chemical and Pharmacological Differences Between the Active Peptide Fractions Fa and Fc Which Are Present in Substance P Preparations Made from Brain Tissue

Feature	Technique or object	Fa	Fc	References
Chemical behavior	Paper electrophoresis	Basic	Acidic	6, 7, 19, 23, 24
Paper chromatography[a]	Biological evaluation	R_f 0.27	R_f 0.63	6, 7, 24
Destroyed by trypsin	Incubation at 37°C	Yes[c]	No	6, 7, 14, 15, 23, 24, 25, 26
Sephadex G-25[b]	Column filtration	R_f 0.5	R_f 0.27	15, 25
Appearance in brain	Human and cattle brains	All areas	Cortex	14, 15, 26
Acetylcholine release	Isolated gut	No	Yes	30
Hypotensive activity	Atropinized rabbit	1	1/18	6, 25
Antagonists	Gut contraction	Not available	Atropine, morphine, cocaine	29, 30

[a] Ascending, n-butanol–acetic acid–water 40:10:50.
[b] Ammonium formate 0.05 M, pH 6.0.
[c] Trypsin-resistant according to Hori.[19]

The peptides we are dealing with cause contractions of isolated gut with blocked receptors for histamine and 5-HT. Fc, in contrast to Fa and Fb, is antagonized by atropine, morphine, and cocaine, which points to an indirect mechanism of action.[29,30] It was shown that Fc releases acetylcholine from isolated gut into the bath fluid.[30] The point of attack or release is the postganglionic nerve fibers, since the ganglionic blocking drug hexamethonium does not prevent the action of Fc. In anesthetized cats it was shown that topically applied Fc releases acetylcholine from brain cortex,[31,32] and that Fb is inactive in this respect.[33] Like crude and highly purified SP preparations, these peptides lower the blood pressure of the atropinized rabbit if injected intravenously, but Fc is eighteen times less active than Fa.[6,25]

The combined biochemical and pharmacological properties shown by Table III clearly differentiate Fa from Fc. Experimentally, it is sufficient to test whether the activity of an extract or a fraction is resistant to trypsin and antagonized by morphine. In this way, it was found that SP preparations made from human amygdala, red nucleus, and hypothalamus yielded during the NaOH elution step on Al_2O_3 chromatography not Fc but Fc', which was destroyed by trypsin and antagonized by morphine.[15]

The SP peptide Fc has great similarity to the highly lipophilic phosphopeptide "nerveside" which is extracted from brain by a 2:1 (v/v) mixture of chloroform and methanol.[34-36] Like Fc, "nerveside" stimulates isolated gut indirectly: this action is antagonized by atropine, the local anesthetic Nupercaine, and morphine[37] but not by the ganglion-blocking drug hexamethonium. "Nerveside" is destroyed by both chymotrypsin and phosphatases, and is partially inactivated by trypsin.[37] Its nature, like that of Fc, is acidic, since it migrates at pH 6,7 to the anode.

Like the other active SP peptides, the component with antistrychnine potency can be dialyzed through cellophane membranes, does not dissolve in dry acetone and ether, and is destroyed by chymotrypsin or alkaline treatment although it resists trypsin.[23] It is therefore probably also a small polypeptide.

C. Differences Between Brain Areas

The SP preparations made from the thirteen human brain areas which are mentioned in Table I were analyzed by means of aluminum oxide column chromatography.[14,15] Fa and Fb were found in all cases, but the trypsin-resistant and acetylcholine-releasing peptide Fc was present only in SP preparations made from the four cortical tissues and absent in those from the nine subcortical areas. Figure 1 demonstrates this finding in detail for precentral gyrus and substantia nigra,[14] the latter yielding with solvent C (0.1 N NaOH) active material which did not show the characteristics of Fc and was therefore considered to be the tail of Fb. Fb of substantia nigra was so potent that its tail contained more activity than was eluted by solvent C for cortex. However, there was no trace of Fc. When, before extracting SP, human brains[15] and cattle brains[26] were divided into cortical and subcortical matter, Fc was again present only in cortical SP preparations and was missing in those made from the rest of the brain. There was no Fa, Fb, and Fc in SP powder prepared from cattle cerebellum.[15]

When searching for Fc in SP preparations made from the human brain areas mentioned in Table I, in each case the material eluted by 0.1 N NaOH from the Al_2O_3 columns was (a) incubated with trypsin and (b) tested on morphine-treated ilea of guinea pig. These eluates from most subcortical areas contained active material which was destroyed by trypsin and was resistant to morphine, and could therefore be considered to be tails of Fb. However, in the cases of amygdala, red nucleus, and hypothalamus this material was destroyed by trypsin and antagonized by morphine, which means it was neither Fb nor Fc. It was concluded that red nucleus and amygdala contain besides Fa and Fb a peptide fraction Fc' with hitherto unknown properties.[15]

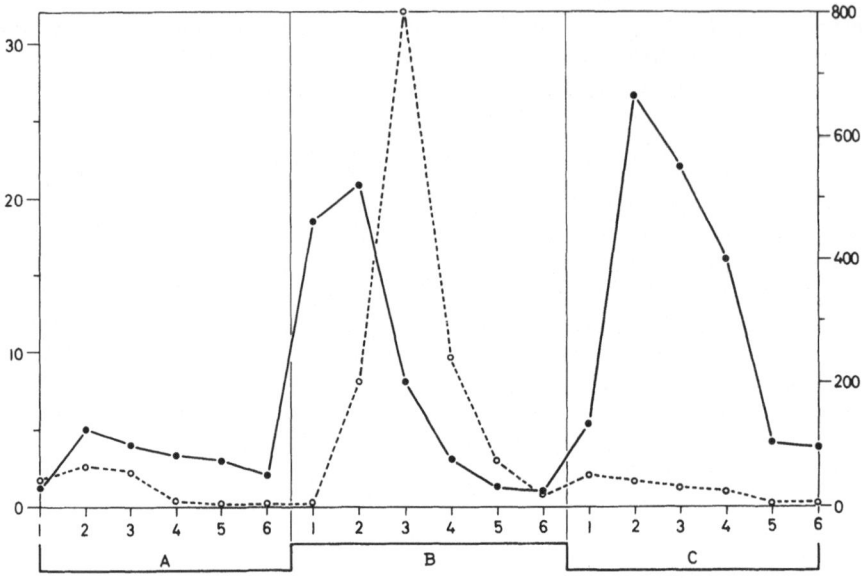

Fig. 1. Differences between SP preparations made from human precentral gyrus (full line) and from human substantia nigra (interrupted line), as revealed by column chromatography on anionotropic aluminum oxide.[14] Ordinates: Biological activity (isolated guinea pig ileum) of the eluates in units per fraction of 5 ml, on the left side for cortex and on the right for substantia nigra. Abscissae: 1. Number of fractions leaving the column, 2. Solvents passing through the column, namely (A) 70% methanol eluting Fa, (B) distilled water yielding Fb, and (C) 0.1 N NaOH eluting Fc. The amounts of SP preparations chromatographed correspond to 107 g of cortical tissue and to 20 g of substantia nigra tissue.

The SP peptide with antistrychnine potency was present in bovine SP preparations made from cortex, brain stem, and cerebellum, and dorsal and ventral halves of spinal cord, but not in that from white matter.[23] The yield was highest from cortex, but there was no parallelism with gut-contracting activity as most clearly seen with cerebellum, which does not contain the other SP peptides.

The phosphopeptide "nerveside" was found to be unevenly distributed within the dog brain.[36] Concentrations were highest in cortical tissue; those of caudate nucleus, diencephalon, and midbrain were only 12–28% of cortical values. The presence of "nerveside" in subcortical structures is important with respect to its possible identity with the SP peptide Fc.

IV. RELEASE OF PEPTIDES FROM PRECURSORS OR STRUCTURES

Several groups have shown that active polypeptides most probably arise by the effect of proteases on inactive precursors. Enzymes used were: trypsin,[18,19,37] plasmin and kallikrein,[19] and a kallikreinlike factor present in brain microsomal fractions.[16,28] The active polypeptides are rapidly destroyed by chymotrypsin and are similar to bradykinin and kallidin, which are also released by trypsin and kallikrein from precursor molecules. Active brain peptides released with trypsin stimulate the chicken rectal cecum preparation[18] and are different from bradykinin, which is inhibitory. On the other hand, the active peptide released by the microsomal factor was very similar to bradykinin or related kinins.[28,38] The materials from which the active peptides were released and which therefore should contain the precursor(s) are crude brain extracts,[38] granules in the press juice of peripheral nerves,[18,19] crude SP preparations,[18] and the SP fractions Fa and Fb.[19] Kininase activity, which destroys the kinin(s) released from precursor(s), was found in all subcellular brain fractions.[28] Its concentration was highest in the soluble and lowest in the microsomal fraction, which, on the other hand, has highest SP-destroying activity.[16]

A peptide that migrates at pH 5.2 to the cathode and contracts the superfused rat uterus is released from the cat's brain cortex.[39–40a] The release of this peptide was increased by direct cortex stimulation, stimulation of the paw, injection of picrotoxin or leptazol. These findings are remarkable since it is expected that SP and the kinin-releasing factors are firmly bound to subcellular particles, requiring unphysiological methods for their release.[16,18,19,28]

The possible appearance in cerebrospinal fluid of a pharmacologically active peptide (kinin) and/or an enzyme which liberates it from globulin, will not be discussed here because of the conflicting observations in this field.[41–43]

V. CONCLUSIONS

At least the following pharmacologically active peptides are now known to be present in crude SP from mammalian brain: (a) Fa peptide in free form and bound as Fb, (2) Fc peptide, (3) the peptide(s) Fc' in human amygdala, red nucleus, and hypothalamus, and (4) the peptide(s) with central depressant and morphine-antagonizing activity. The first three peptides but not the fourth have gut-contracting activity which is, accordingly, not a dependable criterion of a centrally acting peptide. The Fa peptide is very probably identical with that peptide which was as "substance P" highly purified by several workers. The Fc peptide is most interesting because of its exclusive localization in brain cortex, its unusual acidic nature, and acetylcholine-releasing potency. The peptide material with antimorphine activity is more important, as it was possible to induce morphine tolerance by a small protein or peptide extracted from the brain of tolerant animals.[44] Whether or not the phosphopeptide "nerveside" is identical with or related to the SP peptide Fc remains to be clarified.

Pharmacological activity and distribution within the nervous system of these peptides indicate an hitherto unknown role for nervous functions. A highly desirable advance in this challenging field would be the elucidation of the chemical structure of these compounds which in the pure state can be expected to have a very high biological activity.

The brain peptide field was broadened by the findings that there exist in nervous tissue precursors from which trypsin and also an enzymelike brain factor release pharmacologically active peptides. It must be stressed that there is available at present only limited information on this subject.

Finally, there are peptidases and peptide synthesizing enzymes in brain, as shown in other chapters of this volume, but nothing is known about the origin or the fate of the active peptides. Work in this direction together with further knowledge on pharmacological actions of the brain peptides will perhaps allow extension of the concept of the "peptidergic neuron" from neurosecretory areas[45] to the whole brain. Furthermore, this field may become important for the research on acquisition and storage of information, and its transfer by brain extracts containing peptides.[44,46-48] Another possible physiological function of peptides like SP might result from their high vasoactivity and potency to decrease capillary permeability. A high concentration of SP in a given nervous tissue could, perhaps in relation to nervous activity, stimulate the metabolic exchange between capillaries and brain tissue.[49] Similarly, the implication of a protease-polypeptide system in local vasomotor regulation within the central nervous system during health and disease is suggested.[50]

VI. REFERENCES

1. U. S. von Euler and J. H. Gaddum, An unidentified depressor substance in certain tissue extracts, *J. Physiol. (London)* **72**:74–87 (1931).
2. B. Pernow, Studies on substance P. Purification, occurrence and biological actions, *Acta Physiol. Scand.* **29**:Suppl. **105**:1–90 (1953).
3. W. Haefely and A. Hürlimann, Substance P, a highly active naturally occurring polypeptide, *Experientia (Basel)* **18**:297–303 (1962).
4. F. Lembeck and G. Zetler, Substance P: A polypeptide of possible physiological significance, especially within the nervous system, *Int. Rev. Neurobiol.* **4**:159–215 (1962).
5. F. Lembeck and G. Zetler, *in Internat. Encyclopedia of Pharmacol. a. Therap.*, Section 72, Pharmacology of naturally-occurring polypeptides and lipid-soluble acids (J. M. Walker, ed.) (In press.)
6. G. Zetler, Zwei neue pharmakologisch aktive Polypeptide in einem Substanz P-haltigen Hirnextrakt, *Naunyn-Schmiedeberg's Arch. Exp. Path. Pharmakol.* **242**:330–352 (1961).
7. G. Zetler, New pharmacologically active polypeptides present in impure preparations of substance P, *Ann. N.Y. Acad. Sci.* **104**:416–435 (1963).
8. I. Laszlo, Estimation of substance P and adenine nucleotides in the brain of the dog, *J. Physiol. (London)* **184**:17 P (1966).
9. R. A. Boissonnas, J. Franz, and E. Stürmer, On the chemical characterization of substance P, *Ann. N.Y. Acad. Sci.* **104**:376–377 (1963).
10. K. Vogler, W. Haefely, A. Hürlimann, R. O. Studer, R. Lergier, R. Strässle, and K. H. Berneis, A new purification procedure and biological properties of substance P, *Ann. N.Y. Acad. Sci.* **104**:378–389 (1963).

11. H. Meinardi and L. C. Craig, in *Hypotensive Peptides* (E. G. Erdös, N. Back, and F. Sicuteri, eds.), pp. 594–607, Springer-Verlag, New York (1966).

12. H. Zuber, in *Hypotensive Peptides* (E. G. Erdös, N. Back, and F. Sicuteri, eds.), pp. 584–593, Springer-Verlag, New York (1967).

13. G. Bertaccini, J. M. Cei, and V. Erspamer, The action of physalaemin on the systemic arterial blood pressure of some experimental animals, *Brit. J. Pharmacol.* **25**:380–391 (1965).

14. J. Baldauf, P. Harnacke, and G. Zetler, Aktive Peptide in Substanz P-Präparaten aus Cortex, Globus pallidus und Substantia nigra des menschlichen Gehirns, *Naunyn-Schmiedeberg's Arch. Exp. Path. Pharmakol.* **260**:231–241 (1968).

15. J. Baldauf, H. Iven, and G. Zetler, Die Verteilung darmkontrahierender Peptide im menschlichen Gehirn, *Naunyn-Schmiedeberg's Arch. Exp. Path. Pharmakol.* **262**:453–462 (1969).

16. K. Kataoka, The subcellular distribution of substance P in the nervous tissue, *Jap. J. Physiol.* **12**:81–96 (1962).

17. R. W. Ryall, The subcellular distributions of acetylcholine, substance P, 5-hydroxytryptamine, y-aminobutyric acid and glutamic acid in brain homogenates, *J. Neurochem.* **11**:131–145 (1964).

18. U. S. von Euler, Substance P in subcellular particles in peripheral nerves, *Ann. N.Y. Acad. Sci.* **104**:449–463 (1963).

19. S. Hori, Zetler's satellite polypeptides of substance P in subcellular particles of bovine peripheral nerves, *Jap. J. Physiol.* **18**:746–771 (1968).

20. C. A. Baile and H. Meinardi, Action of substance P on the central nervous system of a goat, *Brit. J. Pharmacol.* **30**:302–306 (1967).

21. I. L. Bonta, H. G. Wijmenga, and F. Hohensee, Über die Wirkung von Substanz P und anderen Hirnextrakten auf das Zentralnervensystem, *Acta Physiol. Pharmacol. Neerl.* **10**:114–118 (1961).

22. J. H. Gaddum, M. Randić, and M. W. Smith, An antistrychnine extract from horse intestine, *J. Physiol. (London)* **172**:207–215 (1964).

23. S. Kawaguchi, M. Imaizumi, H. Shio, and K. Kataoka, Substance P and antistrychnine activity, *Naunyn-Schmiedeberg's Arch. Exp. Path. Pharmakol.* **260**:284–297 (1968).

24. G. Zetler, Über das Vorkommen pharmakologisch aktiver Substanzen mit Polypeptid-Charakter in Extrakten aus Darm und Gehirn, *Naunyn-Schmiedeberg's Arch. Exp. Path. Pharmak.* **246**:504–516 (1964).

25. G. Zetler and J. Baldauf, Chromatographische Analyse eines rohen Substanz P-Präparates, *Naunyn-Schmiedeberg's Arch. Exp. Path. Pharmakol.* **256**:86–98 (1967).

26. J. Baldauf and G. Zetler, Darmkontrahierende Hirnpeptide in Cortex und Subcortex, *Naunyn-Schmiedeberg's Arch. Exp. Path. Pharmakol.* **260**:242–253 (1968).

27. U.-H. Knabe, F. Lembeck, and K. Starke, Ausbeute an Substanz P aus dem Hirn verschiedener Tierspecies bei Anwendung zweier Extraktionsverfahren, personal communication (1968).

28. S. Hori, The presence of bradykinine-like polypeptide, kinin-releasing and destroying activity in brain, *Jap. J. Physiol.* **18**:772–787 (1968).

29. G. Zetler, Morphin, Cocain und Atropin als Antagonisten eines darmkontrahierenden Polypeptids aus Rindergehirn, *Naunyn-Schmiedeberg's Arch. Exp. Path. Pharmakol.* **247**: 326 (1964).

30. G. Zetler, in *Hypotensive Peptides* (E. G. Erdös, N. Back, and F. Sicuteri, eds.), pp. 621–630, Springer-Verlag, New York (1966).

31. G. Pepeu, Discussion remark in *Hypotensive Peptides* (E. G. Erdös, N. Back, and F. Sicuteri, eds.), pp. 631–632, Springer-Verlag, New York (1966).

32. G. Pepeu and A. Bartolini, Effect of drugs on acetylcholine (ACh) release from the cerebral cortex of cats with brain stem transection, *Proc. Third Internat. Pharmacological Meeting* (In press.)

33. G. Pepeu, Personal communication (1966).

34. C. C. Toh, Biologically active substances in brain extracts, *J. Physiol.* (*London*) **165**:47–61 (1963).

35. C. C. Toh, Preparation of nerve-stimulating phosphopeptide (substance B, nerveside) in brain extracts and an enzyme in brain which inactivates it, *J. Physiol.* (*London*) **173**:420–430 (1964).

36. C. C. Toh, The regional distribution of the nerve-stimulating phosphopeptide (nerveside, substance B) in the central nervous system. *J. Physiol.* (*London*) **188**:451–456 (1967).

37. C. C. Toh, Personal communication (1968).

38. A. Inouye, K. Kataoka, and T. Tsujioka, On a kinin-like substance in the nervous tissue extracts treated with trypsin, *Jap. J. Physiol.* **11**:319–334 (1961).

39. P. W. Ramwell and J. E. Shaw, The nature of non-cholinergic substances released from the cerebral cortex of cats on direct and indirect stimulation, *J. Physiol.* (*London*) **169**:51–52 P (1963).

40. P. W. Ramwell and J. E. Shaw, The spontaneous and evoked release of noncholinergic substances from the cerebral cortex of cats, *Life Sci.* **6**:419–426 (1963).

40a. J. E. Shaw and P. W. Ramwell, Release of a substance P polypeptide from the cerebral cortex, *Am. J. Physiol.* **215**:261–267 (1968).

41. M. Schachter, *in Recent Advances in Pharmacology* (J. M. Robson and R. S. Stacey, eds.), pp. 156–178, Churchill Ltd., London (1962).

42. E. Habermann, Fortschritte auf dem Gebiet der Plasmakinine, *Naunyn-Schmiedeberg's Arch. Exp. Path. Pharmak.* **245**:230–253 (1963).

43. M. Rocha e Silva, Acquisitions récentes sur la bradykinine, *Actualités Pharmacol.* **15**:162–198 (1963).

44. G. Ungar and M. Cohen, Induction of morphine tolerance by material extracted from brain of tolerant animals, *Int. J. Neuropharmacol.* **5**:183–192 (1966).

45. W. Bargmann, E. Lindner, and K. H. Andres, Über Synapsen an endokrinen Epithelzellen und die Definition sekretorischer Neurone, *Z. Zellforsch.* **77**:282–298 (1967).

46. G. Ungar, *in Protides of the Biological Fluids* (H. Peeters, ed.), pp. 151–162, Elsevier, Amsterdam (1966).

47. G. Ungar and L. N. Irwin, Transfer of acquired information by brain extracts, *Nature* (*London*) **214**:453–455 (1967).

48. F. Rosenblatt, I. T. Farrow, and W. F. Herblin, Transfer of conditioned responses from trained rats to untrained rats by means of a brain extract, *Nature* (*London*) **209**:46–48 (1966).

49. F. Lembeck and K. Starke, Substance P content and effect on capillary permeability of extract of various parts of human brain, *Nature* (*London*) **199**:1295–1296 (1963).

50. L. F. Chapman and H. G. Wolff, Property of cerebrospinal fluid associated with disturbed metabolism of central nervous system, *Science* **128**:1208–1209 (1958).

Chapter 7

BIOLOGICALLY ACTIVE LIPIDS

Leonhard S. Wolfe

The Donner Laboratory of Experimental Neurochemistry
Montreal Neurological Institute and the Department of
Neurology and Neurosurgery, McGill University
Montreal, Canada

I. INTRODUCTION

For many years it has been realized that tissues and body fluids, including the brain and cerebrospinal fluid, contain pharmacologically active substances that have the properties of acidic lipids.[1-7] These substances are most often detected by their ability to elicit contractions of a slow type in smooth muscle that are not affected by atropine or antagonists of 5-hydroxytryptamine and histamine. Their activity is not diminished by preincubation with crystalline chymotrypsin, trypsin, or the broad-spectrum protease from *Streptomyces griseus*, Pronase. Most substances in this class partition into solvents such as diethyl ether, ethyl acetate, or chloroform from acidic but not basic aqueous solutions or extracts. Thus they behave as carboxylic acids. Chemical treatment by carboxyl-complexing reagents (e.g., N,N'-carbo-di-p-tolylimide), hydroxyl-complexing reagents (e.g., phenylisocyanate), and procedures which react with double bonds (e.g., hydrogenation, bromination) abolishes in most cases the smooth-muscle contracting property.[8] These simple procedures suggest compounds which are unsaturated hydroxy fatty acids. Recent investigations have shown that prostaglandins are by far the most important group of physiologically and pharmacologically active substances of this class. Particular members of this new and structurally unique lipid class are among the most powerful smooth-muscle contracting substances known and occur in varying amounts in most if not all tissues. Depending upon the particular *in vitro* preparation used, threshold concentrations for a direct contractile effect are of the order of 1 ng/ml (3×10^{-9} M). An enhancing action on the response to other stimulating agents is produced by concentrations 10–100 times lower. The complex pharmacological actions of the prostaglandins will not be considered in this chapter, or the physiological studies in connection with human reproduction, metabolism of adipose

149

tissue, regulation of intrarenal blood flow, platelet adhesion, etc. (see Refs. 9–15). Prostaglandins either relax or contract spontaneously active smooth muscle and often antagonize the actions of other hormones, particularly adrenaline and noradrenaline. Their effects on tissue *in vivo* often differ from those *in vitro*, and the same tissues in different species also show different responses. The two basic views on the primary site of action of prostaglandins are (1) an action on the cell membrane affecting calcium distribution and fixation on actomyosin and (2) an intracellular biochemical action inhibiting the formation of cyclic 3′,5′-adenosine monophosphate. The precise role of this nucleotide in tissues is not clearly understood, but it has been implicated as an intracellular "second messenger" in the action of several hormones on target cells.[15] Since the discovery that essential polyunsaturated fatty acids are precursors of prostaglandins, it has been suggested that the importance of the essential fatty acids is related to their ability to be converted into prostaglandins. Although still inconclusive, a relationship does exist between those fatty acids which can be converted into biologically active prostaglandins and their ability to reverse the effects of essential fatty acid deficiency. Research in this field is active at present, and rapid advances are to be expected in the elucidation of the mechanism of action of prostaglandins and their physiological importance as local tissue regulatory hormones. In this chapter the central theme is prostaglandins, although other active acidic lipids or extracts will be mentioned when the evidence is good that they do not contain prostaglandins. Several excellent reviews on prostaglandins have appeared, as well as a comprehensive bibliography (see Refs. 11, 12, 14–22).

II. DISCOVERY, CHEMISTRY, AND METABOLISM OF PROSTAGLANDINS

The name "prostaglandin" was given by U. S. von Euler in 1935 to a principle in human seminal plasma and in extracts of sheep vesicular glands which behaved like a lipid-soluble acid, lowered the blood pressure of the rabbit and cat, and stimulated various isolated smooth-muscle preparations.[23,24] Similar effects of seminal fluid were reported independently by Goldblatt.[25] Just over 20 years later Bergström and Sjövall[26] announced the isolation in crystalline form of an unsaturated nitrogen-free hydroxy fatty acid from extracts of sheep seminal vesicles. This compound, called PGF, had a powerful stimulant action on the isolated rabbit duodenum and jejunum but no effect on the rabbit blood pressure. Another single compound less polar and more soluble in lipid solvents, called PGE, was then isolated, and this substance exhibited both vasodepressor and intestinal smooth-muscle contracting properties. In the following few years superb chemical investigations by Bergström, Ryhage, Sjövall, Samuelsson, and others led to the complete formulation of the structure of a number of prostaglandins from sheep seminal vesicles and human seminal plasma.[27–29] In the latter, the richest source of prostaglandins, so far thirteen types have been identified.

Prostaglandins are derivatives of a hypothetical fatty acid given the trivial name of prostanoic acid (Fig. 1).[28] All natural prostaglandins except some metabolites contain a hydroxyl group at C-15 (R) and a trans double bond at C-13-14 (Δ^{13}). Since substitution at C-15 creates an asymmetric center, the S, R notation is used to describe the configuration.[30] In the PGE series the cyclopentane ring is substituted with a ketone group at C-9 and a hydroxyl group at C-11 in the α-position. The PGF series have hydroxyls at C-9 and C-11 both in the α-position. The PGA series, which are dehydration products of the E series and show a peak of absorption at 217 mμ, have a double bond at C-10-11. The PGB series have a double bond at C-8-12 and a

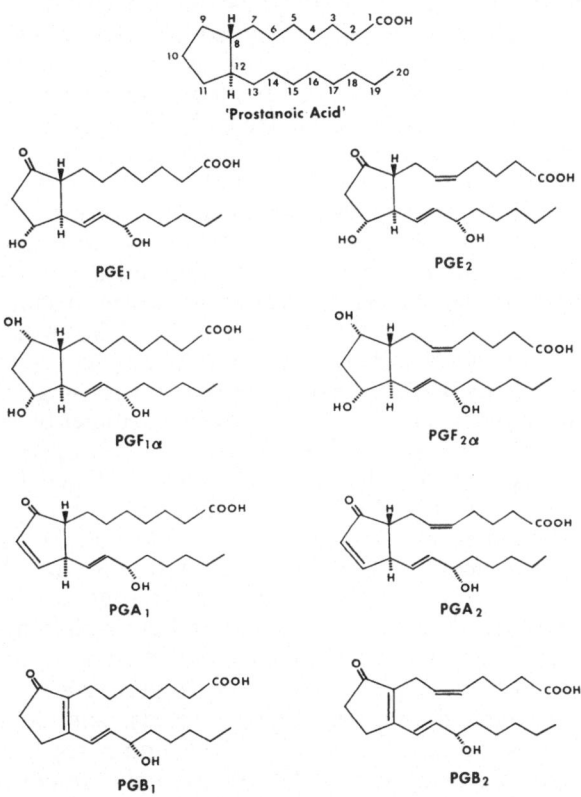

Fig. 1. The absolute configuration of "prostanoic" acid, the hypothetical molecule upon which the chemical names of the prostaglandins are based and the structures of the prostaglandins most commonly encountered in tissue extracts and their dehydrated derivatives.[8] The chemical name of PGE$_1$ is 9-keto-11(α), 15(R)-dihydroxy-prost-13(tr)-enoic acid, PGF$_{1\alpha}$ is 9(α),11(α),15(R)-trihydroxy-prost-13(tr)-enoic acid, etc.

ketone group at C-9. They are formed from the PGE's or PGA's by isomeriz-ation by alkali and absorb strongly at 278 mμ. The suffixes 1-3 in the abbre-viated nomenclature indicate the number of double bonds in the side chains to the ring structure i.e., Δ^{13} trans; Δ^5 cis, Δ^{13} trans; Δ^5 cis, Δ^{13} trans, Δ^{17} cis respectively. The formulas of several prostaglandins are given in Fig. 1.

Soon after the chemical structures of the prostaglandins were deter-mined, two groups simultaneously reported the conversion in high yield of arachidonic acid (all cis eicosa-5,8,11,14-tetraenoic acid) into PGE_2 by aero-bic incubation of homogenates of sheep seminal vesicles.[31,32,37] Subsequent work showed that PGE_1 and PGE_3 were formed from the all cis eicosa-8,11,14-trienoic and eicosa-5,8,11,14,17-pentaenoic acids respectively.[33] The enzyme was microsomal and required a heat-stable factor in the high-speed supernatant which could be substituted by reduced glutathione, tetrahydro-folate, or 6,7-dimethyltetrahydropteridine. It was shown that the three oxy-gens of the prostaglandin molecules originated from molecular oxygen and that the oxygens at C-9 or C-11 came from the same molecule of oxygen. PGE and PGF compounds were formed in about equal amounts if reduced glutathione which facilitated the formation of the E series was not present. The PGE and PGF compounds are not interconvertible (see Refs. 16, 21). With the corresponding polyunsaturated alcohols as substrates, conversion to the prostanols does not take place.[34] The mechanism of the formation of prostaglandins is considered in a little greater detail since it shows a new type of enzyme reaction which most likely occurs generally in tissues containing polyunsaturated fatty acids and also indicates the way in which biologically active unsaturated hydroxy fatty acids other than the prostaglandins may be formed in small amounts in tissues. Recent excellent studies by Hamberg and Samuelsson[35,36] on the mechanism of prostaglandin synthesis in sheep seminal vesicles have shown that the reaction is initiated by a selective removal of a hydrogen from C-13 in the L-configuration of the polyunsatura-ted fatty acid. The evidence is strongly in favor of the initial formation of the 11-peroxy-eicosenoic acid by a lipoxidase type of reaction which can further react at C-9 with cyclization across C-8 and C-12, leading to a rearrangement of the double bond at Δ^{12} trans to Δ^{13} trans and peroxidation at C-15. The resulting endoperoxide is finally transformed into PGE or PGF_α compounds by independent reactions (Fig. 2). The 15,L-hydroperoxy-8,11(cis), 13(trans)-eicosatrienoic acid and the corresponding hydroxy compound which are formed from 8,11,14-eicosatrienoic acid by the action of soy bean lipoxidase are not precursors of prostaglandins. The addition of tetrahydrofolic acid to the high-speed supernatant fraction incubated with seminal vesicle micro-somes and substrate favors the formation of $PGF_{1\alpha}$ and unsaturated mono-hydroxy acids (Fig. 2). It may be that in certain tissues or under particular conditions concentrations of cofactors of this type affect the amounts of these monohydroxy acids which are formed. Fatty acids of this type also exhibit smooth-muscle contracting properties but are in general considerably less powerful than the prostaglandins. They may account for some of the biologi-cal activity of tissue extracts.

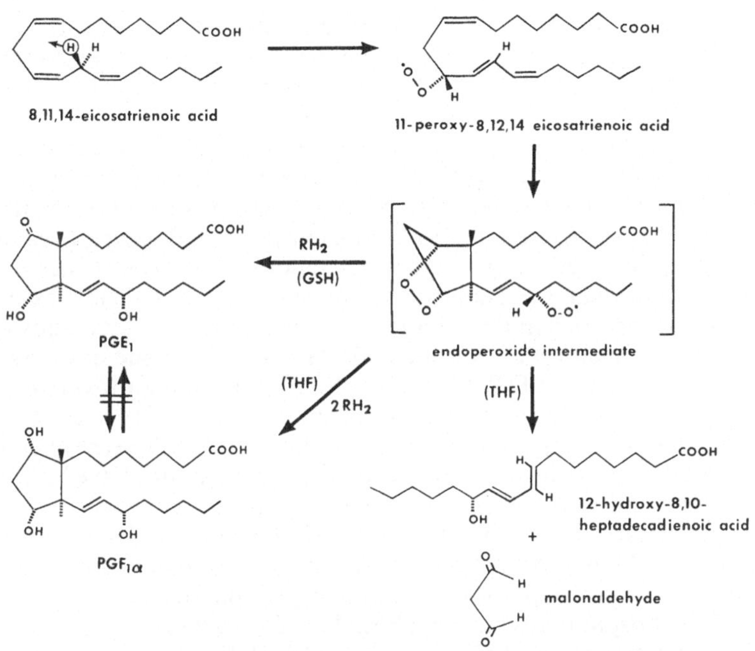

Fig. 2. The mechanism of the conversion of 8,11,14-eicosatrienoic acid (dihomo-γ-linolenic acid) into PGE_1 and $PGF_{1\alpha}$ and by-products formed when the cofactor requirements are limited.[16,35,36]

Homogenates and microsomal fractions of other tissues (lung, intestine, stomach, thymus, kidney, iris, brain) can convert the appropriate precursor fatty acid to the corresponding prostaglandins.[16,17,19,21,37] However, the enzyme capacity is much lower than in the seminal vesicles and only conversions of small percentages of added isotopically labeled precursors have been reported. With cerebral cortex homogenates or microsomal fractions incubated in the presence of oxygen without added substrate, $PGF_{2\alpha}$ increases five- to sevenfold above the unincubated controls with only small increases in $PGF_{1\alpha}$ and the E compounds.[16] The addition of reduced glutathione increased the amounts of PGE_2 formed from endogenous substrates. Experiments with labeled arachidonic acid as substrate clearly indicate that conversion of exogenous fatty acid is small and that formation arose almost entirely from endogenous arachidonic acid formed during the incubation. A similar situation occurs with stomach homogenates or acetone and ether-washed powders of stomach tissue.[38] During homogenization of these tissues even in ice-cold conditions and during subsequent incubation, an active phospholipase A type hydrolysis of phospholipids occurs with a rapid increase in the amounts of free arachidonic acid. Thus the low conversion of added substrate in tissues is not only due to a lower enzyme concentration but also to a rapid dilution of the precursor fatty acid pool when phospholipase activity is present. In addition, there is evidence which suggests that inhibitors of prostaglandin synthesis are formed during incubation but so far their nature is not clear. The size of the total arachidonic acid pool in brain which is almost entirely esterified in membrane phospholipids is much greater than that of the 8,11,14-eicosatrienoic acid, and this accounts for the low amounts of E_1 and $F_{1\alpha}$ formed endogenously. The amounts of free arachidonic acid and $PGF_{2\alpha}$ are low whenever the brain tissue is frozen rapidly in liquid nitrogen. Since both phospholipase A activity and prostaglandin synthesis is rapid and takes place even during homogenization and during brief incubation periods, the levels of prostaglandins reported for some tissues may not necessarily represent the actual amounts in the tissues. The efficiency of conversion to prostaglandins of added labeled substrate fatty acids can be increased by incubation of acetone and ether-washed tissue powders in the presence of 10–20 mM EDTA, reduced glutathione, and antioxidant, conditions which decrease the endogenous phospholipase activity of the tissue.[39] It is likely that an isotope dilution effect by the enzymatic release of endogenous precursors occurs in all tissues in which prostaglandins have been shown to be formed and released and in which insignificant conversions of added labeled precursors have been found. A specific phospholipid may be involved in this action. Thus prostaglandin synthesis may be coupled closely at the cellular level to the activity of membrane phospholipases. Prostaglandins esterified in phospholipids have not been found in tissues.

The metabolism of PGE_1, PGE_2, and PGE_3 has been extensively studied in guinea pig lung tissue *in vitro* as well as in sheep and pig lung and most likely occurs in other species, including human lung.[16,17,29,40–43] Soluble enzymes transform PGE_1 by reduction of the Δ^{13} double bond into dihydro-PGE_1 and then by oxidation of the secondary alcohol at C-15 into 11α-hydroxy-9-15-diketoprostanoic acid (15-keto-dihydro-PGE_1). Metabolism in swine lung differs from that in guinea pig lung in that the enzyme reducing the Δ^{13} double bond is absent and the major metabolite is the 11α-hydroxy-9-15-diketoprost-13-enoic acid (15-keto-PGE_1). Prostaglandins E_2 and E_3 are transformed in analogous fashion. The 15-prostaglandin dehydrogenase has been purified from pig lung and shown to be NAD^+ dependent and highly specific for the C-15 hydroxyl group. The metabolism of the PGF compounds by lung is not completely understood. $PGF_{2\alpha}$ is slowly transformed into less polar metabolites different from those of PGE_2, but they have not been fully characterized. The lower rate of metabolism of $PGF_{2\alpha}$ is interesting since it is the principal lung prostaglandin and more rapid metabolism of the E compounds may have physiological significance. The dihydro compounds of the PGE's have less potent intestinal smooth-muscle contracting activity, but the effect on blood pressure is not significantly different. However, oxidation at C-15 reduces almost completely both the spasmogenic and vasodepressor actions of PGF's. In the whole animal the activity of the metabolic enzymes is such that prostaglandins liberated from tissues into the blood would have a half-life of only a few minutes.[17] Although oxidative metabolism is most active in the lung, it also occurs to some extent in other tissues.

In the liver a different sequence of metabolic reactions occurs. Mitochondrial beta oxidation converts the PGE_1 and $PGF_{1\alpha}$ to the *dinor* derivatives and dihydro-PGE_1, PGE_2 and $PGF_{2\alpha}$ into the *tetranor* compounds. Microsomal omega oxidation of the *tetranor* derivatives also occurs and leads to the formation of polar dicarboxylic acids. Biological activity is completely destroyed by the liver metabolism. These metabolites have been identified in the urine.[16,21,44,69] The E and F compounds are not interconvertible as the C-20 fatty acids; the structures must first be modified by beta oxidation in the liver before interconversions between the two prostaglandin types can take place. In man the inactive urinary metabolites of prostaglandins are derived from the four pathways of metabolism in the lung and liver: (1) 15-dehydrogenation, (2) 13,14 saturation, (3) beta oxidation to C-18 and C-16 monocarboxylic acids and (4) omega oxidation to dicarboxylic acids.

III. DISTRIBUTION OF PROSTAGLANDINS IN TISSUES

Soon after the prostaglandins in human seminal fluid and sheep seminal vesicles had been isolated and chemically characterized, analyses of many other tissues showed that prostaglandins were very widely distributed and it

TABLE I

Distribution of Prostaglandins in Various Tissues

Tissue or fluid	Species	Types[a]
Seminal plasma and seminal vesicles	Human, sheep	$\underline{E_1}, \underline{E_2}, E_3, \underline{F_{1\alpha}}, \underline{F_{2\alpha}}, \underline{A_1}, \underline{A_2}$ $\underline{B_1}, \underline{B_2}$, 19-OH of $\underline{A_1}, \underline{A_2}, \underline{B_1}, B_2$
Menstrual fluid and endometrium	Human	$\underline{F_{2\alpha}}, E_2, F_{1\alpha}, E_1$
Amniotic fluid and umbilical cord blood vessels	Human	$\underline{F_{2\alpha}}, E_2, F_{1\alpha}, E_1$
Iris and anterior chamber fluid	Sheep, rabbit, cat	$\underline{F_{2\alpha}}, E_2$
Lung	Human, monkey, ox, pig, sheep, guinea pig	$\underline{F_{2\alpha}}, E_2, F_{1\alpha}, E_1$ and metabolites
Thyroid	Human	$\underline{F_{2\alpha}}, E_2$
Thymus	Human	$\underline{E_1}$
Submaxillary salivary gland	Human	$\underline{E_2}, F_{2\alpha}$
Stomach and intestine	Rat, sheep, horse, frog	$\underline{F_{2\alpha}}, E_2, E_1, F_{1\alpha}$
Pancreas	Ox	$\underline{F_{2\alpha}}, E_2$
Kidney	Rabbit, pig	$\underline{E_2}, F_{2\alpha}$
Adrenal medulla	Human	$\underline{F_{2\alpha}}, E_2$
Brain and CSF	Ox, cat, dog, chicken, human	$\underline{F_{2\alpha}}, E_2, F_{1\alpha}, E_1$
Spinal cord	Dog, chicken	$\underline{E_2}, F_{2\alpha}$
Phrenic nerve	Human	$\underline{E_1}$
Cervical sympathetic chain	Human	$\underline{E_2}, F_{2\alpha}$
Vagus nerve	Human	$\underline{F_{2\alpha}}, E_2$

[a] The prostaglandin types present in greatest amounts are underlined.

now appears that almost any tissue that is examined contains small amounts of one or another prostaglandin.[11,12,16-18] A great number of human tissues have been examined and found to contain prostaglandins.[45] Table I summarizes these results. The $PGF_{2\alpha}$ and PGE_2 are the main prostaglandin types occurring in tissues and body fluids other than the seminal vesicles. This reflects the relative proportions of the precursor free or esterified polyunsaturated fatty acids in the tissues. In seminal vesicles there is a high concentration of the $C_{20:3}$ $\omega6$ fatty acid as well as PGE_1 and $PGF_{1\alpha}$. Since prostaglandins are present in most tissues in very small amounts (0.05–0.5 μg/g fresh weight), it is not possible at present to chemically characterize each prostaglandin present. Therefore a number of criteria must be satisfied for their identification and measurement which are based on solvent solubility properties, chromatographic behavior in a number of separate systems with pure standards, and biological behavior in several *in vitro* bioassay procedures.[12,46,47] Recoveries can be determined by the use of tritiated PGE_1. Besides thin-layer chromatographic techniques, gas chromatographic procedures have been worked out for the PGF prostaglandins but they are still too insensitive for measurement of amounts found in most tissues.[48,49] A recent interesting finding is that medullary carcinoma of the thyroid, a hormone-secreting tumor, also excessively secretes PGE_2 and $PGF_{2\alpha}$ and that there is an association between this secretion and the diarrhea often present in patients with this neoplasm.[50]

IV. BRAIN PROSTAGLANDINS AND THE EFFECT OF PROSTAGLANDINS ON BRAIN

The existence of smooth-muscle stimulating acidic lipids in mammalian brain with the properties of an unsaturated hydroxy carboxylic acid was reported by Ambache[1,8] before $PGF_{2\alpha}$ was rigorously identified by Samuelsson in ox brain.[51] It is present in amounts of 0.3 μg/g fresh weight determined by isotope dilution. Subsequent studies with ox, cat, rabbit, and rat brain have also shown the definite presence of $PGF_{2\alpha}$ as the major type but not PGE_2 (except in the rabbit) and much smaller amounts of $PGF_{1\alpha}$ and PGE_1.[16,17] In chicken brain, although $F_{2\alpha}$ is present, the predominant type, particularly in the spinal cord is E_2.[52] There is considerable species variation in the amounts of individual prostaglandin types in brain and there may also be regional variations within the brain.

Prostaglandins or prostaglandinlike substances are released into fluids continuously superfusing the cat cerebral cortex, cat cerebellum, and frog spinal cord *in vivo* and also occur in small amounts in ventricular fluids.[16,53-55] A relationship between neuronal activation and prostaglandin release has been reported for the somatosensory cortex of the cat, since direct electrical stimulation, transcallosal stimulation, or stimulation of the contralateral radial nerve (evoked responses) increases significantly the release of prostaglandins.[53] Similarly, electrical stimulation of the frog hind limb increases the release of prostaglandins into superfusates of the frog spinal cord.[56] From the cerebellum of unanesthetized but decerebrate cats the spontaneous release rate was 8–32 ng/hr/cm^2 surface as $PGF_{2\alpha}$.[54] These results suggest that prostaglandins may be central transmitter substances and a preliminary investigation showed some concentration of these lipids in synaptosome subcellular fractions. However, reexamination of this aspect showed no specific subcellular localization and almost all the total tissue prostaglandins were recovered in the high-speed supernatant fraction.[57] Evidence (see below) now suggests that prostaglandins are formed and released at the postsynaptic membrane during synaptic activity.

Pharmacological actions of prostaglandins on the central nervous system of cats and chicks and on the activity of brain stem neurons after iontophoretic application have been examined.[11,12,16,17,58,59] PGE's in doses of 20 μg injected into the ventricles of cats produced sedation and a catatonic type of stupor. $PGF_{2\alpha}$, the main prostaglandin in the cat brain, was without effect. PGE_1 injected intravenously into chicks also caused sedation but $PGF_{2\alpha}$ caused dorsiflexion of the neck and extension of the limbs and potentiated crossed extensor reflexes through a direct action on the spinal cord. The chick spinal cord, as already mentioned, contains much more E_2 than $F_{2\alpha}$. It seems that the prostaglandins pharmacologically active on brain and spinal cord are not those types present to any extent in the tissue. PGE_1 injected intravenously increases gastrocnemius muscle tension in the spinal cat even after dorsal root section and potentiates decerebrate rigidity. This is also due to an action at the spinal cord level. Topical application of PGE_1 to the spinal cord induces contraction of the gastrocnemius muscle. These results suggest that certain prostaglandins can have facilitatory actions on the firing of α-motorneurons. Medullary reticulospinal neurons of the unanesthetized decerebrate cat respond to prostaglandins applied iontophoretically by excitatory or inhibitory responses.[59] With PGE_1 and PGE_2 the predominant response is excitation whereas inhibition was observed more often with $PGF_{2\alpha}$. The responses of single neurons to different prostaglandins and the different responses of different neurons in the brain stem and the spinal cord are similar to the complex pharmacology of these lipids on smooth muscle and other tissues. A general explanation of these results can not be given as yet; it awaits further understanding of the basic mechanisms of prostaglandin actions acting alone or in concert with other central or peripheral neurohormones.

V. RELEASE OF PROSTAGLANDINS FROM ORGANS BY NERVE STIMULATION, VENOMS, AND OTHER TREATMENTS

A. Iris and Irins

Antidromic stimulation of the trigeminal nerve in the rabbit produces an intense and prolonged miosis and vasodilatation which is unrelieved by atropine. Similar miotic spasms can be produced by mechanical stimulation of the iris or other manipulations within the eye, such as collapse of the anterior chamber by paracentesis. Apparently the antidromic reaction is unique to the rabbit but local irritation can elicit an atropine-resistant miosis in other species. A smooth-muscle stimulating principle called irin was extracted from the irides of the rabbit and ox by Ambache and shown to be an unsaturated hydroxy fatty acid quite independently and before the detailed structure of prostaglandins was known.[1,2,17,60] The irin from sheep irides has now been found to contain $PGF_{2\alpha}$ as well as small amounts of other unidentified smooth-muscle contracting factors. The rabbit and cat iris contains PGE_2 as well as $PGF_{2\alpha}$ and an unidentified active material, thus explaining both the miosis and vasodilatation. The pig iris contains the enzyme for transforming polyunsaturated fatty acids to prostaglandins.[61] The fatty acids in the total lipids of the pig iris contain considerable amounts of arachidonic acid but no eicosatrienoic acid, thus explaining the absence of PGE_1 and $PGF_{1\alpha}$. Irins are released into the anterior chamber fluid. It is now generally felt that prostaglandins are mainly responsible for the biological activity of irins but other unidentified biologically active fatty acids are also present.

B. Stomach, Intestine, and Darmstoff

Tyrode solutions bathing the isolated stomach and intestine of the frog acquire an atropine-resistant smooth-muscle stimulating activity with the properties of a dialyzable acidic lipid. This material was called Darmstoff by Vogt.[3–6,16,17] Originally it was thought that frog Darmstoff was a plasmalogenic phosphatidic acid but this is now very doubtful since its biological activity can be accounted for almost entirely by its prostaglandin content (mainly PGF's). Hot aqueous or solvent extracts of horse intestine contain active lipid-soluble acids. These extracts are known to contain phospholipids and lysophospholipids as well as prostaglandins and unsaturated fatty acids.[16,17] Certain phospholipids, particularly those which are acidic at physiological pH's (phosphatidyl serine, phosphatidic acid and its plasmalogen, lysophosphatidic acid plasmalogen), are slow-reacting smooth-muscle stimulants as are also the autooxidation products of polyunsaturated fatty acids (hydroperoxides and hydroxy derivatives). Lysophospholipids desensitize many intestinal smooth muscle preparations to the action of other stimulants, whereas prostaglandins even in picogram amounts enhance the action of stimulants.[3] These various factors must be separated from horse Darmstoff before the quantitative contribution of the prostaglandins present can be assessed. In general microgram amounts of phospholipids, hydroxy and hydroperoxy fatty acids are needed to elicit smooth-muscle contraction but only nanogram amounts of the prostaglandins are required. Prostaglandins must be present in amounts of 0.001–0.0001 % of a milligram of a total lipid or phospholipid fraction for slow-reacting activity to be exhibited. It is the realization of the extraordinary potency of prostaglandins compared with other types of lipids that has cast some doubt on the significance of biological activity of phospholipid and glycolipid preparations. It has been suggested that the prostaglandins in Darmstoff arise through the action of intestinal phospholipase A cleaving off the polyunsaturated fatty acids which become converted to prostaglandins by the intestinal synthetase.[6] The lysocompounds formed may be further hydrolyzed by intestinal phosphatases to yield lysophosphatidic acids or transacylated to the diacylglycerophosphatides. Similarly slow-reacting substances called SRS-C are formed, not liberated, from preformed stores when cobra venoms, bee venom, or purified phospholipase A from the venoms acts on a number of tissues (lung, intestine, stomach).[6] Only a small part of this activity can be accounted for as unsaturated fatty acid hydroperoxides. Prostaglandinlike substances are present which have an activity that can be destroyed by acetylation or reaction with phenylisocyanate.[17] SRS-C extracts do contain PGF and PGE compounds as well as uncharacterized active substances less polar and more polar than the prostaglandins.

In the rat stomach, prostaglandins PGE_1, PGE_2, $PGF_{1\alpha}$ and $PGF_{2\alpha}$ are present and are spontaneously released even after denervation into fluids bathing the serosal surface of unstimulated isolated intact preparations.[47] Stimulation of the vagus nerves or transmural stimulation markedly increases the release of prostaglandins, and the increase is closely dependent upon the rate of stimulation. Stimulation of the periarterial sympathetic nerves does not affect the spontaneous release but inhibits the effect of vagal stimulation. Separation of the individual prostaglandins in the bathing fluids by chromatographic procedures and measurement by bioassay shows that two prostaglandins, PGE_2 and $PGF_{2\alpha}$, are selectively increased after stimulation of the cholinergic fibers.[47] Moreover, the prostaglandins released do not come from preformed stores but are formed during the stimulation period. The accelerated release following nerve stimulation represents an acceleration of biosynthesis in the tissue from precursor fatty acids. These precursors appear to come from membrane phospholipids through the activation of a membrane phospholipase A. Quick-frozen rat stomach contains very little free $C_{20:3} \omega 6$ and $C_{20:4} \omega 6$ fatty acids (6 and 10 μg/g fresh weight, respectively). These acids are present esterified at the β-carbon of phospholipids (114 and 1370 μg/g fresh weight respectively). However, homogenization and incubation at 37°C for short periods causes a rapid increase in these fatty acids in the free form and this is associated with a proportional appearance of the prostaglandins.[38]

The prostaglandins in the rat stomach are formed at sites distal to the intrinsic nervous plexuses. Evidence is accumulating which suggests a relationship between membrane phospholipase activity and prostaglandin formation at certain postsynaptic receptors. This may have general and fundamental physiological significance. A variety of stimuli dependent upon the specific tissue may initiate or accelerate these reactions at the membrane level. Prostaglandins may be regarded as local tissue hormones involved in the regulation of the actions of neurohormones, whether from autonomic endings or central synapses. This regulation can be either at the metabolic level in the adenyl cyclase-catalyzed reactions[19] or at the ionic level in the regulation of calcium movements and binding, or both.[12,13,16]

C. Spleen, Adrenals, and Adipose Tissue

Stimulation of the sympathetic nerves to the isolated spleen of the dog perfused with blood from a second donor animal releases PGE_2 into the venous effluent.[62] Phenoxybenzamine blocks both the contractile response to stimulation and the release of prostaglandins, but the output of nora-drenaline is increased. As in the rat stomach, a postsynaptic endogenous production of PGE_2 occurs in the spleen and the quantities released are greater than the amounts normally contained in the spleen. Perfused adrenal glands release catecholamines in response to acetylcholine and during this reaction prostaglandins are also formed and released.[63] Sympathetic stimu-lation and application of catecholamines release prostaglandins from adipose tissue (rat epididymal fat pad).[16] Thus both adrenergic stimulation and cholinergic stimulation can initiate the formation and release of prostagland-ins in tissues. Knowledge of the specific membrane changes accompanying the coupling of automatic nervous activity to effector responses will do much to clarify the role of prostaglandins. It is also very probable that the forma-tion and release of prostaglandins in the central nervous system is qualita-tively similar to the situation in other organs.

The effects of PGE_1 on the metabolism of the epididymal fat pad have received much attention.[15,16,19] It is mentioned briefly here since cyclic 3',5'-adenosine monophosphate is involved and this nucleotide is present in most tissues with a particularly high concentration in brain. *In vitro* PGE_1 diminishes or inhibits lipolysis whether it is or is not stimulated by catechol-amines and other hormones. Also, PGE_1 has an insulinlike action by in-creasing the incorporation of glucose into the fatty acids of triglycerides. The main metabolic reactions appear to involve cyclic adenosine monophosphate. PGE_1 blocks the catecholamine-stimulated increase in levels of this nucleo-tide in adipose tissue.

VI. OTHER PHARMACOLOGICALLY ACTIVE LIPIDS

Lysolecithin and lysophospholipids in general produce a variety of effects on smooth-muscle preparations.[3–6] Sensitivity to acetylcholine and histamine by guinea pig intestine is decreased, but guinea pig uterus and carotid artery strips are contracted by lysolecithin. Intravenous injection of lysolecithin produces a steep fall in blood pressure but this is most likely mediated by histamine since it is a powerful histamine liberator.

In certain immediate hypersensitivity reactions, interaction of antigen with specific antibodies fixed on the surface of target cells starts an enzymatic sequence leading to the release of histamine, serotonin, kinins, and the formation and release of a slow-reacting substance (SRS-A). The combined effects of these substances contribute to the clinical picture of anaphylaxis. SRS-A has the characteristics of an acidic lipid and contracts isolated bronchial smooth muscle (like $PGF_{2\alpha}$). However, there is good evidence that this activity is not due to prostaglandins.[17,64,65]

Reports of a hemolytic substance in serum and plasma termed G-acid thought to be cis-vaccenic acid (Δ^{11}cis-octadecenoic acid) which stimulates rabbit jejunum requires further investigation.[3] Unsaturated fatty acids do not stimulate smooth muscle unless they are peroxidized or contain hydroxyl groups. It is significant that in the biosynthesis of prostaglandins peroxy intermediates are formed and the homolytic cleavage of the endoperoxide intermediate in certain conditions yields malonaldehyde and an unsaturated hydroxy fatty acid. These substances would have some biological activity and be present to varying degrees in all tissues that synthesize prostaglandins. The presence of these compounds may well account for much of the un-identified unsaturated hydroxy fatty acids other than prostaglandins reported to occur in tissue extracts. Thus Toh's unidentified unsaturated fatty acid, called substance C, from brain extracts[66] may be a prostaglandin or one of the unsaturated hydroxy fatty acids. The α-hydroxy fatty acids (cerebronic and oxynervonic) present in cerebrosides and sulfatides have no biological activity nor do the glycosphingolipids.[8] Earlier reports that gangliosides could contract or relax smooth muscle have been shown to result from im-purities in the preparations which could be removed by extraction with ether or purification on silica gel columns.[67,68] The five major gangliosides in brain in pure form do not contract smooth muscle but they do block the action of bis-quaternary aromatic blocking agents and d-tubocurarine. This results from an ionic interaction between the positive quaternary nitrogen and the negative carboxyl groups of the sialic acids in gangliosides.

VII. REFERENCES

1. N. Ambache, in Comparative Endocrinology (U. S. von Euler and H. Heller, eds.), Vol. 2, pp. 128–158, Academic Press, New York (1963).
2. N. Ambache, Biological characterization of, and structure-action studies on, smooth-muscle contracting fatty acids, Mem. Soc. Endocr. 14:19–28 (1966).
3. W. Vogt, Naturally occurring lipid-soluble acids of pharmacological interest, Pharmacol. Rev. 10:407–435 (1958).
4. W. Vogt, Pharmacologically active acidic phospholipids and glycolipids, Biochem. Pharma- col. 12:415–420 (1963).
5. W. Vogt, in Proc. 2nd Internat. Pharmacol. Meeting (H. W. Raudonat and J. Vanecek, eds.), Vol. 9, pp. 43–51, Pergamon, New York (1965).
6. W. Vogt, in Drugs of Animal Origin (A. Leonardi and J. Walsh, eds.), pp. 29–33, Ferro Edizione, Milan (1967).
7. H. Kirschner and W. Vogt, Pharmakologisch Wirksame Lipoidloslichesauren in Hirn-extrakten, Biochem. Pharmacol. 8:224–234 (1961).

8. N. Ambache, M. Reynolds, and J. Whiting, Investigation of an active lipid in aqueous extracts of rabbit brain, and of some further hydroxy-acids, *J. Physiol. (London)* **166**:251–283 (1963).

9. R. Eliasson, Studies on prostaglandin. Occurrence, formation and biological actions, *Acta Physiol. Scand.* **46**:Suppl. **158**:1–73 (1959).

10. M. Bygdeman, The effect of different prostaglandins on human myometrium *in vitro*, *Acta Physiol. Scand.* **63**:Suppl. **242**:1–78 (1964).

11. E. W. Horton, Biological activities of pure prostaglandins, *Experientia* **21**:113–118 (1965).

12. V. R. Pickles, The prostaglandins, *Biol. Rev.* **42**:614–652 (1967).

13. F. Coceani and L. S. Wolfe, On the action of prostaglandin E_1 and prostaglandins from brain on the isolated rat stomach, *Canad. J. Physiol. Pharmacol.* **44**:933–950 (1966).

14. R. Eliasson and U. S. von Euler, *Prostaglandins—Chemistry and Pharmacology*, Academic Press, New York (1968).

15. R. W. Butcher, Cyclic 3',5'-AMP and the lipolytic effects of hormones on adipose tissue, *Pharmacol. Rev.* **18**:237 (1966).

16. S. Bergström and B. Samuelsson, eds., *The Prostaglandins*, Proc. II Nobel Symp., Almqvist and Wiksell, Uppsala (1967).

17. V. R. Pickles and R. J. Fitzpatrick, Endogenous substances affecting the myometrium, *Mem. Soc. Endocrin.* **14**:3–142 (1966).

18. S. Bergström, L. A. Carlson, and J. R. Weeks, The prostaglandins: A family of biologically active lipids, *Pharmacol. Rev.* **29**:1–48 (1968).

19. S. Bergström, Prostaglandins: Members of a new hormonal system, *Science* **157**:382–391 (1967).

20. S. Bergström, The prostaglandins, *Recent Prog. Horm. Res.* **22**:153–175 (1966).

21. D. Kritchevsky, R. Paoletti, and D. Steinberg, eds., *Progress Biochem. Pharmacol.* **3**:59–138, S. Karger Basel/N.Y. (1967).

22. J. E. Pike, *The Prostaglandins* (Bibliography), Kalamazoo, Michigan, The Upjohn Company (1967).

23. U. S. von Euler, Uber die spezifische blutdrucksenkende Substanz des menslichen Prostata und Samenblasensekretes, *Klin. Wschr.* **14**:1182–1183 (1935).

24. U. S. von Euler, On the specific vasodilating and plain muscle stimulating substances from accessory genital glands in man and certain animals, *J. Physiol. (London)* **184**:837–855 (1936).

25. M. W. Goldblatt, Properties of human seminal plasma, *J. Physiol. (London)* **84**:208 (1935).

26. S. Bergström and J. Sjövall, The isolation of prostaglandin E from sheep prostate glands, *Acta Chem. Scand.* **14**:1693–1701 (1960).

26a. S. Bergström and J. Sjövall, The isolation of prostaglandin F from sheep prostate glands, *Acta Chem. Scand.* **14**:1701–1705 (1960).

27. S. Bergström, R. Ryhage, B. Samuelsson, and J. Sjövall, The structure of prostaglandin E_1, F_1 and F_2, *Acta Chem. Scand.* **16**:501–502 (1962).

28. S. Bergström, R. Ryhage, B. Samuelsson, and J. Sjövall, The structure of prostaglandins E_1, $F_{1\alpha}$ and $F_{1\beta}$, *J. Biol. Chem.* **238**:3555–3564 (1963).

29. S. Bergström and B. Samuelsson, Prostaglandins, *Ann. Rev. Biochem.* **34**:101–108 (1965).

30. R. S. Cahn, C. K. Ingold, and V. Prelog, Specification of asymmetric configuration in organic chemistry, *Experientia* **12**:81 (1956).

31. S. Bergström, H. Danielsson, and B. Samuelsson, The enzymatic formation of prostaglandin E_2 from arachidonic acid, *Biochim. Biophys. Acta* **90**:207–210 (1964).

32. D. A. Van Dorp, R. K. Beerthuis, D. H. Nugteren, and H. Vonkeman, The biosynthesis of prostaglandins, *Biochim. Biophys. Acta* **90**:204–207 (1964).

33. S. Bergström, H. Danielsson, D. Klenberg, and B. Samuelsson, The enzymatic conversion of essential fatty acids into prostaglandins, *J. Biol. Chem.* **239**:PC4006–8 (1964).

34. C. B. Struiyk, R. K. Beerthuis, H. J. J. Pabon, and D. A. Van Dorp, Specificity in the enzymatic conversion of polyunsaturated fatty acids into prostaglandins, *Rec. Trav. Chim.* **85**:1233–1250 (1966).

35. M. Hamberg and B. Samuelsson, On the mechanism of the biosynthesis of prostaglandins E_1 and $F_{1\alpha}$, *J. Biol. Chem.* **242**:5336–5343 (1967).

36. M. Hamberg and B. Samuelsson, Oxygenation of unsaturated fatty acids by the vesicular gland of sheep, *J. Biol. Chem.* **242**:5344–5354 (1967).

37. D. H. Nugteren, R. K. Beerthuis, and D. A. Van Dorp, The enzymatic conversion of *all-cis* 8,11,14-eicosatrienoic acid into prostaglandin E_1, *Rec. Trav. Chim.* **85**:405–419 (1966).

38. C. Pace-Asciak, K. Morawska, F. Coceani, and L. S. Wolfe, *in Proc. Prostaglandin Symposium of the Worcester Foundation for Experimental Biology* (P. W. Ramwell and J. E. Shaw, eds.), pp. 371–378, Interscience, New York (1968).

39. C. Pace-Asciak and L. S. Wolfe, Unpublished observations.

40. E. Änggard and B. Samuelsson, Metabolism of prostaglandin E_1 in guinea-pig lung: the structure of two metabolites, *J. Biol. Chem.* **239**:4097–4102 (1964).

41. E. Änggard, K. Green, and B. Samuelsson, Synthesis of tritium-labeled prostaglandin E_2 and studies on its metabolism in guinea-pig lung, *J. Biol. Chem.* **240**:1932–1939 (1965).

42. E. Änggard and B. Samuelsson, Metabolism of prostaglandin E_3 in guinea-pig lung, *Biochemistry* **4**:1864–1871 (1965).

43. E. Änggard and B. Samuelsson, Purification and properties of a 15-hydroxyprostaglandin dehydrogenase from swine lung, *Ark. Keine (Kemi.?)* **25**:293–300 (1966).

44. E. Granström, U. Inger, and B. Samuelsson, The structure of a urinary metabolite of prostaglandin $F_{1\alpha}$ in the rat, *J. Biol. Chem.* **240**:457–461 (1965).

45. S. M. M. Karim, M. Sandler, and E. D. Williams, Distribution of prostaglandins in human tissues, *Brit. J. Pharmacol. Chemotherap.* **31**:340–344 (1967).

46. E. W. Horton and I. H. M. Main, Identification of prostaglandins in central nervous tissues of the cat and chicken, *Brit. J. Pharmacol. Chemotherap.* **30**:582–602 (1967).

47. F. Coceani, C. Pace-Asciak, F. Volta, and L. S. Wolfe, Effect of nerve stimulation on prostaglandin formation and release from the rat stomach, *Am. J. Physiol.* **213**:1056–1064 (1967).

48. K. Gréen and B. Samuelsson, Thin-layer chromatography of prostaglandins, *J. Lipid Res.* **5**:117–120 (1964).

49. M. Bygdeman and B. Samuelsson, Analyses of prostaglandins in human semen, *Clin. Chim. Acta* **13**:465–474 (1966).

50. E. D. Williams, S. M. M. Karim, and M. Sandler, Prostaglandin secretion by medullary carcinoma of the thyroid, *Lancet* **1**:22–23 (1968).

51. B. Samuelsson, Identification of a smooth-muscle stimulating factor in bovine brain, *Biochim. Biophys. Acta* **84**:218–219 (1964).

52. E. W. Horton and I. H. M. Main, Identification of prostaglandins in central nervous tissues of the cat and chicken, *Brit. J. Pharmacol. Chemotherap.* **30**:582–602 (1967).

53. R. W. Ramwell, J. E. Shaw, The spontaneous and evoked release of prostaglandins from the cerebral cortex of anaesthetized cats, *Am. J. Physiol.* **211**:125–134 (1966).

54. F. Coceani and L. S. Wolfe, Prostaglandins in brain and the release of prostaglandin-like compounds from the cat cerebellar cortex, *Can. J. Physiol. Pharmacol.* **43**:445 (1965).

55. W. Feldberg and R. D. Myers, Appearance of 5-hydroxytryptamine and an unidentified pharmacologically active lipid acid in effluent from perfused cerebral ventricles, *J. Physiol. (London)* **184**:837–855 (1966).

56. P. W. Ramwell, J. E. Shaw, and R. Jessup, Spontaneous and evoked release of prostaglandins from frog spinal cord, *Am. J. Physiol.* **211**:998–1004 (1966).

57. J. M. Hopkin, E. W. Horton, and V. P. Whittaker, Prostaglandin content of particulate and supernatant fractions of rabbit brain homogenates, *Nature (London)* **217**:71–72 (1968).

58. E. W. Horton and I. H. M. Main, Further observations on the central nervous actions of prostaglandins $F_{2\alpha}$ and E_1, *Brit. J. Pharmacol. Chemotherap.* **30**:568–581 (1967).

59. G. L. Avanzino, P. B. Bradley, and J. H. Wolstencroft, Actions of prostaglandins E_1, E_2 and $F_{2\alpha}$ on brain stem neurones, *Brit. J. Pharmacol. Chemotherap.* **27**:157–163 (1966).

60. N. Ambache, Properties of irin, a physiological constituent of the rabbit's iris, *J. Physiol. (London)* **135**:114–132 (1957).

61. D. A. Van Dorp, G. H. Jouvenaz, and C. B. Struijk, The biosynthesis of prostaglandin in pig eye iris, *Biochim. Biophys. Acta* **137**:396–399 (1967).

62. B. N. Davies, E. W. Horton, and P. G. Withrington, The occurrence of prostaglandin E_2 in splenic venous blood of the dog following splenic nerve stimulation, *Brit. J. Pharmacol. Chemotherap.* **32**:127–135 (1968).

63. P. W. Ramwell, J. E. Shaw, W. W. Douglas, and A. M. Poisner, Efflux of prostaglandin from adrenal glands stimulated with acetylcholine, *Nature (London)* **210**:273–274 (1966).

64. W. E. Brocklehurst, Slow reacting substance and related compounds, *Progr. Allergy* **6**:539 (1962).

65. B. Uvnäs, Lipid spasmogens appearing in connection with histamine liberation, *Biochem. Pharmacol.* **12**:439–443 (1963).

66. C. C. Toh, Preparation of a nerve-stimulating phosphopeptide in brain extracts and an enzyme in brain which inactivates it, *J. Physiol.* **173**:420–430 (1964).

67. J. D. Robinson, E. A. Carlini, and J. P. Green, The effects of gangliosides preparations on smooth muscle, *Biochem. Pharmacol.* **12**:1219–1223 (1963).

68. R. I. Irwin and E. G. Trams, Reduction of neuromuscular blocking activity of quaternary compounds by gangliosides, *J. Pharmacol. Exptl. Therap.* **137**:242–248 (1962).

69. P. W. Ramwell and J. E. Shaw, eds., *Prostaglandin Symposium of the Worcester Foundation for Experimental Biology*, pp. 1–402, Interscience, New York (1968).

Chapter 8

HISTIDINE DECARBOXYLASE AND DOPA DECARBOXYLASE

Dorothea Aures

Psychopharmacology Research Laboratories
Veterans Administration Hospital, Sepulveda, California
Department of Pharmacology, College of Medicine
University of California, Irvine, California

Rolf Håkanson

Department of Pharmacology
University of Lund, Lund, Sweden

and

William G. Clark

Psychopharmacology Research Laboratories
Veterans Administration Hospital, Sepulveda, California
Department of Biological Chemistry
University of California, Los Angeles, California

I. INTRODUCTION

The concept of neural transmission is intimately connected with that of the functional significance of the biogenic amines. Although so far relatively few amines have been recognized as transmitter compounds, it is frequently assumed that all amines in the central nervous systems (CNS) have some function related to the control or facilitation of impulse conduction.

Nervous tissue contains a wide variety of amines,[1-4] most of which are still unidentified.[4] Among those which have been identified and extensively studied are the catecholamines (CA), 5-hydroxytryptamine (5-HT), and histamine (Hm). These amines have a characteristic and in some respects rather similar distribution in the brain. 5-HT and noradrenaline (NA) are localized mainly in the phylogenetically older areas of the brain, but their distribution within these areas is not exactly parallel. Apart from the pineal gland,[5] the highest concentrations of 5-HT are in the caudate nucleus and the putamen,[6-8] while the highest concentration of NA is in the rostral and inter-mediate portions of the hypothalamus.[9,10] Dopamine (DA) is localized almost exclusively in the nuclei of the corpus striatum.[11] Hm is found in high concentrations in regions functionally related to the anterior pituitary, i.e., in the hypothalamus and in the median eminence.[12] The presence of Hm in the hypothalamus is perhaps not surprising, since this small area of the brain is well endowed with a variety of active compounds. While the function of hypothalamic Hm is unknown, the presence of Hm in large quantities (up to 100 μg/g) in postganglionic fibers[13] suggests that the amine might partici-pate in the regulation of sympathetic nervous activity. The cellular localiza-tion of neuronal Hm, however, is not well established. The observations of Torp[14] suggest that neuronal Hm is mainly located in mast cells surround-ing the nerve fibers in peripheral nerves. This suggestion has been supported by recent histochemical observations on the cellular localization of Hm.[15]

Kataoka and De Robertis[16] found brain Hm to be much more concen-trated in the gray matter than in the white matter; also the enzymes necessary for its formation were found to be similarly distributed.[17] Subcellular fractionation of rat brain homogenate demonstrated Hm in small nerve endings but not in myelin or free mitochondria. After osmotic shock of mito-chondrial and microsomal fractions, a high concentration of Hm was observed in those fractions containing synaptic vesicles. The presence of Hm in small nerve endings and synaptic vesicles of the brain cortex suggests a possible synaptic role for this biogenic amine. Although little is known about the function of histamine in the CNS, the findings that it is a potent stimulator of neuronal activity when injected either intracarotidally[18] or into the brain[19] suggest that it may be a neurohormone like certain other biogenic amines.

Recently, histochemical techniques have become available which permit the visualization of histamine, indoleamines, and catecholamines at the cellular level. The method of Falck and Hillarp[20-23] for the demonstration of arylalkylamines by fluorescence microscopy has proved to be a valuable tool in studying the localization and physiological significance of these com-pounds in the brain. With this technique central neurons containing DA, NA, and 5-HT have been demonstrated in various parts of the lower brain stem.[24-27] The histochemical method developed for Hm[28-30] still lacks specificity and there are as yet no published studies on the CNS. Conse-quently, it is not yet known to what tissues, vascular, neuronal or others, the occurrence and function of brain Hm is primarily related.

II. AMINE BIOSYNTHESIS

Amines result from decarboxylation of L-α-amino acids. All amino acid decarboxylases seem to be stereospecific and to attack only the L-form of the substrate. Decarboxylation of histidine (Hd) is the sole and hence the rate-limiting step in the formation of Hm. The formation of CA and 5-HT, however, is preceded by hydroxylation of tyrosine to 3,4-dihydroxyphenylalanine (DOPA) and of tryptophan to 5-hydroxytryptophan (5-HTP); the hydroxylation appears to be the rate-limiting step. The stepwise conversion of tyrosine to CA was established by Goodall and Kirshner.[31] Their concept has been amply verified in subsequent studies using radioisotopes of tyrosine and of the various intermediary products.[32-34]

The participation of DOPA decarboxylase in the biosynthesis of CA was postulated by Blaschko[35] in 1939, and since then its role and that of DOPA as an intermediate in CA biosynthesis has been unquestioned. Since DOPA decarboxylase (E.C. 4.1.1.26), which probably also decarboxylates 5-HTP,[36-38] is available in high concentration in most tissues, the initial hydroxylation of tyrosine and tryptophan is claimed to be the rate-limiting step in the biosynthesis of both DA and 5-HT. Consequently, in most tissues the rate of biosynthesis of CA and 5-HT is determined by the rate of hydroxylation rather than by the rate of decarboxylation.

It is very doubtful whether either DOPA or 5-HTP is normally circulating in the blood. Available data suggest that the formation of these amino acids and their conversion into amines takes place in the same cells. Thus, the hydroxylating step is the critical step determining not only the rate of amine formation but also the type of amine which is formed.

The rate of Hm formation, on the other hand, depends entirely upon the Hd decarboxylase activity. The marked adaptability of this enzyme[39,40] is probably of great physiological importance.

Mammalian Hd decarboxylase (E.C. 4.1.1.22) appears to be specific and so far no other substrates have been reported for this enzyme. It is generally accepted that DOPA decarboxylase is nonspecific and capable of decarboxylating a number of aromatic and heterocyclic amino acids, including Hd. Thus, it is sometimes referred to as the "aromatic" amino acid decarboxylase.[38]

III. GENERAL MECHANISM OF AMINO ACID DECARBOXYLATION

With very few exceptions[41] pyridoxal-5'-phosphate (PLP) is the coenzyme for all amino acid decarboxylases of bacterial, as well as mammalian origin.[42-46] Mammalian Hd decarboxylase can be easily resolved into inactive apoenzyme and free coenzyme, and full activity can be restored by recombining the coenzyme and the apoenzyme.[47]

DOPA decarboxylase, however, has proved difficult to resolve completely. The coenzyme seems to be very tightly bound to the apoenzyme[48] and hence it is very difficult to prove that a coenzyme is indispensable for the reaction. It has been shown, however, that addition of PLP to highly purified or semipurified preparations of DOPA decarboxylase causes an increase in enzyme activity. Furthermore, purified rat liver DOPA decarboxylase has an absorption spectrum similar to that of other PLP enzymes.[45,49] These observations provide some evidence that PLP is the coenzyme, also, for DOPA decarboxylase.

PLP acts as a coenzyme in a variety of enzyme reactions involving α-amino acids. It is generally accepted that in PLP enzymes the coenzyme constitutes the catalytic nucleus of the active site and that the surrounding protein structure provides the steric arrangement which accounts for the specificity of the reaction.

A general scheme for PLP-catalyzed reactions was proposed by Braunstein and Shemyakin[50] and by Metzler, Ikawa, and Snell[51] based on their studies of the nonenzymatic interaction of pyridoxal with amino acids. When solutions of pyridoxal or PLP and amino acids are mixed, an immediate yellowing occurs as a result of an imine (Schiff base) formation between the two reactants. The reaction causes a pronounced increase in absorption spectrum in the region from 345 to 430 mμ, the exact position of the maximum, depending upon the pH and the amino acid used. The Schiff base is a highly reactive product. The subsequent transformations of the amino acid that occur are dependent upon the reaction conditions and could involve deamination, racemization, decarboxylation, $\alpha\beta$-eliminations, or cleavage. Addition of appropriate metal ions increases the rate of deamination and $\alpha\beta$-elimination reactions, but inhibits decarboxylation. This may be due partially to the increased rate of other pyridoxal-catalyzed reactions that are promoted by metal ions. Alternatively, it may result from formation of chelates in which the binding of the carboxyl group to metal ion slows its release as carbon dioxide.

General mechanisms have been proposed to explain the above results. All available evidence suggests that the nonenzymatic reactions proceed by mechanisms closely resembling those of the corresponding enzymatic reactions, and, therefore, that PLP provides the active center of each of the many pyridoxal-dependent enzymes. Snell[52] suggested that in enzymatic reactions the apoenzyme plays a role similar to that of the metal ion in nonenzymatic reactions but more effectively.

The original model suggested that the amino acid substrate reacted with the free carbonyl group of enzyme-bound PLP, forming a Schiff base which was subsequently degraded or modified in a specific manner, depending on the stereochemical configuration of the active site. The observations of Bonavita and Scardi[53,54] introduced a slightly different concept, suggesting that PLP was attached to the apoenzyme via an aldimine bond and that the reaction with the substrate involved an initial transaldimination in which the amino acid substrate formed a Schiff base with the enzyme-bound PLP, displacing the apoenzyme amino nitrogen. There are reasons to believe that in some cases the aldimine binding of the coenzyme and the apoenzyme is with the ε-NH$_2$ group of a lysine residue.[55,56] Snell[52] pointed out that the binding of the carbonyl group does not exclude an intermediary azomethine formation between enzyme-bound PLP and substrate: "Such binding may be partially responsible for substrate specificity, since only certain amino compounds would compete successfully with groupings from the protein for the carbonyl group of the coenzyme." Although the formyl and phosphate groups have been emphasized as the primary site of binding, other functional groups of the pyridine ring may be considered. There is no evidence as yet that the phenolic hydroxyl group and the ring nitrogen are directly involved in coenzyme binding. It has been suggested, however, that they are active in facilitating bond cleavage in the substrate.

In the interaction between pyridoxal enzymes and amino acid substrates, the first step is considered to be the transaldimination reaction which involves the enzyme-bound Schiff base of the coenzyme and the amino acid substrate. It has been suggested that the reactants may be particular ionic species of the amino acid and of the enzyme-bound PLP Schiff base.[57] From considerations of nonenzymatic reactions it may be assumed that the free amine form of the amino acid reacts with the protonated form of the enzyme-bound Schiff base,[58] displacing the apoenzyme amino nitrogen through transaldimination. The next step in the reaction scheme is considered to be the weakening of one of the bonds. The system of conjugated double bonds present in the aldimine enables the electrophilic nitrogen atom of the pyridine ring to withdraw electrons from the nitrogen atom which was formerly the α-nitrogen of the amino acid. The bonds (a), (b), and (c) are thus weakened (Fig. 1); cleavage of these bonds then results respectively in the release of a proton, or of carbon dioxide, or of a carbonium ion. Subsequent rearrangements would lead to racemization, deamination, cleavage, or decarboxylation of the amino acid. In the case of decarboxylation, the Schiff base loses

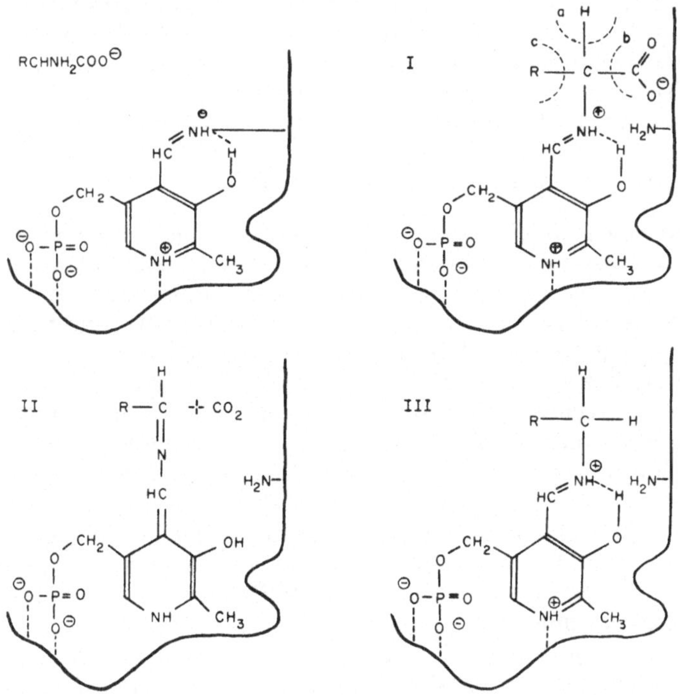

Fig. 1

carbon dioxide to form an intermediate structure (II). Addition of a proton leads to structure (III) which is the Schiff base of the amine produced by decarboxylation of the amino acid. Transaldimination of this Schiff base gives the free amine with regeneration of the holoenzyme (the imine of PLP and apoenzyme).

IV. METHODS FOR THE ASSAY OF AMINO ACID DECARBOXYLASES

Amino acid decarboxylase activity can be estimated in two different ways: (1) estimation of the carbon dioxide produced during the reaction with the amino acid substrate, or (2) estimation of the amine produced during the reaction.

The classical method for the study of amino acid decarboxylases is by manometry. Determination of the amine product became possible when chromatographic, biological, spectrophotometric, and fluorometric methods were introduced.[8,59–62] Recently, the growing interest in the formation and function of biogenic amines has stimulated the development of simple, rapid, and sensitive radiometric methods for measuring very low activities of amino acid decarboxylases.[63–71] The enzyme activity can be estimated by determining the amount of $^{14}CO_2$ released from carboxyl-labeled substrates or by determining the amount of radiolabeled amine produced. During the last decade, such techniques have become available in great variety. Most of them are superior to conventional spectrophotometric and fluorometric techniques in sensitivity, sometimes also in simplicity and versatility.

A. Aromatic L-Amino Acid Decarboxylase

This enzyme is widely distributed in the mammalian organism. Classical manometric techniques are often satisfactory for its assay. In studies of its precise distribution, e.g., in discrete areas of the brain, more sensitive and accurate methods have to be employed. Both fluorometric and radiometric techniques have been successfully applied to such studies.[72,69]

All fluorometric techniques for quantitative estimation of the amine utilize the fluorescence of the native amine, or that of some conjugation or condensation product. For this reason the amount of amine produced can be measured fluorometrically only after adequate separation from the amino acid substrate, which usually interferes with the assay. Such separation can be achieved through ion-exchange chromatography or through selective extraction of the amine with organic solvents. In the method of Bertler and Rosengren[73,37] the enzyme is incubated with L-DOPA or L-5-HTP for a specified period of time at 37°C under nitrogen. The amines are retained on the cation-exchange column, whereas the amino acid substrates pass through. The amines are eluted with hydrochloric acid and quantitated by fluorometry.

Techniques have also been described for anion-exchange chromatography in which the amino acid substrates are retained on the column and the amine products pass into the effluent, for immediate fluorometry.[38]

Extraction with organic solvents has been described for the separation of 5-HT from 5-HTP.[74,8] In this method, the incubate is deproteinized and the clear extract, adjusted to pH 10, is extracted with a mixture of n-butanol and chloroform,[8] or with heptanol.[74] By this procedure 5-HT is almost quantitatively transferred to the organic phase while most of 5-HTP remains in the aqueous phase. The undesirable contamination with 5-HTP can be further minimized by repeated washings with buffer of pH 10. In a final extraction of the butanol–chloroform mixture with dilute hydrochloric acid, 5-HT is transferred into the aqueous phase and subsequently assayed by fluorometry. Procedures of this nature cannot be used to study the decarboxylation of DOPA, for instance, since the CA are labile in an alkaline environment. In studies on tissue homogenates or crude tissue extracts, the presence of large amounts of preformed 5-HT or DA in the tissue makes accurate fluorometric estimation of newly formed amines difficult, especially in tissues with low decarboxylase activity. In this respect, radiometric techniques are superior.

All available radiometric techniques, based on the determination of the amine product, employ modifications of the methods described above for separation of the amine from its precursor.[65,67,69,70] Both ion-exchange chromatography and solvent extraction procedures have been successfully adapted to the special requirements of radiometry. As a rule, radiometric techniques make even greater demands on accurate separation of product from precursor than do the fluorometric techniques and the separation procedures have had to be modified accordingly.

Although simple and sensitive, the measurement of carbon dioxide has not been widely used in the radiometric assay of aromatic amino acid decarboxylase. The reason for this is unclear since the method seems to be well suited for microestimation of this enzyme in very small tissue slices or in minute samples of tissue.[75]

B. L-Histidine Decarboxylase

Hm formation can be determined spectrophotofluorometrically or radiometrically. Also, a method should be mentioned[76,77] in which the Hm content of tissue extracts is determined by bioassay on guinea pig ileum after incubation in the presence and in the absence of Hd. The difference in Hm content at the end of incubation is taken to represent the Hm formed. Fluorometric techniques were developed as a result of the observation of Shore et al.[61] that Hm condenses with o-phthaldialdehyde (OPT) to form a highly fluorescent conjugate. Since Hd forms a similar fluorophor, adequate separation of Hm from its precursor is essential. Selective extraction of Hm can be obtained through ion-exchange chromatography[38,78] or through extraction with organic solvents.[61,79] The most widely used method is one in which the incubation mixture is deproteinized with acid, made alkaline with sodium hydroxide, and extracted with a mixture of n-butanol and chloroform.[78,79] After extraction, Hm is in the organic phase but Hd remains in the aqueous phase. After one washing with dilute sodium hydroxide, the organic solvent mixture is extracted with dilute hydrochloric acid, by which procedure Hm is transferred into the aqueous phase for subsequent fluorometric assay after condensation with OPT. Usually this method is not sufficiently sensitive to detect the small amounts of Hm produced by mammalian Hd decarboxylase in most tissues. The sensitivity of the fluorometric method can be greatly increased if the Hm produced is purified by ion-exchange chromatography and concentrated.[78]

It is still necessary to exercise some caution in the interpretation of data from experiments on tissue homogenates or crude tissue extracts. In most tissues the Hm content is fairly high and the Hd decarboxylase activity comparatively low. This makes it extremely difficult to assess accurately the amount of newly formed Hm against the high tissue blank. An additional difficulty is the nonspecificity of the reaction with OPT.[80,81] A large variety of compounds are known to react with OPT to form highly fluorescent condensation products.[61,80–87] It is thus sometimes difficult to establish that the newly formed OPT-reactive material is in fact Hm. Hence the fluorometric assay for Hd decarboxylase is well suited for studies on highly active, purified preparations of the enzyme, but it is not satisfactory for studies on crude extracts with a high Hm content. For this purpose radiometric techniques are to be preferred and several methods have been described. Schayer[63] introduced a technique in which the Hd is labeled with ^{14}C in the 2-position of the imidazole ring and the resulting radioactive Hm is diluted with nonradioactive carrier and converted to the picrate salt, recrystallized to constant activity, and counted at infinite thickness in a gas-flow counter. An alternative method is to prepare the di-p-iodobenzenesulfonyl derivative (pipsyl Hm) or the dibenzenesulfonyl derivative (BSH).[88] In the latter method the radioactivity can be measured by the liquid scintillation technique, which is more efficient than assay in the gas-flow counter.

The simplest and most sensitive method for measuring Hd decarboxylase activity is that in which Hd labeled with ^{14}C in the carboxyl group is used as substrate. The $^{14}CO_2$ released is trapped in a suitable absorbent medium and its radioactivity is determined directly by liquid scintillation. Several modifications[68,89–91] of the original procedure of Kobayashi[66] have been described. Methods also have been described in which the radioactive Hm formed from ring-labeled Hd is separated as such from the substrate by paper chromatography before counting.[70,92]

C. Critical Evaluation

Spectrophotofluorometric methods for the study of amino acid decarboxylases are of limited usefulness. They are quite satisfactory with purified preparations from tissues rich in the enzyme but not well suited for studies on crude tissue extracts which are (1) rich in the amine and (2) low in enzyme content. There are a number of instances where these methods have failed to detect Hd decarboxylase activity readily measured by radiometric methods.[93] Radiometric techniques are superior since (1) the tissue content of endogenous nonradioactive amine is of no consequence and (2) these methods have a very high sensitivity limited only by the specific radioactivity of the commercially available substrate. Radiometric techniques in which the labeled amine product is measured usually involve laborious separation procedures. Although these techniques have a high degree of sensitivity, specificity, and precision, they all have the disadvantage of being tedious. The $^{14}CO_2$ techniques have a similar, or even higher, sensitivity and offer the advantage of being simple and versatile. These methods have been criticized on the grounds that they may be nonspecific, i.e., $^{14}CO_2$ may be released through breakdown of the amino acid along other pathways than the one expected. This may indeed occur and it is therefore advisable to verify results obtained with these methods by parallel use of some other technique. It should be kept in mind, however, that whereas the recorded enzyme activity according to the CO_2 method may sometimes be too high, it is difficult to envisage a situation in which the enzyme activity registered is too low (unless carbon dioxide trapping systems exist in the extract, which can be easily tested). This is of importance in regard to Hd decarboxylase, since its existence in many tissues has been disputed. A high Hd decarboxylase activity recorded with this technique may have to be verified by some other method but there is no reason to doubt that the inability to register Hd decarboxylase activity with the $^{14}CO_2$ method reflects the absence of the enzyme.

V. GENERAL PROPERTIES OF DOPA DECARBOXYLASE AND HISTIDINE DECARBOXYLASE

Following the discovery of DOPA decarboxylase,[94] which converts DOPA to DA, a close similarity was noted in the distribution of Hd decarboxylase and DOPA decarboxylase activities in mammalian tissues. The possibility was considered that these two activities might reside in a single enzyme.[95] Later, an enzyme catalyzing the decarboxylation of L-5-HTP was detected in mammalian tissues[96] and it was found that the ratio of DOPA decarboxylase to 5-HTP decarboxylase activities in various organs of the rabbit and guinea pig was roughly constant.[97] Thus it seemed possible that the Hd decarboxylase having its pH optimum in the range 8.0–9.5 might also be capable of decarboxylating DOPA and 5-HTP. Finally, when a purified enzyme extract of guinea pig kidney became available, the enzyme was reported to decarboxylate not only these three amino acids but also many other natural and purely synthetic aromatic amino acids.[38] For this reason, it has been suggested that the nonspecific Hd decarboxylase having its pH optimum between 8.0 and 9.5 should be designated "aromatic L-amino acid decarboxylase." It should be noted, however, that objections to this term have been raised because some preparations of DOPA/5-HTP decarboxylase have been reported to be without action on Hd and tryptophan.[98–100] This conflict remains to be resolved. Further research on DOPA decarboxylase should reveal the exact limits of its specificity and establish the precise number of aromatic amino acid decarboxylating isoenzymes in the mammalian organism.

Apart from the similar distribution of the enzyme activities discussed above, further evidence that various aromatic L-amino acids are all decarboxylated by a single enzyme is based on: (1) the failure to dissociate the activities during progressive purification of the enzyme; (2) the demonstration of competitive substrate inhibition; (3) the fact that all the decarboxylations are inhibited by the same inhibitors; and (4) the fact that changes in the ability of a tissue to decarboxylate one substrate causes parallel changes in the ability to decarboxylate the other substrates. Most of this evidence has been obtained by the use of Hd, DOPA, and 5-HTP as substrates. It is important to recognize that another, more specific Hd decarboxylase occurs in some mammalian tissues. This enzyme has very different properties from those of the nonspecific Hd decarboxylase, which has its pH optimum in the range 8.0–9.5 (cf. p. 177).

By fractionating an extract of rabbit kidney cortex on a diethylaminoethyl cellulose column, Rosengren[37] found that the chromatographic distributions of DOPA and 5-HTP decarboxylase activities were very similar. The ratio of the two activities throughout the peak area was almost constant and equal to that observed in crude extracts. The peak also contained all the Hd decarboxylase activity of the extract. Similar results have been reported for the fractionation of guinea pig kidney extracts.[38] Werle and Aures,[98] on the other hand, obtained from this tissue a highly purified preparation which decarboxylated DOPA and 5-HTP but not Hd.

The decarboxylation of DOPA by a rabbit kidney extract was found to be competitively inhibited by 5-HTP, and the decarboxylation of 5-HTP by the same extract was competitively inhibited by DOPA. In agreement with these findings, Rosengren also observed that when o-tyrosine or caffeic acid were tested as inhibitors of the rabbit kidney preparation, the inhibition constants were the same irrespective of whether DOPA or 5-HTP were used as the substrate for the decarboxylase. On the basis of similar inhibition studies with DOPA and 5-HTP, Fellman[101] also concluded that both substrates are decarboxylated by the same enzyme in extracts of ox adrenal glands.

That not only DOPA decarboxylase and 5-HTP decarboxylase, but also Hd decarboxylase (having its optimum activity in the range pH 8.0–9.5) are the same enzyme is also suggested by inhibition studies. α-MethylDOPA, for example, is a potent inhibitor of all three enzyme activities[102–104] and other α-methylated aromatic amino acids behave similarly.[36]

Thus, there is much evidence to suggest that the Hd decarboxylase having its maximum activity in the pH range 8.0–9.5 is a single enzyme which can decarboxylate not only L-Hd, but also L-DOPA and L-5-HTP.* Enzyme preparations which decarboxylate one or more of these three compounds have been found also to decarboxylate other natural and synthetic amino acids,[105,106] thus providing support for the existence of a more general aromatic amino acid decarboxylase. This enzyme is sometimes also referred to as the nonspecific Hd decarboxylase.

Two different Hm-forming enzymes or groups of enzymes have been demonstrated in mammalian tissues.[107,108] As pointed out above, one is possibly identical with the nonspecific aromatic amino acid decarboxylase, also referred to as DOPA decarboxylase. Another enzyme appears to be specific and is probably of greater physiological significance as a catalyst of Hm formation. This enzyme is generally referred to as mammalian Hd decarboxylase or "specific" Hd decarboxylase (E.C. 4.1.1.22). The "specific" Hm-forming enzyme has been found in a variety of tissues.[75,78,109–115] Fetal rat tissue, hamster placenta, and murine mastocytoma are rich sources of this enzyme. Considerable activity has been demonstrated also in gastric mucosa,[75] in the red bone marrow[109,116] and in various transplantable tumors of the rat.[110] In many tissues the enzyme appears to be adaptive[39,40] and consequently the enzyme activity is subject to considerable variation. The Hm-forming enzyme in the brain has not yet been identified.

VI. METHODS FOR THE PURIFICATION OF DOPA DECARBOXYLASE AND HD DECARBOXYLASE

Most methods for the purification of the decarboxylases were reviewed in detail by Schievelbein and Werle[117] and by Schayer;[71] only the main steps used for purification will be described here.

A. L-DOPA Decarboxylase

1. The Method of Fellman[101]

Fresh homogenate of beef adrenals is extracted at pH 6.8, the supernatant is fractionated with ammonium sulfate and the enzyme-containing fraction is treated with calcium phosphate gel to absorb impurities. Glutathione is added and the pH lowered to 4.9. The resulting precipitate is removed, the supernatant is rapidly neutralized, and the enzyme precipitated with ammonium sulfate. The enzyme is rather unstable but can be stored at 4°C for a few days with little loss of activity.

2. The Method of Rosengren[37]

Rabbit kidney cortex is homogenized in phosphate buffer pH 7.5. Proteins precipitated between 35 and 47% ammonium sulfate saturation contain the DOPA/5-HTP-decarboxylase activity. After dialysis, chromatography

* It should be noted that the pH optimum of DA formation, catalyzed by DOPA decarboxylase, is about 6.5.

on DEAE-cellulose, using a phosphate buffer concentration gradient, results in identical peaks for the DOPA, 5-HTP, and Hd decarboxylase activities.

3. The Method of Werle and Aures[98]

Fresh guinea pig kidney extract is exposed to 50–51° C at pH 5.7 for 5 min. The precipitated bulk is spun down and to remove impurities the supernatant containing the enzyme activity is treated with Kieselguhr (Celite) in the presence of 1 M glycerol. The bulk of the enzyme activity is precipitated at 50% ammonium sulfate saturation. Agitation in a 35% saturated ammonium sulfate solution at pH 4.9 removes inactive proteins and in a 26% saturated ammonium sulfate solution at pH 5.6, dissolves the enzyme. A further purification is achieved by column chromatography on anionotropic aluminum oxide. DOPA and 5-HTP decarboxylase activities are found to coincide during the purification, but the Hd decarboxylating activity is removed.

B. L-Histidine Decarboxylase

1. The Method of Håkanson[78]

This method was originally developed for purification of Hd decarboxylase from fetal rat tissues and hamster placenta[115] but it has been found useful also for other tissues.[113,114] As a rule tissues can be stored at −30° C for several weeks without loss of enzyme activity; this permits the collection and pooling of large amounts of material. The tissue is homogenized in 2 vol of 0.1 N sodium acetate–acetic acid buffer, pH 4.5–5.5, and centrifuged. Heating for 5 min at 55°C removes impurities and the enzyme is then precipitated with ammonium sulfate. The procedure results in a 200-fold purification of Hd decarboxylase in a homogenate of fetal rat tissues. More than 50% of the enzyme activity of the initial homogenate is preserved in the final extract. In frozen condition the enzyme is stable for only a few days. The purification procedure causes an almost complete resolution of the holoenzyme, and PLP must be added to the incubation mixture to restore the enzyme activity.

2. The Method of Aures[118,71]

This method involves the extraction of a lyophilized murine mastocytoma followed by a heat step for 5 min at 55°C to remove impurities. The bulk of the decarboxylase activity is collected in an ammonium sulfate fraction between 25 and 50% saturation. The precipitate is redissolved and further fractionated by repeated stepwise pH fractionation. The most active material is usually obtained in the protein fraction precipitated at between pH 5.1 and 4.7. Addition of the coenzyme, PLP, is indispensable for the restoration of full enzyme activity.

The final fraction has a greatly increased specific activity for Hd, for which the decarboxylating rate is much higher than for DOPA. The purification increase for the Hd decarboxylating activity is approximately 1800-fold, compared with the initial homogenate. This highly active fraction is stable for a few days only, but about 10% of its activity remains for several weeks. This procedure was successfully adapted for the purification of Hd decarboxylase from various tissues such as fetal rat, fetal rat liver, rat stomach mucosa, and rat bone marrow.[108,71]

3. General Comments

The process of purification may be directed toward (1) high purity in absolute terms or (2) high purity in relative terms, i.e., successful separation from contaminating Hm-forming isoenzymes. These objectives are not always identical. The purpose of the study will be decisive in the selection of the enzyme source. Hamster placenta, which is low in DOPA decarboxylase, is the superior source of enzyme for extensive purification and detailed characterization of Hd decarboxylase. Murine mastocytoma, which is rich in both Hd decarboxylase and DOPA decarboxylase, is an excellent source of enzyme for studies on the separation of the two isoenzymes.

As yet, column chromatographic procedures have had only limited use in the purification of mammalian Hd decarboxylase.[119-121] The major problem is the extreme lability of the enzyme in its highly purified state.

VII. KINETIC PROPERTIES

Several Hm-forming enzymes, or groups of enzymes with different properties, may occur in mammalian tissues. Until precise chromatographic procedures are available for the separation and identification of the various enzyme species, their kinetic characteristics will remain the most important criterion in distinguishing between them. It is not enough to demonstrate that a particular Hm-forming enzyme is different from DOPA decarboxylase by means of inhibitors or by its being unaffected by the addition of benzene. At present, the only reliable method for positive identification of mammalian Hd decarboxylase is through kinetic analysis of partially purified enzymes. Available data indicate that mammalian Hd decarboxylase obtained from fetal rat tissues, hamster placenta, murine mastocytoma, and rat stomach mucosa has the following characteristics[78,111-115,75]: (1) with very high substrate concentration, maximal reaction velocity is observed at pH 5.5–6.0; (2) the pH optimum of the reaction is inversely related to the substrate concentration: at very low Hd concentration the optimal pH is close to 7.0; and (3) the apparent Michaelis–Menten constant decreases markedly with increasing pH. In contrast, the nonspecific Hm-forming enzyme (possibly identical with DOPA decarboxylase) has its pH optimum at about 9 and requires very high Hd concentrations for saturation at this pH. These differences have some important practical consequences. It is possible, and advisable, to select incubation conditions that minimize the interference of DOPA decarboxylase with the Hm-forming capacity; for instance, by choosing a pH 6.0–6.9 range and by restricting the Hd concentration to that which is saturating or near saturating at that particular pH value.

VIII. DISTRIBUTION AND PROPERTIES OF HISTIDINE DECARBOXYLASE AND DOPA DECARBOXYLASE IN NERVOUS TISSUE

Many reports have presented data on the formation of Hm and aryl-alkylamines in various regions of the brain and in peripheral nerves. But little has been done to purify and characterize the enzyme responsible for the amine-forming capacity of these tissues. With few exceptions all information available on the purification procedures and on the properties of Hd de-carboxylase and DOPA decarboxylase are obtained from tissue sources other than the brain and it cannot be excluded that the neuronal enzymes are different from those of nonneuronal tissue. There is little doubt, however, that in nervous tissue as well as in other tissues DOPA and 5-HTP are decarboxy-lated by the same enzyme.

A. L-Histidine Decarboxylase

High Hd decarboxylase activity has been observed in splenic nerves and sympathetic ganglia, less activity in the phrenic nerve and vagus, and low activity in the spinal cord, brain stem, and cerebral cortex.[122] The distribu-tion of DOPA decarboxylase was similar. Formation of Hm from Hd by brain tissue has been demonstrated *in vitro* in cattle, rat, cat, pig, man, and dog.[63,122–124] Injected Hm penetrates slowly, or not at all, from the blood into the brain.[125] This indicates that the Hm present in the brain is formed there. Formation of Hm in the brain has also been demonstrated *in vivo* in cats after intraventricular administration of radioactive Hd.[17] In cat, pig, dog, and man the Hm-forming capacity is much higher in the hypothalamus than in other parts of the brain.[124] This distribution corresponds closely to that of Hm itself.[12]

The demonstration of a Hm-forming capacity does not necessarily imply the presence of "specific" Hd decarboxylase. It should be noted that the nonspecific aromatic amino acid decarboxylase has a distribution in the brain somewhat similar to that of the Hm-forming enzyme. The Hm-forming enzyme in nervous tissue remains to be characterized.

B. Aromatic L-Amino Acid Decarboxylase

DOPA decarboxylase and 5-HTP decarboxylase have been found in the brain of all animals tested. Recently, however, Robins *et al.*,[126] using both fluorometric and radiometric techniques, found very low levels of aromatic amino acid decarboxylase in human brain tissue, obtained by autopsy within 6 hr after death, or by biopsy, in contrast to other species. The ratio 5-HTP/ DOPA decarboxylase activity in various brain regions has been found to be constant, which supports the concept of a single enzyme. There is a fairly good correlation between 5-HT content and decarboxylase activity in various parts of the mammalian brain, but not in the avian brain.[127] The highest decarboxylase activity in mammalian brain is found in the caudate nucleus;

fairly high values have been recorded in the medulla, the thalamus and superior colliculus.[128] The enzyme activity of the cortex is low.

The administration of DOPA or 5-HTP results in an accumulation of DA and 5-HT, respectively, in various regions of the brain. A substantial portion of these accumulated monoamines is found in the endothelial cells of the brain but not in the peripheral capillaries, which have been shown to be rich in aromatic amino acid decarboxylase.[129–131] It has been suggested that these endothelial cells represent a functionally important storage site for this enzyme in the brain, and also act as a blood–brain barrier to the substrate amino acids. It has been shown[132,133] that capillary DOPA decarboxylase can be selectively inhibited, leading to an increased penetration of the amino acid into the brain and elevated levels of brain DA. It is important to realize that not all aromatic amino acid decarboxylase in the brain has a neuronal localization. Possibly brain Hd decarboxylase has a similar dual distribution. Udenfriend[134] has reported that a large fraction of brain DOPA decarboxylase is recovered in the synaptosomes and has suggested that the entire biosynthetic machinery necessary for neuronal monoamine formation is contained in the amine-storing vesicles.

Sectioning of the medial forebrain bundle within the lateral hypothalamus of the rat reduces both the 5-HT content and the 5-HTP decarboxylase activity in the brain.[135] The reduction in 5-HT and 5-HTP decarboxylase extended to all parts of the telencephalon as well as to areas in which there is no histological evidence of degeneration. It was therefore proposed that there must be a transynaptic change in neuron metabolism and it was concluded that the integrity of the median forebrain bundle is necessary for the maintenance of normal enzyme activity in the telencephalon.

The pineal gland of several species has a high 5-HT content[5,136–139] and conspicuously high activities of aromatic amino acid decarboxylase,[63,140,141] most of which seems to reside in the pinealocytes.[141] The functional significance of 5-HT in this organ is unknown.

IX. INHIBITION OF DECARBOXYLASES

Metabolic blocking agents are commonly used to study the physiological role of body constituents. The agent is studied as an inhibitor in test experiments using isolated enzyme systems. Also, the mechanism of inhibition may be investigated. An inhibitor found powerful *in vitro* is then applied *in vivo* to control the formation of the biological compound. But the actions of drugs *in vivo* are more complex than simplified *in vitro* experiments can show. A drug known to act as an inhibitor of the enzymatic formation of a hormone from its direct precursor may act in many other ways in the test animal. Catabolizing as well as biosynthetic pathways of the hormone may be blocked and as a consequence little or no change in tissue stores may be observed. Or, if other pathways using the immediate precursor are blocked, thus making more substrate available, and if the supply of substrate is the rate-limiting factor in the formation, the enzyme reaction may supply sufficient hormone in spite of inhibition. Since nature works with many "safety valves," an enzyme can be inhibited to a large extent, but still produce hormone levels adequate for normal function. Also, permeability changes as a direct or indirect effect of drug administration (as well as effects on other natural regulators) increase the difficulty of explaining *in vivo* drug experiments. Ideally all variants in the formation and metabolism of an endogenous compound of interest should be carefully considered to draw meaningful conclusions. Considerations for exercising caution in translating data obtained with purified enzymes *in vitro* into corresponding metabolic terms were expressed by Srere[142] quite generally for enzyme regulatory studies. More recent calculations of localized enzyme concentrations in tissue resulted in values which were much higher than the usual assay concentrations, and were considered to be more truly physiological. Also, at high enzyme concentrations regulatory behavior may appear or disappear in ways that escape observation in *in vitro* assay concentrations. Nevertheless, simplified measurements can shed light on a problem and as research progresses, they may prove to be relevant to understanding and learning to control the mechanism. Thus, the development of monoamine oxidase inhibitors, which interfere with the oxidative deamination of monoamines, e.g. CA, 5-HT, or 1,4-methylhistamine, led to the development of therapeutic drugs for mental disorders. Excellent reviews have been written of investigations on inhibition of the decarboxylases.[143-148] Here, only a few compounds will be mentioned, selected because of their importance as experimental tools in studies on the formation and function of biogenic amines.

The inhibitors of the amino acid decarboxylases can be roughly classified into compounds acting on the coenzyme PLP and compounds acting on the apodecarboxylase. Hundreds of compounds have been tested as inhibitors for DOPA decarboxylase, and some as inhibitors of the specific Hd decarboxylase. Only a few were investigated as to their mechanism of action. Possibly the most potent compounds inhibit in a mixed fashion, by interference with both the apoenzyme and the coenzyme. A comparison of inhibitor potency reported by different authors can only give a crude lead, because the many factors which influence enzyme activity vary under different conditions of enzyme assay. Examples: the pH and ionic strength of the incubate influence; the ionization constant of enzyme protein, coenzyme, substrate and inhibitor; the concentration of substrate, type of substrate, and concentration of coenzyme are obviously important, if the inhibitor is of the competitive type; and the source of the enzyme and its purity will play a role, since substances introduced by the tissue can modify the reaction. Also, the timing and sequence of addition of the reactants to the incubation mixture, as well as the incubation time, are factors strongly modifying the degree and the type of inhibition.

Several recent reports have indicated a variety of pharmacological effects ascribed to the inhibition of the specific Hd decarboxylase, following administration of 4-bromo-3-hydroxybenzyloxyamine (NSD-1055). After its injection into rats, the Hm content of the stomach, heart, and urine was markedly lowered.[149] Gastric secretion in the rat, which is believed to be regulated by Hm,[150] was drastically lowered after injection of NSD-1055.[151] Human subjects suffering from mastocytosis, a disease involving abnormal proliferation of mast cells and excessive Hm production, have had their symptoms relieved by oral administration of NSD-1055.[152] An anti-inflammatory effect of NSD-1055 was reported in a histological study of experimentally induced granulomas and contact dermatitis caused by 2,4-dinitrobenzene.[153] These effects had been attributed to the ability of the NSD-1055 to inhibit specific Hd decarboxylase.[152]

NSD-1055 is also a strong inhibitor of DOPA decarboxylase, glutamic acid decarboxylase, and other PLP enzymes. Leinweber[154] investigated the mechanism of the inhibition of specific Hd decarboxylase by a group of O-substituted hydroxylamines. Also, he reacted these compounds with PLP and tested the resulting oximines (abbreviated [Oxyamine]·[PLP]) as to their action on Hd decarboxylase. The [NSD-1055]·[PLP] was found to be competitive with the PLP and noncompetitive with the Hd, whereas the NSD-1055 itself was considered competitive with the substrate Hd. These data were obtained by mixing all the reactants and initiating the decarboxylation reaction by adding the enzyme. Various degrees of inhibition were obtained, when the Hd decarboxylase apoenzyme was reacted with the PLP and/or the NSD-1055 prior to the decarboxylation reaction. We found the inhibition of Hd-, DOPA-, and glutamic acid decarboxylases by NSD-1055 to be competitive with PLP, if the apoenzyme was first reacted with the PLP.[155] However, these results varied with the time of reaction. The mechanism of inhibition is undoubtedly complex. From available data it can be concluded that the [Oxyamine]·[PLP] and also other PLP–Schiff bases (including the Schiff bases with PLP and the substrates or with the corresponding amines) can transaldiminate with the lysine-bound PLP of the holoenzyme.[47,57] Some PLP–Schiff bases with amino acids and amines undergo an irreversible transformation under physiological conditions forming tetrahydroisoquinoline type compounds, thus causing depletion of coenzyme and substrate,[43,156] and/or the tetrahydroisoquinoline itself acts as an inhibitor of the enzyme.[157] The possible reactions are given in the schema:

Enzyme + PLP $\rightleftharpoons$ [Enzyme]·[PLP]

Substrate + PLP $\rightleftharpoons$ [Substrate]·[PLP] → irreversible transformation

Inhibitor* + PLP $\rightleftharpoons$ [Inhibitor]·[PLP] → irreversible transformation

[Enzyme]·[PLP] + Substrate $\rightleftharpoons$ [Enzyme]·[PLP]·[Substrate]

[Enzyme]·[PLP] + Inhibitor $\rightleftharpoons$ [Enzyme]·[PLP]·[Inhibitor]

[Enzyme]·[PLP]·[Substrate] + [Inhibitor]·[PLP] $\rightleftharpoons$
                                               [Enzyme]·[PLP]·[Inhibitor] + Substrate

* Inhibitor can be any amino, hydrazino, or oxyamino compound capable of forming a Schiff base with PLP.

Thus, considering all the possible reactions, the degree of inhibition by compounds which can react with the carbonyl group of PLP depends upon the equilibrium constants of all these possible reactions, the concentration of the reactants, the sequence of the addition of the reactants, and on the time of reaction.

As Snell pointed out (cf. p. 168), the binding of PLP to the ε-amino group of the lysine in the apoenzyme may be partially responsible for substrate specificity. An inhibitor specificity may be postulated on a similar principle. Schott and Clark[156] noted that the formation of tetrahydroisoquinoline structures was favored if the aromatic amino acids had a hydroxyl group in m-position to the side chain.

The fact that the specific Hd decarboxylase, but not the DOPA decarboxylase, is easily resolved into apoenzyme and coenzyme may be responsible for an inhibitor specificity with amine derivatives. Also, specific substrate-enzyme binding sites are involved in the binding of inhibitors. The decarboxylation of either DOPA, 5-HTP, or Hd by the general amino acid decarboxylase is more strongly inhibited by compounds structurally related to DOPA than by compounds structurally similar to Hd. Compounds similar to Hd inhibit the specific Hd decarboxylase more readily. Inhibitors related to either Hd or DOPA have been used to assess the nature of the Hd decarboxylating enzymes. Mackay and Shepherd[104] tested the inhibitory potency of various compounds on the decarboxylation of Hd by guinea pig kidney enzyme, which represents the general decarboxylase, and by rat hepatoma, which more closely represents the specific Hd decarboxylase. Substances which demonstrate the most striking differentiation in inhibitor action are presented in Table I. A group of O-substituted hydroxylamines, which represent the most

TABLE I

Inhibition of Histidine Decarboxylases

Compound	$C_{50} \times 10^{4a}$	
	Guinea pig	Rat hepatoma
DL-α-MethylDOPA	0.01	100
L-DOPA	0.2	9
Catechol	1.8	65
DL-5-HTP	0.65	75
DL-α-Methyl 5-HTP	0.075	1
DL-α-Methyl Hd	150	15
SNR-1610 (Imidazole-4(5)ylmethoxyamine)	0.065	0.0021
NSD-1024 (3-Hydroxybenzyloxyamine)	0.0052	0.006
NSD-1034 (N'-Methyl-'N'-(3-hydroxybenzyl) hydrazine	0.0025	0.0098
NSD-1055 (3-Hydroxy-4-bromobenzyloxyamine)	0.0027	0.0014

^{a} C_{50} is the molar concentration, which is required to reduce the initial velocity of inhibited decarboxylation reaction by 50% under the conditions of incubation of Mackay and Shepherd.[104]

potent inhibitors for both decarboxylases, are also listed. The concentration of α-methylDOPA required for 50 % inhibition of the specific Hd decarboxylase is 10,000 times greater than the concentration required for 50 % inhibition of the general decarboxylase.

In contrast, the inhibitor concentration requirement of α-methylhistidine is higher for the general decarboxylase than for the specific Hd decarboxylase. α-MethylDOPA and other α-substituted amino acids were found to inhibit DOPA decarboxylase competitively. Although α-methylDOPA is known to react with the coenzyme of the decarboxylases, its action as an inhibitor lies primarily in its ability to compete with the aromatic amino acids for the apoenzyme. Putter and Kroneberg[158] demonstrated that D-α-methylDOPA weakly inhibits the decarboxylase *in vitro* through binding of free pyridoxalphosphate. The L-form, however, forms in addition an amino acid-coenzyme-apoenzyme-complex. It is decarboxylated to form α-methyldopamine which is β-hydroxylated *in vivo* to α-methyl-NA. The DOPA decarboxylase inhibitor α-methyl-*m*-tyrosine is also decarboxylated to the corresponding amine, which is further metabolized *in vivo* to form metaraminol.[159] Both the methyl-NA and metaraminol play major roles in investigations of nerve transmission, as "false transmitters," by displacing neuronal NA from synaptosomes. The α-hydrazino analogue of Hd (MK-785) strongly inhibits the specific Hd decarboxylase of the fetal rat and the α-hydrazino-analogue of α-methylDOPA (MK 485) (α-hydrazino-DL-α-methylDOPA) strongly inhibits the guinea pig kidney enzyme.

Hartman *et al.*[160] listed a series of substrate analogues as DOPA decarboxylase inhibitors. The most effective ones were cinnamoyl derivatives, having the structure R—CH=CH—CO—R', where R is 3-hydroxyphenyl or 3,4-dihydroxyphenyl or 5-hydroxyindole, and R' is OH or O-alkyl or aryl. These were found to be competitive with the substrate *in vitro*, and thus could make excellent tools to deplete the monoamines, but they had no effect *in vivo* on brain amine levels.

X. ACTIONS AND SIDE ACTIONS OF ANTIDECARBOXYLASES *In Vivo*

Administration of decarboxylase inhibitors to the animal affects the tissue levels of the amines and/or the urinary excretion of the metabolites. As already mentioned (cf. p. 181), NSD-1055 reduces the Hm content of various tissues *in vivo*. MK-785 depletes tissue Hm maximally within 3–6 hr in animals but the levels return to normal within 24 hr. Moreover, it is toxic in man. The α-methyl-5-HTP, although it is an efficient inhibitor *in vitro* for both decarboxylases, does not lower *in vivo* Hm levels.[161,162] The NSD-1034 (*N'*-methyl-*N'*-[3-hydroxybenzyl] hydrazine), one of the most potent decarboxylase inhibitors, was recently reported by Hempel and Männel[163] to completely block oxidative tyrosine metabolism, initiated by transamination of tyrosine to 3,5-*p*-hydroxyphenylpyruvic acid. As a result the tyrosine content was elevated in adrenals, brain stem, and venous blood. Also, MK-485 was found to block tyrosine catabolism.[164] The strong DOPA decarboxylase inhibitor RO 4-4602 (N-[DL-seryl]-N'-[2,3,4-trihydroxybenzyl]hydrazine) produced an interesting effect upon administration to rats. DOPA decarboxylase was found to be strongly inhibited, but cardiac NA was only reduced 50%.[165] Since the conversion of DOPA to DA is not the rate-limiting step in the biosynthesis of CA, a high degree of decarboxylase inhibition may hardly affect the CA stores, especially if the conditions are such that there is only a minimal requirement for the amines. After exposure to stress situations, however, the amine stores are depleted faster in the presence of decarboxylase inhibitors due to the increased usage of CA. Indeed, Johnson and Pritzker[166] showed in experiments with cold-stressed rats that inhibition of DOPA decarboxylase with RO 4-4602 did not influence the levels of CA in rats kept at 27°C, but did significantly diminish the cold-induced increase in NA secretion. Pretreatment with RO 4-4062 also prevented the characteristic behavioral abnormalities in animals and man after a loading with L-tryptophan.[167] 5-HT levels in guinea pig brain rise after L-tryptophan treatment, but not when the tryptophan load followed treatment with RO 4-4062. This favors the conclusion that the excitation seen after L-tryptophan is caused by the amine metabolite and not by the amino acid itself.

A series of α-substituted amino acids emerged from a program directed toward the discovery of DOPA decarboxylase inhibitors.[144,168,169] Some of them play a major role in modern medicine because of their action on the CNS. The in vivo inhibition of DOPA decarboxylase by α-methylDOPA (Aldomet) becomes evident when exogenous DOPA or certain other amino acids are administered. The usual accumulation of DA in the test animal under these conditions can be minimized by treatment with α-methylDOPA or α-methyl-m-tyrosine, but not methyl-3-methoxyphenylalanine. Both α-methylDOPA and α-methyltyrosine produce symptoms attributed to an effect on the CNS. α-MethylDOPA is a primary depressant, whereas α-methyl-m-tyrosine causes overt stimulant effects. Both agents were found to cause a short-term reduction of brain 5-HT and DA and a more prolonged depletion of brain NA. It was noted by early investigators[170,171] that in animals which had received α-methylDOPA, tissue NA concentrations remained depressed long after DA and 5-HT concentrations had returned to normal. Sourkes[172] championed the idea that the amines formed during metabolism of α-methyl-DOPA, principally α-methyl-NA, replace NA at binding sites in the brain and in peripheral autonomic neurons. Other displacements of the NA by metaraminol have been reported by several investigators[173–175] and the amine metabolites of α-methylDOPA have been shown to prevent uptake of NA by tissue.[176] As already mentioned, the term "false transmitters" has been coined for the amines derived from α-methyl-amino acids. The vast literature on the biochemistry and pharmacology of methylDOPA and related structures has been extensively reviewed by Stone and Porter.[177] The pharmacological action of this series of decarboxylase inhibitors is complex, and their mechanism of action is still far from understood. Experimental results must be viewed in the light of all known metabolic pathways involved. The interaction of the α-methyl-amino acids and their metabolites with the natural component by replacement contributes to their pharmacological effects, as does the inhibition of various enzymes, e.g., the ring hydroxylating enzyme, which catalyzes the conversion of phenylalanine to tyrosine, of tryptophan to 5-HTP,[178,179] and of tyrosine to DOPA.[180] The β-hydroxylase which converts DA to NA is weakly inhibited by α-methylDOPA[181] and α-methyl-m-tyramine. The methylated amines are also converted to the α-hydroxylated derivative by the same enzyme, forming the so-called "false transmitters."[182–184] α-Methylated NA is probably more tightly bound in the neuron than is NA itself, thus the former is less potent as a direct sympathomimetic[185,186] and less susceptible to destruction by monoamine oxidase. Catechol-O-methyltransferase is also competitively inhibited, the methylated derivatives being about as good substrates as DA for this enzyme, which converts DA to the pharmacologically inactive O-methylated derivative.[187] The main pathway of tyrosine degradation,[164] as well as D-amino acid oxidase, is inhibited by α-methylDOPA. Fat mobilization in response to adrenal medullary hormones is blocked by α-methylamino acid analogues.[188,189]

Some of the cinnamoyl compounds introduced as decarboxylase inhibitors[146,190] were shown to have antidecarboxylase activity in $vivo$. Moran and Sourkes[164] claimed that the sole action of 5-(3-hydroxycinnamoly)-salicylic acid is its antidecarboxylase effect in $vivo$, in contrast to other decarboxylase inhibitors. The authors measured the alveolar $^{14}CO_2$ derived from ^{14}C-amino acids injected into rats. Comparison of data from substrate labeled in the carboxyl position with data from substrate labeled elsewhere, together with a study of the effect of selective administration of inhibitors led to the above conclusion.

It remains doubtful whether the cinnamoyl derivatives interfere significantly with the endogenous synthesis of monoamines, e.g., at nerve endings, since they did not diminish the amine concentration in the nictitating membrane even following repeated electrical stimulation.[146,190] A rapid detoxication mechanism was assumed, to explain the weak effect. Although the cinnamic acid derivative did effectively counteract the rise in monoamine levels and blood pressure due to exogenous aromatic amino acids, the 5-(3,4-dihydroxycinnamoyl)-salicylic acid was found to reduce tremors and motor excitation induced by 5-HTP or DOPA, and to potentiate barbiturate hypnosis.[191,192]

All these findings call for caution in ascribing pharmacological effects observed only to "antidecarboxylase" activity of decarboxylase blocking agents. Multiple inhibitory effects are most likely, especially if the drug can interact with PLP.

XI. BEHAVIORAL SIGNIFICANCE OF AMINO ACID DECARBOXYLASES

Levels of biogenic amines have in certain cases been found to be correlated with behavior patterns. Some mental deficiencies have been linked to abnormal balances in biogenic amines and their precursors and metabolites. Instances in which abnormal levels of decarboxylase activities have been considered as contributing to a behavioral disorder are discussed below.

In phenylketonuria, three different theories have been proposed to explain the reduced serum levels of 5-HT and the diminished excretion of 5-hydroxyindoleacetic acid. There is an inherited deficiency of the liver enzyme which hydroxylates phenylalanine to tyrosine, and tryptophan to 5-HTP.[193] The metabolic products of phenylalanine (viz., phenylpyruvic, phenyllactic, and phenylacetic acid) are abnormally high in phenylketonuria and inhibit 5-HTP decarboxylase.[194] Phenylalanine and certain other L-amino acids depress the uptake of 5-HTP by brain tissue *in vitro* and *in vivo*; thus the 5-HT synthesis is impaired by a shortage of substrate for the decarboxylase.[195]

In children with histidinemia, a disease with speech disorder as a symptom, unusually high levels of Hd in blood and urine are found. A metabolic block of the histidase appears to be responsible for this. As a consequence the large Hd pool is apparently shunted into the transamination pathway, since imidazolpyruvic acid and imidazolelactic acid are found in these patients, but not in normals.[196] Snyder *et al.*[197] speculated on the possibility that the large Hd pool is also shunted into the Hm pathway in these patients, but no evidence for or against this assumption has been shown.

Abnormality in the levels of amines and their metabolites in schizophrenia is still disputed. Some 5-HT antagonists are hallucinogens,[198] but in psychotics, tryptamine, 5-hydroxyindole acetic acid, and indole-3-acetic acid levels seem to be usually within the normal range. However during acute psychotic episodes these three compounds are excreted at a higher rate.[199]

In patients with Parkinson's syndrome, the concentration of 5-HT in certain areas of the brain is reduced.[200] Also there is a reduction of brain DA and NA and diminished urinary excretion of DA,[201] although the latter has not been confirmed. Whether this reflects slower synthesis, increased turnover, or impaired storage, is not known.

Mice, which have high brain levels of DOPA/5-HTP decarboxylase[126] showed an increase in 5-HTP decarboxylase activity in the frontal cortex during aggression as compared to "unfought controls."[202] This may shed some doubt on the concept that the decarboxylase is not the rate-limiting step in the 5-HT and/or CA biosynthesis.

Many workers have attempted to use drugs to better understand behavior. Different neurohumors have been manipulated by drugs (cf. Refs. 198, 203, 204, 205). The large variety of biochemical actions of these drugs is bewildering. In pharmacological correlations compatible with CA and 5-HT hypotheses of affective disorders, drugs which specifically act as inhibitors of the decarboxylation step in the amine biosynthesis have not yet been recognized clinically. The ideal tool, if it can be found, for correlating effect and function is the "absolutely specific inhibitor" for a particular enzyme system. Since this is more a fiction than a possibility, only the most sophisticated intradisciplinary group efforts will be able to dissect, integrate, and correlate the wealth of present and upcoming information in this field.

ACKNOWLEDGMENT

We wish to thank Dr. H. F. Schott for valuable suggestions and help in preparing this manuscript.

The manuscript was prepared while holding grant support from National Institutes of Health, U.S. Public Health Service (MY-03663), U.S. Army (DA-HC19-67-0004), and Swedish Medical Research Council (K68-12X-1007-03).

XII. REFERENCES

1. T. L. Perry, A. Hansen, J. G. Foulks, and G. M. Ling, Aliphatic and aromatic amines of cat brain, *J. Neurochem.* **12**(5):397–405 (1965).
2. T. L. Perry, S. Hansen, and L. MacDougall, Amines of human whole brain, *J. Neurochem.* **14**(7):775–782 (1967).
3. T. Nakajima, F. Wolfgram, and W. G. Clark, Identification of 1,4-methylhistamine, 1,3-diaminopropane, and 2,4-diaminobutyric acid in bovine brain, *J. Neurochem.* **14**(12):1113–1118 (1967).
4. N. Seiler and M. Wiechmann, Zum Nachweis von Aminen im 10^{-10}-Molmassstab. Trennung von 1-Dimethylaminonaphthalin-5-sulfonsäure-amiden auf Dünnschichtchromatogrammen, *Experientia* **21**(4):203–204 (1965).
5. W. B. Quay, J. Ariens-Kappers, and J. F. Jongkind, Innervation and fluorescence histochemistry of monoamines in the pineal organ of a snake, *J. Neuro-Visceral Relations* **31**:11–25 (1968).
6. A. H. Amin, T. B. B. Crawford, and J. H. Gaddum, *XIX International Physiological Congress.* Abstracts. pp. 165–166 (1953).
7. B. Twarog and J. H. Page, Serotonin content of some mammalian tissues and urine and a method for its determination, *Am. J. Physiol.* **175**(1):157–161 (1953).
8. D. F. Bogdanski, A. Pletscher, B. B. Brodie, and S. Udenfriend, The distribution and assay of serotonin in brain, *J. Pharmacol. Exp. Therap.* **117**(1):82–88 (1956).
9. M. Vogt, The concentration of sympathin in different parts of the central nervous system under normal conditions and after the administration of drugs, *J. Physiol.* **123**(3):451–481 (1954).
10. Å. Bertler and E. Rosengren, Occurrence and distribution of catecholamines in brain, *Acta Physiol. Scand.* **47**(4):350–361 (1959).
11. Å. Bertler and E. Rosengren, Occurrence and distribution of dopamine in brain and other tissues, *Experientia* **15**(1):10 (1959).
12. G. W. Harris, D. Jacobsohn, and G. Kahlson, The occurrence of histamine in cerebral regions related to the hypophysis, *Ciba Foundation Colloquia on Endocrinology*, Vol. 4, pp. 186–194, London, Churchill (1952).
13. U. S. von Euler, *in Histamine* (E. W. Wolstenholme and C. M. O'Connor, eds.), pp. 235–241, London, Churchill (1956).
14. A. Torp, Histamine and mast cells in nerves, *Med. Exp.* **4**(3):180–182 (1961).
15. R. Håkanson, Ch. Owman, and B. Sporrong, To be published.

16. K. Kataoka and E. De Robertis, Histamine in isolated small nerve endings and synaptic vesicles of rat brain cortex, *J. Pharmacol. Exp. Therap.* **156**(1):114–125 (1967).
17. T. White, Formation and catabolism of histamine in cat brain *in vivo*, *J. Physiol.* **152**:299–308 (1960).
18. P. L. McGeer, in *Comparative Neurochemistry* (D. Richter, ed.), pp. 387–391, Pergamon Press, New York (1964).
19. R. G. Heath and F. Verster De Balbian, Effects of chemical stimulation to discrete brain areas, *Am. J. Psychiat.* **117**(11):980–990 (1961).
20. B. Falck, Observations on the possibilities of the cellular localization of monoamines by a fluorescence method, *Acta Physiol. Scand.* **56**:Suppl. **197**:1–24 (1962).
21. B. Falck, N. Å. Hillarp, G. Thieme, and A. Torp, Fluorescence of catecholamines and related compounds condensed with formaldehyde, *J. Histochem. Cytochem.* **10**(3):348–354 (1962).
22. H. Corrodi and N. Å. Hillarp, Fluoreszenzmethoden zur histochemischen Sichtbarmachung von Monoaminen. I. Identifizierung der fluoreszierenden Produkte aus Modellversuchen mit 6,7-Dimethoxyisochinolinderivaten und Formaldehyd, *Helv. Chim. Acta* **46**(6):2425–2430 (1963).
23. H. Corrodi and N. Å. Hillarp, Fluoreszenzmethoden zur histochemischen Sichtbarmachung von Monoaminen. Identifizierung des fluoreszierenden Produktes aus Dopamin und Formaldehyd, *Helv. Chim. Acta* **47**(3):911–918 (1964).
24. N. E. Andén, A. Carlsson, A. Dahlström, K. Fuxe, N. Å. Hillarp, and K. Larsson, Demonstration and mapping out of nigro-neostriatal dopamine neurons, *Life Sci.* **3**(6):523–530 (1964).
25. N. E. Andén, A. Dahlström, K. Fuxe, and K. Larsson, Further evidence for the presence of nigro-neostriatal dopamine neurons in the rat, *Am. J. Anat.* **116**(1):329–334 (1965).
26. A. Dahlström and K. Fuxe, Evidence for the existence of monoamine-containing neurons in the central nervous system. I. Demonstration of monoamines in the cell bodies of brain stem neurons, *Acta Physiol. Scand.* **62**:Suppl. **232**:1–55 (1964).
27. A. Dahlström and K. Fuxe, Evidence for the existence of monoamine neurons in the central nervous system; experimentally induced changes in the intraneuronal amine levels of bulbospinal neuron systems, *Acta Physiol. Scand.* **64**:Suppl. **247**:1–35 (1964).
28. B. Ehinger and R. Thunberg, Induction of fluorescence in histamine-containing cells, *Exptl. Cell Res.* **47**(1–2):116–122 (1967).
29. R. Håkanson and Ch. Owman, Concomitant histochemical demonstration of histamine and catecholamines in enterochromaffin-like cells of gastric mucosa, *Life Sci.* **6**:759–766 (1967).
30. B. Ehinger, R. Håkanson, Ch. Owman, and B. Sporrong, Histochemical demonstration of histamine in paraffin sections by a fluorescence method, *Biochem. Pharmacol.* **17**:1997–1998 (1968).
31. M. Goodall and N. Kirshner, Biosynthesis of adrenaline and noradrenaline *in vitro*, *J. Biol. Chem.* **226**:213–221 (1957).
32. S. Udenfriend and J. B. Wyngaarden, Precursors of adrenal epinephrine and norepinephrine *in vivo*, *Biochim. Biophys. Acta* **20**(1):48–52 (1956).
33. G. Rosenfeld, L. C. Leeper, and S. Udenfriend, Biosynthesis of norepinephrine and epinephrine by the isolated perfused calf adrenal, *Arch. Biochem. Biophys.* **74**(1):252–265 (1958).
34. M. Goodall and N. Kirshner, Biosynthesis of epinephrine and norepinephrine by sympathetic nerves and ganglia, *Circulation* **17**:366–371 (1958).
35. H. Blaschko, The specific action of L-dopa decarboxylase, *J. Physiol.* **96**:50–51 (1939).
36. A. Yuwiler, E. Geller, and S. Eiduson, Studies on 5-hydroxytryptophan decarboxylase. I. *In vitro* inhibition and substrate interaction, *Arch. Biochem.* **80**:162–173 (1959).
37. E. Rosengren, Are dihydroxyphenylalanine decarboxylase and 5-hydroxytryptophan decarboxylase individual enzymes? *Acta Physiol. Scand.* **49**(4):364–369 (1960).

38. W. Lovenberg, H. Weissbach, and S. Udenfriend, Aromatic L-amino acid decarboxylase, *J. Biol. Chem.* **237**(1):89–93 (1962).
39. R. W. Schayer and O. H. Ganley, Adaptive increase in mammalian histidine decarboxylase activity in response to nonspecific stress, *Am. J. Physiol.* **197**(3):721–724 (1959).
40. G. Kahlson, E. Rosengren, D. Svahn, and R. Thunberg, Mobilization and formation of histamine in the gastric mucosa as related to acid secretion, *Am. J. Physiol.* **174**(3):400–416 (1964).
41. J. Rosenthal, B. M. Guiarard, G. W. Chang, and E. E. Snell, Purification and properties of histidine decarboxylase from lactobacillus 30a, *Proc. Nat. Acad. Sci. U.S.A.* **54**(1):152–158 (1965).
42. I. C. Gunsalus, W. D. Bellamy, and W. W. Umbreit, Phosphorylated derivative of pyridoxal as the coenzyme of tyrosine decarboxylase, *J. Biol. Chem.* **155**(2):685–686 (1944).
43. D. Heyl, E. Luz, S. A. Harris, and K. Folkers, Phosphates of the vitamin B_6 group. I. The structure of codecarboxylase, *J. Am. Chem. Soc.* **73**:3430–3433 (1951).
44. A. M. Rothschild and R. W. Schayer, Characterization of histidine decarboxylase from rat peritoneal fluid mast cells, *Biochim. Biophys. Acta* **34**:392–398 (1959).
45. J. Awapara, R. P. Sandman, and C. Hanley, Activation of dopa decarboxylase by pyridoxal phosphate, *Arch. Biochem.* **98**:520–525 (1962).
46. S. Ono and P. Hagen, Pyridoxal phosphate: a coenzyme for histidine decarboxylase, *Nature* **184**:1143–1144 (1959).
47. R. Håkanson, Mammalian histidine decarboxylase: Interaction between apoenzyme and pyridoxal-5'-phosphate, *Europ. J. Pharmacol.* **1**:(5):383–390 (1967).
48. S. Udenfriend, Amino acid decarboxylation step in the biosynthesis of norepinephrine, serotonin and histamine, *Vitamins and Hormones* **22**:445–450 (1964).
49. W. T. Jenkins, P. A. Yphantis, and J. W. Sizer, Glutamic-aspartic transaminase. I. Assay, purification and general properties, *J. Biol. Chem.* **234**:51–57 (1959).
50. A. E. Braunstein and M. M. Shemyakin, A theory of amino acid metabolic processes catalyzed by pyridoxal-dependent enzymes, *Biokhimiia* **18**:393–411 (1953).
51. D. E. Metzler, M. Ikawa, and E. E. Snell, A general mechanism for vitamin B_6-catalyzed reactions, *J. Am. Chem. Soc.* **76**:648–652 (1954).
52. E. E. Snell, Chemical structure in relation to biological activities of vitamin B_6, *Vitamins and Hormones* **16**:77–125 (1958).
53. V. Bonavita and V. Scardi, On the role of the 4-formyl group of the pyridoxal-5'-phosphate in the activation of apotransaminase, *Experientia* **14**(1):7 (1958).
54. V. Bonavita and V. Scardi, The interaction between pyridoxal-5-phosphate and arginine apodecarboxylase, *Experientia* **14**(1):133 (1958).
55. E. H. Fischer, A. B. Kent, E. R. Snyder, and E. G. Krebs, The reaction of sodium borohydride with muscle phosphorylase, *J. Am. Chem. Soc.* **80**(11):2906–2907 (1958).
56. R. C. Hughes, W. T. Jenkins, and E. H. Fischer, The site of binding of pyridoxal-5'-phosphate to heart glutamic-aspartic transaminase, *Proc. Nat. Acad. Sci. U.S.A.* **48**(9):1615–1618 (1962).
57. R. Håkanson, Pyridoxal-5'-phosphate enzymes. Influence of substrate concentration on the pH optimum of enzyme reactions involving transaldimination, *Z. Physiol. Chemie* **348**(12):1730–1733 (1967).
58. W. P. Jencks and E. Cordes, in *Chemical and Biological Aspects of Pyridoxal Catalysis* (E. E. Snell, P. M. Fasella, A. Braunstein, and A. Rossi Fanelli, eds.), pp. 57–67, Pergamon Press, Oxford (1963).
59. T. P. Waalkes and S. Udenfriend, A fluorometric method for the estimation of tyrosine in plasma and tissues, *J. Lab. Clin. Med.* **50**(5):733–736 (1957).
60. S. Hess and S. Udenfriend, A fluorometric procedure for the measurement of tryptamine in tissues, *J. Pharmacol. Exp. Therap.* **127**:175–177 (1959).

61. P. A. Shore, A. Burkhalter, and V. H. Cohn, A method for the fluorometric assay of histamine in tissues, *J. Pharmacol. Exp. Therap.* **127**:182–186 (1959).

62. A. Carlsson and B. Waldeck, A fluorometric method for the determination of dopamine (3-hydroxytyramine), *Acta Physiol. Scand.* **44**:293–298 (1958).

63. R. W. Schayer, Formation and binding of histamine by rat tissues *in vitro*, *Am. J. Physiol.* **187**:63–65 (1956).

64. G. Kahlson, E. Rosengren, H. Westling, and T. White, The site of increased formation of histamine in the pregnant rat, *J. Physiol.* **144**(3):337–348 (1958).

65. H. Weissbach, W. King, A. Sjoerdsma, and S. Udenfriend, Formation of indole-3-acetic acid and tryptamine in animals, A method for estimation of indole-3-acetic acid in tissues, *J. Biol. Chem.* **234**:81–86 (1959).

66. Y. Kobayashi, Determination of histidine decarboxylase activity by liquid scintillation counting of $C^{14}O_2$, *Anal. Biochem.* **5**:284–290 (1963).

67. S. Snyder and J. Axelrod, Inhibition of histamine methylation *in vivo* by drugs, *Biochem. Pharmacol.* **13**(5):805–808 (1964).

68. D. Aures and W. G. Clark, A rotating diffusion chamber for $C^{14}O_2$ determination as applied to inhibitor studies on mouse mast cell tumor histidine decarboxylase, *Anal. Biochem.* **9**:35–47 (1964).

69. R. E. McCaman, M. W. McCaman, J. M. Hunt, and M. S. Smith, Microdetermination of monoamine oxidase and 5-hydroxytryptophan decarboxylase activities in nervous tissues, *J. Neurochem.* **12**(1):15–24 (1965).

70. R. Håkanson, Radiometric micromethods for the study of some amino acid decarboxylases, *Acta Pharmacol. Toxicol.* **24**(2–3):217–231 (1966).

71. R. W. Schayer, in *Methods of Biochemical Analysis* (David Glick, ed.), Vol. 16, pp. 273–291, Interscience (Wiley) New York, London (1968).

72. R. Kuntzman, P. A. Shore, D. F. Bogdanski, and B. B. Brodie, Microanalytical procedures for fluorometric assay of brain dopa-5HTP decarboxylase, norepinephrine and serotonin and a detailed mapping of decarboxylase activity in brain, *J. Neurochem.* **6**(3):226–232 (1961).

73. Å. Bertler and E. Rosengren, On the distribution in brain of monoamines and of enzymes responsible for their formation, *Experientia* **15**(10):382–383 (1959).

74. M. Mirolli, Advantages of *n*-heptanol in the extraction of 5-hydroxytryptamine (5-HT), *Experientia* **22**(12):788 (1966).

75. D. Aures, R. Håkanson, and A. Schauer, Histidine decarboxylase and dopa decarboxylase in the rat stomach. Properties and cellular localization, *Europ. J. Pharmacol.* **3**(3):217–234 (1968).

76. N. G. Waton, Studies on mammalian histidine decarboxylase, *Brit. J. Pharmacol.* **11**(2):119–127 (1956).

77. N. G. Waton, Histamine as an impurity in samples of histidine, *J. Pharm. Pharmacol.* **15**:574–578 (1963).

78. R. Håkanson, Histidine decarboxylase in the fetal rat, *Biochem. Pharmacol.* **12**(11):1289–1296 (1963).

79. A. Burkhalter, The formation of histamine by fetal rat liver, *Biochem. Pharmacol.* **11**:1–8 (1962).

80. T. O. Turner and S. L. Wightman, Ortho-phthalaldehyde as a spray reagent for thin layer chromatograms, *J. Chromatog.* **32**(1):315–322 (1968).

81. D. Aures, R. Fleming, and R. Håkanson, Separation and detection of biogenic amines by thin layer chromatography. Micro-analysis of tissue amines and of enzyme involved in their metabolism, *J. Chromatog.* **33**(3,4):480–493 (1968).

82. J. J. Pisano, J. B. Wilson, L. Cohen, D. Abraham, and S. Udenfriend, Isolation of γ-amino-butyrylhistidine (homocarnosine) in brain, *J. Biol. Chem.* **236**(2):499–502 (1961).

83. V. H. Cohn and P. A. Shore, A microfluorometric method for the determination of agmatine, *Anal. Biochem.* **2**:237–241 (1961).
84. S. Udenfriend, *Fluorescence Assay in Biology and Medicine*, Academic Press, New York, London (1962).
85. L. T. Kremzner and C. C. Pfeiffer, Identification of substances interfering with the fluorometric determination of brain histamine, *Biochem. Pharmacol.* **15**(2):197–200 (1966).
86. B. Elliott and J. A. Michaelson, Improved method for the fluorometric estimation of spermidine, *Anal. Biochem.* **19**(1):184–186 (1967).
87. R. P. Maickell and F. P. Miller, Fluorescent products formed by reaction of indole derivatives and *o*-phthalaldehyde, *Anal. Chem.* **38**(13):1937 (1967).
88. R. W. Schayer, Histidine decarboxylase of rat stomach and other mammalian tissues, *Am. J. Physiol.* **189**(3):533–536 (1957).
89. R. J. Levine and D. E. Watts, A sensitive and specific assay for histidine decarboxylase activity, *Biochem. Pharmacol.* **15**(7):841–849 (1966).
90. R. D. Smith and C. F. Code, Histamine formation, *Mayo Clin. Proc.* **42**(2):105–111 (1967).
91. F. J. Leinweber and L. A. Walker, Isotopic determination of histidine decarboxylase: a disposable assay vial, *Anal. Biochem.* **21**(1):131–134 (1967).
92. J. F. Skidmore and M. W. Whitehouse, Inhibition of histamine formation catalysed by substrate-specific mammalian histidine decarboxylases. Drug antagonism of aldehyde binding to protein amino groups, *Biochem. Pharmacol.* **15**(12):1965–1983 (1966).
93. G. Kahlson, Nascent histamine and methods of its estimation, *Int. Physiol. Congr. Proc.* **1**:856–862 (1962).
94. P. Holtz, R. Heise, and K. Ludtke, Fermentativer Abbau von 3,4-Dioxyphenylalanin (DOPA) durch Niere, *Arch. Exp. Pathol. Pharmakol.* **191**(1):87–118 (1938).
95. H. Blaschko, The amino acid decarboxylases of mammalian tissue, *Advan. Enzymol.* **5**:67–85 (1945).
96. S. Udenfriend, C. T. Clark, and S. Titus, 5-Hydroxytryptophan decarboxylase: A new route of metabolism of tryptophan, *J. Am. Chem. Soc.* **75**(2):501–502 (1953).
97. E. Westermann, H. Balzer, and J. Krell, Hemmung der Serotoninbildung durch α-Methyldopa, *Arch. Exp. Pathol. Pharmakol.* **234**:194–205 (1958).
98. E. Werle and D. Aures, Ueber die Reinigungen und Spezifität der Dopa-decarboxylase, *Z. Physiol. Chem.* **316**(1–2):45–60 (1959).
99. J. Awapara, T. L. Perry, C. Hanley, and E. Peck, Substrate specificity of dopa decarboxylase, *Clin. Chim. Acta* **10**(3):286–289 (1964).
100. P. Hagen, Observations on the substrate specificity of dopa decarboxylase from ox adrenal medulla, human phaeochromocytoma and human argentaffinoma, *Brit. J. Pharmacol.* **18**(1):175–182 (1962).
101. J. H. Fellman, Purification and properties of adrenal l-dopa decarboxylase, *Enzymologia* **20**:366–376 (1959).
102. T. L. Sourkes, Inhibition of dihydroxyphenylalanine decarboxylase by derivatives of phenylalanine, *Arch. Biochem. Biophys.* **51**(2):444–456 (1954).
103. S. E. Smith, The pharmacological actions of 3,4-dihydroxyphenyl-α-methylalanine (α-methyldopa), an inhibitor of 5-hydroxytryptophan decarboxylase, *Brit. J. Pharmacol.* **15**(2):319–327 (1960).
104. D. Mackay and D. M. Shepherd, A study of potential histidine decarboxylase inhibitors, *Brit. J. Pharmacol.* **15**(2):552–556 (1960).
105. V. Erspamer, A. Glasser, C. Pasini, and G. Stoppari, *In vitro* decarboxylation of tryptophans by mammalian decarboxylase, *Nature* **189**:483 (1961).
106. R. Ferrini and A. Glasser, *In vitro* decarboxylation of new phenylalanine derivatives, *Biochem. Pharmacol.* **13**:798–801 (1964).

107. P. O. Ganrot, A. M. Rosengren, and E. Rosengren, On the presence of different histidine decarboxylating enzymes in mammalian tissues, *Experientia* **17**:263 (1961).
108. H. Weissbach, W. Lovenberg, and S. Udenfriend, Characteristics of mammalian histidine decarboxylating enzymes, *Biochim. Biophys. Acta* **50**(1):177–179 (1961).
109. R. Håkanson, Histidine decarboxylase in the bone marrow of the rat, *Experientia* **20**(4):205–206 (1964).
110. R. Håkanson, Histidine decarboxylase in experimental tumors, *J. Pharm. Pharmacol.* **18**:769–774 (1966).
111. D. Aures and W. W. Noll, Specific histidine decarboxylase from different mammalian tissues, *Fed. Proc.* **25**(2) Part 1:291 (1966).
112. W. W. Noll, Some properties of mouse mastocytoma histidine decarboxylase, Thesis, Yale University School of Medicine, New Haven. 30 p. (1966).
113. R. Håkanson and Ch. Owman, Distribution and properties of amino acid decarboxylases in gastric mucosa, *Biochem. Pharmacol.* **15**:489–499 (1966).
114. R. Håkanson, Histamine-forming isoenzymes in the fetal and adult mouse, *Europ. J. Pharmacol.* **1**(1):34–41 (1967).
115. R. Håkanson, Kinetic properties of mammalian histidine decarboxylase, *Europ. J. Pharmacol.* **1**(1):42–46 (1967).
116. D. Aures, G. Winqvist, and E. Hansson, Histamine formation in the blood and bone marrow of the guinea pig, *Am. J. Physiol.* **208**(1):186–189 (1965).
117. H. Schievelbein and E. Werle, Aminosäuredecarboxylasen und verwandte Aminosäure-lysen, in *Handbuch der Physiologisch-und Pathologischen Analyse*, Vol. VI C:503–607 (Hoppe-Seyler/Thierfelder, eds.), Springer-Verlag, Berlin, Heidelberg, New York (1966).
118. D. Aures, To be published.
119. E. Werle and W. Lorenz, Histamin und Histidin-decarboxylasen in Schilddrüse und Thymus, *Biochem. Pharmacol.* **15**(8):1059–1070 (1966).
120. R. Håkanson and L. Josefsson, To be published.
121. D. Aures, To be published.
122. P. Holtz and E. Westermann, Über die Dopa-decarboxylase und Histidindecarboxylase des Nervengewebes, *Arch. Exp. Pathol. Pharmakol.* **227**:538–546 (1956).
123. E. Werle and A. Schauer, Histamin in Nerven, *Z. Exp. Med.* **127**:16–21 (1956).
124. T. White, Formation and catabolism of histamine in brain tissue *in vitro*, *J. Physiol.* **149**:34–42 (1959).
125. B. N. Halpern, Th. Neven, and C. W. M. Wilson, The distribution and fate of radioactive histamine in the rat, *J. Physiol.* **147**(4):437–449 (1959).
126. E. Robins, J. M. Robins, A. B. Croninger, S. G. Moses, S. J. Spencer, and R. W. Hudgens, The low level of 5-hydroxytryptophan decarboxylase in human brain, *Biochem. Med.* **1**(3):240–251 (1967).
127. G. R. Pscheidt and B. Haber, Regional distribution of dihydroxyphenylalanine and 5-hydroxytryptophan decarboxylase and of biogenic amines in the chicken, *J. Neurochem.* **12**(7):613–618 (1965).
128. R. E. McCaman and M. H. Aprison, The synthetic and catabolic enzyme systems for acetylcholine and serotonin in several discrete areas of the developing rabbit brain, *Prog. Brain Res.* **9**:220–233 (1964).
129. Å. Bertler, B. Falck, and E. Rosengren, The direct demonstration of a barrier mechanism in the brain capillaries, *Acta Pharmacol.* **20**(4):317–321 (1964).
130. Å. Bertler, B. Falck, Ch. Owman, and E. Rosengren, The localization of monoaminergic blood–brain barrier mechanisms, *Pharmacol. Rev.* **18**(1):369–385 (1966).
131. Ch. Owman and E. Rosengren, Dopamine formation in brain capillaries—an enzymic blood–brain barrier mechanism, *J. Neurochem.* **14**(5):547–550 (1967).

132. J. Constantinidis, G. Bartholini, R. Tissot, and A. Pletscher, Accumulation of dopamine in the parenchyma after decarboxylase inhibition in the capillaries of brain, *Experientia* **24**(2):130–131 (1968).

133. G. Bartholini, W. P. Burkhard, A. Pletscher, and H. M. Bates, Increase of cerebral catecholamines caused by 3,4-dihydroxyphenylalanine after inhibition of peripheral decarboxylase, *Nature* **215**(5103):852 (1967).

134. S. Udenfriend, in *Mechanisms of Release of Biogenic Amines* (U. S. von Euler, S. Rosell, and B. Üvnas, eds.), pp. 103–108, Pergamon Press, London, New York (1966).

135. A. Heller, L. S. Seiden, W. Porcher, and R. Y. Moore, 5-Hydroxytryptophan decarboxylase in rat brain: effect of hypothalmic lesions, *Science* **147**(3660):887–888 (1965).

136. N. J. Giarman and M. Day, Presence of biogenic amines in the bovine pineal body, *Biochem. Pharmacol.* **1**(3):235 (1958).

137. N. J. Giarman, D. X. Freedman, and L. Picard-Ami, Serotonin content of the pineal glands of man and monkey, *Nature* **186**(4723):480–481 (1960).

138. N. Prop and J. Ariends-Kappers, Demonstration of some compounds present in the pineal organ of the albino rat by histochemical methods and paper chromatography, *Acta Anat.* **45**(1–2):90–109 (1961).

139. W. B. Quay and A. Halevy, Experimental modification of the rat pineal content of serotonin and related indole amines, *Physiol. Zool.* **35**(1):1–7 (1962).

140. R. Håkanson and Ch. Owman, Effect of denervation and enzyme inhibition on dopa decarboxylase and monoamine oxidase activity of rat pineal gland, *J. Neurochem.* **12**:417–429 (1965).

141. R. Håkanson and Ch. Owman, Pineal dopa decarboxylase and monoamine oxidase activities as related to the monoamine stores, *J. Neurochem.* **13**:597–605 (1966).

142. P. A. Srere, Enzyme concentration in tissues, *Science* **158**(3803):936–937 (1967).

143. W. G. Clark, in *Metabolic Inhibitors* (R. M. Hoechster and J. H. Quastel, eds.), Vol. 1, pp. 315–381, Academic Press, New York (1963).

144. T. L. Sourkes and A. D'Iorio, in *Metabolic Inhibitors* (R. M. Hoechster and J. H. Quastel, eds.), Vol. II, pp. 79–98, Academic Press, New York (1963).

145. D. M. Shepherd and D. Mackay, in *Progress in Medicinal Chemistry* (G. P. Ellis and G. B. West, eds.), Vol. V, pp. 199–250, Plenum Press, New York (1967).

146. W. G. Clark, Studies on inhibition of l-dopa decarboxylase *in vitro* and *in vivo*, *Pharmacol. Rev.* **11**:330–349 (1959).

147. R. W. Schayer, in *Handbook of Experimental Pharmacology* (O. Eichler and A. Farah, eds.), Vol. 18, Part 1, pp. 688–725, Springer-Verlag, Berlin (1966).

148. A. Pletscher, K. F. Gey, and W. P. Burkhard, in *Handbook of Experimental Pharmacology* (O. Eichler, A. Farah, and V. Erspamer, eds.), Vol. 19, pp. 592–635, Springer-Verlag, New York (1966).

149. R. J. Levine, T. L. Sato, and A. Sjoerdsma, Inhibition of histamine synthesis in the rat by α-hydrazino analog of histidine and 4-bromo-3-hydroxybenzyloxyamine, *Biochem. Pharmacol.* **14**(2):139–150 (1965).

150. C. F. Code, Histamine and gastric secretion: a later look, 1955–1965, *Fed. Proc.* **24**(6):1311–1321 (1965).

151. R. J. Levine, Effect of histidine decarboxylase inhibition on gastric acid secretion in the rat, *Fed. Proc.* **24**(6):1331–1333 (1965).

152. R. J. Levine, Histamine synthesis in man: Inhibition by 4-bromo-3-hydroxybenzyloxyamine, *Science* **154**(3752):1017–1019 (1966).

153. A. Nikulin, P. Stern, and Z. Zeger-Vidovic, Die Beteutung des Histamins im Entzündungsprozess, *Arch. Int. Pharmacodyn. Ther.* **166**(2):305–312 (1967).

154. F. J. Leinweber, Mechanism of histidine decarboxylase inhibition by NSD-1055 and related hydroxylamines, *Mol. Pharmacol.* **4**:337–348 (1968).

155. D. Aures, L. E. A. Rodrigues, and S. F. Laws, To be published.
156. H. F. Schott and W. G. Clark, Dopa decarboxylase inhibition through the interaction of coenzyme and substrate, *J. Biol. Chem.* **196**(1):449–462 (1952).
157. P. Holtz and E. Westerman, Hemmung der Glutaminsäuredecarboxylase des Gehirns durch Brenzcatechinderivate, Naunyn-Schmiedeberg's *Arch. Exp. Pathol. Pharmakol.* **231**:311–332 (1957).
158. J. Putter and G. Kroneberg, Untersuchungen über die Stereospezifität der decarboxylase-hemmenden Wirkung von α-methyldopa, *Arch. Exp. Pathol. Pharmakol.* **249**(5):470–478 (1964).
159. A. Carlsson and M. Lindquist, In vivo decarboxylation of α-methyl dopa and α-methyl-metatyrosine, *Acta Physiol. Scand.* **54**(1):87–94 (1962).
160. W. J. Hartman, R. J. Akawie, and W. G. Clark, Competitive inhibition of 3,4-dihy-droxyphenylalanine (dopa) decarboxylase in vitro, *J. Biol. Chem.* **216**(2):507–529 (1955).
161. B. Robinson and D. M. Shepherd, The inhibition of the l-histidine decarboxylases of guinea-pig kidney and rat hepatoma, *J. Pharmacol.* **14**(1):9–15 (1962).
162. G. R. Marschke and G. N. Beall, The effects of two histidine decarboxylase inhibitors in guinea pig anaphylaxis, *Biochem. Pharmacol.* **14**(2):192–194 (1965).
163. K. Hempel and H. F. K. Männel, Inhibition of tyrosine degradation in vivo by the dopa decarboxylase blocking agent NSD-1034, *Experientia* **24**(5):429–430 (1968).
164. J. F. Moran and T. L. Sourkes, Multiple inhibitory effects of α-methylamino acid on the metabolism of C^{14}-labeled tyrosine, dopa and α-methyldopa, *J. Pharmacol. Exp. Therap.* **148**(2):252–261 (1965).
165. A. Pletscher and K. F. Gey, The effect of a new decarboxylase inhibitor on endogenous and exogenous monoamines, *Biochem. Pharmacol.* **12**:223–228 (1963).
166. G. E. Johnson and K. Pritzker, The influence of the dopa decarboxylase inhibitor Ro4-4602 on the urinary excretion of catecholamines in cold stressed rats, *J. Pharmacol. Exp. Therap.* **152**(3):432–438 (1966).
167. J. V. Hodge, J. A. Oates, and A. Sjoerdsma, Reduction of the central effects of tryptophan by a decarboxylase inhibitor, *Clin. Pharmacol. Therap.* **5**(2):149–155 (1964).
168. E. J. Glamkowski, G. Gal, M. Sletzinger, C. C. Porter, and L. S. Watson, A new class of potent decarboxylase inhibitors, *J. Med. Chem.* **10**:852–855 (1967).
169. A. Parulkar, A. Burger, and D. Aures, Some nuclear-substituted derivatives of α-methyldopa, *J. Med. Chem.* **9**(5):738–741 (1966).
170. S. M. Hess, R. H. Connamacher, M. Ozaki, and S. Udenfriend, Effects of α-methyldopa and α-methyl-m-tyrosine on metabolism of norepinephrine and serotonin in vivo, *J. Pharmacol. Exp. Therap.* **134**(2):129–138 (1961).
171. C. C. Porter, J. A. Totaro, and C. M. Leiby, Some biochemical effects of α-methyl-3,4-dihydroxyphenylalanine and related compounds in mice, *J. Pharmacol. Exp. Therap.* **134**(2):139–145 (1961).
172. T. L. Sourkes, The action of alpha-methyldopa on the brain, *Brit. Med. Bull.* **21**(1):66–69 (1965).
173. C. C. Porter, J. A. Totaro, and A. Burcin, The relationship between radioactivity and norepinephrine concentrations in the brains and hearts of mice following administration of labeled methyldopa or 6-hydroxydopamine, *J. Pharmacol. Exp. Therap.* **150**(1): 17–22 (1965).
174. R. Lindman and E. Muscholl, α-Methylnoradrenalin in das isolierte Kaninchenherz und seine Freisetzung durch Reserpin und Guanethidin in vivo, *Arch. Exp. Pathol. Pharmakol.* **249**(6):529–548 (1965).
175. N. E. Andén, On the mechanism of noradrenaline depletion by α-methyl metatyrosine and metaraminol, *Acta Pharmacol. Toxicol.* **21**(3):260–271 (1964).

176. R. Imaizumi, M. Oka, and T. Ohuche, Mechanism of the antihypertensive effect of α-methyldopa, *Nature* **203**(4948):982 (1964).

177. C. A. Stone and C. C. Porter, in *Advances in Drug Research* (N. J. Harper and A. Simmond, eds.), Vol. 4, pp. 71–93, Academic Press, New York (1967).

178. W. G. Burkard, K. F. Gey, and A. Pletscher, Inhibition of the hydroxylation of tryptophan and phenylalanine by methyldopa and similar compounds, *Life Sci.* **3**(5):27–33 (1964).

179. B. E. Roos and B. Werdenius, The effect of methyldopa on the metabolism of 5-hydroxytryptamine in brain, *Life Sci.* **2**:92–96 (1963).

180. T. Nagatsu, M. Levitt, and S. Udenfriend, The initial step in norepinephrine biosynthesis, *J. Biol. Chem.* **239**(9):2910–2917 (1964).

181. C. R. Creveling, J. W. Daly, B. Witkop, and S. Udenfriend, Substrates and inhibitors of dopamine β-oxidase, *Biochim. Biophys. Acta* **64**(1):125–134 (1962).

182. E. Muscholl and L. Maitre, Release by sympathetic stimulation of α-methylnoradrenaline stored in the heart after administration of α-methyldopa, *Experientia* **19**(12):658–659 (1963).

183. R. Lindmar, E. Muscholl, and K. H. Rahn, Effects of rest and physical activity on the urinary excretion of noradrenaline and α-methylnoradrenaline in human subjects treated with α-methyldopa, *Europ. J. Pharmacol.* **2**(4):317–319 (1968).

184. W. Haeffely, A. Hurlimann, and H. Thoenen, Adrenergic transmitter changes and response to sympathetic nerve stimulation after differing pretreatment with α-methyldopa, *Brit. J. Pharmacol.* **31**:105–119 (1967).

185. M. D. Day and M. J. Rand, Some observations on the pharmacology of α-methyldopa, *Brit. J. Pharmacol.* **22**(1):72–85 (1964).

186. A. Carlsson, A. Dahlström, K. Fuxe, and H. Å. Hillarp, Failure of reserpine to deplete noradrenaline neurons of α-methylnoradrenaline formed from α-methyldopa, *Acta Pharmacol. Toxicol.* **22**(3):270–276 (1965).

187. J. Axelrod and R. Tomshick, Enzymatic o-methylation of epinephrine and other catechols, *J. Biol. Chem.* **233**(3):702–705 (1958).

188. K. Stock and E. Westermann, Effect of α-methyldopa and α-methyl-m-tyrosine on the mobilization of free fatty acids, *Experientia* **20**(9):495–496 (1964).

189. E. Westermann and K. Stock, Wirkung von α-methyldopa und α-methyl-m-tyrosine auf den Fettstoffwechsel der Ratte, *Arch. Exp. Pathol. Pharmakol.* **247**(4):299–300 (1964).

190. W. G. Clark and R. S. Pogrund, Inhibition of dopa decarboxylase *in vitro* and *in vivo*, *Circulation Res.* **9**:721–733 (1961).

191. J. Cahn, M. Herold, C. Helbecque, T. Lasjuamias, and A. M. Juillard, Effects comparés de différents inhibiteurs de la l-dopa décarboxylase chez le rat et la souris, *Compt. Rend. Soc. Biol.* (Paris) **156**(6):1094–1096 (1962).

192. M. Herold, J. Cahn, C. Helbecque, T. Lasjaumias, and A. M. Juillard, Action de quelques inhibiteurs de la l-dopa décarboxylase sur la durée de la narcose barbiturique chez la souris, *Compt. Rend. Soc. Biol.* (Paris) **156**(7):1270–1272 (1962).

193. T. L. Perry, Urinary excretion of amines in phenylketonuria and mongolism, *Science* **136**:879–880 (1962).

194. A. N. Davison and M. Sandler, Inhibition of 5-hydroxytryptophan decarboxylase by phenylalanine metabolites, *Nature* **181**:186–187 (1958).

195. A. Yuwiler and R. T. Louttit, Effects of phenylalanine diet on brain serotonin in the rat, *Science* **134**:831–832 (1961).

196. R. C. Baldridge and V. H. Auerbach, The metabolism of histidine: VI. Histidemia and imidazolepyruvic aciduria, *J. Biol. Chem.* **239**(5):1557–1559 (1964).

197. S. H. Snyder, P. Myron, M. W. Kies, and S. Berlow, Metabolism of 2-C^{14}-labeled l-histidine in histidinemia, *J. Clin. Endocr.* **23**(6):595–597 (1963).

198. D. W. Wooley, *The Biochemical Basis of Psychosis*, Wiley, New York (1962).
199. G. Bruce and G. R. Pscheidt, Correlations between behavior and urinary excretion of indole amines and catecholamines in schizophrenic patients as affected by drugs, *Fed. Proc.* **20**(4):889–893 (1961).
200. O. Hornykiewicz, Die topische Lokalisation und das Verhalten von Noradrenalin und Dopamin in der Substantia nigra des Normalen und parkinsonkranken Menschen, *Wien. Klin. Wochschr.* **75**(18):309–312 (1963).
201. O. Hornykiewicz, Dopamine (3-hydroxytryamine) and brain function, *Pharmacol. Rev.* **18**:925–964 (1966).
202. B. Elefteriou and R. L. Church, Brain 5-hydroxytryptophan decarboxylase in mice after exposure to aggression and defeat, *Physiol. Behav.* **3**:323–325 (1968).
203. A. B. Rothballer, in *Aggression and Defense. Neural Mechanisms and Social Patterns* (C. Clemente and D. B. Lindsley, eds.), *Brain Function* **5**:135–170 (1967), Univ. Calif. Press, Berkeley and Los Angeles.
204. J. J. Schildkraut and S. S. Kety, Biogenic amines and emotion, *Science* **156**(3771):21–30 (1967).
205. S. S. Kety and F. E. Samson, Jr. (eds.), Neural properties of the biogenic amines, *Neurosci. Res. Program Bull.* **5**(1):1–119 (1967).

Chapter 9

METABOLISM OF CATECHOLAMINES

Leslie L. Iversen

Department of Pharmacology
University of Cambridge
Cambridge, England

I. DISTRIBUTION OF CATECHOLAMINES

The three naturally occurring catecholamines are dopamine (3,4-dihydroxy-β-phenylethylamine; 3-hydroxytyramine) (DM), L-noradrenaline (arterenol; norepinephrine) (NA) and L-adrenaline (epinephrine) (ADR). NA is the transmitter substance released from the terminals of postganglionic neurons of the sympathetic nervous system. It is present in all parts of such neurons, but is markedly more concentrated in their fine terminal branches. In addition NA and DM occur in the brain and spinal cord of most vertebrates (Table I).

TABLE I

Occurrence of Catecholamines in CNS of Vertebrates

Species	Brain catecholamine content ($\mu g/g$)		
	Noradrenaline	Dopamine	Adrenaline
Rat	0.49	0.60	Approx 0.02
Cat	0.22	0.28	—
Dog	0.16	0.19	—
Pig	0.14	0.22	Approx 0.01
Guinea pig	0.38	0.34	Approx 0.01
Rabbit	0.29	0.32	—
Pigeon	0.38	Approx 0.40	Approx 0.05
Frog (*Rana pipiens*)	v. low	—	2.10
Toad (*Bufo marinus*)	v. low	—	2.40

For references see 85.

Within the CNS, bioassay and chemical analysis have revealed a characteristic regional distribution of NA which parallels the occurrence of specialized catecholamine-containing neurons visualized in tissue sections by a fluorescence histochemical technique. The application of this histochemical technique has led to a greatly improved understanding of the detailed morphological distribution of catecholamine-containing neurons in the CNS in recent years.[1] Histochemical and biochemical studies have also revealed the presence of separate neuronal systems in the CNS in which the predominant catecholamine is dopamine, rather than NA. While NA-containing nerve terminals are widely distributed throughout the CNS, with a particular abundance in the hypothalamus, DM-containing neurons have a more circumscribed distribution, being particularly abundant in the striatum.[2] The occurrence of NA and DM in specific neuronal systems lends strong support to the hypothesis that these substances function as neurotransmitter substances in the CNS. ADR is present in only very small amounts in the mammalian CNS, although it is the predominant catecholamine in the CNS of certain amphibians (Table I).

All three catecholamines are also found in the chromaffin tissue of the adrenal medulla.[3] While CNS and most peripheral tissues contain only microgram quantities of catecholamines, the adrenal medulla represents a massive local concentration of the catecholamines and the related enzymes. As such it has proved an extremely valuable tissue for studies of catecholamine metabolism. NA and ADR are also found in extraadrenal chromaffin tissue, and DM occurs in local abundance in specialized mast cells found in the peripheral tissues of ruminant animals.[4]

II. BIOSYNTHESIS OF CATECHOLAMINES

A. Overall Pathway

Catecholamine-containing cells synthesize their endogenous amine content from the plasma amino acid L-tyrosine by a pathway first suggested by Blaschko in 1939 :[5]

L-tyrosine → L-DOPA → dopamine → L-noradrenaline → L-adrenaline

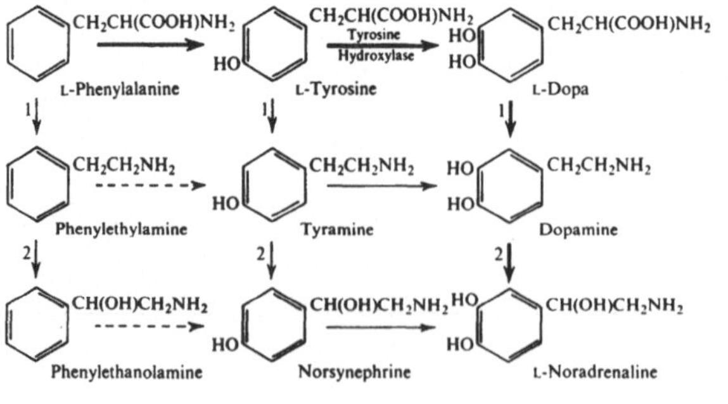

Fig. 1. Biosynthetic pathways for catecholamines.

Heavy lines indicate major pathway *in vivo*. 1 = Aromatic-L-amino acid decarboxylase, 2 = Dopamine-β-hydroxylase. Thin lines indicate reactions which have been demonstrated to occur both *in vivo* and *in vitro*, dotted lines reactions demonstrated only *in vitro*.

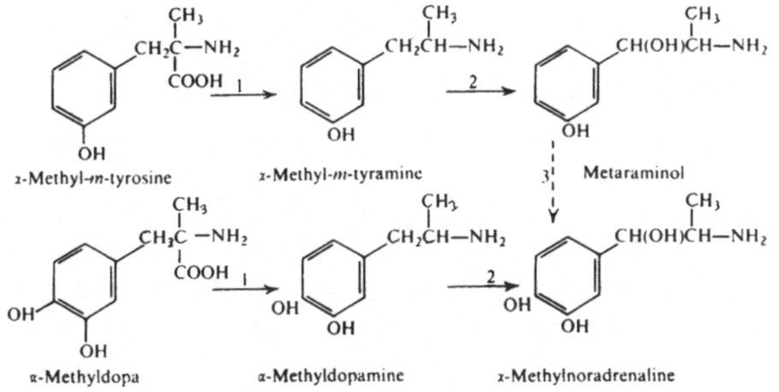

Fig. 2. Biosynthesis of false transmitters.

1 = Aromatic-L-amino acid decarboxylase, 2 = Dopamine-β-hydroxylase, 3 = liver microsomal hydroxylase (?).

The first direct evidence for this pathway came from studies of the synthesis of catecholamines in the adrenal medulla, which was shown to be capable of converting radioactively labeled tyrosine or DOPA into DM, NA, and ADR both *in vivo* and *in vitro*.[6-9] The same pathway can be demonstrated, in brain tissue or in tissues innervated by the sympathetic nervous system when these are exposed to radioactively labeled tyrosine or DOPA.[10-13] The broad specificity of many of the enzymes involved in the biosynthetic pathway implies that several alternative pathways are in principle available for the synthesis of NA from L-Tyr (Fig. 1). It seems unlikely, however, that these are of any quantitative importance under normal conditions. However, the finding that synthetic substrates, such as α-methylDOPA and α-methyl-*m*-tyrosine, can be converted by the biosynthetic machinery *in vivo* to form false sympathetic transmitter substances (Fig. 2) is of considerable pharmacological importance.[14]

B. Properties of Individual Enzymes

1. Tyrosine Hydroxylase

(L-Tyrosine, tetrahydropteridine: oxygen oxidoreductase (3-hydroxylating))

The conversion of L-Tyr to L-DOPA is catalyzed by the enzyme tyrosine hydroxylase, which occurs in the adrenal medulla, brain, and sympathetically innervated tissues.[15,16] The properties of the partially purified enzyme from bovine adrenal medulla are described in recent reviews.[17,18] Tyrosine hydroxylase is a mixed function oxidase, i.e., it catalyzes a reaction of the type:

$$RH + O_2 + XH_2 \rightarrow ROH + H_2O + X$$

The bovine adrenal medullary enzyme utilizes either tetrahydrofolic acid with a ferrous salt, or a synthetic tetrahydropteridine as cofactor (XH_2). The naturally occurring cofactor is thought to be dihydrobiopterin (Fig. 3) or a derivative. Tyrosine hydroxylase appears to be the most specific of the

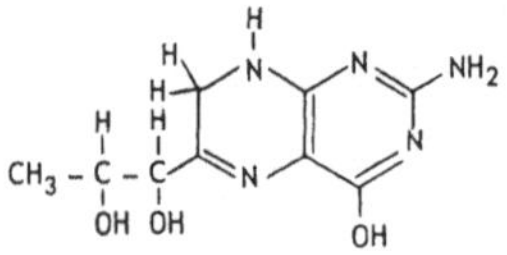

Fig. 3. Dihydrobiopterin.

enzymes involved in catecholamine biosynthesis; the partially purified enzyme from bovine adrenal gland catalyzes a small conversion of L-Phe to L-Tyr,[19] but other related amino acids such as D-Tyr, L-Tyr, DL-*m*-Tyr, and 3-iodo-Tyr are inactive as substrates. L-α-Methyl-Tyr, a competitive inhibitor of the enzyme, is a poor substrate, but a significant conversion of this substance to α-methyl-L-DOPA can, nevertheless, be demonstrated *in vitro* and *in vivo*.[20,21]

TABLE II

Substrate Specificity of Aromatic Amino Acid Decarboxylase[a]

Substrate	Relative specific activity
L-DOPA	6400
5-Hydroxytryptophan (5-HTP)	1000
L-Tryptophan	220
L-Phenylalanine	104
p-Tyrosine	34
α-MethylDOPA	32
α-Methyl-5-HTP	13
L-Histidine	10
α-Methyl-*m*-tyrosine	4

[a] Data from Lovenberg *et al.*[24] Partially purified enzyme from guinea pig kidney. All substrates tested at optimal concentrations or V_{max} values were extrapolated from Lineweaver–Burke plots. Incubations were carried out in presence of iproniazid $(10^{-3}M)$ and pyridoxal phosphate $(7 \times 10^{-5}M)$.

2. Aromatic L-Amino Acid Decarboxylase

The decarboxylation of L-DOPA to DM is reviewed by Sourkes[22] and by Aures, Håkanson and Clarke in this volume.[23] A single enzyme appears to be responsible for the decarboxylation of L-DOPA, L-5-HTP, L-His and other aromatic amino acids such as Phe and Tyr (Table II). This is now commonly referred to as aromatic L-amino acid decarboxylase after the suggestion of Lovenberg *et al.*[24] As with other amino acid decarboxylase, pyridoxal phosphate is required as a cofactor.

3. Dopamine β-hydroxylase

(3,4-dihydroxyphenylethylamine, ascorbate: oxygen oxidoreductase (hydroxylating) E.C. 1.14.2.1)

TABLE III

Substrate Specificity of Dopamine-β-Hydroxylase[a]

Substrate	Relative activity
p-Tyramine	100
m-Tyramine	60–80
o-Tyramine	0
Dopamine	85–90
Epinine (N-methyldopamine)	25
N-methyltyramine	25
α-methyltyramine	60–70
β-methyltyramine	5–10
3-Methoxytyramine	50–70
3,5-Dimethoxytyramine	25–35
3,4,5-Trimethoxyphenylethylamine(Mescaline)	2–5
β-Phenylethylamine	10–15
p-Methoxyphenylethylamine	2–5
3,4-Dimethoxyphenylethylamine	2–5
p-Hydroxy-β-phenylacetic acid	0

[a] Data from Creveling et al.[27] Purified enzyme from bovine adrenal medulla. Substrates were tested at 10^{-2} M and 2×10^{-2} M in the presence of saturating concentrations of ascorbate and fumarate.

The final stage in the biosynthesis of NA involves a second mixed function oxidase, dopamine-β-hydroxylase. The purified enzyme from bovine adrenal medulla is a copper-containing protein of mol wt = 290,000. Ascorbic acid is required as a cofactor and enzyme activity is stimulated by catalytic amounts of fumarate and by certain other dicarboxylic acids.[25,26] Dopamine-β-hydroxylase is so far the only enzyme involved in catecholamine metabolism which has been isolated in a highly purified state.[25] There have been extensive studies of the substrate specificity of dopamine-β-hydroxylase,[27,28] which show it to be capable of catalyzing the β-hydroxylation of a variety of phenylethylamine derivatives (Table III).

4. Phenylethanolamine-N-methyl transferase

(S-adenosyl methionine:phenylethanolamine N-methyl transferase)

In adrenal and extraadrenal chromaffin cells NA may be further converted to ADR by the enzyme phenylethanolamine-N-methyl transferase. There is little evidence, however, that this enzyme exists in any significant amounts in the mammalian sympathetic nervous system or CNS, although it does exist in these tissues in certain amphibians. The partially purified enzyme from monkey adrenal glands requires S-adenosyl-L-methionine as methyl donor, and will N-methylate a variety of β-phenylethanolamines (Table IV).[29,30] Although only β-hydroxylated phenylethylamines appear to act as substrates

TABLE IV

Substrate Specificity of Phenylethanolamine
N-Methyl Transferase[a]

Substrate	Relative activity
L-Normetanephrine	100
3-Methoxytyramine	0
DL-Phenylethanolamine	72
β-Phenylethylamine	0
DL-Norsynephrine (octopamine)	18
p-Tyramine	0
DL-m-Hydroxyphenylethanolamine	60
L-Noradrenaline	21
D-Noradrenaline	15

[a] Data from Axelrod.[29] Partially purified enzyme from
monkey adrenal. All substrates were tested at 3.3 ×
10^{-3}M in the presence of ^{14}C-Methyl-S-adenosyl Met
(3.3 × 10^{-5}M).

for the enzyme, other β-phenylethylamines may act as competitive inhibitors.[31] The enzyme is not absolutely stereospecific for L-NA; D-NA
also acts as a substrate, but with a lower affinity than L-NA.[31] ADR and
related secondary amines can be further converted to the respective N-
dimethyl derivatives by the enzyme. Small amounts of N-dimethyl ADR
occur naturally in the adrenal medulla.[32]

C. Subcellular and Cellular Distribution of Biosynthetic Enzymes

The enzymes involved in the biosynthesis of NA are present in peripheral tissues, in sympathetic ganglia, in the adrenal medulla, and in the
brain. In all these tissues it is likely that tyrosine hydroxylase and dopamine-
β-hydroxylase are present exclusively in those cells which normally contain
catecholamines (adrenal medullary chromaffin cells, postganglionic sympathetic neurons, and catecholamine-containing neurons of the CNS).[33]
This is not the case with aromatic-L-amino acid decarboxylase, which since it
is involved in the synthesis of other aromatic amines such as serotonin must
presumably be present in other amine-containing cells. There is evidence that
this enzyme also occurs in abundance in certain glial cells in the CNS which
are intimately associated with the cerebral vasculature. In these cells the
enzyme is thought to play an important role in the blood–brain barrier which
largely prevents circulating DOPA from gaining access to the brain.[34]

The results of a recent systematic study of the intraneuronal distribution
of the biosynthetic enzymes in splenic nerve homogenates show that only the
last step in NA biosynthesis is specifically associated with intraneuronal NA

storage particles.[35] This confirms previous findings that purified preparations of such particles contain dopamine-β-hydroxylase activity (see Ref. 36). Aromatic amino acid decarboxylase activity is predominantly localized in the cytoplasm of splenic nerve and other tissues, although there is evidence that some of the enzyme activity may be loosely associated with intraneuronal storage particles.[35] Tyrosine hydroxylase activity does not appear to be associated at all with neuronal storage particles,[35] but in many tissues part of the tyrosine hydroxylase activity is associated with a particulate fraction which can be sedimented by centrifugation at 15–20,000 × g.[35,15] The nature of the enzyme binding sites in this heavy particulate fraction remains obscure. Suggestions that a single intracellular particle exists with which all three biosynthetic enzymes are associated[18] do not seem to be well founded. The finding that brain homogenates contain particles containing all three biosynthetic enzymes and capable of synthesizing NA from Tyr[18] does not necessarily support this contention. It is well known that brain homogenates contain large numbers of complex synaptosomal particles, and such particles have been shown to contain tyrosine hydroxylase and aromatic amino acid decarboxylase.[37]

D. Regulation of Catecholamine Biosynthesis

Knowledge of the rate-limiting step in a biosynthetic pathway is essential for an understanding of the regulation of the pathway. It is at this point that the most effective physiological and pharmacological control may be exerted. Recent evidence suggests that the first step in the pathway from Tyr to NA is the rate-limiting reaction: the conversion of L-Tyr to L-DOPA catalyzed by tyrosine hydroxylase. The evidence for this proposal has been reviewed recently[38] and is briefly as follows. In isolated guinea pig hearts perfused with labeled DOPA or DM, rates of NA synthesis greatly in excess of the normal *in vivo* rate may be attained, and the rate of NA synthesis increases with increasing concentrations of DOPA or DM. With labeled L-Tyr as substrate, however, the maximum rate of NA synthesis which can be attained is much lower and approximates to the normal *in vivo* rate. Furthermore, the rate of NA synthesis does not continuously increase as the concentration of L-Tyr is increased but reaches a saturation level at high L-Tyr concentrations. The overall K_m for the conversion of Tyr to NA in this system is 2×10^{-5} M, which is close to the K_m for the conversion of L-Tyr to L-DOPA by the enzyme tyrosine hydroxylase *in vitro* (Table V). The activity of tyrosine hydroxylase is always much lower than that of the other synthetic enzymes in tissue extracts. In the adrenal medulla, for example, tyrosine hydroxylase activity is more than one thousand times less than that of aromatic amino acid decarboxylase or dopamine-β-hydroxylase. The maximum activity of tyrosine hydroxylase in tissue extracts corresponds closely to the maximum rate of NA synthesis observed *in vivo*. The concentration of L-Tyr in plasma and tissues is of the order of 10^{-4} M, while only negligible amounts of L-DOPA or dopamine are usually found. There is thus considerable evidence

TABLE V

K_m Values for Synthetic Enzymes

Enzyme	Source	Substrate	K_m (μM)	Cofactor	K_m (μM)	Reference
Tyrosine hydroxylase	Bovine adrenal	L-Tyr	50	DMPH$_4$[a]	200	(18)
Aromatic amino acid Decarboxylase	Guinea pig kidney	L-DOPA	400	—	—	(24)
Dopamine-β-hydroxylase	Bovine adrenal	Dopamine	6000	Ascorbate	600	(25)
Phenylethanolamine N-methyl transferase	Rabbit adrenal	L-NA	10	S-Adenosyl-L-methionine	10	(44)

[a] DMPH$_4$ = 2-amino-4-hydroxy-6,7-dimethyltetrahydropteridine.

that tyrosine hydroxylase represents the rate-limiting step in NA biosynthesis. Since plasma concentrations of L-Tyr are normally high in comparison with the K_m of the enzyme for L-Tyr, it seems probable that the enzyme is normally operating under saturating conditions of substrate. The possibility that the transport of L-Tyr from the plasma or CSF into catecholamine-synthesizing cells might be the rate-limiting step in the synthetic process has, however, not yet been ruled out. Assuming that tyrosine hydroxylase is the rate-limiting step, the observation that NA, DM, and other catechol compounds inhibit the enzyme *in vitro* is highly significant.[39] The nature of this inhibitory action remains obscure. NA and other catechols (Table VI) do not compete with L-Tyr, but appear to compete with the pteridine cofactor. Other hydroxylases utilizing pteridine cofactors, such as Phe and Try hydroxylase, are also inhibited by NA and related catechol compounds. Although the catechol inhibition of tyrosine hydroxylase appears to be mediated by competition or complexing with the pteridine cofactor, the data do not rule out the alternative possibility that NA interacts with a separate regulatory or allosteric binding site on the enzyme. In any event, these data suggest a possible physiological mechanism for the regulation of catecholamine biosynthesis,[40,41] in which the rate of DM or NA synthesis is controlled by the now familiar type of negative feedback system in which the concentration of end product regulates the rate of an early step in the metabolic pathway. Recent studies have shown that the rate of NA synthesis in sympathetic nerves is related to nerve activity, and may be increased by electrical stimulation of such nerves.[42,43]

The synthesis of ADR from NA in the adrenal medulla, catalyzed by phenylethanolamine N-methyl transferase also appears to be under quite complex regulation *in vivo*. The N-methyl transferase is quite potently inhibited by its product, L-ADR (K_i = 50–100 μM)[44]; since the endogenous concentration of ADR in the adrenal medulla is extremely high, this probably represents a very effective control mechanism *in vivo* to limit the rate of ADR synthesis. In addition, a longer term control mechanism exists in that the synthesis of the N-methyl transferase enzyme in the adrenal medulla appears to be dependent on steroid hormones secreted by the surrounding adrenal cortex. Enzyme activity falls after hypophysectomy, and can be restored by the administration of ACTH or glucocorticoids.[45] This intriguing finding may help to explain the evolutionary significance of the close association between the tissues of the adrenal cortex and medulla.

E. Inhibitors of Catecholamine Biosynthesis

Although many effective inhibitors of aromatic amino acid decarboxylase and dopamine β-hydroxylase have been discovered (see Refs. 46,27), very few of these compounds proved to be effective inhibitors of NA synthesis *in vivo*. Since these enzymes, however, are not rate-limiting under normal conditions, and are present in great excess in most tissues, with hindsight this is no longer surprising. Tyrosine hydroxylase is clearly the point at which to approach the pharmacological control of the biosynthetic pathway, and in the last few years intense efforts have been made to find an effective long-lasting inhibitor of this enzyme. Currently available inhibitors of tyrosine hydroxylase, however, leave much to be desired. Although many inhibitors of the enzyme are now known[39] (Table VI), no potent irreversible inhibitors have so far been described. Inhibitors of tyrosine hydroxylases are either

TABLE VI

Inhibitors of Tyrosine Hydroxylase[a]

Tyrosine analogues	K_i (μM)
3-Iodo-α-methyl-L-Tyr	0.2
3-Iodo-L-Tyr	0.4
3,5-Diiodo-L-Tyr	9.3
α-Methyl-L-Tyr	17.0

Catechol compounds	Concentration for 50% inhibition—(mM)
Catechol	5
L-NA	1
L-DOPA	2
D-DOPA	4
L-α-MethylDOPA	1.5
3,4-Dihydroxyphenylpropyl-acetamide (H-22/54)	0.02

[a] From Udenfriend *et al.*[39] Partially purified enzyme from bovine adrenal medulla.

analogues of L-Tyr, which act as competitive inhibitors, or catechol compounds which appear to act competitively with the pteridine cofactor. Many
of these compounds, such as the iodo-tyrosines and L-α-methyl-Tyr have
been found to be effective inhibitors of catecholamine biosynthesis *in vivo*, and
can lead to a depletion of catecholamine stores in the intact animal.[47,48]

III. THE CATABOLISM OF CATECHOLAMINES

A. Metabolic Pathways

The availability of high specific activity tritium-labeled catecholamines
and studies of the metabolism of these compounds *in vivo* have led to great
advances in understanding of the metabolism of catecholamines.[49] NA,

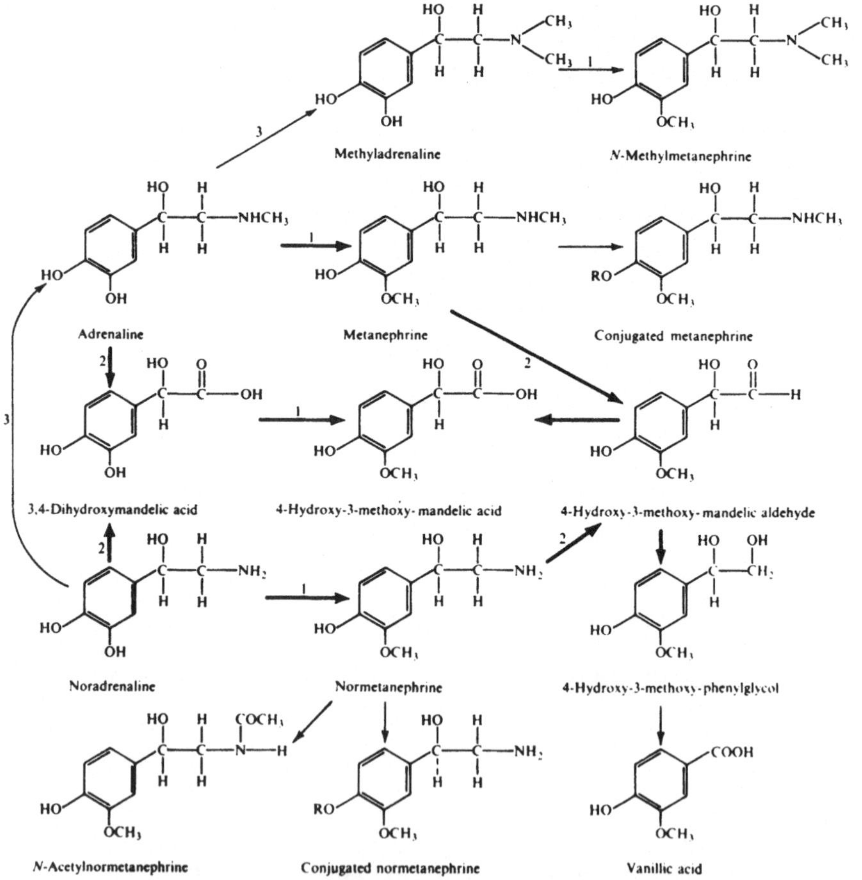

Fig. 4. Catabolic pathways for adrenaline and noradrenaline.

Thick lines indicate major pathways *in vivo*. 1 = COMT, 2 = MAO, 3 = phenylethanolamine-N-methyl transferase. After Axelrod.[49]

ADR, and DM are metabolized by a series of reactions involving 3-O-methylation catalyzed by the enzyme catechol-3-O-methyl transferase (COMT) and oxidative deamination catalyzed by the enzyme monoamine oxidase (MAO) outlined for ADR and NA in Fig. 4. Both O-methylation and deamination can occur in all tissues which contain catecholamines. In addition MAO and COMT occur in great abundance in the liver, kidney, and gastrointestinal tract tissue. The high enzyme activities of the latter tissues probably constitute a very important detoxication barrier, which normally prevents a variety of pharmacologically active amines absorbed from the intestine from gaining access to the general circulation. The liver and kidney enzymes also play an important role in the metabolism of catecholamines (particularly ADR) released into the circulation from the adrenal medulla. Axelrod and co-workers[52–53] showed that the major metabolic pathway for circulating catecholamines involved a primary O-methylation step, followed by extensive deamination of the O-methylated metabolites to yield 3-methoxy, 4-hydroxymandelic acid (VMA), and 3-methoxy, 4-hydroxyphenyl-glycol (MHPG). The urinary metabolites of injected DL-ADR-^{3}H in rat and man are summarized in Table VII.

TABLE VII
Metabolic Fate of Adrenaline-^{3}H[a]

Metabolite	Urine content as percentage of administered dose	
	Man	Rat
Free and conjugated ADR	6.5	14.1
3,4-Dihydroxymandelic acid + corresponding glycol:		
Free	0.8	1.4
Conjugated	0.8	1.2
Metanephrine:		
Free	5.0	4.8
Conjugated	33.7	28.3
3-Methoxy, 4-hydroxymandelic acid (VMA)	39.2	6.2
3-Methoxy, 4-hydroxyphenylglycol	6.8	17.6
Total recovered in urine	94.0	73.6

[a] Data from LaBrosse et al.[51] and Kopin et al.[52] DL-adrenaline-^{3}H was administered intravenously to humans and rats, and urine was collected for the following 54 hr (human) or 48 hr (rat) and analyzed for labeled metabolites of ADR. Results are mean values for 12 normal human subjects and 8 rats. In the rat significant amounts of the injected dose were excreted in the feces during this period.

Studies of the metabolism of catecholamines in the CNS have been hampered by the existence of a blood–brain barrier which prevents circulating labeled catecholamines from gaining access to the CNS. The metabolism of labeled catecholamines in the intact brain has been studied, however, after the injection of small amounts of labeled catecholamine into the CSF.[53,54] The metabolism of labeled NA and DM has also been studied in brain slices incubated *in vitro*.[55] Both of these methods confirm that NA and DM are metabolized by both *O*-methylation and deamination in all regions of the brain, as in the periphery. An interesting feature of NA metabolism in the brain is that the major end products are not acids, but the reduced alcohol derivatives, 3-methoxy, 4-hydroxyphenylethylglycol, and 3,4-dihydroxy-phenylglycol[55] (Table VIII). Although this is true for NA metabolism in both the intact brain and in tissue slices, labeled DM in the same tissue slices is metabolized principally to the acid metabolites 3-methoxy, 4-hydroxy-phenylacetic acid (homovanillic acid), and 3,4-dihydroxyphenylacetic acid, with only minor amounts of the corresponding alcohols,[55] (Table VIII).

TABLE VIII

Metabolism of Noradrenaline and Dopamine by Slices of Rabbit Cerebral Cortex[a]

NA-^{3}H		DM-^{14}C	
Individual metabolites as percentage of total ^{3}H metabolites		Individual metabolites as percentage of total ^{14}C metabolites	
3,4-Dihydroxymandelic acid	16.0	3,4-Dihydroxyphenyl-acetic acid	57.7
3-Methoxy, 4-hydroxy-mandelic acid (VMA)	6.2	3-Methoxy, 4-hydroxy-phenylacetic acid	23.2
Normetanephrine	16.2	3-Methoxytyramine	7.1
3,4-Dihydroxyphenyl-glycol	26.2	3,4-Dihydroxyphenyl-ethanol	0.9
3-Methoxy, 4-hydroxy-phenylglycol	35.4	3-Methoxy, 4-hydroxy-phenylethanol	0.6
		Noradrenaline	6.1
		Noradrenaline metabolites	4.2

[a] Data calculated from Rutledge and Jonason.[55] Cortex slices were incubated for 30 min in a medium containing 1.5×10^{-6}M, DL-NA^{3}H or DM-^{14}C. Metabolites were assayed in slices + medium. Results are mean values for several experiments. Note: At the end of the 30 min incubation, 38.0% of the added NA-^{3}H and 46.5% of the added DM-^{14}C had been metabolized.

B. Individual Enzymes

1. Catechol-O-Methyl Transferase

(*S*-adenosyl-methionine: catechol *O*-methyl transferase. E.C. 2.1.1.6)

The partially purified enzyme from rat liver utilizes *S*-adenosyl-L-methionine as methyl donor, and has an absolute requirement for magnesium or other divalent cation.[56] Catecholamines are methylated principally on the *meta* position, but the *para* position can also be methylated to some extent depending on its nucleophilic characteristics. 3,4-Dihydroxyacetophenone, with a strongly nucleophilic *para*-phenolic hydroxyl group, yields about 50% each of *para* and *meta* O-methylated products, while dopamine yields 90% *meta* and 10% *para* O-methylated products.[57] A variety of catechol compounds and also trihydroxy phenolic compounds may be O-methylated by COMT. Monophenols, however, are not substrates. While O-methylation may occur in either *meta* or *para* positions with catechol substrates, there is no evidence that 3,4-dimethoxy products can be formed by this enzyme. When COMT is incubated with 3-methoxy, 4-hydroxyphenylethylamine, or 3-hydroxy, 4-methoxyphenylethylamine, a partial interconversion of the two isomers takes place, but no 3,4-dimethoxyphenylethylamine (DMPEA) can be detected.[58] This finding is of interest in view of recent speculation that DMPEA may be formed as an abnormal metabolite of catecholamines in schizophrenia. K_m values for catechol substrates such as NA and ADR are in the range $1-4 \times 10^{-4}$ M. The K_m values for catechol substrates of this, and other transferase enzymes (cf. Table V), vary according to the fixed concentration of second substrate (S-adenosyl Met) used. In a careful study of the kinetic properties of partially purified COMT from rat brain, D'Iorio and Mevrides[59] showed that a series of K_m values for ADR or for S-adenosyl-Met could be obtained at various fixed concentrations of the second substrate. From these values it was possible to estimate that the K_s for ADR (dissociation constant for the ADR-binding site) was 0.5×10^{-4} M. K_m values for S-adenosyl-Met ranged from 4×10^{-5} M to 8×10^{-5} M according to the concentration of ADR present.

2. Monoamine Oxidase

(Monoamine: oxygen oxidoreductase deaminating, E.C. 1.4.3.4)

Despite its relatively long history and considerable physiological and pharmacological importance, monoamine oxidase (MAO) remains a very poorly characterized enzyme. Indeed it is not yet clear whether the term monoamine oxidase describes the activity of a single enzyme or of several different enzymes or isozymes. Differences in the substrate specificity and sensitivity to inhibitors of MAO in different tissues and within the same tissue have been reported.[60] Gorkin[60] has also reported a partial chromatographic separation of what appear to be different forms of the enzyme from rat liver mitochondria.

In all tissues MAO activity is firmly associated with the lipoprotein membranes of mitochondria and has proved remarkably resistant to many attempts to obtain a soluble purified preparation. Considerable advances have been made in recent years, however, and the properties of soluble purified preparations from liver, kidney, and brain have been described.[61-63] Even the most highly purified preparations, however, appear to contain micellar or membranous structures.[62] Purified preparations of MAO contain enzyme-bound flavine groups, (FAD), but enzyme-bound copper does not seem to be a requisite for enzyme activity.[62,63a] MAO will catalyze the oxidative deamination of a wide variety of monoamines according to the reaction:

$$R \cdot CH_2 \cdot NH_2 + H_2O + O_2 \rightarrow R \cdot CHO + NH_3 + H_2O_2$$

The resulting aldehydes are further metabolized either by reduction to the corresponding alcohol (catalyzed by alcohol dehydrogenase), or by oxidation to the acid. Little attention has been paid to the latter steps. The oxidation of the aldehyde products is thought to be catalyzed by aldehyde dehydrogenase, an enzyme of broad specificity which is present in many tissues.[64] K_m values of pig brain MAO for catecholamine substrates are shown in Table IX. The O-methylated metabolites of NA and ADR have lower K_m values for pig brain MAO than the parent catecholamines. Despite conflicting reports, it seems clear that MAO exhibits stereochemical selectivity in favor of the naturally occurring L-anantiomers of optically active amines such as NA, ADR, and related compounds.[65] Many of the wide range of drugs which are structural analogues of NA are deaminated by MAO, tyramine and β-phenylethylamine being particularly good substrates. Analogues with α-methyl or other substituents, such as amphetamine, are not substrates but may act as competitive inhibitors.

TABLE IX

K_m Values for Monoamine Oxidase Purified from Pig Brain[a]

Substrate	K_m ($M \times 10^{-4}$)
Tyramine	1.20
Tryptamine	1.15
5-Hydroxytryptamine	2.78
L-Noradrenaline	0.65
L-Adrenaline	0.88
DL-Normetanephrine	0.25
DL-Metanephrine	0.40

[a] All K_m values were determined at saturating oxygen concentrations. Data kindly made available by Dr. K. Tipton (unpublished).

C. Cellular and Subcellular Localization of MAO and COMT

Although MAO is thought to occur in adrenergic neurons, the MAO activity of extraneuronal tissues is usually very much greater than that associated with the sympathetic innervation, so that little or no change in the MAO activity of peripheral organs is generally found after surgical sympathectomy or immunosympathectomy.[66] In a few tissues (rabbit, iris, rat pineal gland),[67] however, this is not the case, and a marked drop in MAO activity has been observed in these organs after surgical sympathectomy. Histochemical studies confirm that there is no specific association of enzyme activity with the peripheral sympathetic nervous system or with catecholamine-containing neurons in the CNS.[68] Precisely which cells other than sympathetic neurons contain MAO is not yet clear. In the CNS not only neurons but also glial cells associated with cerebral blood vessels appear to contain MAO.[69] In all tissues MAO activity appears to be entirely localized in mitochondria. Thus in brain homogenates most of the MAO activity is recovered in a purified fraction of free mitochondria, although small amounts of the enzyme may also be present in mitochondria entrapped within synaptosome particles.[70] Recent studies in which liver and brain mitochondria were subfractionated show that MAO is predominantly localized in the outer membranes of the mitochondrion.[71,72] When sympathetically innervated smooth muscle tissues or peripheral sympathetic nerves are homogenized, a considerable amount of MAO activity may be recovered in a microsomal fraction on differential or density-gradient centrifugation.[73] The "microsomal" enzyme activity, however, is most probably associated with fragments of mitochondrial membranes formed as an artifact during homogenization.[74]

The cellular localization of COMT is also obscure. High enzyme activity is present in the liver and kidney, but lower activity is also associated with all parts of the peripheral sympathetic nervous system and the CNS. In the brain the regional distribution of COMT correlates fairly well with that of endogenous catecholamines.[75] In the CNS and in peripheral tissues the activity of COMT is generally lower than the MAO activity. This may be related to the fact that whereas COMT is exclusively related to catecholamine metabolism, a variety of other amines such as 5-hydroxytryptamine are important substrates for MAO. Although COMT is probably specifically related to catecholamine metabolism, the enzyme does not appear to be present in catecholamine-containing cells since the COMT activity of peripheral tissues is not reduced after sympathectomy.[76] A major part of the COMT activity in homogenates of peripheral tissues or CNS is associated with the cell sap fraction. However, in brain homogenates some COMT activity is also associated with synaptosomal fractions.[77]

D. Inhibitors of MAO and COMT

Because of the clinical usefulness of MAO inhibitors in the treatment of angina, hypertension, and certain forms of depression, a very large number of such substances has been developed (see Ref. 46). Some representatives of the main types of MAO inhibitor are illustrated in Fig. 5. While reversible competitive inhibitors, such as harmaline and amphetamine, are known, most of the clinically useful drugs produce an irreversible inhibition of the enzyme. In many cases irreversible inhibition of the enzyme follows an initially reversible binding of the drug to the enzyme, or metabolic conversion of the inhibitor

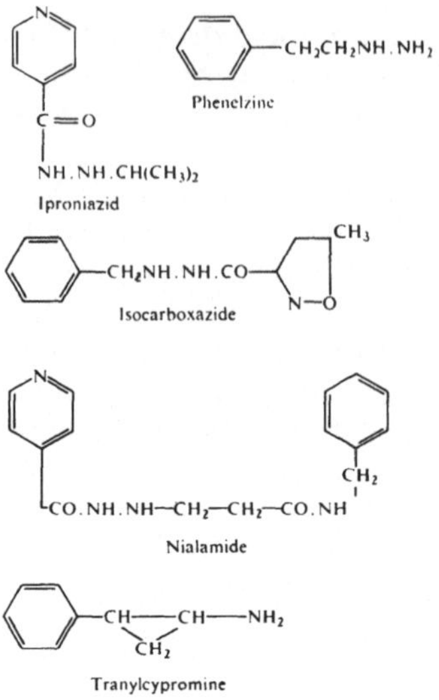

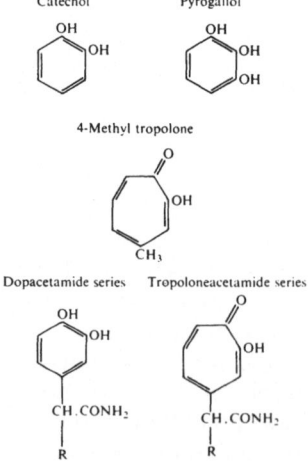

Fig. 5. Some monoamine oxidase inhibitors.

to an active derivative.[78] Repeated therapeutic doses of the long-lasting MAO inhibitors can produce a near total inhibition of MAO in man or experimental animals.

Few effective inhibitors of COMT are available. Pyrogallol and several related compounds are quite potent competitive inhibitors, but their effects are not long lasting *in vivo*. COMT is also inhibited by a variety of tropolone derivatives, papaverine derivatives, thyroxines, and dopacetamides (Fig. 6).[59,79,80] The latter group includes H-22/54 (3,4-dihydroxyphenyl-propylacetamide) mentioned previously as an inhibitor of tyrosine hydroxylase.

Fig. 6. Some catechol *O*-methyl transferase inhibitors.

E. Role of MAO and COMT in Adrenergic Transmission

Neither MAO or COMT appear to play a role at adrenergic nerve endings analogous to that of acetylcholinesterase at cholinergic synapses. As discussed above, there is no evidence that either enzyme is specifically localized at adrenergic synapses. Nor does inhibition of either or both enzymes produce a marked potentiation or prolongation of adrenergic responses, as found in the cholinergic system with anticholinesterases.[81] NA released at sympathetic nerve terminals appears to be rapidly inactivated not by metabolism, but by a transport mechanism which transfers NA from its site of action in the synaptic cleft back into the nerve terminals.[82] The operation of this uptake process is probably responsible for the finding that very little overflow of NA is found in the venous effluent of peripheral organs during sympathetic nerve stimulation.[83] While inhibitors of the uptake system may increase this overflow, inhibitors of COMT or MAO have little or no effect.[84] Although uptake into sympathetic nerves represents a major pathway for the inactivation of extracellular catecholamines in both peripheral tissues and in the brain,[85] some metabolism of extracellular catecholamines does occur at extraneuronal sites. The major metabolic products formed during the metabolism of extracellular catecholamines are the O-methylated and O-methylated/deaminated derivatives. This is the case, for example, when labeled NA is injected either into the circulation or into the CSF, or when NA is released from sympathetic nerve endings by nerve stimulation or by indirectly acting sympathomimetic amines such as tyramine.[86] In order to be metabolized by MAO and COMT in extraneuronal sites, the catecholamines have to penetrate into the cells which contain these enzymes. Recent evidence suggests that the entry of catecholamines into such extraneuronal cells may also involve a mediated transport system, since the extraneuronal metabolism of labeled NA can be blocked by a variety of drugs (such as phenoxybenzamine, imipramine, normetanephrine) which are not themselves inhibitors of either MAO or COMT.[87,88] Within adrenergic neurons, in the periphery and in the CNS, MAO appears to play some role in regulating the storage level of endogenous catecholamines. After treatment with MAO inhibitors, the catecholamine content of these neurons increases, often very markedly.[89] Further evidence that MAO plays a role in the intraneuronal metabolism of catecholamines is provided by the effects of reserpine and similar drugs on catecholamine metabolism. Reserpine and related drugs are thought to act by destroying binding sites within the intraneuronal storage particles of adrenergic neurons. This in turn leads to an almost complete depletion of the normal catecholamine content of these neurons, both in the sympathetic nervous system and in the CNS. However, the depletion of catecholamine content by reserpine does not lead to a release of free catecholamines from the cell, but is accompanied by a marked increase in deaminated metabolites of the catecholamines, and particularly in the catechol-deaminated metabolites (3,4-dihydroxymandelic acid, 3,4-dihydroxyphenylacetic acid, and the corresponding glycol).[10] This suggests that when NA or DM are released intraneuronally from their normal storage sites by reserpine or similar drugs, they are metabolized by intraneuronal MAO and the resulting catechol-deaminated metabolites are then released. Indeed there is reason to suppose that catechol-deaminated metabolites arise largely from the activity of intraneuronal MAO, whereas extraneuronal deamination occurs only after a primary O-methylation step to give the O-methylated/ deaminated metabolites.

In summary, MAO appears to play a role in regulating the intraneuronal storage levels of NA and DM in catecholamine-containing neurons in the periphery and in the CNS. The functional significance of MAO and COMT in extraneuronal sites, however, is still not clear, although it seems unlikely that extraneuronal metabolism plays a primary role in terminating the actions of neurally released catecholamines.

IV. OTHER PATHWAYS OF CATECHOLAMINE METABOLISM

A. Alternative Biosynthetic Pathways

As indicated in Fig. 1, a number of alternative pathways exist, in principle, for the biosynthesis of catecholamines. An enzyme system has been described in the microsomal fraction of liver which is capable of hydroxylating monophenolic substrates such as tyramine or octopamine to form catechol derivatives.[91,92] Since tyramine can be formed from L-Tyr by the action of aromatic amino acid decarboxylase, this suggests the possibility of the pathways:

$$\text{L-Tyr} \rightarrow \text{tyramine} \rightarrow \text{norsynephrine} \rightarrow \text{NA}$$

or: $$\text{L-Tyr} \rightarrow \text{tyramine} \rightarrow \text{dopamine} \rightarrow \text{NA}$$

However, although very small amounts of labeled NA can be detected after the administration of labeled tyramine *in vivo*,[93] this is unlikely to be a pathway of any quantitative significance under normal conditions.

B. Oxidation of Catecholamines to Adrenochromes and Melanin

An enzyme has been described in cat salivary glands which catalyzes the oxidation of ADR to adrenochrome.[94] The enzyme is most active in the cat parotid gland, but is also present in the salivary glands of other species and in other tissues, such as skin and lung. The presence of this enzyme may be related to the finding that small amounts of tritium-labeled water are formed in salivary glands and saliva after the injection of NA-^{3}H.[95]

Rat brain contains an enzyme catalyzing the oxidative conversion of dopamine (but not NA or ADR) to melanin.[96] This finding may be related to the occurrence of melanin deposits in regions of the brain such as the substantia nigra, in which dopamine-containing neurons are concentrated.

V. COMPARATIVE ASPECTS OF CATECHOLAMINE METABOLISM

While there have been few detailed studies of comparative aspects of catecholamine metabolism, the major biosynthesis and catabolic pathways outlined above appear to apply to all vertebrate species. An important difference between amphibians and other classes, already noted, is that ADR rather than NA is present in the sympathetic nerves and CNS, implying that phenylethanolamine-N-methyl transferase activity is widely distributed in the tissues of such animals. The COMT and MAO activities of peripheral tissues vary considerably among vertebrate species (Table X), but there have been few quantitative studies of the relative importance of O-methylation and deamination in the catabolism of NA and ADR in different species. There are interesting species differences in the role of MAO in regulating the catecholamine content of CNS neurons. Whereas in most species MAO inhibition leads to a marked increase in the brain level of NA, this is not the case in some species, such as cat and dog.[97]

Cardiac MAO in the rat shows a remarkable continuing increase in activity throughout the life of the animal. In addition to this, adult female rats have almost twice as high cardiac MAO activity as males of the same age.[98] The MAO activity of other rat tissues does not show these changes with age and sex.*

TABLE X

Comparison of Monoamine Oxidase and Catechol O-Methyl Transferase Activity in the Hearts of Various Vertebrate Species[a]

Species		MAO activity (µmoles tyramine/hr/g protein)	COMT activity (µmoles 3,4-DHBA/hr/g protein)
Rat	Atria	104	7.5
	Ventricles	242	5.1
Mouse	Atria	99	9.6
	Ventricles	35	13.3
Guinea pig	Atria	80	4.3
	Ventricles	80	4.3
Pigeon	Atria	24	5.9
	Ventricles	13	5.5
Toad	Atria	49	1.1
(*Bufo marinus*)	Ventricles	56	3.8

[a] MAO was assayed using tyramine-^{3}H as substrate; COMT with 3,4-dihydroxybenzoic acid (3,4-DHBA) and ^{3}H-methyl-S-adenosyl-methionine as substrates. Results are mean values for four determinations. Data kindly made available by Mr. B. Jarrott (unpublished observations).

* The material reviewed in this chapter does not, in general, include information published since the end of 1967.

VI. REFERENCES

1. K. Fuxe, Evidence for the existence of monoamine neurons in the central nervous system. IV. Distribution of monoamine nerve terminals in the central nervous system, *Acta Physiol. Scand.*, **64**:Suppl. **247** (1965).
2. O. Hornykiewicz, Dopamine (3-hydroxy-tyramine) and brain function, *Pharmacol. Rev.* **18**:925–964 (1966).
3. U. S. Euler, *Noradrenaline*, Charles C. Thomas, Springfield, Ill. (1956).
4. B. Falk, T. Nystedt, E. Rosengren, and J. Stenflo, Dopamine and mast cells, *Acta Pharmacol. Tox.* **21**:51–81 (1964).
5. H. Blaschko, The specific action of L-DOPA decarboxylase, *J. Physiol. (London)* **96**, 5–51P. (1939).
6. D. J. Demis, H. Blaschko, and A. D. Welch, The conversion of dihydroxyphenylal-anine-2-C^{14} (DOPA) to norepinephrine by bovine adrenal medullary homogenates. *J. Pharmacol. Exp. Ther.* **113**:14–15 (1956).
7. S. Udenfriend and J. B. Wyngaarden, Precursors of adrenal epinephrine and norepinephrine *in vivo*, *Biochim. Biophys. Acta* **20**:48–52 (1956).
8. J. Pellerin and A. D'Iorio, Metabolism of DL-3,4-dihydroxyphenylalanine-α-C^{14} in bovine adrenal homogenate, *Canad. J. Biochem. Physiol.* **35**:151–156 (1957).
9. P. Hagen, Biosynthesis of norepinephrine from 3,4-dihydroxyphenylethylamine (dopamine), *J. Pharmacol. Exp. Therap.* **116**:26–27 (1956).
10. McC. Goodall and N. Kirshner, Biosynthesis of epinephrine and norepinephrine by sympathetic nerves and ganglia, *Circulation* **17**:366–371 (1956).
11. D. T. Masuoka, H. F. Schott, and L. Petriello, Formation of catecholamines by various areas of cat brain, *J. Pharmacol. Exp. Therap.* **139**:73–76 (1963).
12. J. M. Musacchio and M. Goldstein, Biosynthesis of norepinephrine and norsynephrine in the perfused rabbit heart, *Biochem. Pharmacol.* **12**:1061–1063 (1963).
13. S. Spector, A. Sjoerdsma, P. Zaltzmann-Nirenberg, M. Levitt, and S. Udenfriend, Norepinephrine synthesis from tyrosine-C^{14} in isolated perfused guinea pig hearts, *Science* **139**:1299–1301 (1963).
14. C. A. Stone and C. C. Porter, Biochemistry and pharmacology of methyldopa and some related structures, *Adv. Drug Res.* **4**:71–93 (1967).
15. T. Nagatsu, M. Levitt, and S. Udenfriend, The initial step in norepinephrine biosynthesis, *J. Biol. Chem.* **239**:2910–2917 (1964).
16. S. P. Bagchi and P. L. McGeer, Some properties of tyrosine hydroxylase from the caudate nucleus, *Life Sciences* **3**:1195–1200 (1964).
17. S. Udenfriend, Tyrosine hydroxylase, *Pharmacol. Rev.* **18**:43–52 (1966).
18. S. Udenfriend, Biosynthesis of the sympathetic neutrotransmitter, norepinephrine, *Harvey Lectures, 1964–65*:57–84 (1966).
19. M. Ikeda, M. Levitt, and S. Udenfriend, Hydroxylation of phenylalanine by purified preparations of adrenal and brain tyrosine hydroxylase, *Biochem. Biophys. Res. Comm.* **18**:482–488 (1965).
20. S. Udenfriend, P. Zaltzmann-Nirenberg, and T. Nagatsu, Inhibitors of purified beef adrenal tyrosine hydroxylase, *Biochem. Pharmacol.* **14**:837–845 (1965).
21. L. Maître, Presence of α-methyl-DOPA metabolites in heart and brain of guinea pigs treated with α-methyl-tyrosine, *Life Sciences* **4**:2249–2256 (1965).
22. T. L. Sourkes, DOPA decarboxylase: substrates, coenzyme, inhibitors, *Pharm. Rev.* **18**:53–60 (1956).
23. D. Aures, R. Håkanson, and W. C. Clark, Histidine decarboxylase and DOPA decarboxylase, this volume, pp. 165–196.

24. W. Lovenberg, H. Weissbach, and S. Udenfriend, Aromatic L-amino acid decarboxylase, J. Biol. Chem. 237:89–93 (1962).
25. S. Kaufman and S. Friedman, Dopamine-β-hydroxylase, Pharmacol. Rev. 17:71–100.
26. M. Goldstein, in The Biochemistry of Copper, pp. 443–453, Academic Press, New York (1956).
27. J. B. Van der Schoot and C. R. Creveling, Substrates and inhibitors of dopamine-β-hydroxylase, Adv. Drug Research. 2:47–88 (1965).
28. M. Goldstein and J. F. Contrera, The substrate specificity of phenylamine-β-hydroxylase, J. Biol. Chem. 237:1898–1902 (1962).
29. J. Axelrod, Purification and properties of phenylethanolamine-N-methyl transferase, J. Biol. Chem. 237:1657–1660 (1962).
30. J. Axelrod, Methylation reactions in the formation and metabolism of catecholamines and other biogenic amines, Pharmac. Rev. 18:95–113 (1966).
31. R. W. Fuller and J. M. Hunt, Substrate specificity of phenethanolamine N-methyl transferase, Biochem. Pharmacol. 14:1896–1897 (1965).
32. J. Axelrod, N-Methyladrenaline, a new catecholamine in the adrenal gland, Biochim. Biophys. Acta 45:614–615 (1960).
33. L. L. Iversen, The Uptake and Storage of Noradrenaline in Sympathetic Nerves, pp. 56–57, Cambridge University Press, London (1967).
34. C. Owman and E. Rosengren, Dopamine formation in brain capillaries—an enzymic blood–brain barrier mechanism, J. Neurochem. 14:547–550 (1967).
35. L. Stjärne and F. Lishajko, Localization of different steps in noradrenaline synthesis to different fractions of a bovine splenic nerve homogenate, Biochem. Pharmacol. 16:1719–1728 (1967).
36. L. T. Potter, Storage of norepinephrine in sympathetic nerves, Pharm. Rev. 18:439–451 (1966).
37. P. L. McGeer, S. P. Bagchi, and E. G. McGeer, Subcellular localization of tyrosine hydroxylase in beef caudate nucleus, Life Sci. 4:1859–1867 (1965).
38. M. Levitt, S. Spector, A. Sjoerdsma, and S. Udenfriend, Elucidation of the rate-limiting step in norepinephrine biosynthesis in the perfused guinea pig heart, J. Pharmacol. Exp. Therap. 148:1–8 (1965).
39. E. G. McGeer and P. L. McGeer, In vitro screen of inhibitors of rat brain tyrosine hydroxylase, Can. J. Biochem. 45:115–131 (1967).
40. S. Spector, R. Gordon, A. Sjoerdsma, and S. Udenfriend, End-product inhibition of tyrosine hydroxylase as a possible mechanism for regulation of norepinephrine synthesis, Mol. Pharmacol. 3:549–555 (1967).
41. E. Costa and N. H. Neff, this volume, pp. 45–90.
42. R. H. Roth, L. Stjärne, and U. S. von Euler, Acceleration of noradrenaline biosynthesis by nerve stimulation, Life Sci. 5:1071–1075 (1966).
43. R. Gordon, J. V. O. Reid, A. Sjoerdsma, and S. Udenfriend, Increased synthesis of norepinephrine in the rat heart on electrical stimulation of the stellate ganglia, Mol. Pharmacol. 2:606–613 (1966).
44. R. W. Fuller and J. M. Hunt, Inhibition of phenethanolamine N-methyl transferase by its product, epinephrine, Life Sci. 6:1107–1112 (1967).
45. R. J. Wurtman and J. Axelrod, Control of enzymatic synthesis of adrenaline in the adrenal medulla by adrenal cortical steroids, J. Biol. Chem. 241:2301–2305 (1966).
46. A. Pletscher, K. F. Gey, and W. P. Burkard, Inhibitors of monoamine oxidase and decarboxylase of aromatic amino acids, Handb. d. Exp. Pharmacol. Vol. XIX, pp. 593–735, Springer-Verlag, Berlin (1964).
47. S. Spector, A. Sjoerdsma, and S. Udenfriend, Blockade of endogenous norepinephrine synthesis by α-methyl-tyrosine, an inhibitor of tyrosine hydroxylase, J. Pharmacol. Exp. Therap. 147:86–95 (1966).

48. S. Spector, R. O. Mata, A. Sjoerdsma, and S. Udenfriend, Biochemical and pharmacological effects of iodo-tyrosines; relation to tyrosine hydroxylase inhibition *in vivo*, *Life Sci.* **4**:1307–1311 (1965).

49. J. Axelrod, The metabolism, storage and release of catecholamines, *Recent Prog. Horm. Res.* **21**:597–622 (1965).

50. J. Axelrod, J. K. Inscoe, S. Senoh, and B. Witkop, *O*-Methylation, the principal pathway for the metabolism of epinephrine and norepinephrine in the rat, *Biochim. Biophys. Acta* **27**:210–211 (1958).

51. E. H. La Brosse, J. Axelrod, I. J. Kopin, and S. S. Kety, Metabolism of 7-H^3-norepinephrine-d-bitartrate in normal young man, *J. Clin. Invest.* **40**:253–260 (1961).

52. I. J. Kopin, J. Axelrod, and E. K. Gordon, The metabolic fate of H^3-epinephrine and C^{14}-metanephrine, *J. Biol. Chem.* **236**:2109–2113 (1961).

53. E. Mannarino, N. Kirshner, and B. S. Nashold, Jr. The metabolism of C^{14}-noradrenaline by cat brain, *J. Neurochem.* **10**:373–379 (1963).

54. J. Glowinski, I. J. Kopin, and J. Axelrod, Metabolism of H^3-norepinephrine in the rat brain, *J. Neurochem.* **12**:25–30 (1965).

55. C. O. Rutledge and J. Jonason, Metabolic pathways of dopamine and norepinephrine in rabbit brain *in vitro*, *J. Pharm. Exp. Therap.* **157**:493–502 (1967).

56. J. Axelrod and R. Tomchick, Enzymatic *O*-methylation of epinephrine and other catechols, *J. Biol. Chem.* **233**:702–705 (1958).

57. S. Senoh, J. Daly, and J. Axelrod, Enzymatic *p-O*-methylation by catechol *O*-methyl transferase, *J. Am. Chem. Soc.* **81**:6240–6245 (1959).

58. J. W. Daly, J. Axelrod, and B. Witkop, Dynamic aspects of enzymatic *O*-methylation and demethylation of catechols *in vitro* and *in vivo*. *J. Biol. Chem.* **235**:1155–1159 (1960).

59. A. D'Iorio and C. Mavrides, Actions of the thyroid hormones and analogues *in vitro* on catechol *O*-methyltransferase, *Biochem. Pharm.* **12**:1307–1313 (1963).

60. V. Z. Gorkin, Monoamine oxidase, *Pharmacol. Rev.* **18**:115–120 (1966).

61. S. Nara, B. Gomes, and K. T. Yasunobu, Amine oxidase VII. Beef–liver mitochondrial monoamine oxidase, a copper containing protein, *J. Biol. Chem.* **241**:2774–2779.

62. V. G. Erwin and L. Hellerman, Mitochondrial monoamine oxidase—I. Purification and characterization of the bovine kidney-enzyme, *J. Biol. Chem.* **242**:4230–4238 (1967).

63a. K. F. Tipton, The prosthetic groups of pig brain mitochondrial monoamine oxidase, *Biochim. Biophys. Acta* **159**:451–459 (1968).

63b. K. F. Tipton, The purification of pig brain mitochondrial monoamine oxidase, *European J. Biochem.* **4**:103–107 (1968).

63c. T. L. Sourkes, Properties of the monoamine oxidase of rat liver mitochondria, *Adv. Pharmacol.* **6A**:61–69 (1968).

64. R. A. Deitrich, Tissue and subcellular distribution of mammalian aldehyde-oxidizing capacity, *Biochem. Pharm.* **15**:1911–1922 (1966).

65. A. Giachetti and P. Shore, Optical specificity of monoamine oxidase, *Life Sci.* **5**:1373 (1966).

66. L. L. Iversen and J. Glowinski, The physiological disposition and metabolism of norepinephrine in immunosympathectomized animals, *J. Pharmacol. Exp. Therap.* **151**:273–284 (1966).

67. S. H. Snyder, J. Fischer, and J. Axelrod, Evidence for the presence of monoamine oxidase in sympathetic nerve endings, *Biochem. Pharmacol.* **14**:363–365 (1965).

68. G. B. Koelle and A. de T. Valk, Physiological implications of the histochemical localization of monoamine oxidase, *J. Physiol.* (*London*) **126**:434–447 (1954).

69. A. Bertler, B. Falck, C. Owman, and E. Rosengren, The localization of monoaminergic blood–brain barrier mechanisms, *Pharmacol. Rev.* **18**:369–385 (1966).

70. G. R. De Lores Arnaiz and E. De Robertis, Cholinergic and non-cholinergic nerve endings in the rat brain : II. Subcellular localization of monoamine oxidase and succinate dehydrogenase, *J. Neurochem.* **9**:503–508 (1962).

71. C. Schnaitman, V. G. Erwin, and J. W. Greenwalt, The submitochondrial localization of monoamine oxidase, *J. Cell. Biol.* **32**:719–735 (1967).

72. K. F. Tipton, The sub-mitochondrial localization of monoamine oxidase in rat liver and brain, *Biochim. Biophys. Acta* **135**:910–920 (1967).

73. R. H. Roth and L. Stjärne, Monoamine oxidase activity in the bovine splenic nerve granule preparation, *Acta Physiol. Scand.* **68**:342–346 (1966).

74. B. Jarrott and L. L. Iversen, The subcellular distribution of monoamine oxidase activity in rat liver and vas deferens. *Biochem. Pharmacol.* **17**:1619–1625 (1968).

75. J. Axelrod, R. W. Albers, and C. D. Clements, Distribution of catechol *O*-methyl transferase in the nervous system and other tissues, *J. Neurochem.* **5**:68–72 (1959).

76. L. T. Potter, T. Cooper, V. L. Willman, and D. E. Wolfe, Synthesis, binding, release and metabolism of norepinephrine in normal and transplanted dog hearts, *Circulation Res.* **16**:468–481 (1965).

77. M. Alberici, G. R. De Lores Arnaiz, and E. De Robertis, Catechol *O*-methyl transferase in nerve endings of rat brain, *Life Sci.* **4**:1951–1960 (1965).

78. A. Pletscher, Monoamine oxidase inhibitors, *Pharmacol. Rev.* **18**:121–129 (1966).

79. S. B. Ross and O. Haljasmaa, Catechol *O*-methyl transferase inhibitors, *In vitro* inhibition of the enzyme in mouse brain extract, *Acta Pharmacol. Tox.* **21**:205–214 (1964).

80. A. Carlsson, M. Lindqvist, S. Fila-Hromadko, and H. Corrodi, Synthese von Catechol-*O*-Methyl Transferase-hemmenden Verbindurgen, *Helv. Chim. Acta* **45**:270–276 (1962).

81. J. R. Crout, Effect of inhibiting both catechol *O*-methyltransferase and monoamine oxidase on cardiovascular responses to norepinephrine, *Proc. Soc. Exp. Biol. Med.* **108**:482–484 (1961).

82. L. L. Iversen, *The Uptake and Storage of Noradrenaline in Sympathetic Nerves*, Cambridge University Press, London (1967).

83. G. L. Brown, The release and fate of the transmitter liberated by adrenergic nerves, *Proc. Roy. Soc. Ser.* B **162**:1–19 (1965).

84. R. H. Stinson, Electrical stimulation of the sympathetic nerves of the isolated rabbit ear and the fate of the neurohormone released, *Can. J. Biochem.* **39**:309–316 (1961).

85. J. Glowinski and R. J. Baldessarini, Metabolism of norepinephrine in the central nervous system, *Pharm. Rev.* **18**:1201–1238 (1966).

86. J. Axelrod, in *Mechanisms of Release of Biogenic Amines*, pp. 189–209, Pergamon Press, Oxford (1966).

87. A. J. Eisenfeld, J. Axelrod, and L. Krakoff, Inhibition of the extraneuronal accumulation and metabolism of norepinephrine by adrenergic blocking agents, *J. Pharmacol. Exp. Therap.* **156**:107–113 (1967).

88. A. J. Eisenfeld, L. Landsberg, and J. Axelrod, Effect of drugs on the accumulation and metabolism of extraneuronal norepinephrine in the rat heart, *J. Pharmacol. Exp. Therap.* **158**:378–385 (1967).

89. B. B. Brodie, S. Spector, and P. A. Shore, Interaction of drugs with norepinephrine in the brain, *Pharmacol. Rev.* **11**:548–564 (1959).

90. J. Glowinski, L. L. Iversen, and J. Axelrod, Storage and synthesis of norepinephrine in the reserpine-treated rat brain, *J. Pharmacol. Exp. Therap.* **151**:385–399 (1966).

91. J. Axelrod, J. K. Inscoe, and J. Daly, Enzymatic formation of *O*-methylated dihydroxy derivatives from phenolic amines and indoles, *J. Pharmacol. Exp. Therap.* **149**:16–22 (1965).

92. L. Lemberger, R. Kuntzman, A. H. Conney, and J. J. Burns, Metabolism of tyramine to dopamine by liver microsomes, *J. Pharmacol. Exp. Therap.* **150**:292–297 (1965).

93. C. R. Creveling, M. Levitt, and S. Udenfriend, An alternative route for the biosynthesis of norepinephrine, *Life Sci.* **1**:523–526 (1962).

94. J. Axelrod, Enzymic oxidation of epinephrine to adrenochrome by the salivary gland, *Biochim. Biophys. Acta* **85**:247–254 (1964).

95. J. Suko, O. Linet, and G. Hertting, Tritiumwasser in speicheldrussen im Speichel und Plasma der Katze nach Injektion von DL-7-H³-Noradrenaline, *Arch. Pharmakol. u. Exp. Path.* **256**:439–449 (1967).

96. C. Vanderwende, Studies on the oxidation of dopamine to melanin by rat brain, *Arch. Int. Pharmacodyn.* **152**:433–444 (1964).

97. S. Sanan and M. Vogt, Effect of drugs on the noradrenaline content of brain and peripheral tissues and its significance, *Brit. J. Pharmacol.* **18**:109–127 (1962).

98. A. Horita, Cardiac monoamine oxidase in rat, *Nature* **215**:411–412 (1967).

Chapter 10

HISTAMINE

Jack Peter Green*

Department of Pharmacology
Mount Sinai School of Medicine
New York, New York

Histamine has attracted less attention than most other biogenic amines in the nervous system, probably because a specific and sensitive chemical method for its determination has only recently become available. It is likely that future work will place histamine in that group of biogenic amines for which neural function is suspected. Even now there are some observations on histamine analogous to those which on other biogenic amines have stimulated hypotheses about function.

I. MEASUREMENT OF HISTAMINE IN NEURAL TISSUES

Unless special care is taken, both the bioassay and all the chemical methods yield erroneously high levels of brain histamine. This risk is especially high with brain because of the relatively low levels of histamine compared to the high levels of interfering substances.

In the bioassay, which is performed on the isolated guinea pig ileum (always in the presence of drugs that block the action of acetylcholine and 5-hydroxytryptamine), the interfering materials are "slow-reacting substances," and substances with antihistaminic activity.[1] The slow-reacting substances are the troublesome ones. They are found in many tissues but in notably high activity in brain. In low concentrations, slow-reacting substances cause the ileum to contract slowly, but in high concentration they are as "fast-reacting" as is histamine and therefore cause a contraction not obviously different from that induced by histamine. In a carefully conducted bioassay, there is no ambiguity because the activity of histamine is prevented by antihistaminic drugs in doses that do not block the action of slow-reacting substances. Nevertheless, a quantitative bioassay requires that histamine be

* Supported by grants from the National Institute of General Medical Sciences (GM 14278-02) and the National Science Foundation (GB-6248).

separated from the interfering substances, and ion-exchange chromatography (though not the common solvent extractions) serves this purpose. Histamine can be accurately determined by bioassay of an acid extract of brain (and associated structures) after chromatography on carboxylic acid ion-exchange resins.[2] The effect of these purified extracts in causing contraction of the guinea pig ileum is totally blocked by an antihistaminic drug. The extracts also contain methylhistamine (see Section X), but its low concentration in brain coupled with its slight activity on ileum (1/200th or less than that of histamine) make it a negligible source of error.

The fluorometric assay of histamine is based on condensation of the amine with o-phthaldialdehyde.[3] Butanol extraction readily removes interfering materials from all tissues studied except brain[1,4] and blood.[5] Brain, even after this extraction, contains along with histamine, spermidine which reacts with o-phthaldialdehyde to produce a fluorophor with spectral properties very nearly like those of histamine.[6,7] Histamine and spermidine must therefore be separated, and chromatography[6,7] or a modified extraction[142] effects this separation.

The history of recent work on brain histamine serves both to emphasize the caution necessary in assaying brain for histamine and to clarify the confused literature. The original fluorometric method[3] appeared to show a concentration of 250 ng of histamine/g of wet weight of whole brain (rat, guinea pig, rabbit, and dog) and bioassay appeared to confirm this value. The validity of this fluorometric method also appeared to be supported by treating both the brain extract and authentic histamine with diamine oxidase: the amount of apparent histamine in the brain extract and true histamine were destroyed by this enzyme at the same rate.[3]

But this fluorometric method seemed to show that histamine was uniformly[3] distributed in brain in contrast with nonuniformity[2] observed with bioassay. Furthermore, the spectral properties of the fluorophor produced from brain extracts differed slightly but significantly from that produced by authentic histamine.[3,8]

A specific bioassay showed that whole wet brain of the guinea pig contained 53 ng of histamine/gm.[1] This value was confirmed by fluorometric measurements of brain extracts in which interfering materials were removed by chromatography[6,7,9,10] and was also confirmed by an enzymatic-isotopic method[11] for measuring histamine. Isotope dilution experiments[12] further showed the lack of specificity of the original fluorometric method when applied to brain (although, as emphasized, this original method is specific for all tissues studied except brain and blood).

The apparent and seemingly confirmatory value of 250 ng obtained by the original fluorometric method and bioassay was unhappily fortuitous. By accident both methods gave spuriously high values. The interfering materials in both methods are different, but by a perverse coincidence, they gave the same values of apparent histamine. The fact that diamine oxidase caused the same rate of destruction of true histamine and of apparent brain histamine shows only that the extract contained histamine and may also reflect the fact that histamine is not a specific substrate for diamine oxidase. Thus did a series of odd circumstances and bizarre coincidences cause confusion that still persists in some minds.

In the search for the material contributing to the spurious histamine values in the original fluorometric assay of brain histamine, Green showed in unpublished work that of a group of diamines and polyamines, only spermidine was both extracted by the solvents used and also reacted to yield a fluorophor with spectral properties like those of histamine. The molar fluorescent yield after spermidine and histamine were carried through the extraction procedure was eleven times higher (or even greater) for histamine than for spermidine, but since the molar concentration of spermidine in brain is at least 500 times that of histamine,[13] it seemed likely that spermidine could be the interfering substance. That spermidine was indeed the substance in brain causing the spuriously high values of histamine was clearly shown.[6,7] Spermidine is almost uniformly distributed in brain,[14] a fact that may account for the mistaken claim that histamine is uniformly distributed, for the unmodified fluorometric method measured more spermidine than histamine, and also for this reason estimates of apparent "histamine" in mammalian brain have ranged from 1700 to 50 ng/g.

In peripheral nerve, the estimates of histamine were made before full awareness of the presence of interfering materials, and the values are suspect (see Section III).

II. CONCENTRATION AND DISTRIBUTION OF HISTAMINE IN BRAIN AND ASSOCIATED STRUCTURES

A. Concentration in Whole Brain

Table I shows the concentrations of histamine in whole brains of different species. The concentration, about 0.5 nmole, which is significantly lower than that of 5-hydroxytryptamine and catecholamine, does not differ much among the mammals that have been studied. It is likely that almost all this histamine, at least in the mammals listed in Table I, is in neural tissue and not in mast cells; mast cells are not found in the central nervous tissue of those species that have been examined (though they are found in associated structures as described below)[2,145] except in the hedgehog which has mast cells in its thalamic structures.[19] Gray matter is richer than white matter in histamine.[35]

B. Regional Distribution in Brain

Table II shows the regional distribution of histamine in the brains of several species. High concentrations of histamine (higher than those of 5-hydroxytryptamine, dopamine, and norepinephrine)[20] are found in all three areas of the hypophysis, especially in the hypophyseal stalk. It is not commonly recognized that histamine was shown in the hypophysis almost 50

TABLE I

Concentration of Histamine in Whole Brain

Species	Concentration (ng/g wet weight)	Method
Guinea pig[1]	53	Bioassay
Guinea pig[18]	70	Bioassay
Guinea pig[6]	57	Fluorometry
Guinea pig[7]	68[7]	Fluorometry
Guinea pig (pregnant)[15]	61	Fluorometry
Fetal guinea pig[15]	0	Fluorometry
Rat[9]	52	Fluorometry
Rat[6]	76	Fluorometry
Rat[16]	55	Fluorometry
Rat[11]	60	Enzymatic-isotopic assay
Rat[10]	46	Fluorometry
Rat[17]	57	Bioassay
Rat[18]	70	Bioassay
Baby rat[15]a		
3–5 days old	223	Fluorometry
14 days old	104	Fluorometry
24 days old	68	Fluorometry
11 weeks old	31	Fluorometry
Rabbit[16]	56	Fluorometry
Rabbit[6]	65	Fluorometry
Rabbit[18]	130	Bioassay
Baby rabbit[15]		
3 days old	43	Fluorometry
9–10 weeks old	37	Fluorometry
Mouse[18]	60	Bioassay
Chick[18], 1–21 days old	120	Bioassay
Frog[11]	350	Enzymatic-isotopic assay
Frog[18]	380	Bioassay

a Others have also noted that levels decline with development.[144]

years ago by Abel and Kubota,[131] and was rediscovered much later.[24] Although there is a gross correlation between the concentration of histamine and number of mast cells in the three portions of the hypophysis,[2] there is reason to think that the histamine in the anterior lobe of the hypophysis is held in cells different from mast cells, whereas histamine in the posterior lobe and stalk is found, at least in part, in mast cells. Compound 48/80, a chemical that releases histamine from mast cells, did not affect the concentration of histamine in the anterior lobe (cat) while reducing it in the stalk (by 70 %) and posterior lobe (by 60 %).[21] It is relevant that the rabbit, which is deficient in mast cells, has less histamine in the posterior lobe of its hypophysis than have other species.

It would appear, too, that histamine in the anterior lobe of the hypophysis, though not in mast cells, is held differently from histamine in the brain

TABLE II

Distribution of Histamine in the Central Nervous System and Associated Structures[k]

	Means, expressed as ng/g wet weight					
	Dog[(2)a]	Cat[(21)a]	Cat[(22)b]	Monkey[(15)c]	Rabbit[(18)a]	Frog[(11)d]
Spinal cord	< 40	< 100		63	< 100	
Pons and medulla longata					140	180
Pons				84		
Medulla				64		
Floor of fourth ventricle	55	70			180	
Vestibular area	< 30					
Region of nuclei X and XII	< 40					
Region of reticular formation	60	50				
Brachium pontis	< 15	< 30				
Nucleus cuneatus	< 30	< 30				
Nucleus gracilis	< 30	< 30				
Area postrema	920	e				
Cerebellum	< 50	< 20	100	15	60	60
Mesencephalon						
Colliculi	220			217		
Superior colliculus	218	150			210	
Inferior colliculus	200	120			170	
Central gray matter	220	160			280	
Region of red nucleus	215	100				180
Optic lobe						310
Diencephalon						
Thalamus			340	196		
Medial	260	250			270	
Lateral thalamus						
Ventrolateral		75				
Dorsolateral	220	350				
Medial geniculate body	270	370				
Lateral geniculate body	140	180				
Hypothalamus	610		930	1,850	700	760
Corpora mammillaria	740	1,150				
Ventral hypothalamus	900	800				
Dorsal hypothalamus	460	480				
Supraoptic nucleus	340					
Preoptic area		430				

Table II—*continued*

	Dog[2]a	Cat[21]a	Cat[22]b	Monkey[15]c	Rabbit[18]a	Frog[11]d
	Means, expressed as ng/g wet weight					
Telencephalon						
Cerebral cortex		90	100		110	290
Occipital cortex				50		
Temporal cortex				54		
Parietal cortex				74		
Frontal cortex		68[23]a		71		
Cortical area 4	80					
Cortical area 17	70					
Cerebral white	<55					
Caudate nucleus	140	50	340	103	150	
Hippocampus	80	60		81	90	
Corpus callosum	<50	<80				
Amygdala	140					
Olfactory bulb	50					
Other						
Optic nerve	<50	<60				
Optic chiasma				167		
Associated structures						
Hypophysis g	900g	2,400g	1,800g	580		
Anterior lobe g	750g	2,400g			650	
Posterior lobe g	1,060g	1,700g			400g	
Hypophyseal stalk and infundibulum g	14,500g	5,200g				
Median eminence g	15,100g	12,500g				
Pineal body h	500h	380h,i				
Choroid plexus j	6,500j	210j			610	

a Bioassay.   b Spectroscopic assay (dinitrophenyl derivative).
c Fluorometric assay.   d Enzymatic-isotopic assay.
e Values of two samples from cat were 1300 and 1500; three others contained < 300.
f The rat d hypothalamus has 170 ng/g,[11] the pig a 300.[24]
g It is estimated that of the histamine in the hypophysis of the dog,[2] 57 % is in the anterior lobe, 24 % in the posterior lobe, and 19 % in the stalk and infundibulum; the corresponding percentages[21] for the hypophysis of the cat are 54, 26, and 20. The mean weight of whole hypophysis of the dog is twice that of the cat; the anterior lobe of the cat's hypophysis is relatively larger than that of the dog. In both species, it appears that most of the histamine in the anterior lobe is not in mast cells, whereas most of the histamine in the posterior lobe, stalk, and infundibulum is in mast cells. The concentrations in pig hypophysis are: anterior lobe, 380; posterior lobe, 3300; median eminence, 6300. The relatively low level of histamine in the posterior lobe of the rabbit hypophysis may reflect the paucity of mast cells in this species.[25]
h In human and cattle pineal bodies which contain respectively 76,400 and 6300 ng/g, histamine content correlated with the number of mast cells. This correlation was suggested in 15 other species; in rabbit, pig, rat, and horse, the pineal bodies contain no histamine and no mast cells. However, release of histamine from the cattle pineal body was refractory to compound 48/80, behavior uncharacteristic of mast cells. And mast cells were not found in dog or cat pineal bodies, both of which contain histamine.[26]
i One animal.
j Contains mast cells.
k Additional studies were published after this table was assembled.[142,144,145]

because reserpine (and chlorpromazine and iproniazid) did not change the levels of histamine in the hypophysis while altering the levels in brain[21] (see Section XII, B1).

Not certain is the type of cell holding histamine in the pineal body. A correlation between mast cell number and content of histamine among pineal bodies from different species does not prove that all the histamine is in mast cells[26]: studies with drugs, as noted above, showed that a suggestive correlation between mast cell count and histamine content among three parts of the hypophysis could have been misleading. The inability of compound 48/80 to release histamine from the bovine pineal body may suggest that the histamine is not in mast cells, but it must be noted that negative experiments with compound 48/80 do not always rule out mast cells. In some organs of some species, these cells are notably refractory to compound 48/80,[26-28] (see Section IV) though sensitive to other histamine releasers such as octylamine. Octylamine does release 30% of the histamine from the bovine pineal body.[26] It can be tentatively concluded that at least some of the histamine in the pineal body is in mast cells.

With the exception of the pineal body, portions of the hypophysis, and choroid plexus, the structures listed in Table II lack mast cells,[2] and their histamine must be held in neural cells, or supporting cells, or both. The con-concentration of histamine is highest in the hypothalamus, less in the thalamus and midbrain, and least in the cerebral cortex. The regional distributions of histamine, 5-hydroxytryptamine, and norepinephrine have been tabulated,[2] and the similarities are clear.

C. Subcellular Distribution in Brain

Carlini and Green[1] found the microsomal fraction of guinea pig whole brain to be especially rich in histamine (Table III). This distribution was confirmed in rat brain, though not in guinea pig midbrain. The different

TABLE III

Subcellular Distribution of Histamine in Brain

	Percentages of total histamine		
	Whole guinea pig brain[1]a	Rat cortex[17]a	Guinea pig midbrain[29]b
Nuclei and debris	18	8	40
Mitochondrial fraction	18	21	60
Microsomal fraction	44	51	0
Soluble fraction	20	20	0

a Bioassay.
b Fluorometric assay.

distribution in guinea pig midbrain is probably not due to different methods of assay since the observed total concentration of histamine did not differ.

Most of the apparent histamine in cow splenic nerve was found in the fraction containing microsomes and soluble material[30] (see Section III).

After intraventricular injection into rats, radioactive histamine was found in the fraction containing nerve endings and in a mixed fraction containing microsomes, myelin, and supernatant fluid.[31] It is not known that the exogenous histamine mixed with the endogenous pool. Other experiments have shown that endogenous and exogenous histamine do not freely mix in all tissues,[32] brain as well.[33] Histamine is well taken up *in vitro* by both mitochondrial and microsomal fractions of brain,[34] but this uptake does not indicate with which fraction histamine is associated in intact brain.

In confirming the work of Carlini and Green on brain, Katuoka and DeRobertis[17] (Table III) also carried out electron microscopy on the particulate fractions. The histamine-rich microsomal fraction was shown to contain small nerve endings. The subcellular distribution of histamine in brain differs from that of norepinephrine, 5-hydroxytryptamine, and slow-reacting substances,[1] but all are found in fractions containing particles resembling nerve endings.

III. CONCENTRATION OF HISTAMINE IN PERIPHERAL NERVOUS SYSTEM AND ASSOCIATED STRUCTURES

Histamine has been measured by bioassay in peripheral nerve trunks and in tissue of the autonomic nervous system from different species, and the levels have been tabulated.[35-38] There is not sufficient agreement among the different studies to prompt a table here. These differences could be attributable to a lack of specificity in the methods. Euler has pointed out that unlike authentic histamine, the histaminelike material in the nerves was destroyed by heating in alkaline solution.[36] New studies of histamine in these structures are necessary. Even with these equivocations, it is probably safe to conclude[35-37] that the postganglionic sympathetic nerves are rich in histamine.*

IV. MAST CELLS IN PERIPHERAL NERVE: A SOURCE OF HISTAMINE[145]

Histamine content of peripheral nerves is grossly correlated with mast cell count.[39] While such a correlation cannot alone establish that all the histamine in these structures is in mast cells (see Section II, B), other work shows that mast cells account for much of the histamine in the nerves.

From studies on the subcellular distribution of histamine in splenic nerve,[30] which showed histamine to be mainly in the fraction containing microsomes and soluble material, one may suggest that most of the histamine is not in mast cells because in mast cells histamine is found in the crude

* Recent fluorometric studies support this conclusion.[147]

mitochondrial fraction.[27] But the observations on histamine in splenic nerve are equivocal since the method may not have measured only histamine[36] (see Section III).

Mast cells are found in the peripheral nerves of all species studied.[40] An intensive study has been made of mast cells in peripheral nerve, primarily but not exclusively in the sciatic nerve of the rat.[41–45] The epineurium, perineurium, and endoneurium all contain mast cells. Only in the epineurium and perineurium are the mast cells associated with blood vessels. Section of the sciatic nerve is followed by disruption of mast cells in the epineurium, perineurium, and endoneurium at the site of the lesion. But in the distal part of the sectioned nerve the number of endoneurial mast cells increases, an increase that precedes axonal outgrowth. The increase in mast cells could very well explain the increase in histamine that follows sectioning of sciatic nerves of the rat and cat.[35]

Compound 48/80, after repeated injections (daily intraperitoneal injections to rats for 14 days) or after direct application to the nerves, disrupts most of the mast cells in the epineurium and perineurium (of the sciatic, peroneal, tibial and sural nerves and of the brachial plexus and dorsal root ganglion).[42,43] But the endoneurial mast cells remain intact after both kinds of treatment with compound 48/80, as do the nerve branches in skin. Compound 48/80 also releases histamine from the superior cervical ganglion of the cat.[46]

A diffusion barrier protects the endoneurial mast cells from compound 48/80 since after the compound was injected directly into the endoneurium the mast cells there are disrupted. Interestingly, this diffusion barrier is broken after section of the nerve.[42,43]

This work not only shows that mast cells are a significant source of histamine in peripheral nerve, but also shows, once again, that not all mast cells react to compound 48/80. The presence of mast cells in peripheral nerve may have even broader implications. Mast cells in the epineurium, perineurium, and endoneurium are disrupted by trauma mild enough to cause no immediate microscopic damage to the axons or myelin sheath.[44] These highly vulnerable cells are full of substances with potent pharmacological effects,[25] such as histamine and heparin (in mast cells of all species studied), 5-hydroxytryptamine in the mouse and rat, dopamine in some species, and slow-reacting substances in mast cells of the rat and almost surely of other species. The release of these and perhaps other, unidentified substances in nerve injury could have profound effects. For one, the release of heparin, which in turn releases lipases, including phospholipase from tissues, could well be implicated in the process of Wallerian degeneration.[41] The peripheral neuropathy induced by isonicotinic acid hydrazide is associated with an increase in number of endoneurial mast cells.[45]

What other roles mast cells may play in peripheral nerve is a subject more suitable for further work than for speculation, but it should be noted that mast cell histamine, especially that in the endoneurium, could function in normal (see Section XII, B, 2) as well as pathological neural activity.

V. THE SOURCE OF HISTAMINE IN THE NERVOUS SYSTEM: THE FORMATION OF HISTAMINE

Relatively little histamine is taken up by whole brain from blood.[47,48] More is taken up after intraventricular perfusion,[31,49,50] by which route histamine was shown to be taken up by gray matter in preference to white matter[49] and to be bound to particulate material[31] (see Section II, C). From the ventricles, histamine is absorbed into the bloodstream.[49]

In specific areas, uptake from blood may account for some of the histamine.[51] The anterior lobe of the hypophysis took up histamine from the blood, as did the posterior lobe to a lesser extent, and neither seemed to form histamine. The hypothalamus did not take up histamine but formed it.[50,51] Experiments with minced tissue[52] showed that the rate of histamine formation was greatest by far in hypothalamic tissue; the area postrema was less, and cerebral cortex showed slight but detectable activity. Homogenates[53] of cerebral cortex, brain stem, and spinal cord showed very little histidine decarboxylase activity. The phrenic and vagal nerves had more, and high activity was found in the splenic nerve and the stellate ganglion.

There are at least two histidine decarboxylases, both of which require pyridoxal and are found in the soluble portion of the cell. The nature of the one in neural tissue is not clear. Probably both a specific histidine decarboxylase and the nonspecific L-amino acid decarboxylase were measured with homogenates in the experiments noted above. In peripheral nerve it is likely that at least some of the decarboxylase is in the associated mast cells (see Section IV) and therefore is a specific histidine decarboxylase. Why histamine formation was not detectable in the mast cell-rich hypophysis is not clear.

VI. BINDING OF HISTAMINE BY NEURAL TISSUE

After histamine is formed in the soluble portion of the cell by histidine decarboxylase, histamine must be sequestered in order to escape the catabolizing enzyme(s), at least one of which, histamine methyltransferase, is found in the soluble portion of the cell.[54] The histamine is sequestered by particulate material in brain (see Section II, C). No information is available on a mechanism for conveying histamine from the soluble cytoplasm to the particulate material. It is possible that in the soluble cytoplasm, histamine is in complex with acidic material which prevents histamine catabolism.[27,48] Or histamine may be unprotected, subject to ready catabolism, in which circumstance one of the determinants of the steady-state levels of histamine in brain would be how much of the histamine formed in the soluble cytoplasm by chance escapes catabolism before being bound. Another possibility is that the decarboxylation occurs at or near the surface of the particulate material. These problems are not peculiar to histamine (or to neural tissue) but exist for other biogenic amines. Studies on uptake of 5-hydroxytryptamine led to the suggestion that the soluble cytoplasm may influence the extent of 5-hydroxytryptamine binding to particulate material.[34]

Both *in vitro*[34] and during intraventricular perfusion,[31] histamine is taken up by the microsomal fraction of brain and the fraction containing particles resembling nerve endings. Particles have a capacity for uptake of histamine (and of 5-hydroxytryptamine) approaching that of commercial cation-exchange resins.[34] Uptake was slightly greater at 37°C than at 0°C.

To what compound or compounds histamine is bound in particulate material of brain is not known. Histamine forms complexes with acidic lipids and other acidic substances,[27,48] including sulfomucopolysaccharides to which histamine in mast cells is believed to be bound.[27,55] Like histamine, sulfomucopolysaccharides are present in gray matter and in the microsomal fraction of brain.[56] In the mast cells found in nerve fibers and in structures associated with brain, it may be assumed that histamine is bound as in other mast cells.

VII. ASPECTS OF THE CHEMISTRY OF HISTAMINE PERTINENT TO ITS BINDING AND ITS BIOLOGICAL ACTIVITIES[48,55]

Histamine (Fig. 1) exists in two tautomeric forms which have not been separated. Recently reviewed evidence suggests that A is the dominant compound, but the presence of B can also be inferred from studies showing metabolites of 1-methyl-5(β-aminoethyl)imidazole in urine. Histamine forms chelates in which the primary amino group on the side chain and the pyridinelike nitrogen in the ring bind the cation; complexes of 1:1 or 2:1 are possible. Salt formation occurs through the primary amino group which is protonated at physiological *p*H. Imidazole forms hydrogen bonds through both the pyridinelike and pyrole nitrogens, and can form oligomers by intermolecular hydrogen bonding; histamine should have similar capacity for hydrogen bonding. The 5-carbon in histamine (tautomer A) is strongly nucleophilic, as suggested by molecular orbital calculations (Table IV) and shown by experiment; the ease of cyclization of histamine, for example, rests on the nucleophilicity of this carbon. The 2-carbon (tautomer A) is electron deficient and could attract nucleophilic groups. The calculated energies of the highest occupied and lowest empty molecular orbitals suggest that the imidazole ring does not engage in charge-transfer reactions. Imidazoles can function as esterases and are part of the active sites of many enzymes.

These molecular characteristics could prompt numerous guesses on how histamine can interact with tissues, but none of these has yet been shown to be pertinent to the action of histamine on nervous or any other tissue.

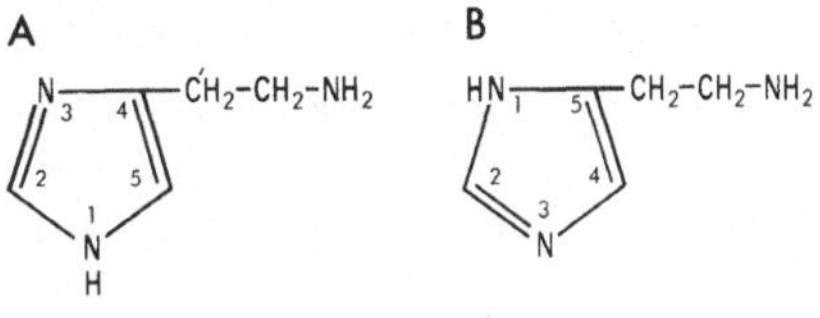

Fig. 1. The two tautomers of histamine.

TABLE IV

Reactivity Indexes of the Atoms in the Imidazole Ring in Histamine (Fig. 1A), Calculated by the Hückel Approximation, Confined to π-Electrons[48,55]

Position	Reactivity index[a]				
	q	f^E	f^N	S^E	S^N
1	1.55	0.04	0.44	1.09	0.39
2	0.99	0.58	0.81	1.15	0.84
3	1.30	0.07	0.20	1.08	0.51
4	1.06	0.49	0.05	1.17	0.60
5	1.10	0.72	0.49	1.38	0.70

[a] q = π-electron density; f^E = frontier electron density for an electron-donating reaction; f^N = frontier electron density for an electron-accepting reaction; S^E = superdelocalizability for an electron-donating reaction; S^N = superdelocalizability for an electron-accepting reaction.

VIII. PATHWAYS FOR THE CATABOLISM OF HISTAMINE

Figure 2 shows the main pathways for catabolism of histamine,[57] not all of which have been shown in brain. There are significant species differences (see Section IX). An important pathway in most species, including man, is direct oxidative deamination to imidazole acetaldehyde. The enzyme catalyzing this reaction is called histaminase, and it is not agreed that this is the same as diamine oxidase.[58] Almost all of the aldehyde is converted to imidazoleacetic acid by aldehyde dehydrogenase. A small portion of the aldehyde is reduced by alcohol dehydrogenase to histidol, also called imidazole-ethanol[59] (i.e., 4(5)-imidazoyl-ethane-2-ol). It has not been shown that the relative extent of oxidation or reduction of this aldehyde is determined by the relative levels of NADP and NADPH in tissues, as has been shown for analogous reactions in the catabolism of 5-hydroxytryptamine.

Some of the imidazoleacetic acid is converted to the ribotide, i.e., 5'-phosphoribosylimidazoleacetic acid, which is then dephosphorylated to the riboside, ribosylimidazoleacetic acid.[60] The formation of the ribotide requires 5'-phosphoribosyl-1-pyrophosphate and, oddly, stoichiometric amounts of ATP; the role of ATP in the reaction is now known. The riboside is formed by action of a phosphatase on the ribotide.

The more important pathway for histamine metabolism is methylation[57] to 1-methyl-4(β-aminoethyl)imidazole, usually called methylhistamine or sometimes 1,4-methylhistamine to differentiate it from the 1,5-isomer. S-Adenosylmethionine is the methyl donor. Methylhistamine is oxidatively deaminated to form methylimidazole acetaldehyde. From this, by the action

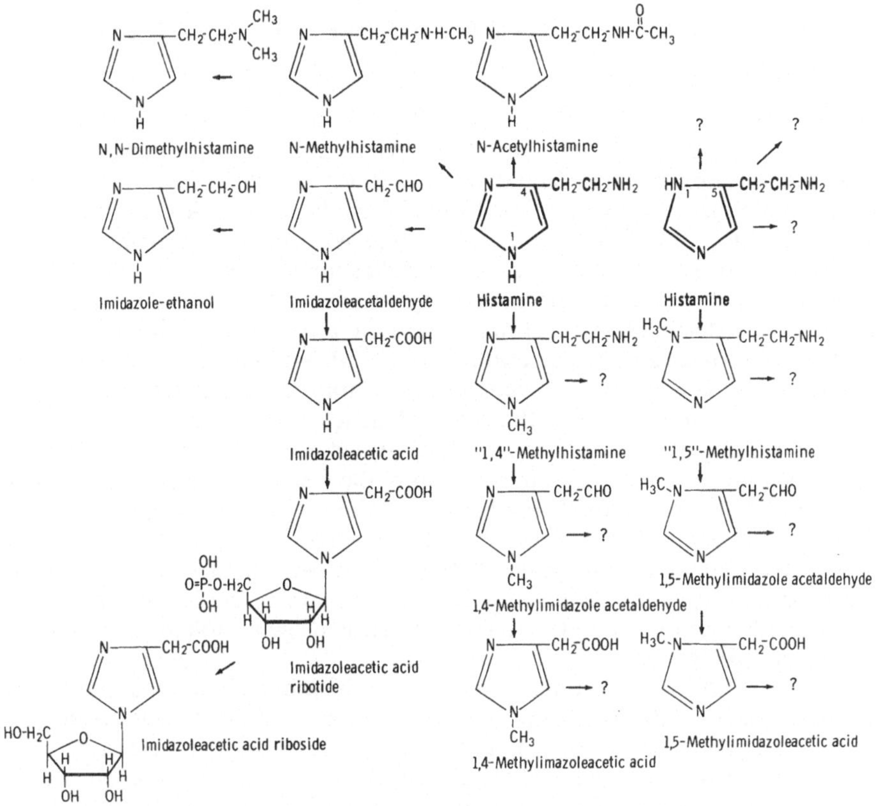

Fig. 2. The catabolism of histamine.

of aldehyde dehydrogenase, results the major urinary product of histamine catabolism, methylimidazoleacetic acid. Conceivably, methylimidazole acetaldehyde could be enzymatically reduced to methylhistidol, in analogy with the formation of histidol from imidazole acetaldehyde, but no attempts to show this reaction have been published. Nor have the ribotide or riboside of methylimidazoleacetic acid been described.

Other pathways have been described[57]: acetylation of the primary amino group, methylation of that group, and methylation of the ring nitrogen of the tautomeric to form 1-methyl-5-(β-aminoethyl)imidazole (i.e., 1,5-methylhistamine) which goes to 1,5-methylimidazoleacetic acid. Histamine (and some other amines) form amides with palmitic and probably other fatty acids, a reaction catalyzed by the microsomal fraction of liver and probably of brain.[61] This amide has not been searched for in brain or other tissues, but it is formed by the same enzyme system that forms the amide of ethanolamine which is found in high amounts in brain of rat and guinea pig (36 and 45 μg/g, respectively).[62]

IX. CATABOLISM OF HISTAMINE BY NEURAL TISSUE

Not all the reactions described above have been shown to occur in brains of all species. The relative activities of the catabolizing enzymes can vary among species.[57] The activity of histamine methyltransferase is very high in guinea pig brain and relatively low in rat brain.[54]

No imidazoleacetic acid was formed from histamine or histidine by the perfused cat brain.[50] But when histamine was injected into the cerebral ventricle of the rat, most of it was recovered as imidazoleacetic acid.[31] Paradoxically, with putrescine or cadaverine as substrates, no diamine oxidase activity was detected in the brain of rat or any other mammalian brain examined (mouse, guinea pig, and rabbit).[63] If both groups of observations are correct, one is tempted to suggest that histaminase and diamine oxidase are two different enzymes in the brain of, at least, rat. Imidazoleacetic acid must be taken up by brain, since it has central effects (see Section XI), but it is unlikely that the histamine injected into the ventricles was converted to imidazoleacetic acid outside the brain and then taken up by the brain. Imidazoleacetic acid ribotide and riboside were shown in rat brain after the injection of radioactive histidine.[64] It is not known whether the ribotide and riboside are formed in the brain or are carried there by the blood; imidazoleacetic acid ribotide synthetase has so far been shown (looked for ?) only in liver and kidney.[60]

Ring methylation is a major pathway of histamine catabolism in brains of some species. After perfusion of cerebral ventricles in cats with histamine, both methylhistamine and methylimidazoleacetic were detected.[50] The histamine formed endogenously (during histidine perfusion) was less readily methylated than exogenous histamine.[33] This difference may be due to more ready accessibility of the exogenous histamine to the methylating enzyme in the supernatant fraction. Pools for histamine have been seen in other tissues.[32]

Minces of both cortex and hypothalamus of cat methylated histamine, the latter showing greater activity.[52] The regional distribution of this enzyme in monkey brain, measured in homogenates,[65] is shown in Table V.

The hypothalamus, which in the monkey is especially rich in histamine (Table II), showed high activity. The hypophysis, which has less histamine than does the hypothalamus (Table II), has the greatest activity of all the areas measured.

X. CATABOLITES OF HISTAMINE IN NEURAL TISSUE

The concentration of methylhistamine in guinea pig brain is 72 ng/g,[66] in cat brain 60 ng/g,[67] about the same molar concentration (0,5 nmole/g) as histamine (Table I). Examination of the regional distribution of methylhistamine[22] (Table VI) showed highest concentration in the hypophysis, and next highest in the hypothalamus, an order reminiscent of the distribution of both histamine (Table II) and histamine methyltransferase (Table V).

TABLE V

Histamine Methyltransferase in the Nervous System and Associated Structures of the Monkey[65]

	μmoles of methylhistamine formed from histamine per hour per gram of tissue
Peripheral nervous system	
Sympathetic chain	0.79
Saphenous nerve	0.55
Femoral nerve	0.48
Vagus nerve	0.46
Splanchnic nerve	0.40
Sural nerve	0.36
Central nervous system	
Spinal cord	0.54
Pons and medulla oblongata	
Pons	0.41
Area postrema and dorsal medulla	0.54
Cerebellum (vermis)	0.36
Mesencephalon	
Superior colliculus	0.49
Diencephalon	
Dorsal thalamus	0.45
Hypothalamus	1.00
Habenula	0.61
Telencephalon	
Cerebral cortex	
Postcentral gyrus	0.57
Precentral gyrus	0.38
Caudate nucleus	0.62
Hippocampus	0.76
Pyriform cortex and amygdala	0.88
Fornix	0.56
Corpus callosum	0.55
Olfactory bulb	0.78
Olfactory tract	0.80
Septum	
Dorsal	0.97
Basal	0.66
Other	
Optic chiasma	0.48
Associated structures	
Hypophysis	
Anterior lobe	3.83
Posterior lobe	0.93
Pineal body	1.05

Histamine methyltransferase is highly specific, no other imidazole or other substance examined serving as substrate.[54] But obviously no exact correlation between the regional distribution of the enzyme and of methylhistamine is expected, since the steady-state levels of methylhistamine will also be influenced by the presence of other histamine-catabolizing enzymes and the activity of monoamine oxidase, which destroys methylhistamine. The relative rates at which various regions of the brain oxidize methylhistamine are not known, but oxidation of tyramine and 5-hydroxytryptamine by monoamine oxidase was most rapid in the hypothalamus and thalamus (dog, cow, and man).[69]

Methylhistamine is found mostly in the crude mitochondrial fraction (Table VII).[66] The insensitivity of the method precluded examination of subfractions of the mitochondrial fraction.

Neither brain[21] nor peripheral nerve[70] nor ganglia[70] showed any derivatives of histamine that were methylated on the primary amino group of side chain. The riboside and ribotide of imidazoleacetic acid were shown in rat brain after intraperitoneal injection of radioactive histidine,[64] but neither the source nor the concentrations of these catabolites is known (see Section IX). Traces of N-acetylhistamine were observed in the stellate ganglion.[70]

XI. ARE THE CATABOLITES OF HISTAMINE "INACTIVATION PRODUCTS"?

Both histamine and methylhistamine are found in particles containing nerve endings (Tables III and VII), and the amines have a similar regional distribution (Tables II and VI). These analogies suggest that methylhistamine should not be dismissed as a mere catabolite of histamine. Methylhistamine may have a function. Its casual dismissal as an inactivation product of histamine rests on its having less activity than histamine on the test systems used to measure histaminelike activity.[71-75,139] Analogous observations could have led to the inference that melatonin and epinephrine are merely inactivation products of 5-hydroxytryptamine and norepinephrine. Methylation of many naturally existing substances results in products with enhanced activity which can be manifest as qualitative differences between the original substance and the methylated products; or the methylated products can antagonize or potentiate the action of the parent compound (see references).[76]

What role, if any, methylhistamine may play in function is not known. It is worth noting that it accumulates after treatment with monoamine oxidase inhibitors,[22] including after isocarboxazid was given to patients in therapeutic doses.[76] Methylhistamine was shown to be released with histamine during reflex vasodilatation, after lumbar sympathectomy, and after neural stimulation of the submaxillary gland (see Section XII, B, 2).

No catabolite of histamine (or any biogenic amine) can be presumed to lack biological activity. Both N-acetylhistamine and N,N-dimethylhistamine, like histamine, produce an arousal action in pentobarbital-treated rabbits,

TABLE VI

Regional Distribution of Methylhistamine (μg/g) in Cat Nervous System and Associated Structures[22]

Area	Methylhistamine (μg/g)
Cerebellum	0.1
Mesencephalon	0.27
Diencephalon	
Thalamus	0.30
Hypothalamus	0.56
Telencephalon	
Cerebral cortex	0.3
Caudate nucleus	0.4
Hypophysis	2.2
Cerebrospinal fluid[68]	<0.002

the dimethyl derivative being more effective than histamine.[75]* Histidol accumulates in the presence of disulfiram and could contribute to the effects of the drug.[59] Imidazoleacetic acid is reported to protect against anaphylaxis[77]; to have hypnotic and possibly analgesic action[78]; to be more active than γ-aminobutyric acid on the crayfish stretch receptor neuron, an effect of imidazoleacetic acid that is blocked by chlorpromazine and dibenzyline (neither of which affect γ-aminobutyric acid)[79]; and to be as potent as γ-aminobutyric acid in inhibiting cerebral cortical neurons of cats.[103,149] It has been suggested that imidazoleacetic acid (and maybe other catabolites of biogenic amines) function in the nervous system of animals low in the evolutionary scale[80]; this notion is in accord with the observation[50,63] (which is controversial[31]) that diamine oxidase is found in vertebrate but not in mammalian brain.[63] One may also wonder if future work will reveal imidazoleacetic acid as a neural "depressant," counteracting the "excitant" action of histamine.

TABLE VII

Subcellular Distribution of Methylhistamine in Whole Guinea Pig Brain[66]

Fraction	Percent
Nuclei and debris	19
Mitochondrial faction	55
Microsomal and supernatant faction	26

* It was recently suggested that N,N-dimethylhistamine may function in gastric secretion in the guinea pig.[148]

XII. HISTAMINE AND NEURAL ACTIVITY

One group of observations that has stimulated the idea that histamine may function in the brain is the relatively low incidence of allergy in schizophrenic patients. A brief review of this work has appeared,[80] and will not be discussed here.

A. The Effect of Exogenous Histamine on Neural Tissues

Just as there are marked differences among species in the relative importance of the catabolic pathways for histamine, in the distribution of the amine in the body, in the effect of drugs and hormones on histamine levels, and in the relative sensitivity to histamine,[80] there are species differences in the neural actions of histamine.

Histamine stimulates the superior cervical ganglion of the cat.[81,82] This action is blocked by antihistamine drugs but not by hexamethonium or atropine. The effect of histamine, like that of other agonists, is enhanced by prior preganglionic nerve stimulation or by nicotine and is blocked by morphine and cocaine. Tachyphylaxis occurs as it does with other agonists, and there is cross tachyphylaxis between histamine and some other agonists.[83]

Histamine releases catecholamines from the adrenal medulla, and this effect is prevented specifically by antihistamine drugs.[84] In the cat, histamine acted directly on the adrenal medullary cells; but in the dog, the effect of low doses of histamine was prevented by ganglionic blocking agents, implying that part of the action is dependent on reflex pathways and on direct stimulation of the nerve endings in the adrenal medulla.

Release of catecholamines from chick intestine may explain the relaxant effect of histamine on this preparation.[85] This effect is prevented by pretreatment with doses of reserpine that deplete the tissue of catecholamine, and can be restored by treating the depleted tissue with epinephrine or norepinephrine. Both α- and β-receptors are involved in the relaxant effect of histamine, as well as receptors sensitive to antihistamine drugs, and, surprisingly, to hexamethonium, which suggests that histamine may stimulate the presynaptic terminals of cholinergic synapse. Histamine is known to lower the threshold of peripheral nerve (crab) to stimulation[86] and to stimulate mammalian peripheral afferent fibers.[87,88]

The several sites of inotropic action of histamine on the chick intestine have parallels in the rat stomach.[89] Here histamine also has both a direct action and actions that are prevented by hexamethonium and scopolamine. It was inferred that histamine stimulates preganglionic cholinergic fibers and adrenergic fibers. On human stomach, neither hexamethonium nor scopolamine affected the contractile effect of histamine,[90] but both inhibited histamine-induced gastric secretion.[91,92]

Suggesting the possibility of subtle relationships between histamine and the adrenergic system (see review)[93] is the observation that epinephrine potentiates the depressant action of histamine in baby chicks.[94] The effect of an intraventricular injection of histamine in raising blood pressure in cats may be due to stimulation of the sympathetic ganglia.[73,95,96]

Histamine increases cyclic 3',5'-AMP in rabbit brain slices.[114] The increase is tenfold greater than that produced by norepinephrine or 5-hydroxy-tryptamine.

As noted above, histamine stimulates peripheral nerve.[86–88] Histamine activated the cerebellum too.[101] About 50% of neurons of the isolated brain of the snail were stimulated by histamine, about 15% inhibited, and the rest were unresponsive; both the excitatory and inhibitory effects were blocked by an antihistamine.[139] Inhibitory actions of histamine on the cortex have been described[102,103,149] and less consistently on the spinal cord of cats.[133,134,150] Histamine induced sleepiness, muscular weakness, salivation, and increased respiratory rate after intraventricular injection into cats[104] and catatonia when introduced into the cisterna magna of dogs[105] or into the septal region of monkeys.[106] It increased avoidance response of rats, an effect antagonized by an antihistamine.[140]

LSD, reserpine, and hydroxyzine[112] increase resistance of rats to histamine. In mice, histamine, like chlorpromazine, potentiated the depressant effects of pentobarbital and chloral hydrate.[97] In rats[98] and rabbits,[75,99] intravenous histamine reversed the action of pentobarbital and in mice, that of chlorpromazine.[100] In high concentrations, chlorpromazine, brom-LSD, bufotenine, 5-hydroxytryptamine, and substances like toluidine blue inhibit the methylation of histamine.[54,109–111,135] In mice histamine potentiated the catatonic effects of reserpine, tetrabenazine, papaverine, mescaline, and 3,4-dimethoxyphenylethylamine[107]; 3,4-dimethoxyphenylethylamine blocked the action of histamine on blood pressure.[108] In a series of papers[107] it was shown that some drugs having catatonic effects that are potentiated by histamine are inhibitors of diamine oxidase (see Section IX). (These observations recall the parallelism in potency of some compounds as antihistamines and as inhibitors of both diamine oxidase and histamine methyltransferase, suggesting a common characteristic among the receptor and the two enzymes[141].) Perhaps the most mystifying interaction of histamine with another substance is the interaction with lead acetate. The subcutaneous administration of histamine shortly after, or along with, the intravenous administration of lead acetate is followed by calcification of the vagus nerve and part of the sympathetic nervous system.[25]

The results with exogenous histamine, introduced into different species in different doses by different routes, cannot be critically integrated into an hypothesis about the function of histamine. They show only that histamine, too, has effects on the nervous system. More relevant to the possible role of histamine in neural function are studies of endogenous histamine.

B. Endogenous Histamine and the Nervous System

1. Brain

Specific methods for measuring brain histamine showed that some psychotropic drugs cause changes in histamine levels.

Drugs were compared for their effects on histamine levels in the hypophysis (anterior lobe, posterior lobe, and stalk), hypothalamus (mammillary body, ventral and dorsal hypothalamus, and preoptic region) and medial thalamus of the cat.[21] Compound 48/80 lowered levels to about 40 % of normal in the posterior lobe and stalk of the hypophysis, the two regions rich in mast cells (see Section IV), without affecting levels in the other areas. Reserpine in the dose range 0.5 to 10 mg/kg did not affect histamine levels in the hypophysis but lowered it in other areas, in hypothalamus to about 40 %, in thalamus to 64 % of controls. With a lower dose (0.1 or 0.25 mg/kg) the effect was seen only in the mammillary body and the ventral hypothalamus. The maximum fall in the hypothalamus occurred at 18 hours after a single dose of 0.5 mg/kg, and recovery was still not complete after 120 hours.

Chlorpromazine, when given three times at a dose of 50 mg/kg, increased histamine levels in the hypothalamus by about 50 % and in the medial thalamus by 32 %, by inhibiting histamine methyltransferase. When the dose was lowered to 25 mg/kg, there was no effect; when the same dose was given five times, the histamine concentration rose by more than threefold in the mammillary body and ventral hypothalamus, and more than twofold in other parts of the hypothalamus, and in the medial thalamus. Two other phenothiazines were also shown to increase histamine in the hypothalamus. The phenothiazines failed to raise histamine levels in the hypophysis, perhaps because most of the histamine is in the mast cells which do not methylate histamine at a measurable rate.[132]

Iproniazid treatment for 5 days (25 mg/kg) increased histamine levels in all parts of the hypothalamus by 46–74 %, and in the medial thalamus by 28 %. Bulbocapnine appeared to increase the concentration in the hypothalamus, and pentobarbital in the preoptic region and in the medial thalamus. Morphine was without effect.

A correlation between endogenous levels of histamine and behavior remains to be shown. Following infusion of histidine into rabbits,[18] histamine levels rose, mainly in the midbrain (central gray) and hypothalamus, these high levels persisted for 16–32 hours. There was no obvious effect on behavior such as is seen after infusions of 3,4-dihydroxyphenylalanine or 5-hydroxytryptophan. If endogenous histamine has a function, subtle methods of measurement are needed to reveal it.

Experiments with drugs[115,116] that have a relatively greater inhibitory effect on the specific than on the nonspecific histidine decarboxylase could be useful. And the development of histochemical and other techniques for measuring small amounts of histamine in brain should stimulate experiments analogous to those that lend credence to the idea that 5-hydroxytryptamine functions in the brain.[136] Yet another approach to relating function to changes in histamine levels may be the use of anticerebral antibodies: specific antibodies to the caudate nucleus and to the hippocampus selectively increased histamine levels only in those homologous portions of the brain of cats.[23] It would be interesting, too, to learn if certain conditions provoke histamine synthesis in brain as they do in other organs.[137,138]

2. Peripheral Nervous System

The early idea of vasodilating histaminergic nerves[143] has been recently revived. It has been suggested that the vasodilatation that follows elevation of blood pressure may be due to a liberation of histamine from the sympathetic nerves.[117–119,151] As mentioned previously, histamine appears to be found in high concentration in postganglionic sympathetic nerves along with histidine decarboxylase and histamine methyltransferase. Direct experimental evidence could not be obtained that endogenous histamine is released during reflex vasodilatation,[118,151] perhaps because the methods were not sensitive enough (see Section I, A). The evidence for release of endogenous histamine rests on the ability of antihistamine to block the reflex vasodilatation, and of some histamine depletors such as octylamine to reduce vasodilatation.

Studies with the exogenous histamine, however, are in accord with the idea that histamine is the vasodilator. After the gracilis muscle was perfused with radioactive histamine, the reflex vasodilatation produced by intravenous norepinephrine (but not the vasodilatation produced locally by glyceryl trinitrate) was associated with the appearance in venous blood of increased amounts of radioactive histamine and methylhistamine. A similar increase occurred after sectioning the lumbar sympathetic chain.[118]

The hypotensive response evoked by stimulating a zone above the median raphe in the posterior medulla of the cat was blocked by either atropine or an antihistamine and potentiated by physostigmine or diamine oxidase inhibitors.[119–122] After the cat was loaded with radioactive histidine, radioactive histamine was released during the hypotensive response. Compound 48/80 was also observed to release radioactive histamine. The vasodilatation following direct stimulation of the sympathetic trunk or during direct or reflex stimulation of the carotid sinus nerve was also attributed to the release of histamine, an inference derived from studies with drugs including compound 48/80 which reduced the dilator responses. Loading experiments with radioactive histidine showed directly that histamine was released by stimulation of the sympathetic trunk, carotid sinus, or sciatic nerve. These observations led to the conclusion that histamine is a neuromediator for the baroreceptor depressor reflex. Other substances may be involved, such as prostaglandin E and the polypeptide substance P, which mimic a component of reflex vasodilatation. It may be relevant that stimulation of the vagus nerve to the frog's stomach released material that was antagonized by an antihistamine, atropine, and an antagonist of polypeptides. As pointed out by the author of these studies,[119] the concept of sympathetic histaminergic nerves requires more subtle thinking than that of adrenergic or cholinergic nerves: the release of histamine was not related to the change in vascular resistance; the release was short-lived, but not attributable to depletion of histamine and the disappearance of histamine from the effluent was not associated with a change in perfusion pressure.*

These recent observations recall early ones. Stimulating the cut end of nerves releases histaminelike material.[35,36] If stimulation is repeated, no

* The idea of histamine-mediated vasodilatation has been recently challenged again.[152]

release is seen unless a fresh cut is made. These observations led Euler[36] to suggest that histamine is not a mediator but only another substance that is allowed to escape during stimulation. Perhaps explaining a release that depends on cutting the nerve are recent studies showing that sectioning of peripheral nerve (including those nerves with vasodilator fibers described above) disrupts the mast cells at the site of the lesion (see Section IV). The disruption would release histamine from the mast cell and make it subject to escape during stimulation.

In the light of present evidence, there is no reason to reject out of hand the possibility that the mast cells are at least one source of the histamine released during reflex vasodilatation or during stimulation of the sympathetic nerves. As pointed out before (see Section IV), mast cells in all portions of the peripheral nerve (epineurium, perineurium, and endoneurium) are disrupted by conditions mild enough to cause no evident damage to the axons or myelin sheath. In the experiments described above, mast cells could account for the uptake and release of radioactive histamine by the vasodilating nerves and for the formation of histamine.[123] The effects on vasodilatation of the mast cell depletors, octylamine and compound 48/80, further implicate mast cells.* It is conceivable that during reflex vasodilatation, some event occurs in the vasodilator fibers which causes the release of histamine from mast cells in the fibers. Such a mechanism in no way makes the vasodilatation less neural. Rather, it suggests a source for the histamine that is released by neural activity and could account for some of the nuances observed, such as the short-lived nature of the release and it could account for the effects of the chemicals that release histamine; at the same time this hypothesis integrates these new physiological experiments with older ones and with recent histological observations (see Section IV).

Observations on the isolated stomach suggest that neural or muscular activity releases histamine.[89] Histamine output from stomach (guinea pig, kitten) could be effected by any procedure that changed the tome of the stomach: transmural stimulation, vagal stimulation, drugs (neostigmine, LSD) or mechanical distention. Release of histamine was inconsistent, recalling the inconsistency noted in histamine release during reflex vasodilatation. The impression was that mechanical deformation of the stomach caused histamine release, a suggestion in accord with the idea that the source of histamine may be the mast cell, which easily releases histamine (see Section IV). It was also considered that the acetylcholine that is released from nerves in the stomach might stimulate mast cells to release histamine.[89] There is no evidence that acetylcholine has such action, but other naturally existing compounds, especially polymers, have. It should be recalled that substance

---

* It is fair to note here that not all the effects of compound 48/80, certainly not all of its immediate effects, are attributable to the release of histamine. The action of compound 48/80 on the superior cervical ganglion was not attributable to histamine release.[46] Compound 48/80, which has pharmacological effects,[124] including neural effects,[125] different from those of histamine, antagonizes the effects of angiotensin, oxytocin, bradykinin, 5-hydroxytryptamine, and acetylcholine.[126]

P, a polypeptide, mimics a component of reflex vasodilatation,[117] and something like it, along with histamine, is released by stimulation of the frog's vagus nerve.[127] Whatever the substance or substances may be that release histamine (from the mast cell or other storage site) it need not be peculiar to nerve, for contraction of the dog's bladder, which has few nerves in its wall, also releases histamine.[128]

The idea that cholinergic stimulation is associated with the release of histamine has greatest support from studies, extensively reviewed, showing that vagal stimulation leads to a release of histamine from stomach. Many,[91] though not all,[92] believe that histamine is the final mediator of gastric secretion.* Recent work indicates that histamine may have an analogous role in salivary secretion.[129,130] From histological studies and other evidence, it is clear that the source of histamine in stomach for gastric secretion is not mast cells. For salivary glands, the evidence for the nonmast cell source of histamine rests on experiments with compound 48/80,[130] which though suggestive are equivocal because not all mast cells are disrupted by, or equally accessible to, compound 48/80 (see Section IV).

Stimulation of the skin and cornea released histaminelike material.[87] These and other observations such as the pain-producing effect of histamine, the analgesic effect of locally administered antihistamine drugs and of compound 48/80 have suggested that histamine is the mediator for cutaneous pain,[87] a controversial idea.[88]

XIII. SUMMARY

There has been growing interest in histamine in the nervous system, attributable in part to the availability of specific and sensitive chemical methods for its measurement. Histamine is present in the brains of all species examined, together with enzymes for its synthesis and catabolism. Its high concentration in the phylogenetically old parts of the brain and its association with subcellular fractions containing nerve endings make it a candidate for function. Its catabolites are also found in brain and it is suggested that they may have activity. Injected histamine and some of its catabolites have profound and varied effects on the nervous system and interact with some drugs that affect neural tissue, including psychotropic drugs. The concentration of endogenous histamine in various parts of the brain is influenced by psychotropic drugs. Provocative work on peripheral nerve and exocrine glands supports the idea that histamine merits consideration as functioning in these tissues; it is suggested that some of the histamine that appears to function in peripheral nerves may derive from mast cells. Important in the evaluation of histamine in the nervous system is an awareness of marked differences among species in their sensitivity to histamine, in the relative content of histamine in their organs, and in the relative importance of the catabolic pathways.

* Species differences may explain the different opinions.[153,154]

XIV. REFERENCES

1. E. A. Carlini and J. P. Green, The subcellular distribution of histamine, slow-reacting substance and 5-hydroxytryptamine in the brain of the rat, *Brit. J. Pharmacol.* **20**:264–277 (1963).
2. H. M. Adam, in *Regional Neurochemistry* (S. S. Kety and J. Elkes, eds.), pp. 293–306, Pergamon Press, London (1961).
3. P. A. Shore, A. Burkhalter, and V. H. Cohn, A method for the fluorometric assay of histamine in tissues, *J. Pharmacol. Exp. Therap.* **127**:182–186 (1959).
4. E. A. Carlini and J. P. Green, The measurement of histamine in brain and its distribution, *Biochem. Pharmacol.* **12**:1448–1449 (1963).
5. H. Graham, J. A. D. Scarpellini, B. P. Hubka, and O. H. Lowry, Measurement and normal range of free histamine in human blood plasma, *Biochem. Pharmacol.* **17**:2271–2280 (1968).
6. L. T. Kremzner and C. C. Pfeiffer, Identification of substances interfering with fluorometric determination of brain histamine, *Biochem. Pharmacol.* **14**:1189–1195 (1965).
7. I. A. Michaelson and P. Z. Coffman, An improved ion-exchange purification procedure for the fluorometric assay of histamine, *Anal. Biochem.* **27**:257–261 (1969).
8. V. H. Cohn, Panel Discussion, *Fed. Proc.* **23**:1112–1116 (1964).
9. H. Green and R. W. Erickson, Effect of some drugs upon rat brain histamine content, *Intern. J. Neuropharmacol.* **3**:315–320 (1964).
10. M. Medina and P. A. Shore, Increased sensitivity in a specific fluorometric method for brain histamine, *Biochem. Pharmacol.* **15**:1627–1629 (1966).
11. S. H. Snyder, R. J. Baldessarini, and J. Axelrod, A sensitive and specific enzymatic isotopic assay for tissue histamine, *J. Pharmacol. Exp. Therap.* **153**:544–549 (1966).
12. R. J. Levine, T. L. Sato, and A. Sjoerdsma, Inhibition of histamine synthesis in the rat by α-hydrazino analog of histidine and 4-bromo-3-hydroxy benzyloxamine, *Biochem. Pharmacol.* **14**:139–149 (1965).
13. H. Tabor and C. W. Tabor, Spermidine, spermine and related amines, *Pharmacol. Rev.* **16**:245–300 (1964).
14. H. Shimizu, Y. Kakimoto, and I. Sano, The determination and distribution of polyamine in mammalian nervous system, *J. Pharmacol. Exp. Therap.* **143**:199–204 (1964).
15. V. Iliev and C. Pfeiffer, Personal communication (1968).
16. J. Crossland, G. N. Woodruff, and Judith H. Woodruff, The histamine content of brain during bulbocapnine-induced catalepsy, *Life Sci.* **5**:193–197 (1966).
17. K. Kataoka and E. DeRobertis, Histamine in isolated small nerve endings and synaptic vesicles of rat brain cortex, *J. Pharmacol. Exp. Therap.* **156**:114–125 (1967).
18. H. Adam, Personal communication (1967).
19. D. J. Campbell and J. A. Kiernan, Mast cells in the central nervous system, *Nature* **210**:756–757 (1966).
20. A. Björklund, B. Falck, and E. Rosengren, Monamines in the pituitary gland of the pig, *Life Sci.* **6**:2103–2110 (1967).
21. H. M. Adam and H. K. A. Hye, Concentration of histamine in different parts of brain and hypophysis of cat and its modification by drugs, *Brit. J. Pharmacol.* **28**:137–152 (1966).
22. T. White, Histamine and methylhistamine in cat brain and other tissues, *Brit. J. Pharm.* **26**:494–501 (1966).
23. Lj. Mihailović, B. D. Janković, B. Beleslin, K. Mitrović, and Lj. Kržalić, Effect of intraventricularly injected anticerebral antibodies on the histamine-like substance and potassium content of various regions of the brain of the cat, *Nature* **203**:763–765 (1964).
24. G. W. Harris, D. Jacobsohn, and G. Kahlson, in *Ciba Foundation Colloquia on Endocrinology* (G. E. W. Wolstenholme, ed.), Vol. 4, pp. 186–194, Churchill Press, London (1952).
25. H. Selye, *The Mast Cells*, Butterworth Press, London (1965).

26. A. B. M. Macahado, L. C. M. Faleiro, and W. D. Da Silva, Study of mast cell and histamine contents of the pineal body, *Zellforsch. Mikroskop. Anat. Abt. Histochem.* **65**:521–529 (1965).

27. J. P. Green, in *Advances in Pharmacology* Vol. 1, pp. 349–422, Academic Press, New York (1962).

28. L. Enerbach, Mast cells in rat gastrointestinal mucosa, *Acta Pathol. Microbiol. Scand.* **66**:313–322 (1966).

29. I. A. Michaelson and P. Z. Coffman, The subcellular localization of histamine in guinea pig brain—A re-evaluation, *Biochem. Pharmacol.* **16**:2085–2090 (1967).

30. U. S. von Euler, The presence of the adrenergic neurotransmitter in intraaxonal structures, *Acta Physiol. Scand.* **43**:155–166 (1958).

31. S. H. Snyder, J. Glowinski, and J. Axelrod, The physiologic disposition of H^3-histamine in the rat brain, *J. Pharmacol. Exp. Therap.* **153**:8–14 (1966).

32. A. V. Furano and J. P. Green, Differences in the disposition of endogenous and exogenous substances by cells, *Nature* **199**:380–381 (1963).

33. B. Jonson and T. White, Histamine metabolism in the brain of conscious cats, *Proc. Soc. Exp. Biol. Med.* **115**:874–876 (1964).

34. J. D. Robinson, J. H. Anderson, and J. P. Green, The uptake of 5-hydroxytryptamine and histamine by particulate fractions of brain, *J. Pharmacol. Exp. Therap.* **147**:236–243 (1965).

35. H. Kwiatkowski, Histamine in nervous tissue, *J. Physiol.* **102**:32–41 (1943).

36. U. S. von Euler, in *Ciba Foundation Symposium on Histamine* (G. E. W. Wolstenholme, ed.), pp. 235–241, Little, Brown, Boston (1956).

37. E. Werje, in *Ciba Foundation Symposium on Histamine* (G. E. W. Wolstenholme, ed.), pp. 264–269, Little, Brown, Boston (1956).

38. I. Vugman and M. Rocha e Silva, in *Handbook of Experimental Pharmacology* (O. Eichler and A. Farah, eds.), Part 1, pp. 81–115, Springer-Verlag, New York (1966).

39. A. Torp, Histamine and mast cells in nerves, *Medicina Experimentalis* **4**:180–182 (1961).

40. H. J. Gamble and S. Goldby, Mast cells in peripheral nerve trunks, *Nature* **189**:766–767 (1961).

41. L. Enerbäck, Y. Olsson, and P. Sourander, Mast cells in normal and sectioned peripheral nerve, *Zellforsch. Mikroskop. Anat. Abt. Histochem.* **66**:596–608 (1965).

42. Y. Olsson, The effect of the histamine liberator, compound 48/80 on mast cells in normal peripheral nerves, *Acta Pathol. Microbiol. Scand.* **68**:563–574 (1966).

43. Y. Olsson, The effect of the histamine liberator, compound 48/80 on mast cells in sectioned peripheral nerves, *Acta Pathol. Microbiol. Scand.* **68**:575–584 (1966).

44. Y. Olsson, Degranulation of mast cells in peripheral nerve injuries, *Acta Neurologica Scandinavica* **43**:365–374 (1967).

45. Y. Olsson, Mast cell changes in INH-induced neuropathy in the rat, *Acta Pathol. Microbiol. Scand.* **69**:1–10 (1967).

46. S. B. Gertner, The effect of compound 48/80 on ganglionic transmission, *Brit. J. Pharmacol.* **10**:103–109 (1955).

47. S. H. Snyder, J. Axelrod, and H. Bauer, The fate of C^{14}-histamine in animal tissues, *J. Pharmacol. Exp. Therap.* **144**:373–379 (1964).

48. J. P. Green, Uptake and binding of histamine, *Fed. Proc.* **26**:211–218 (1967).

49. M. Draškoci, W. Feldberg, K. Fleischauer, and P. S. R. K. Haranath, Absorption of histamine into the blood stream on perfusion of the cerebral ventricles, and its uptake by brain tissue, *J. Physiol.* **150**:50–66 (1960).

50. T. White, Formation and catabolism of histamine in cat brain *in vivo*, *J. Physiol.* **152**:299–308 (1960).

51. H. M. Adam, H. K. A. Hye, and N. G. Waton, Studies on uptake and formation of histamine by hypophysis and hypothalamus in the cat, *J. Physiol.* **175**:70–71P (1964).

52. T. White, Formation and catabolism of histamine in brain tissue *in vitro*, *J. Physiol.* **149**:34–42 (1959).
53. P. Holtz and E. Westermann, Über die Dopadecarboxylase und Histidindecarboxylase des Nervengewebes, *Arch. Exp. Pathol. Pharmakol.* **227**:538–546 (1956).
54. D. D. Brown, R. Tomchick, and J. Axelrod, The distribution and properties of a histamine-methylating enzyme, *J. Biol. Chem.* **234**:2948–2950 (1959).
55. J. P. Green, in *Mechanisms of Release of Biogenic Amines* (U. S. von Euler, S. Rosell, and B. Uvnas, eds.), pp. 125–145, Pergamon Press, London (1966).
56. J. D. Robinson, Jr. and J. P. Green, Sulfomucopolysaccharides in brain, *Yale J. Biol. Med.* **35**:248–257 (1962).
57. R. W. Schayer, Catabolism of physiological quantities of histamine *in vivo*, *Physiol. Rev.* **39**:116–126 (1959).
58. F. Buffoni, Histaminase and related amine oxidases, *Pharm. Rev.* **18**:1163–1199 (1966).
59. T. Nakajima and I. Sano, A metabolite of histamine: 4(5)-imidazoyl-ethane-2-Ol, *Biochim. Biophy. Acta* **82**:260–265 (1964).
60. G. M. Crowley, The enzymatic synthesis of 5'-phosphoribosylimidazoleacetic acid, *J. Biol. Chem.* **239**:2593–2601 (1964).
61. N. R. Bachur and S. Udenfriend, Microsomal synthesis of fatty acid amides, *J. Biol. Chem.* **241**:1308–1313 (1966).
62. N. R. Bachur, K. Masek, K. L. Melmon, and S. Udenfriend, Fatty acid amides of ethanolamine in mammalian tissues, *J. Biol. Chem.* **240**:1019–1024 (1965).
63. W. P. Burkard, K. F. Gey, and A. Pletscher, Diamine oxidase in the brain of vertebrates, *J. Neurochem.* **10**:183–186 (1963).
64. J. D. Robinson, Jr. and J. P. Green, Presence of imidazole-acetic acid riboside and ribotide in rat tissues, *Nature* **203**:1178–1179 (1964).
65. J. Axelrod, P. D. MacLean, R. Wayne Albers, and H. Wiessbach, in *Regional Neurochemistry* (S. S. Kety and J. Elkes, eds.), pp. 307–311, Pergamon Press, London (1961).
66. D. H. Fram and J. P. Green, Methylhistamine in guinea pig brain, *J. Neurochem.* **15**:597–602 (1968).
67. T. L. Perry, S. Hansen, J. G. Foulks, and G. M. Ling, Aliphatic and aromatic amines of cat brain, *J. Neurochem.* **12**:397–405 (1965).
68. T. L. Perry, S. Hansen, and L. C. Jenkins, Amine content of normal human cerebrospinal fluid, *J. Neurochem.* **11**:49–53 (1964).
69. N. Weiner, The distribution of monoamine oxidase and succinic oxidase in brain, *J. Neurochem.* **6**:79–86 (1960).
70. E. Werle and D. Palm, Histamin in nerven, *Biochem. Z.* **323**:255–264 (1952).
71. H. M. Lee and R. G. Jones, The histamine activity of some β-aminoethyl heterocyclic nitrogen compounds, *J. Pharmacol. Exp. Therap.* **95**:71–78 (1949).
72. H. Westling, Observations on the action of histamine and related substances on the bronchial resistance in the guinea pig, *Acta Physiol. Scand.* **40**:75–82 (1957).
73. T. White, Some effects of histamine and two histamine metabolites on the cat's brain, *J. Physiol.* **159**:198–202 (1961).
74. T. M. Lin, R. S. Alphin, F. G. Henderson, D. N. Benslay and K. K. Chen, The role of histamine in gastric hydrochloric acid secretion, *Ann. N. Y. Acad. Sci.* **99**:30–44 (February 28, 1962).
75. L. Goldstein, C. C. Pfeiffer, and C. Munoz, Quantitative EEG Analysis of the Stimulant Properties of Histamine and Histamine Derivatives, *Fed. Proc.* **22**:424 (1963).
76. D. H. Fram and J. P. Green, Methylhistamine excretion during treatment with a monoamine oxidase inhibitor, *Clin. Pharmacol. Therap.* **9**:355–357 (1968).
77. C. L. Fox and S. E. Lasker, Protection by histamine and metabolites in anaphylaxis, scalds and endotoxin shock, *Am. J. Physiol.* **202**:111–113 (1962).

78. E. Roberts and D. G. Simonsen, A hypnotic and possible analgesic effect of imidazoleacetic acid in mice, *Biochem. Pharmacol.* **15**:1875–1877 (1966).

79. E. G. McGeer, P. L. McGeer, and H. McLennan, The inhibitory action of 3-hydroxytyramine, gamma-aminobutyric acid (GABA) and some other compounds toward the crayfish stretch receptor neuron, *J. Neurochem.* **8**:36–49 (1961).

80. J. P. Green, Histamine and the nervous system, *Fed. Proc.* **23**:1095–1102 (1964).

81. U. Trendelenburg and A. Jones, Facilitation of ganglionic responses after a period of preganglionic stimulation, *J. Pharmacol. Exp. Therap.* **147**:330–335 (1965).

82. L. C. Iorio and R. J. McIsaac, Comparison of the stimulating effects of nicotine, pilocarpine and histamine on the superior cervical ganglion of the cat, *J. Pharmacol. Exp. Therap.* **151**: 430–437 (1966).

83. G. P. Lewis and E. Reit, The action of angiotensin and bradykinin on the superior cervical ganglion of the cat, *J. Physiol.* **179**:538–553 (1965).

84. J. Staszewska-Barczak and J. R. Vane, The release of catechol amines from the adrenal medulla by histamine, *Brit. J. Pharmacol.* **25**:728–742 (1965).

85. S. D. Everett and S. P. Mann, Catecholamine release by histamine from the isolated intestine of the chick, *Eur. J. Pharmacol.* **1**:310–320 (1967).

86. G. A. Kerkut and M. A. Price, The effects of drugs on the threshold of peripheral nerve, *Life Sci.* **2**:722–724 (1963).

87. S. R. Rosenthal, Histamine as the chemical mediator for cutaneous pain, *Fed. Proc.* **23**:1109–1111 (1964).

88. A. Iggo, *in Ciba Foundation Study Group No. 1* (G. E. W. Wolstenholme, ed.), Vol. 1, pp. 41–59, Little, Brown, Boston (1959).

89. W. D. M. Paton and J. R. Vane, An analysis of the responses of the isolated stomach to electrical stimulation and to drugs, *J. Physiol.* **165**:10–46 (1963).

90. A. Bennett and B. Whitney, A pharmacological investigation of human isolated stomach, *Brit. J. Pharmacol.* **27**:286–298 (1966).

91. C. F. Code, Histamine and gastric secretion: a later look, 1955, 1965, *Fed. Proc.* **24**:1311–1321 (1965).

92. C. G. Clark, V. J. Curnow, J. G. Murray, F. O. Stephens, and J. H. Wyllie, Mode of action of histamine in causing gastric section in man, *Gut* **5**:537–545 (1964).

93. U. S. von Euler, *in Handbook of Experimental Pharmacology* (O. Eichler and A. Farah, eds.), Part 1, pp. 318–333, Springer-Verlag, New York (1966).

94. F. J. Rauzzino and J. Seifter, Potentiation and antagonism of biogenic amines, *J. Pharmacol. Exp. Therap.* **157**:143–148 (1967).

95. U. Trendelenburg, Stimulation of sympathetic centers by histamine, *Circulation Res.* **5**:105–110 (1957).

96. T. White, Peripheral vascular effects of histamine administered into the cerebral ventricles of anaesthetized cats, *Experientia* **21**:132–133 (1965).

97. N. L. Sadre and N. M. Tiwari, Prolongation of chloral hydrate and pentobarbitone sleeping time by chlorpromazine and histamine in mice, *Arch. Intern. Pharmacodyn.* **163**:6–10 (1966).

98. D. Bovet, R. Kohn, M. Marotta, and B. Silvestrini, Some effects of histamine in the normal and *Haemophilus pertussis* vaccinated rat, *Brit. J. Pharmacol.* **13**:74–83 (1958).

99. M. Monnier, M. Fallert, and I. C. Bhattacharya, The waking action of histamine, *Experientia* **23**:21–25 (1967).

100. F. J. Rosenberg and P. J. Savarie, Histamine and the reversal of chlorpromazine-induced depression, *J. Pharmacol. Exp. Therap.* **146**:180–185 (1964).

101. J. Crossland and J. F. Mitchell, The effect of the electrical activity of the cerebellum of a substance present in cerebellar extracts, *J. Physiol.* **132**:391–405 (1956).

102. A. S. Marrazzi, E. R. Hart, and T. M. Gilfoil, A potential histaminogenic (allergic) mechanism for psychosis, *Recent Adv. Biol. Psych.* **3**:164 (1961).

103. K. Krnjevic and J. W. Phillis, Actions of certain amines on cerebral cortical neurones, *Brit. J. Pharmacol.* **20**:471–490 (1963).

104. W. Feldberg and S. L. Sherwood, Injections of drugs into the lateral ventricle of the cat, *J. Physiol.* **123**:148–167 (1954).

105. C. Fazio and V. Sacchi, in *Psychotropic Drugs* (S. Gurattini and V. Ghetti, eds.), p. 104, Elsevier, Amsterdam (1957).

106. R. G. Heath, Schizophrenia: biochemical and physiologic aberrations, *Int. J. Neuropsychiat.* **2**:597–610 (1966).

107. L. C. Cesare, R. S. Carlini, and E. A. Carlini, Influence of histamine on the catatonia induced in mice by tetrabenazine and reserpine, *Arch. Intern. Pharmacodyn.* **169**:26–34 (1967).

108. C. VanderWende and J. C. Johnson, Species variation and the antihistaminic properties of 3,4-dimethoxyphenylethylamine, *Life Sci.* **6**:2345–2352 (1967).

109. A. Gustafsson and G. P. Forshell, Purification of a N-methyltransferase, *Acta Chem. Scand.* **17**:541–542 (1963).

110. S. H. Snyder and J. Axelrod, Inhibition of histamine methylation *in vivo* by drugs, *Biochem. Pharmacol.* **13**:536–537 (1964).

111. T. White, Inhibition of the methylation of histamine in cat brain, *J. Physiol.* **159**:191–197 (1961).

112. A. S. Weltman and A. M. Sackler, Effect of lysergic acid diethylamide (LSD-25) on growth metabolism and the resistance of male rats to histamine stress, *J. Pharm. Sci.* **54**:1382–1384 (1965).

113. A. M. Sackler, A. S. Weltman, and R. R. Sackler, Effect of tranquillizing agents on the resistance of rats to histamine stress, *Nature* **183**:896–897 (1959).

114. S. Kakiuchi and T. W. Rall, The influence of chemical agents on the accumulation of adenosine 3′,5′-phosphate in slices of rabbit cerebellum, *Mol. Pharmacol.* **4**:367–378 (1968).

115. J. D. Reid and D. M. Shepherd, Inhibition of histidine decarboxylases, *Life Sci.* **1**:5–8 (1963).

116. G. Kahlson and E. Rosengren, New approaches to the physiology of histamine, *Physiol. Rev.* **48**:155–196 (1968).

117. L. Beck, A. A. Pollard, S. O. Kayaal, and L. M. Weiner, Sustained dilatation elicited by sympathetic nerve stimulation, *Fed. Proc.* **25**:1596–1606 (1966).

118. M. J. Brody, Neurohumoral mediation of active reflex vasodilatation, *Fed. Proc.* **25**:1583–1592 (1966).

119. R. S. Tuttle, Physiological release of histamine-^{14}C in the pyramidal cat, *Am. J. Physiol.* **213**:620–624 (1967).

120. R. S. Tuttle, Relationship between blood histamine and centrally evoked hypotensive response, *Am. J. Physiol.* **209**:745–750 (1965).

121. R. S. Tuttle, Histaminergic component in the baroreceptor reflex of the pyramidal cat, *Fed. Proc.* **25**:1593–1595 (1966).

122. R. S. Tuttle, Evidence for histaminergic nerves in the pyramidal cat, *Am. J. Physiol.* **211**:903–910 (1966).

123. J. P. Green, A new neoplastic mast cell grown in culture, *Eur. J. Pharmacol.* **3**:68–73 (1968).

124. C. A. Papacostas, E. R. Loew, and B. G. West, Studies on the toxicology of a histamine liberator, compound 48-80, *Arch. Intern. Pharmacodyn.* **120**:353–362 (1959).

125. M. Rocha e Silva, Central effects produced by injection of 48/80 into the cerebral ventricles of mice, *Brit. J. Pharmacol.* **14**:243–245 (1959).

126. T. B. Paiva and A. C. M. Paiva, The inhibition of cationic myotropic drugs by compounds 48/80 and 46/108, *Biochem. Pharmacol.* **15**:1303–1308 (1966).

127. I. Singh and S. I. Singh, Histaminergic transmission in a frog vagus stomach muscle preparation, *Arch. Intern. Biochim.* **74**:365–373 (1966).

128. N. Ambache and G. S. Barsoum, The release of histamine by isolated smooth muscles, *J. Physiol.* **96**:139–145 (1939).

129. E. Werle and W. Lorenz, Speicheldrüsensekretion nach Pilocarpin, Histamin, und Kininen, *Arch. Intern. Pharmacodyn.* **161**:477–488 (1966).

130. F. Erjavec, M. A. Beaven, and B. B. Brodie, Uptake and release of ^{3}H-histamine in cat submixillary gland, *Fed. Proc.* **26**:237–240 (1967).

131. J. J. Abel and S. Kubota, Presence of histamine (4-imidazoleethylamine) in the hypophysis cerebri and other tissues of the body and its occurrence among the hydrolytic decomposition products of proteins, *J. Pharmacol. Exp. Therap.* **13**:243–299 (1919).

132. A. V. Furano and J. P. Green, The uptake of biogenic amines by mast cells of the rat, *J. Physiol.* **170**:263–271 (1964).

133. J. W. Kissel and E. F. Domino, The effects of some possible neurohumoral agents on spinal cord reflexes, *J. Pharmacol. Exp. Therap.* **125**:168–177 (1959).

134. D. R. Curtis, J. W. Phillis, and J. C. Watkins, Cholinergic and non-cholinergic transmission in the mammalian spinal cord, *J. Physiol.* **158**:296–323 (1961).

135. C. R. Merril, S. H. Snyder, and D. F. Bradley, Inhibition of histamine methyltransferase by serotonin and chlorpromazine derivatives: electronic aspects, *Biochim. Biophys. Acta* **118**:316–324.

136. R. Levi and J. P. Green, 5-Hydroxytryptamine and the central nervous system, *Annual Reports in Medicinal Chemistry*, 1966, pp. 273–285 (C. K. Cain, ed.), Academic Press, New York (1967).

137. R. W. Schayer, Histamine and circulatory homeostasis, *Fed. Proc.* **24**:1295–1297 (1965).

138. G. Kahlson and E. Rosengren, New approaches to the physiology of histamine, *Physiol. Rev.* **48**:155–196 (1968).

139. G. A. Kerkut, R. J. Walker, and G. N. Woodruff, The effects of histamine and other naturally occurring imidazoles on neurones of helix aspersa, *Brit. J. Pharmacol.* **32**:241–252 (1968).

140. M. C. Gerald and W. C. Stern, Interactions of histamine and antihistamines with behavioral systems, *Fed. Proc.* **27**:273 (1968).

141. V. H. Cohn and W. Wynn, Inhibition of histamine metabolism by antihistaminic drugs, *Fed. Proc.* **27**:243 (1968).

142. A. H. Anton and D. F. Sayre, A modified fluorometric procedure for tissue histamine and its distribution in various animals, *J. Pharmacol. Exp. Therap.* **166**:285–292 (1969).

143. G. Ungar, Demonstration de la mise en liberté des substances histaminiques. Transmission neuro-humorale histaminergique, *J. Physiol. Path. Gén.* **34**:77–91 (1936).

144. L. A. Pearce and S. M. Schanberg, Histamine levels during brain development, *Fed. Proc.* **28**:353 (1969).

145. Y. Olsson, Mast cells in the nervous system, *Int. Rev. Cytol.* **24**:27–70 (1968).

146. I. A. Michaelson, P. Z. Coffman, and D. F. Vedral, The regional distribution of histamine in brain of the Rhesus monkey (Macaca mulatta), *Biochem. Pharmacol.* **17**:2435–2441 (1968).

147. M. J. Ryan and M. J. Brody, Histamine (H) in sympathetic nerves of the dog, *Fed. Proc.* **28**:353 (1969).

148. E. M. Kovacs and S. Heisler, The role of N,N-dimethylhistamine in the cortisone induced acid hypersecretion of guinea-pigs, *Fed. Proc.* **28**:353 (1969).

149. J. W. Phillis, A. K. Tebēcis, and D. H. York, Histamine and some antihistamines: their actions on cerebral cortical neurones, *Brit. J. Pharmacol.* **33**:426–440 (1968).

150. J. W. Phillis, A. K. Tebēcis, and D. H. York, Depression of spinal motoneurones by noradrenaline, 5-hydroxytryptamine, and histamine, *Eur. J. Pharmacol.* **4**:471–475 (1968).

151. J. A. Levin, J. D. Bartlett, Jr., and L. Beck, Active reflex vasodilatation induced by intravenous epinephrine or norepinephrine in primates, *J. Pharmacol. Exp. Therap.* **161**:262–270 (1968).

152. G. Glick, A. S. Wechsler, and S. E. Epstein, Mechanisms of reflex vasodilation: assessment of the role of neural reuptake of norepinephrine and release of histamine, *J. Clin. Invest.* **47**:511–520 (1968).

153. R. Håkanson, Ch. Owman, and N.-O. Sjöberg, Three different systems of monoamine storing cells in the gastrointestinal tract of fetal and neonatal rats, *Acta Physiol. Scand.* **75**:213–220 (1969).
154. R. Håkanson, B. Lilja, and Ch. Owman, Cellular localization of histamine and monoamines in the gastric mucosa of man, *Histochemie* **18**:74–86 (1969).

Chapter 11

SEROTONIN

Irvine H. Page and Arvid Carlsson

*Research Division of the Cleveland Clinic Foundation
Cleveland, Ohio; and the Department of Pharmacology
University of Göteborg, Sweden*

I. INTRODUCTION

The literature on the distribution and function of serotonin in brain has grown enormously in the past 15 years. The authors do not consider it possible within limited space to document adequately all the evidence, hence they elect to give certain general references from which the reader can, with relative ease, get to the source. For those interested in tables of critical values, the *Handbook of Experimental Pharmacology*[11] will suffice.

After the discovery of serotonin in blood in 1948 during a routine survey of its distribution within the body, it was found in the brain; prior to this there had been no reason to suspect its presence there. It soon became clear that the blood–brain barrier was not penetrated by this amine and that the brain was dependent on synthesis and degradation within itself for its metabolism.

Serotonin, like other humoral agents, seems to serve many quite different functions in the body and the number suggested for the brain alone has been legion. The chief of these are (1) that it acts as a transmitter for serotonergic neurons and (2) that it functions in the control of mood.

II. DISTRIBUTION

High concentrations of serotonin are found in the hypothalamus while those in the neopallium and cerebellum are low. The development of a histochemical fluorescence method for the demonstration of serotonin and catecholamines by Hillarp and Falck has made possible extraordinary detailed examinations by the Swedish school of Carlsson, Fuxe, Dahlström, Andén, etc. They showed the existence of central serotonergic neurons whose concentration was low in the cell bodies and high in the terminal portions.

Serotonin is localized mainly in beadlike enlargements of the nerve endings, so-called varicosities, where it is stored in granular vesicles.

The serotonergic neuron cell bodies are localized mainly in the raphé nuclei of the lower brain stem and some of them are also found surrounding the pyramidal tract and in the medioventral part of the caudal tegmentum. Almost none occur in the diencephalon, telencephalon, or spinal cord. Electron microscopic studies of the cells in the nucleus raphé dorsalis show well-developed granular reticulum and prominent Golgi apparatus around the nucleus. Granular vesicles are seen mainly in the serotonin-rich zone surrounding the nucleus, probably representing a storage site. The varicosities of the terminals seem to be presynaptic structures. The largest number of serotonergic nerve terminals are found in the lumbar enlargement and the sacral portion of the cord. Large numbers of them are found in the autonomic sacral nucleus, the sympathetic lateral column, and in some visceral efferent nuclei of the pons and the medulla. The lower brain stem is rich in them. Terminals are also diffusely scattered in most parts of the reticular formation. The fact that the serotonergic nerve terminals are found in both sympathetic and parasympathetic nuclei shows that they are not concerned alone with regulation of parasympathetic function.

There have been many detailed studies by the Swedish school of other portions of the brain but they need not be recounted here as the literature is readily available.

III. SYNTHESIS AND DEGRADATION OF SEROTONIN

A. 5-Hydroxylation of Tryptophan in Brain

Hydroxylation of tryptophan in the five position is the first and probably the rate-limiting reaction in the formation of serotonin. Brain homogenates are able to hydroxylate tryptophan, and the same is true *in vivo* as measured by the synthesis of serotonin from tryptophan infused into the brain.

There has been much uncertainty as to whether such a specific hydroxylase exists in brain, but with the development of a sensitive radioassay, the conversion of tryptophan to 5-hydroxytryptophan showed the presence of this enzyme in a variety of tissues. Beef and rat pineal tissue contained the highest activity; rat and rabbit brain stem, mouse mast cell tumor, and human carcinoid cells also showed easily measurable amounts. Both hydroxylase and decarboxylase tend to be localized in nerve endings.

The enzymatic reaction requires a reduced pteridine and oxygen; it is inhibited by *p*-chlorophenylalanine. The latter specific serotonin depletor when given to rats also inhibits the hydroxylase. The enzyme inhibition is correlated with cerebral serotonin depletion, this presumably being the cause. *p*-Chlorophenylalanine is a competitive inhibitor *in vitro* but causes irreversible inhibition *in vivo*. The parallelism between the serotonin content and tryptophan hydroxylase activity in brain after administration of *p*-chlorophenylalanine suggests that cerebral serotonin results from the action of tryptophan hydroxylase within the brain. The parallelism also is evidence that tryptophan hydroxylation is the rate-limiting enzymatic step in serotonin biosynthesis. The enzyme does not normally appear to be fully saturated with substrate, hence the rate of synthesis of serotonin may be partially dependent upon the availability of tryptophan.

The striking similarity between the enzymatic systems responsible for synthesis and degradation of serotonin and those for the catecholamines is clear. Indeed, catecholamines seem to be good inhibitors of beef tryptophan hydroxylase. But their distribution is not the same, suggesting that the need for serotonin occurs in places different from catecholamines. Further, inhibition of tyrosine hydroxylase by the methyl ester of α-methyl-*p*-tyrosine causes the catecholamines but not the serotonin in rat brain and spinal cord to disappear almost completely within 24 hr. There was little difference among the various regions in rate and degree of depletion of the terminals and cell bodies, but after cord transection the loss of norepinephrine was prevented below the lesion but not above it.

B. Breakdown of Serotonin

Serotonin is degraded chiefly by monoamine oxidase, and for this reason inhibitors of this enzyme are commonly used to increase its concentration in tissues. The first step is formation of 5-hydroxyindole acetaldehyde, followed by 5-hydroxyindole acetic acid resulting from the action of aldehyde dehydrogenase. The latter acid is widely used as a measure of serotonin metabolism. The 5-hydroxyindole acid content of urine is, however, a fallacious measure of the serotonin within the brain.

C. Methylation of Serotonin

There is chromatographic evidence of the presence of methylated derivatives of serotonin in some urine samples. The problem is far from solved, both as to the identification and significance of several derivatives.

One of the more interesting metabolic alterations of serotonin is its *N*-methylation. An enzyme was separated from lung that catalyzes conversion to the psychotomimetics bufotenine and *N,N*-dimethyltryptamine. Whether such a metabolic change is a causal mechanism in psychotic states is unknown. Methylation in the *l*-position has also produced some *l*-methylindoles which have strong psychotropic effects on conditioned rats. These may prove of much interest if they can be shown to occur naturally. The possibility of the formation of the highly active 5-methoxy-*N*:*N*-dimethyltryptamine from such a substrate as serotonin has been studied.

Interest in *O*-methylation was greatly furthered by Lerner's demonstration that the *O*-methyl derivative of *N*-acetylserotonin is a pineal skin-lightening hormone which he named "melatonin."

D. Tryptophol

Kveder, Iskeric, and Keglevic[19] in 1962 first described formation of 5-hydroxytryptophol from serotonin *in vivo* in rats. The alcohol appears to be a major metabolite of serotonin in platelets. Recently it was shown that both human and rat brain homogenates convert serotonin into 5-hydroxytryptophol and 5-hydroxyindoleacetic acid via 5-hydroxyindoleacetaldehyde. The conversion to the alcohol is effected by an alcohol dehydrogenase that requires reduced nicotinamide adenine dinucleotide phosphate as coenzyme.

IV. OTHER PSYCHOTROPIC DRUGS AND SEROTONIN

The literature on this problem is vast and would suffer from being touched lightly. We cannot do better than to use Pletscher's orientating schema to expose the problem.[24] He outlines the following five types of action of these drugs:

1. Decrease of storage capacity for monoamines—*Rauwolfia* alkaloids and benzoquinolizine derivatives.
2. Decrease of synthesis—α-propyldopacetamide, α-methylDOPA, *p*-chlorophenylalanine.
3. Inhibition of monoamine oxidase—iproniazid, nialamide, harmaline. The excitation induced by monoamine oxidase inhibitors appears to be largely due to accumulation of serotonin, as shown by experiments with specific inhibitors of synthesis.
4. Interference with monoamine penetration—chlorpromazine. However, this effect is probably secondary to hypothermia.
5. Increase of sensitivity of central nervous receptors—imipramine and amitriptyline. This "sensitization" is apparent only; it is due to blockade of the "amine pump" at the level of the cell membrane of monoaminergic neurons, leading to an increased serotonin level at receptor sites. The sensitivity of the receptors themselves is not increased but, if anything, decreased.

There are many drugs that affect these various aspects of both serotonin and catecholamine metabolism. Let us consider just a few examples.

A. Imipramine

Since Sigg's discovery that imipramine potentiates the action of norepinephrine and sympathetic nerve stimulation in the peripheral nervous system, the view was favored that a similar augmenting effect on central norepinephrine might account for its antidepressive action. The peripheral potentiation is probably due to blockade of an amine-concentrating mechanism in the cell membranes of adrenergic neurons. While derivatives of imipramine proved active inhibitors of this mechanism in the central norepinephrine neurons, imipramine itself was found surprisingly weak by Carlsson, Fuxe, and Ungerstedt despite its good antidepressive action. They have recently found that in the central serotonergic neuron a reserpine-resistant uptake concentrating mechanism exists, which is blocked by imipramine.

Thus, intraventricular injections of serotonin were made into the lateral ventricles of rats pretreated intraperitoneally with reserpine and nialamide. Imipramine was given intraperitoneally 15 min before the serotonin. This resulted in marked blockade of the reserpine-resistant concentrating mechanism for serotonin in the serotonergic cell bodies, nonterminal axons, and terminals lying close to the ventricles. This, as well as other evidence, suggests the importance of this action for the antidepressive action of imipramine.

There seem to be quite different structural requirements for blockade between the noradrenergic and serotonergic neurons. Protriptyline and desipramine are much more active inhibitors of the membrane pump of central noradrenergic neurons than imipramine and amitriptyline. But for the serotonergic neurons the reverse holds true. This led to the tentative conclusion that stimulation of the noradrenergic neurons causes behavioral "activation," i.e., release of the inhibition often seen in depressed patients, whereas stimulation of the serotonergic neurons causes elevation of mood. This may be related to the observation that in depressed patients the level of 5-hydroxyindoleacetic acid in the cerebrospinal fluid is frequently low.

The different antidepressive agents on central adrenergic and serotonergic neurons have been compared and found highly variable in their effectiveness. Protriptyline and desipramine were especially potent on the adrenergic and imipamine, amitriptyline and nortriptyline on the serotonergic. Comparing these effects with clinical experience on antidepressive activity, the impression was gained that the adrenergic neurons cause activation and the serotonergic, mood elevation.

B. Reserpine and Serotonin—Kinetics

In 1955, Pletscher, Shore, and Brodie[24] made the remarkably interesting observation that reserpine caused a sharp decrease in concentration of serotonin in many tissues by blocking the storage process. The depletion but not the actual content of cerebral serotonin was related to the tranquilizing effect of reserpine. This was shortly followed by the observation of others that the catecholamines were similarly affected. These observations set off a blizzard of papers, which although now somewhat abating, still continues. In a short space, we cannot do justice to the debate.

In essence, the chief problem is whether specific depletion of brain serotonin causes the sedation resulting from reserpine.

There have been many studies attempting selectively to deplete either serotonin or norepinephrine but these have proved difficult to affect one without the other and, further, it is not clear how much depletion is necessary to elicit physical signs. Brodie suggested that there is a direct association between functional recovery from reserpine and recovery of the cell membrane pump which—in his opinion—is normally responsible for the uptake and storage of serotonin. He applied steady-state kinetics, showing that serotonin, norepinephrine, and dopamine stores, though present in constant amount, are in continual flux, with synthesis balancing efflux. This integration is controlled by "neurochemical transducers" that translate electrical impulses into the discharge of precise amounts of neurohormones to the receptor. Brodie considers them the "primary units of behavior," since the organism responds to environmental change only because these units control the quantity of free neurohormones at nerve endings.

The central effect of reserpine has been found to be a function of the initial rate of release as expressed by the proportion of serotonin stores that disappear per minute rather than the final degree of depletion. These results do not prove that the central action of reserpine is caused by change in serotonin, but show two associations requiring exploration: (1) the close relationship between the initial rate of serotonin release and the intensity of the reserpine action; (2) the association between the recovery mechanism in brain that takes up serotonin and recovery of animals from sedation. Brodie's kinetic analysis of serotonin release suggests that the primary action of reserpine is to inhibit the carrier process at nerve endings. He is convinced that reserpine sedation is associated with changes in cerebral serotonin rather than with catecholamines. This is supported by Spector, Sjoerdsma, and Udenfriend's finding that α-methyltyrosine which blocks catecholamine synthesis does not elicit sedation until catecholamines are reduced 90 % or more, while reserpine produces it in doses that deplete stores of serotonin and norepinephrine about 55 %.

Koe and Weissman discovered that p-chlorophenylalanine lowers cerebral serotonin specifically and prevents the rise after administration of monoamine oxidase inhibitors. The drug inhibits the rate-limiting tryptophan hydroxylase. But pretreatment with it does not prevent reserpine sedation! Brodie and Reid counter this with the observation that after administration of maximal doses of p-chlorophenylalanine, the synthesis of serotonin-^{14}C from tryptophan-^{14}C injected directly into the hypothalamus is about 30 % of normal. It may be that the reduction of serotonin in the brain is not sufficient to block the central action of reserpine by analogy with the decline to 10 % in norepinephrine necessary to produce chloropromazinelike signs when α-methyltyrosine is given.

Thus, Brodie's studies seem to indicate that the tranquilizing effect of reserpine and chlorpromazine involve different neural pathways.

Carlsson takes a different view. He does not believe that reserpine acts on the cell membrane pump but on the intraneuronal storage granules. By blocking the amine-concentrating mechanism of these granules, reserpine causes depletion of transmitter and blockade of transmission, as demonstrated in the peripheral adrenergic system. This is in contrast to Brodie, who believes that reserpine, by blocking the cell membrane pump, causes an increase in the concentration of serotonin at postsynaptic receptor sites. According to the Swedish school, there is no close correlation between amine levels and function. Functional recovery from reserpine sets in at persistent low amine levels but coincides with recovery of a small essential transmitter pool of the granules, as demonstrated with isotope techniques.

Furthermore, Carlsson believes that depletion of catecholamines rather than serotonin is responsible for the gross reserpine syndrome. This view is based on (1) the ability of DOPA rather than 5-hydroxytryptophan to antagonize this syndrome, and (2) the presence of the syndrome in animals whose serotonin stores have been selectively protected against reserpine, and the absence of the syndrome in animals whose catecholamine stores have been selectively protected. Carlsson admits that serotonin depletion may be involved in more subtle components of the reserpine syndrome.

Although some important studies have been made only on peripheral adrenergic neurons, the relatively recent work of the Swedish school leaves little doubt that central adrenergic and serotonergic neurons behave similarly. Many clinical phenomena will find their explanation when the different central amine-containing neurons are traced. For example, years ago when we first used reserpine in high doses for treatment of hypertension, we noted the occurrence of Parkinson's syndrome in some of our patients. We had no idea why. In 1964 Andén et al.[1] showed that dopamine occurs in at least three different central neuronal systems. One of these, a pathway originating in the *substantia nigra* and ascending through the internal capsule to the caudate nucleus-putamen, when blocked by reserpine elicits Parkinsonism, just as the spontaneous variety occurs when these neurons are destroyed.

V. SEROTONIN AS A NEURAL TRANSMITTER

The facts that serotonin is both synthesized and degraded in the brain; that it is present in specific neurons, and concentrated in synaptic vesicles; and that in lower animals it has some functions similar to acetylcholine, have led to the belief that it functions as a transmitter and that there are *serotonergic* as well as *adrenergic* pathways.

The rate of ciliary beating in the gill of the muscle *Mytilus edulis* was greatly increased by electrical stimulation of the branchial nerve. If nerve activity causes release of serotonin, depletion of it by reserpine or blockade by bromolysergic acid diethylamide should diminish the cilio-excitatory effect. This occurred: the rate of ciliary beating was, however, unrelated to the total concentration of serotonin in the gills. Rather, it seemed dependent on the concentration of free serotonin, which in turn may have been determined by branchial nerve activity. Perhaps somewhat similar phenomena are encountered in the bioluminescence of the animal *Meganyetiphases*.

One of the more convincing experiments suggesting the importance of serotonin as a transmitter is that in which electrodes were implanted in the area of the medial mammillary nucleus of the posterior hypothalamus of dogs. The dogs were trained to press a lever for electrical self-stimulation, and the thresholds were measured. Drugs such as α-methylDOPA, which lower cerebral norepinephrine, dopamine, and serotonin, resulted in marked depression of self-stimulation response. This inhibiting effect correlated best with serotonin levels. If serotonin was elevated and norepinephrine unchanged, threshold was also lowered. The serotonin antagonist bromolysergic acid diethylamide lowered the threshold although high doses raised it. A monoamine oxidase inhibitor which elevated brain serotonin lowered the threshold. In this self-reward system serotonin seems to play an important part.

Marrazzi and Hart were convinced that serotonin played a highly important part in neurohumoral synaptic inhibition, overshadowing epinephrine and norepinephrine by virtue of its greater potency. When given by arterial injection into a common carotid artery, the serotonin was believed to penetrate the blood–brain barrier to inhibit the ipsilateral cortical synapses, as indicated by reduction in cortically recorded action potentials. The high potency of serotonin appeared to be characteristically cerebral, since the synaptic inhibition they had recorded in the ciliary ganglion required about 75 times the dose. Marrazzi has shown that many chemical psychotogens are synaptic inhibitors and suggests that serotonin or similar substances are likely candidates for the role of endogenous psychotogens acting by distorting synaptic equilibria.

Evidence gathered by Andén, Carlsson, and Hillarp has left little doubt in their minds that serotonin is a transmitter. They cite (1) the similar distribution in *nonadrenergic* neurons to that of norepinephrine, e.g., in peripheral adrenergic neurons, (2) the release and increased synthesis of serotonin caused by electrical stimulation of the axon, (3) the origin of serotonergic neurons in the lower brain stem and termination in various parts of the brain and spinal cord innervating, for example, spinal sympathetic centers in the intermediolateral nucleus, (4) the fundamentally similar actions of reserpine on catecholamines and serotonin, i.e., blockade of storage mechanism, as clear evidence in favor of this hypothesis. It should be recalled that only a few years ago most investigators emphatically took the opposite view.

While the evidence is fragmentary, still it adds up to the probability that the serotonin in the serotonergic nerve has a transmitter function, in part inhibitory and especially prominent in lower animals. The high concentration of serotonin in the pineal and hypothalamic areas suggests association with other rhythmic functions. What it comes to is that the function and metabolism of the cerebral indolealkylamines are only dimly perceived but even though dim, their vast importance can hardly be mistaken.

VI. LYSERGIC ACID DIETHYLAMIDE (LSD)

In animals LSD elicits similar effects to those from drugs which cause increased serotonin levels in the central nervous system. It seems possible that LSD interferes with serotonergic nerve transmission, especially since it had been shown that it increases the serotonin content of rat's brain, within 10 min of its administration. LSD *in vivo* or *in vitro* does not change the net synthesis of serotonin from 5-hydroxytryptophan, especially of the particulate fraction. For these reasons, the effect of LSD on the activity of the central monoamine-containing neurons was studied utilizing monoamine synthesis inhibitors. Amine depletion obtained with such inhibitors is highly dependent on nerve impulse flow. The effects of LSD and 5-hydroxytryptophan on hind limb reflexes of *spinal* rats were examined as well.

LSD decreased greatly the rate of amine depletion in serotonergic neurons after inhibition of serotonin synthesis by α-propyldopacetamide, indicating that it decreases the activity of these neurons. This was deduced since it had been shown that the serotonergic terminals caudal to a spinal cord section, which lack nerve impulses, were not depleted after inhibition of synthesis.

The functional studies on spinal reflexes showed LSD and 5-hydroxytryptophan to have similar actions. Since the central action of LSD appeared also in reserpine-α-propyldopacetamide pretreated rats, these effects of LSD may be partly due to direct stimulation of the serotonin receptors of the postsynaptic neurons and not to release of neuronal serotonin. They also suggest that such direct stimulation evokes a negative feedback on the presynaptic serotonergic neurons, thus resulting in decreased neuronal activity as indicated by chemical studies. These results, then, are taken to mean that serotonin receptor stimulation may be responsible for some pharmacological effects of LSD and possibly the hallucinogenic ones as well. This hypothesis is supported by the fact that nonhallucinogenic lysergic acid derivatives do not produce either the chemical or functional changes noted after LSD.

VII. SEROTONIN AND TEMPERATURE CONTROL

Feldberg and Myers found evidence that serotonin and catecholamines have a reciprocal relationship to temperature control. Rectal temperature of anesthetized cats was sharply elevated by injecting serotonin into the cerebral ventricles, whereas catecholamines lowered it. This action appeared to be mediated by the hypothalamus. It is not known whether this is a physiological phenomenon or not.

VIII. SEROTONIN AND MENTAL DISEASE

Any suggestion concerning the cause of mental disease inevitably receives attention no matter how meager its justification. Such is true for serotonin in this relationship. The facts are that serotonin can be produced, stored, and destroyed in the brain. On isolated uterus its contractile action is antagonized by LSD, but whether the psychotomimetic action of LSD is affected by serotonin in intact brain is unknown. Some indolealkylamines are hallucinogens and produce behavioral disturbances. There is some evidence that tryptophan metabolism may be abnormal in certain mental patients.

There has been much fancy, the chief bit probably that conjured by the great pharmacologist, Gaddum, who made the much-quoted and totally unsupported epigram, "It is possible that the serotonin in our brains plays an essential part in keeping us sane." This was so much nonsense and Gaddum knew it! The best case for the participation of serotonin in mental disease was made by Woolley in his book, "The Biochemical Bases of Psychoses or the Serotonin Hypothesis about Mental Diseases." The brilliant discoveries of A. Hoffmann at the Sandoz Laboratories in Basel of both LSD and psilocybin have contributed greatly to the basic interest if not understanding of this aspect of cerebral metabolism.

The idea of involvement of amino acids in the mechanisms of mental disease has been tested in a number of ways. One was to administer large amounts while the patient was receiving monoamine oxidase inhibitor. Methionine and tryptophan changed the behavior of schizophrenic patients. There appears to be a relationship between behavioral worsening and rise in urinary tryptamine. Recently, clinical evidence suggests that tryptophan (5–7 g/day) was as effective in treating severe unremitting depression as electroconvulsive therapy. Addition of monoamine oxidase inhibitor somewhat improved on tryptophan alone.

Since N,N-dimethyltryptamine has psychotomimetic actions, it has been sought in the body fluids of schizophrenic patients. Some evidence suggesting the presence of serotonin, N-methylserotonin, and bufotenine in pooled human urine was found. Subsequently, there has been much conflicting evidence concerned with the occurrence of such methylated serotonins in body fluids of both normal persons and schizophrenics. The most recent and probably the most carefully controlled study clearly demonstrated the presence of bufotenine in the urine of schizophrenic patients, identification being made by paper, thin-layer, and gas-liquid chromatography. Exogenous sources of preformed catechol- and indoleamines were excluded from the diet, although it was not possible to exclude their formation within the gut. Administration of tranylcypromine, a monoamine oxidase inhibitor, increased the excretion of bufotenine. It was noted that worsening of the behavioral symptoms started about 2 wk after bufotenine increased in the urine.

Much has been rightly made of the association of phenylalanine excess in the blood with the occurrence of mental disease. Woolley has insisted that experiments carried out in mice with induced phenylketonuria show that the cause of this mental defect is deficiency of serotonin imposed early in infancy. Associated with the presence of serotonin deficiency was a loss of mental alertness which disappeared when the deficiency was prevented, even though the subjects continued to have experimental phenylketonuria.

Recent evidence has supported the view that mental state and behavior are more closely associated with the norepinephrine content of brain than with its serotonin content, but many competent investigators do not agree. The introduction of effective chemical antagonists is beginning to throw some light on this vexing problem.

IX. REFERENCES

1. N. Andén, A. Dahlström, K. Fuxe, and K. Larsson, Mapping out of catecholamine and 5-hydroxytryptamine neurons innervating the telencephalon and diencephalon, *Life Sci.* **4**:1275–1279 (1965).

2. N. Andén and T. Magnusson, An improved method for the fluorometric determination of 5-hydroxytryptamine in tissues, *Acta Physiol. Scand.* **69**:87 (1967).

3. J. Axelrod, in *Mechanisms of Release of Biogenic Amines*, p. 189, Pergamon Press, New York (1966).

4. F. Berti and P. A. Shore, A kinetic analysis of drugs that inhibit the adrenergic neuronal membrane amine pump, *Biochem. Pharmacol.* **16**:2091–2094 (1967).

5. B. B. Brodie, M. S. Comer, E. Costa, and A. Olabac, The role of brain serotonin in the mechanism of the central action of reserpine, *J. Pharmacol. Exp. Therap.* **152**:340–349 (1965).

6. A. Carlsson, in *Handbook of Experimental Pharmacology* (O. Eichler and A. Farah, eds.), Springer-Verlag, New York (1965).

7. A. Carlsson, B. Falck, and N. Hillarp, Cellular localization of brain monoamines, *Acta Physiol. Scand.* **56**:Suppl. 196 (1962).

8. A. Carlsson, K. Fuxe, B. Hamberger, and M. Lindqvist, Biochemical and histochemical studies on the effects of imipramine-like drugs and (+)-amphetamine on central and peripheral catecholamine neurons, *Acta Physiol. Scand.* **67**:481–497 (1966).

9. E. Costa, The role of serotonin in neurobiology, *Int. Rev. Neurobiol.* **2**:175 (1960).

10. A. Dahlström and K. Fuxe, Evidence for the existence of monoamine neurons in the central nervous system. II. Experimentally induced changes in the intraneuronal amine levels of bulbospinal neuron systems; and K. Fuxe, IV. Distribution of monoamine nerve terminals in the central nervous system, *Acta Physiol. Scand.* **64**:Suppl. 247 (1965).

11. *Handbook of Experimental Pharmacology* (O. Eichler and A. Farah, eds.), Springer-Verlag, New York (1965).

12. K. Engelman, W. Lovenberg, and A. Sjoerdsma, Inhibition of serotonin synthesis by *para*-chlorophenylalanine in patients with the carcinoid syndrome, *New Eng. J. Med.* **277**:1103–1108 (1967).

13. V. Erspamer, Pharmacology of indolealkylamines, *Pharmacol. Rev.* **6**:427–587 (1954).

14. V. Erspamer, in *Progress in Drug Research* (E. Jucker, ed.), pp. 151–367, Interscience Publ., New York (1961).

15. V. Erspamer, ed. *Handbook of Experimental Pharmacology*, Springer-Verlag, New York (1965).

16. U. S. von Euler, S. Rosell, and B. Uvnäs, *Mechanisms of Release of Biogenic Amines*, Pergamon Press, Oxford (1966).

17. S. Garratini and L. Valzelli, *Serotonin*, Elsevier, Amsterdam (1965).

18. B. Falck, in *Progress in Brain Research* (H. E. and W. A. Himwich, eds.), Vol. 8, pp. 28–44, Elsevier, Amsterdam (1964).

19. S. Kveder, S. Iskrić, and D. Keglević, 5-Hydroxytryptophol: a metabolite of 5-hydroxytryptamine in rats, *Biochem. J.* **85**:447–449 (1962).

20. W. Lovenberg, E. Jequier, and A. Sjoerdsma, Tryptophan hydroxylation: measurement in pineal gland, brainstem and carcinoid tumor, *Science* **155**:217–219 (1967).

21. T. E. Mansour, Effect of hormones on carbohydrate metabolism of invertebrates, *Fed. Proc.* **26**:1179–1185 (1967).

22. I. H. Page, Serotonin (5-hydroxytryptamine), *Physiol. Rev.* **34**:563–588 (1954).

23. I. H. Page, Serotonin (5-hydroxytryptamine); the last four years, *Physiol. Rev.* **38**:277–335 (1958).

24. A. Pletscher, P. A. Shore, and B. B. Brodie, Serotonin as a mediator of reserpine action in brain, *J. Pharmacol. Exp. Therap.* **116**:84–89 (1956).

25. W. B. Quay, Circadian rhythm in rat pineal serotonin and its modifications by estrous cycle and photoperiod, *Gen. Comp. Endocrinol.* **3**:473–479 (1963).
26. G. P. Quinn, P. A. Shore, and B. B. Brodie, Biochemical and pharmacological studies on Ro 1-9569 (tetrabenazine) a non-indole tranquilizing agent with reserpine-like effects, *J. Pharmacol. Exp. Therap.* **127**:103 (1959).
27. A. Sjoerdsma, Serotonin, *New Eng. J. Med.* **261**:181 (1959).
28. L. Sollero, *Serotonina e Substâncias Antitriptamínicas*, Universidade do Brasil, Rio de Janeiro (1963).
29. P. Stark, E. S. Boyd, and R. W. Fuller, A possible role of serotonin in hypothalamic self-stimulation in dogs, *J. Pharmacol. Exp. Therap.* **146**:147–153 (1964).
30. J. R. Vane, The use of isolated organs for detecting active substances in the circulating blood, *Brit. J. Pharmacol. Chemotherap.* **23**:360 (1964).

Chapter 12

ACETYLCHOLINE, CHOLINE ACETYLTRANSFERASE, AND ACETYLCHOLINESTERASE

Lincoln T. Potter

Biophysics Department
University College, London, England

The intention of this chapter is to summarize established information about acetylcholine (ACh), choline acetyltransferase (ChAc), and acetylcholinesterase (AChE); and to provide pertinent references to original and review literature which permit access to detailed data. The neurobiology of cholinergic synapses, and techniques employed in synaptic biochemistry are emphasized.

I. ACETYLCHOLINE

A. Acetylcholine as a Chemical Transmitter

1. History

Dale's book[1] and a recent review[2] give readable accounts of the discovery and significance of ACh in cholinergic nerves, which may be summarized as follows. ACh was first synthesized by Baeyer in 1867. DuBois-Reymond was apparently the first to suggest, in 1877, that motor nerves might release a chemical which excited muscles; the specific suggestion of adrenaline as such an agent was made by Elliot in 1904. Hunt and his co-workers then discovered the pharmacological potency of ACh. This work was greatly extended in 1914 by Dale, who found ACh in ergot (with Ewins; the actual source was probably microorganisms), compared its action to that produced by parasympathetic nerves, and suggested that its brevity of action in animals was due to its hydrolysis by tissue esterases. In 1921 Loewi convincingly demonstrated what had been claimed by Dixon in 1906, that perfused hearts release an ACh-like substance upon stimulation of their vagus nerves. Loewi's subsequent pharmacological experiments, and Dale and Dudley's chemical identification of acetylcholine in spleen tissue in 1929,

TABLE I

Tissue Distribution of ACh, ChAc, and AChE[a]

Tissue	ACh (mμmoles)	ChAc (μmoles/min)	AChE (μmoles/min)
Brains (insects)		47 (31)	
(fly head extract)		4 → 73 (37)	
(squid ganglia)		→ 1840 (34)	550–730 (6)
(guinea pig)		ca.0.5 → 46 (38)	
(rat)	139 (22)	0.95 → 727 (21)	47 (18)
Cortex (cerebral)	121–247 (31)	0.2–1.0 (31,32)	4.2–18.2 (2)
(cerebellar)	6–17 (31)	0–0.005 (31)	42 (2)
Caudate nucleus		2.8–3.7 (31)	154 → ? (2,43)
Sup. colliculus	94 (31)	0.4 (31)	29–44 (2)
Retina (fowl)		4.0–5.9 (31)	
(mammal)	275–330 (31)	0.9–3.2 (31)	
Optic nerve	0–17 (31)	0–0.005 (31)	0.4 (2)
Spinal gray matter	82 (31)	0.5 (31)	24 (2)
Spinal roots (dorsal)	0–14 (31)	0–0.005 (31)	0.5–2.2 (2)
(ventral)	495–990 (31)	2.0–4.6 (31)	1.4–8.0 (2)
Peripheral nerves (mixed)	220–495 (2,31)	0.3–1.0 (31)	2.4–18.7 (47)
Sup. cervical ganglion	990–2420 (31)	1.5–1.8 (31)	61–88 (2,47)
Eel electric organ		1.4–2.3 (6)	→ 12.5 × 10^6 (54)
Human placenta	1540 (2)	ca.25 → 2340 (39)	
Lactobacillus pl. extract		518 (25)	

[a] All values are per gram of protein. Tissues are from laboratory mammals unless otherwise noted; data were recalculated assuming, where necessary, that the weight of acetone-dried tissue was 20%, and protein was 10% of fresh weight. ChAc activity is ACh synthesized, and AChE activity is ester (ACh, acetyl-β-methylcholine or acetylthiocholine) hydrolyzed at 37°C for vertebrate tissues and 25–30° for other tissues. Arrows show purification → final activity, expressed as V_{max} if possible. References are in parentheses. Data from different laboratories should not be rigidly compared.

identified ACh as the transmitter. Subsequent bioassays by Feldberg, with Dale, Vogt and Brown, in the early thirties made it clear that ACh was released at neuromuscular junctions as well as at parasympathetic endings, and that the substance was present in many nerves within the spinal cord and brain as well. Further information on acetylcholine synthesis, storage, turnover, and metabolism then accumulated rapidly.[3] The distribution of ACh in some tissues and species is summarized in Table I.

2. Present Status

Most investigators now consider that present knowledge limits the neurophysiological functions of ACh to chemical transmission at synapses. The details of how ACh is released and how it acts have been largely worked out by microelectrode studies of neuromuscular junctions.[4] For a summary of experimental techniques, results and references in great clarity, see Katz's book.[5] In brief: an impulse moves along a nerve because of the cable properties of the axon, and because the axonal membrane depolarizes and repolarizes in such a way as to use the electrochemical energy of ionic concentration gradients across it, for boosting the impulse to full strength at every point it passes; depolarization of the nerve terminal is accompanied by a local movement of ions, including calcium, the entry of which is essential for the transmitter release process; after a measurable delay, packages or "quanta" of ACh (probably 5000–50,000 molecules each) are released from specific spots on the presynaptic membrane; the ACh diffuses across the synaptic cleft and combines with highly specific "receptor molecules" on the outside of the postsynaptic cell membrane; the combination initiates a permeability change to small ions which depolarizes the postsynaptic cell, and thereby leads to muscle contraction, gland secretion, or further nerve impulses; the ACh is hydrolyzed to acetate and choline by AChE in the synaptic membranes. The attractive hypothesis that quanta of released ACh correspond to the transmitter held in synaptic vesicles within cholinergic nerve terminals is supported in the discussion of ChAc.

Nachmansohn[6] has promoted the concept that a local release and turnover of ACh are responsible for the membrane permeability changes of excitable membranes, including axonal membranes. The arguments for this view have been countered by facts, e.g., ChAc and ACh are not apparent in certain sensory nerves,[7] and AChE activity in axons and muscles can be abolished without affecting action potentials.[5,8] Another hypothesis, by Burn and Rand,[9] is that nerves which release norepinephrine do so because they first release ACh. The indirect evidence for this has been repeatedly challenged,[10] and there is no direct evidence for the presence of ChAc or ACh in noradrenergic nerves.

Although the function of ACh at cholinergic synapses is well established, there is little information about the function of ACh in certain nonneural tissues and biological substances. These include primate placentas,[11] some plants and micro-organisms, eggs, venoms, and honey.[2] Neural ChAc can synthesize propionylcholine almost as easily as ACh,[12] but recent tracer studies do not show such synthesis in intact ganglia[13] or in isolated cortical nerve terminals[14] during active synthesis of ACh. The site of propionylcholine synthesis in ox spleens[15] has not been established. In addition, the butyryl-, α-aminobutyryl-, urocanyl-, senecioyl-, sinapyl- and acrylyl-esters of choline have been found in some organisms, where their function is unknown.[15]

B. Chemistry

Whittaker has handsomely reviewed this subject through 1962.[15]

1. Structure

ACh is the acetyl ester of the strong quaternary base, choline. Its chemical structure is $CH_3CO-OCH_2CH_2\overset{+}{N}(CH_3)_3$, molecular weight 146.2. Its crystalline structure and infrared spectrum have been determined, and the melting points of a large number of salts are known.[15] In solution, the quaternary nitrogen group is charged at all pH values. The diffusion coefficient is 8×10^{-6} cm^2/sec at 18°C, and 14.8×10^{-6} cm^2/sec at 37°C. ACh lowers the surface tension of water, e.g., -2 dynes/cm at 25°C and 40 mM, and has pronounced although nonspecific electrogenic effects at polar–nonpolar interfaces.[2] By itself, ACh does not absorb ultraviolet or visible light, but Mn(ACh)$_2$Cl$_4$ is intensely fluorescent.[16] After its precipitation from absolute ethanol or acetone, this complex absorbs light maximally at 350 and 436 mμ and fluoresces at 518 mμ (Potter, unpublished, uncorrected values); unfortunately the fluorescence disappears in solution. A similar cobalt-ACh complex has weak fluorescence at 410 mμ when absorbing at 346 mμ. ACh and choline form nonfluorescent salts with at least 14 strongly anionic substances; the salts of dipicrylamine in water or acetone have a molar absorbance of about 36,000 at 420 mμ.[30]

The most convenient salts for storage are the iodides and perchlorates, which are much less hygroscopic than the chlorides and bromides.

2. Solubility

All the common salts of ACh and choline are very soluble in water, progressively less so in alcohols up to hexanol, and insoluble in nonpolar solvents. The perchlorates may be made by crystallizing them from ethanol upon the addition of concentrated perchloric acid, and any of the salts may be crystallized from reasonably dry ethanol by the addition of ether. Crystallization in the latter manner does not fully separate choline from ACh even though the choline salts, especially the acid tartrate, are the less soluble.

Many water-insoluble salts have been used for isolation and identification purposes.[15] Of these the phosphotungstate is the heaviest, and the chloroplatinate probably the least soluble although most expensive. The reineckate has been widely used and can be recommended for acid precipitations, but it should be crystallized on ice from a hot saturated solution before use, and splitting of ACh-reineckate requires the use of silver salts or ion-exchange resins.[15] The tetraphenylborate is comparably effective but it complexes potassium too well for many uses. Dipicrylamate salts are preferred by the author because ACh regeneration requires only acidification— saturated solutions of the acid dipicrylamate at pH 2 are approximately micromolar. The salts of ACh (and choline) with any of these anions have very similar solubilities: their saturated water solutions are of the order of millimolar, they are slightly more soluble in cold alcohols, freely soluble in ketones (including sulfoxides), and are relatively insoluble in nonpolar solvents.

3. Stability

In the absence of catalysts, particularly esterases, ACh is stable for years at pH 4 and may be sterilized by boiling at this pH; stock solutions are conveniently made in millimolar acetic acid or NaH_2PO_4. Acid-catalyzed hydrolysis of the ester linkage is rarely important experimentally, but warm acids below pH 1–2 should be avoided. Base-catalyzed hydrolysis is more rapid. The constants given by Tammelin[17] indicate that the rate of splitting of 0.1 M ACh at 25°C in moles/sec is 0.427 times the OH^- concentration; the half-life at pH 12 is therefore about 17 sec, and at pH 7, about 20 days. The rate is a function of both temperature and concentration. Under assay conditions suitable for AChE assays, i.e., 2 mM ACh at pH 7 and 37°C, the initial rate of hydrolysis is 0.1 % of the substrate per hour.[18] The boiling of tissue extracts in alkali has been a standard procedure for the elimination of ACh.

4. Assays

ACh (0.01–5 μmole) is very readily assayed by Hestrin's application[19] of the general ferric hydroxamate reaction for esters. At pH 12–14, ACh reacts with hydroxylamine to give choline and acetylhydroxamate; at pH 1.2, three molecules of the latter product combine with one ferric ion to form a complex which absorbs at 500 mμ. More sensitive procedures include gas chromatography,[15] and enzymatic assay[20] of ethanol produced from the acetate moiety of ACh. Physiological amounts of choline may be assayed with ChAc, by using the other components of a specific radiometric microassay in excess[12,21]; by the same procedure 10^{-10} moles of ACh can be determined after its hydrolysis with AChE.

For most laboratory purposes, bioassays of ACh are still the most suitable for the amounts found in tissues and biological fluids. Whittaker[15] has reviewed the use of various test preparations, including the frog rectus abdominus muscle, guinea pig ileum, cat blood pressure, clam hearts, and leech dorsal muscle. By using very thin muscle strips in small baths, and suitable amplification of the output of strain gages or other transducers, threshold amounts of 10^{-12} moles are detectable with any of the above muscles. Many biological fluids can be studied without purification, and tissue extracts in cold 10 % trichloracetic acid may be assayed after removal of the acid by ether extraction.[22]

5. Chromatography

ACh and its analogues may be isolated and tentatively identified by any of four chromatographic techniques. Of these, paper electrophoresis is the most generally useful because it is convenient and quick, gives very sharp and reproducible separations (e.g., separation of labeled ACh and choline by 1 cm gives 99.8 % complete isolation of ACh), and is not affected by hygroscopic impurities or quantities of mixed salts below 1 μmole.[23] Paper and thin-layer (cellulose) chromatography are excellent for ACh[15] when prior purification methods have left the same anion as that used in the chromatography system, as after elution of ion-exchange resins or regeneration of dipicrylamate complexes with acids. Systems which give fairly high R_f values, such as butanol–ethanol–acetic acid–water (8 : 2 : 1 ; 3) are generally the most useful.

Spots are located in iodine vapor, or as esters, with the ferric hydroxamate reaction. In dilute solution, ACh, like other amines, may be concentrated or isolated on cation-exchange resins of the carboxylic acid type. ACh is best adsorbed at pH 4.35 and eluted with 0.1 M NaH_2PO_4.[15] Since the pK of the COO^- groups on the resin is about 4–5, the material is readily stripped of adsorbed materials with 0.1 N HCl.

Concentration of tissue extracts for chromatography is facilitated if they are made in cold ethanol -2% acetic acid,[22] especially if the tissue is previously freeze-dried. Concentration of trace amounts of ACh from biological salt solutions can be effected with dipicrylamine at pH 7.[30] Enough lithium dipicrylamate is added to the solution to complex all the ACh and potassium, and these are extracted into 0.01–0.1 volumes of 2-octanone. The organic solvent is then shaken with enough 0.1 N perchloric acid to precipitate K perchlorate and H dipicrylamate and to give ACh perchlorate in solution. Small quantities of the ACh complex may also be regenerated on paper in the acid buffer used for electrophoresis.[23]

II. CHOLINE ACETYLTRANSFERASE

A. Biology

1. Distribution in Tissues and Cells

In general, ChAc in multicellular animals is found normally only in nerves. Table I gives some representative values for its distribution.

The only unequivocal exception is the placenta of higher primates which, although not innervated, contains considerably more ACh and ChAc than most nervous tissues.[2,11] This enzyme appears to have a different reaction mechanism[24] from the ChAc in brains.[12,21] Another exception appears to be the gill plates of the mussel, *Mytilus edulis*, where the movement of cilia is controlled by ACh.[2,7] Mammalian cardiac muscle, the spleen, intestinal muscle, and corneal epithelium have also been suggested as sites where ACh is produced by nonneural tissues, but these tissues have cholinergic nerves which may account for all or a part of the observed synthesis. Attempts to confirm early reports that red blood cells synthesize ACh have not been successful;[7] in the author's laboratory no ChAc was found in lysed or whole human cells (unpublished) when a specific and very sensitive assay method was used.[21] ChAc is inducible in *Lactobacillus plantarum*[25] and is presumably present in the other organisms where ACh is found, including nettles, Malayan jackfruit, *Paramoecium*, and *Trypanosome*; the function of the enzyme in these organisms is not known. ChAc has not been detected in the venom sac of hornets, although the venom contains large amounts of ACh.[2]

Denervated vertebrate muscles retain a few percent of their normal content of ACh and a few percent of normal ChAc activity in the original end plate region.[26] At this time a slow trickle of ACh continues from the tissue, a few miniature end plate potentials are seen in frog muscles, and a few vesicles the size of synaptic vesicles are present in Schwann cells.[27] In this abnormal situation, ACh synthesis and turnover apparently occur in nonneural cells, presumably in Schwann cells, since these are known to engulf the degenerating cholinergic nerve terminals. Still, most of the ChAc disappears from the distal part of cut nerves, leaving some in proximal regenerating tips.[2] Nothing is known about what regulates the synthesis of ChAc in nerves, and it would be of great interest to know the details of ChAc appearance in embryonic cells.

ChAc is present in high concentration in the axons of ventral spinal roots,[2] and is therefore probably present in their motor horn cell bodies as well. In the rat brain more than half of the total enzyme is present in nerve endings.[21] The richest sources of the enzyme are the ventral roots, caudate nuclei of the brain, and parts of the optic nervous system, including the retina (especially of birds) and the predominant lobes of the brains of insects and squid (Table I).

Three new procedures for the localization of ChAc and ACh are under study (unpublished) in different laboratories. These include application of radiometric assays to the study of single cells, a specific fluorescence technique for ACh in cells, and electron microscopic procedures for ChAc. Our knowledge of the location of these substances should be greatly extended in the next few years.

2. Intracellular Localization

Reports on the intracellular site of ChAc have been contradictory, since Whittaker and co-workers have found the enzyme in supernatant fractions after centrifugal analysis of lysed nerve endings, whereas DeRobertis and co-workers have demonstrated that much of the enzyme can be isolated in association with microsomal particles like the vesicles which hold ACh.[28] Fonnum[29] found that ChAc required considerable salt for its solubilization, and suggested that its apparent association with microsomes was an artifact produced by the very dilute salt conditions required for subcellular fractionation. Direct examination of this suggestion has unequivocally demonstrated that ChAc is found in solution if rat brains are disintegrated by means which lyse nerve endings in 200 mM KCl, and that it reversibly and progressively adsorbs to microsomal particles if the salt concentration is lowered, especially at low pH.[21] ChAc is a strongly cationic enzyme which adsorbs to negatively charged particles unless there is sufficient salt to prevent such association. After purification, however, the enzyme remains in true solution in dilute buffers.[12,21] Thus, as far as can be experimentally determined, ChAc is a cytoplasmic enzyme.

Under the same conditions used for localizing ChAc in brain homogenates, 60% of the total amount of newly synthesized ACh is found in microsomal fractions containing synaptic vesicles; labeled ACh in the medium is not trapped in this manner.[21] It must be concluded that ACh is made in the cytoplasm of nerve endings and is either actively or passively concentrated in vesicles. The mechanism for such concentration is being studied in several laboratories.

3. Regulation of Acetylcholine Synthesis

When isolated ganglia[3] or diaphragms[30] are exposed to physostigmine (which can inhibit intracellular esterases), the level of ACh in the tissues rises until it is approximately twice normal. The initial rate of accumulation is rapid, which suggests that the rate of ACh synthesis in unstimulated nerves is set higher than required to balance ACh losses; the rate then becomes progressively slower, as if an equilibrium state were being approached. In the presence of physiological amounts of choline, ACh synthesis keeps pace with its release even during short bursts of nerve impulses at high frequency;[3,30] at such times the synthesis rate is at least twice the highest rate seen at rest in physostigmine.[30] Thus synthesis is at least partially regulated in response to ACh release.

Any of four mechanisms could regulate the activity of ChAc in nerves: limiting amounts of substrates or of active enzyme, product inhibition, or mass action. There is reason to believe that the amount of active enzyme present in nerve terminals is more than adequate for transmitter requirements, and that choline is not a limiting substrate.[3,14,21,26,30] The free level of acetyl-CoA in nerve terminals is not known and could be limiting; moreover CoA competes with acetyl-CoA.[12] But a regulatory mechanism for ACh synthesis based upon another for acetyl-CoA and/or CoA seems needlessly complicated. ACh, even at 50 mM concentrations, does not significantly inhibit ChAc.[12,21] At present, therefore, the most plausible guess is that the rate and amount of ACh synthesis depend upon the equilibrium position of ChAc with its substrates and products.[21] The equilibrium constant for bovine brain ChAc, $K = (ACh)(CoA)/(choline)(acetyl-CoA)$, is about 40.[12] Thus, for example, if cytoplasmic choline is roughly 50 μM, and the ratio of CoA to acetyl-CoA is roughly 20 (as in whole brains and the liver), then levels of free ACh near 0.1 mM would be sufficient to reach equilibrium. Such levels could reasonably be in equilibrium with ACh in vesicles, where the concentration has been estimated from direct, quantal, and theoretical considerations to be in the range 100–1000 mM.[28]

B. Enzymology

1. Assay Methods

In order to measure the total amount of ChAc in tissues, it is best to lyse nerve endings and to obtain the enzyme in solution. This can be done by extraction of acetone powders with isotonic salt solutions,[31] or by treatment of homogenates in salt solutions with solvents or detergents.[29,32] Vigorous disintegration of tissues in isotonic KCl can be equally effective, but if hypo-tonic lysis of nerve endings is used, ChAc adsorbs strongly to particles and should be desorbed with both isotonic salts and a solvent or detergent; 1 % butanol is convenient.[21] EDTA or another agent which helps to protect sulfhydryl groups on the enzyme should be included in the extraction medium. Unless a generating system for acetyl-CoA is used, cysteine and other SH-compounds should not be used, since they are transacetylated non-enzymatically by acetyl-CoA; NaCN has been recommended instead.[33] ChAc is reasonably stable when these precautions are taken and enzyme dilution is avoided. Dialysis tubing should be boiled in EDTA.

ACh produced in reaction mixtures may be bioassayed[31] but much more convenient and accurate assays are performed by measuring the rate of acetylation of choline with radioactive acetyl-CoA. Addition of the acetyl-CoA is necessary for defined conditions, but the reaction rate is usually constant for a shorter period than with systems which generate acetyl-CoA. The product, labeled ACh, may be isolated by precipitation as the reineckate or tetraphenylborate salt, by ion-exchange or by paper electrophoresis.[21,32] The latter procedure is preferred by the author, since it is specific for ACh, gives minimal blanks, the acetylation of competitive substrates can be followed simultaneously, and choline, acetyl-CoA, and ACh are all separated for equilibrium measurements.

Enzyme preparations free of deacylases can be assayed by determining the rate of appearance of CoA thiol groups from acetyl-CoA in the presence of choline (forward reaction) or the rate of disappearance of CoA in the presence of ACh (backward reaction). The free thiol is most easily measured[21,24,34] with Ellman's reagent.[35]

2. Purification

ChAc from squid head ganglia[34,36] and from fly brains[37] has been partially purified by the use of ammonium sulfate and protamine. Greater purification was achieved from guinea pig brains with ammonium sulfate and by removing inactive proteins at pH 4.5.[38] The enzyme from mature and immature human placentas has been purified moderately with ammonium sulfate, calcium phosphate adsorption, and by chromatography on Sephadex and DEAE-cellulose.[24,39] The most active preparations from vertebrate nervous tissues have been obtained from rat brains and the striatum of cow brains, by precipitation and extraction steps at pH 5, and the use of CM-Sephadex, ammonium sulfate, and Sephadex G-200.[12,21] The activity of these preparations is summarized in Table I.

3. Physical Properties

Highly purified enzyme preparations are stable for months at 4°C, and may be dialyzed against dilute buffers with little loss of activity.[12,21] The bovine brain enzyme is stable between pH 5 and 10.5 and at temperatures below 40°.[12] The molecular weights of ChAc from rabbit,[39] rat,[21] and cow brains,[12] and from human placentas[39] have been estimated as between 50,000 and 67,000, of which the higher estimates appear most accurate. Acrylamide gel electrophoresis shows that the bovine enzyme is much more cationic than most tissue proteins,[12] which may explain the tendency of ChAc to adsorb to negatively charged membranes in nonionic media.[21,29]

ChAc preparations from different species show different adsorption affinities for membranes,[29] and require different concentrations of ammonium sulfate for enzyme precipitation.[12,34,38,39] These observations indicate that the enzymes differ appreciably in their ionic properties, and this may prove the reason for the variable effects of salts in activating or inactivating the soluble enzymes.[21,34] Subunit aggregation or disaggregation could also explain some of the results obtained with salts.

4. Kinetics and Thermodynamics

An ionic strength of about 50 mM is necessary for the activity of all ChAc preparations, and is sufficient for full activity with the cow brain enzyme; the purified enzymes from many mammalian brains,[21] squid head ganglia,[34] and placentas[24] are further activated by salt concentrations up to about 300 mM. Salts increase the Michaelis constant for choline with squid ChAc but appear to decrease it with the placental enzyme. All ChAc preparations are inhibited by copper and other trace metals, and are protected or reactivated by EDTA, sulfhydryl compounds, or $NaCN$[33]; apparently the enzyme has essential sulfhydryl groups. The pH optimum for the bovine enzyme is broad, with nearly constant activity between pH 7 and 9.5.

The kinetics of ChAc from bovine brain striata have been studied most completely; assays were performed at pH 7 and in 150 mM KCl so as to approximate the intracellular ionic environment of the enzyme.[12] Lineweaver–Burk plots for choline and acetyl-CoA at several different concentrations of the other substrate intersect at the baseline, whereas those for ACh and CoA intersect above it. Indicated Michaelis constants are: choline, 450 μM; acetyl-CoA, 11 μM; ACh, 1–8 mM; and CoA, 22–200 μM. (There is a range of values for the products which depends upon the concentration of the other in the reaction.) V_{max} for the forward reaction is four times that for the backward reaction, and the equilibrium constant is approximately 40. Choline, acetyl-CoA and ACh do not show substrate or product inhibition, but CoA competes with acetyl-CoA with a K_I of about 16 μM.

With placental ChAc, Lineweaver–Burk plots for choline and acetyl-CoA, at several concentrations of the other substrate, appear parallel.[24] It has been suggested, therefore, that this nonneural enzyme is acetylated by acetyl-CoA prior to the binding of choline. Such a "ping pong" mechanism clearly does not occur with ChAc from vertebrate brains, to which both substrates (or products) must bind before the appearance of products. Michaelis constants for placental ChAc in 0.3 M NaCl are: choline, 1.1 mM; acetyl-CoA, 20 μM; ACh, 2 mM; and CoA, 20 μM. V_{max} for ACh synthesis was about three times that for its hydrolysis, and the equilibrium constant was

145 in 0.3 M NaCl and 5100 without it. NaCl slightly increased the pH optimum, lowered the temperature optimum, and did not affect the apparent energy of activation for the forward reaction.

5. Substrates and Inhibitors

Most studies of substrates for ChAc have been done with relatively impure enzyme preparations or with only a few concentrations of each test substance. However, the following data for bovine brain ChAc[12] generally confirm earlier qualitative results.[36,40,39] Approximate Michaelis constants (mM) and relative V_{max} values (choline = 100) for N-substituents on aminoethanol are: monomethyl 4 mM, $V_{max} = 1$; dimethyl 3,9; monoethyl 3,1; diethyl 3,2; triethyl 13,10. For the uncharged analogue of choline, 3,3-dimethylbutanol, the values are 400 and 1. Propionyl-CoA substitutes for acetyl-CoA with an apparent K_m of 15 μM and the same V_{max}. The enzyme is thus highly specific for choline, but not for acetyl-CoA.

There are few potent inhibitors of ChAc, and none of these are known to act *in vivo*. The above substrates in concentrations 10–100 times that of choline appear as competitive inhibitors, as do many quaternary amino compounds at concentrations of 5–100 mM.[36,40] High concentrations of most salts, especially iodides and divalent cations above 100–200 mM, are inhibitory.[21,34,36] The most potent compound yet developed is a *bis*-quaternary amine which inhibits brain ChAc 50 % at 0.9 μM.[41]

III. ACETYLCHOLINESTERASE

This discussion is oriented towards the AChE of synaptic membranes. Other cholinesterases (ChE, often known as pseudo-ChE, plasma ChE, or butyryl-ChE) and simple esterases are considered for comparative purposes. The distinguishing characteristics of these three groups of enzymes are summarized in Section III, B, 1.

A. Biology

1. Distribution in Tissues; Biological Significance

AChE has been found in the innervated tissues of all vertebrates and invertebrates which have been studied.[6] Some representative values for its distribution are given in Table I. The cytological techniques discussed in the next section indicate that enzyme activity is maximal in the synaptic membranes of cholinergic synapses. The specialized junctional tissue of electric organs represents, therefore, an unusual source. The significance of AChE at postsynaptic membranes is demonstrated by the prolonged duration of ACh effects seen after esterase inhibitors.[5] The importance of the enzyme for organisms in general is further emphasized by the effectiveness and destructiveness of the irreversible AChE inhibitors; nearly every major insecticide and chemical warfare agent belongs to this class.

The significance of AChE in neuronal membranes, other than at post-synaptic sites, is unknown. In many locations, e.g., synapses in the cat superior cervical ganglion, the amount of enzyme in the presynaptic membranes exceeds that of the receptor cells.[8] It has been hypothesized that such AChE indicates that ACh acts directly on the nerve endings from which it is released, thereby causing further release.[8] However, at neuromuscular junctions the quantal release of ACh does not cause further quantal release.[4] It has been demonstrated that an esterase inhibitor which can pass membranes causes an appreciable increase in the total ACh of cat superior cervical ganglia.[3] It seems likely, therefore, that AChE in cholinergic nerves does hydrolyze some of the ACh which is synthesized throughout the same fibers, although the physiological merit of this action is unknown.

The presence of AChE in nonneural membranes raises further questions about the functions of the enzyme. In the placenta there are generous quantities of ChAc, AChE, and ACh, and it has been suggested that ACh could function in some way to permit antibody transfer through membranes.[11] In red cells the presence of AChE is not universal, and there is little reason to believe that significant ACh metabolism occurs.[7,8] Observations that ACh increases the ionic permeability of erythrocytes, and that such permeability is decreased or reversed by esterase inhibitors, do not, therefore, provide evidence of a physiological phenomenon. AChE inhibitors also alter the permeability of frog skin, frog skeletal muscles, and crab gills, but the concentrations required seem well beyond those necessary for enzyme inhibition.[8]

2. Cellular and Subcellular Localization; Histochemical Studies

Three techniques have helped to define the localization of AChE: classical subcellular fractionation of tissues, microanalysis of single cells, and histochemistry.

Almost all the AChE of tissues remains associated with membrane fragments during centrifugal fractionation in sucrose solutions; and its solubilization requires high concentrations of salts, organic solvents, or detergents.[28,42,43] This is the best evidence that AChE is localized in membranes *in vivo*, and it permits further interpretation of data obtained with diffusible histochemical stains, than would otherwise be possible. Early fractionation studies of brain tissue[28] showed that AChE was associated with many membrane fragments, particularly microsomes (except synaptic vesicles). Subsequent studies show that much of the enzyme is associated with isolated nerve endings which have attached postsynaptic membranes, and that the enzyme can be isolated with a paired complex of pre- and postsynaptic membranes after lysis of the endings.[14,44] This complex can be separated with EDTA, and fractions rich in postsynaptic membranes have been obtained with AChE specific activities up to 340 times that of the original tissue (Potter, unpublished).

Within the limits of dissection, the ultramicroassays of AChE by Giacobini[45] and others, using Cartesian divers as small as 0.005 μliters, have greatly extended our knowledge of the quantitative amounts of enzyme in cells and parts of cells. The technique is based upon the minute volume changes which occur when acetate, released by enzymatic hydrolysis of ACh, reacts with bicarbonate buffer to liberate CO_2. These studies show that the amount of enzyme in end plates is at least 50 times that in extrajunctional muscle; they demonstrate high concentrations in cholinergic cells like motor horn cells; and they emphasize the wide variation of enzyme levels between different juxtaposed cells, e.g., 10–15 % of dorsal root ganglion cells are highly active whereas the others have little or no enzyme. In general, AChE activity did not correlate with cell size or surface volume, and the further finding of considerable "cytoplasmic" (interpreted as microsomal) enzyme confirms (see below) that AChE, in cell bodies, is as much a property of intracellular membranes as of surface membranes.

Histochemical methods for the demonstration of AChE and ChE have been developed by Koelle, Couteaux, and their co-workers.[8] In the most widely used of several procedures, partially formalin-fixed tissues are incubated with acetylthiocholine in the presence of copper ions. The enzymatic product thiocholine reacts with copper to give a precipitate of copper thiocholine; this in turn is reacted with a sulfide to give brown copper sulfide which is microscopically visible in the vicinity of the enzyme reaction. By the use of suitable pH's and inhibitors, AChE and ChE can be distinguished. The results provide a qualitative measure of the enzyme. Very extensive applications of the method[8] permit the following conclusions. AChE is widely distributed in nerves, particularly known cholinergic ones, and it is present in dendrites, cell bodies, axons, and at the surface of nerve terminals. It is maximal at muscle end plates and is seen in red cells. ChE, in contrast, is absent from nerves, except in their nuclei, but is concentrated in glia, vascular and connective tissues, Schwann cells, liver, pancreas, and intestinal sites. Small amounts are present at end plates but there is more elsewhere in muscles.

3. Enzyme Metabolism; Effects of Innervation and Denervation

The factors which regulate the amount of AChE in nervous tissues are unknown. The total activity of the enzyme generally increases with growth, both during embryological life[8] and in the proximal regenerating tips of cut nerves.[2] It would be of great interest to know whether the AChE level in a given cell follows the appearance or level of ACh. The accumulation of enzyme in regenerating nerves, and observations of its rapid loss in the distal parts of nerves after sectioning, support the general belief that most proteins are synthesized in cell bodies and are transferred down axons. An apparently contrary observation is that after "irreversible" inhibition of AChE with organophosphorus compounds, the enzyme reappears uniformly in axons[46] rather than as a gradient from cell to ending,[47] as though some synthesis of AChE is possible in the axons.

The presence of AChE at postsynaptic sites appears to be regulated by a factor from nerves. At the time of the original innervation of muscles, AChE becomes concentrated at end plates.[8] After nerve section both end plate and extrajunctional AChE decrease after 3 days by about 50 % and continue to fall slowly unless reinnervation takes place; reimplantation leads to a slow return to near-normal esterase levels at the end plate.[48] Since blockade of ACh action in glands with an atropine analogue, and blockade of ACh release at end plates with botulinum toxin do not cause as much decrease in AChE as does denervation,[49] it has been argued that the neural effect is not mediated by ACh. Further work on this point would be of great interest, to determine whether or not ACh is responsible for AChE "induction." Muscle activity does not seem to affect the enzyme, since tenotomy does not cause a fall in AChE, and electrotherapy does not prevent the AChE decrease which follows denervation.[48]

4. Comparison of Acetylcholinesterase with Acetylcholine Receptors

Many authors have considered the possibility that part or all of the molecular configuration of ACh receptors in membranes is composed of AChE. Too often the "evidence" given for this concept is that AChE is present wherever ACh receptors are found, and that both have structures which specifically fit ACh. Careful attempts to demonstrate overlapping of the surfaces of receptors and the enzyme, as deduced from the effects of ACh analogues and suitable inhibitors have so far failed to provide convincing evidence. On the other hand, there are several reasons to believe that the effective active sites are distinct. Inhibitors of AChE do not alter responses to iontophoretically applied carbachol, which shows that the enzyme can be blocked when ACh receptors are not.[5,50] Receptor responses are blocked by concentrations of atropine, hexamethonium, and curare (at parasympathetic, ganglionic, and neuromuscular junctions, respectively) which do not inhibit AChE. Finally, there are spatial differences between the location of AChE and receptor molecules even in cholinergically innervated tissues. In muscles, AChE is more restricted to end plates than are the receptors, and the receptors are present in low concentration near muscle-tendon junctions where ACh hydrolysis is relatively ineffective.[50] Receptor sensitivity increases 100–1000-fold over denervated muscles[50] at a time when the level of end plate and extrajunctional AChE is falling.[48]

B. Enzymology

1. Assay Methods

Measurement of the activity of a relatively pure cholinesterase preparation is an easy matter for which there are a large number of effective and convenient methods.[51] However, measurements involving tissue samples from most sources are complicated by the need to distinguish the separate activities of mixed amounts of AChE, ChE, and simple esterases. The only suitable way to do so is with the use of relatively specific substrates and inhibitors; although the literature is a guide, accurate measurements require considerable study of the preparations at hand.[18] *In general:* AChEs hydrolyze several substrates at different rates (ACh > acetyl-β-methylcholine $\gg$ butyrylcholine) but not benzoylcholine, and are readily inhibited by *bis*-quaternary compounds like ambenonium; ChEs hydrolyze many esters including ACh and benzoylcholine, but not acetyl-β-methylcholine, and are inhibited by low concentrations of organophosphates, notably diisopropyl-fluorophosphate (DFP), Mipafox, and iso-OMPA; AChE and ChE are inhibited by 10^{-5} M physostigmine whereas simple esterases are not.[51] Optimal substrate concentrations are 1–10 mM for AChE, and 10–100 mM for ChE; the optima for substrates, pH, temperature, salt concentration, and inhibitor concentration are interdependent.[51,52] In practice most of the authors cited have used ACh and DFP, or acetyl-β-methylcholine, for assay of AChE, and butyrylcholine for ChE, in each case subtracting the effect of physostigmine-resistant esterases. ACh is the preferable substrate for AChE because it is the natural one and its analogues are split at a different rate. Use of the organophosphates requires care because of their toxicity and instability in water; because they combine with many substances, the concentration required for inhibition of ChEs depends upon the purity of the preparation.

For occasional use, the procedure of Hestrin,[19] which measures the amount of unreacted ester by the ferric hydroxamate method, is probably the most convenient, since any reasonable substrate, pH, medium, and temperature may be used. Another widely used method is that of Ellman *et al.*[35] in which the free thiol of hydrolyzed acetylthiocholine is measured with a sulfhydryl reagent; although limited to thiocholine esters as substrates, the method is otherwise highly adaptable and sensitive. Most other methods for AChE[51] measure the amount of acetate produced, by CO_2 production from bicarbonate buffer, by titration, pH change, electrometric or indicator method, or by assay of radioactive acetate. Microassay modifications of all these procedures have been developed; for references and a radiometric assay of comparable sensitivity to that of Cartesian diver methods see Ref. 18.

2. Purification and Crystallization

Initial attempts to solubilize AChE from caudate nuclei (the best mammalian source) proved very difficult. A method was recently developed[43] which made use of Triton X-100 for solubilization, followed by extensive purification by chromatography on benzyl-DEAE-cellulose, Sephadex G-25 and G-200. The enzyme of red cells has been moderately purified after solubilization with butanol.[51] Soluble ChE from horse serum has been purified 14,000-fold with ammonium sulfate, ultracentrifugation, and electrophoresis to a homogeneous protein by electrophoretic and ultracentrifugal analysis.[53]

AChE has been purified and crystallized from eel electric organs. Kremzner and Wilson stored pieces of the tissue under toluene for 6 weeks to remove mucin, and solubilized the enzyme with 5 % ammonium sulfate. The enzyme was then purified with ammonium sulfate and chromatography on columns of benzyl-DEAE-cellulose, Sephadex G-200, cellulose phosphate, and DEAE cellulose.[42] A large-scale purification of 10 Kg of tissue by similar techniques[54] has yielded 60 mg of enzyme homogeneous by disc electrophoresis and ultracentrifugation. Large crystals were obtained in 35 % ammonium sulfate; the final activity was 12,500 μmoles ACh hydrolyzed min^{-1} mg protein^{-1}.

3. Physical Properties

AChE is generally quite stable during purification and storage, but high dilution and many organic solvents promote its denaturation. Studies of the eel enzyme at a specific activity of 11,000[42] showed a single peak during ultracentrifugation with $S_{20} = 10.8 S$, and a diffusion rate of 4.3×10^{-7} cm^2/ sec at 20°; the molecular weight was calculated as 230,000. By gel filtration the weight appeared about 250,000 and no evidence of larger or smaller units was obtained. Titration with an irreversible inhibitor indicated that the activity per mole of active sites (at 88 % purity) was 610,000 μmoles/min. From this there appeared to be four active sites per molecule, suggesting subunits of molecular weight 57,000. AChE from *Torpedo*[55] sediments more slowly in 0.1–0.3 M Na, K, or Mg chlorides than in dilute solutions, as though combination and dissociation of subunits were occurring. It has been suggested that part of the activating effects of salts on AChE depends upon such disaggregation. The eel enzyme[54] crystallized as hexagonal rods, and amino acid analyses of the pure enzyme have been performed. AChE from human caudate nuclei[43] has a similar molecular weight (230,000) by ultracentrifugation and gel filtration, and a turnover number of 420,000/min. Pure horse serum ChE[53] has a sedimentation rate of $S_{20} = 9.9 S$ indicating a molecular weight comparable to that of AChE. One mole of DFP was bound per 84,000 g of enzyme, suggesting two to four active sites to account for the sedimentation rate. The isoelectric point is at pH 3.1.

4. Anionic and Esteratic Sites on the Enzyme; Reactions with Substrates and Inhibitors

See reviews by Wilson[56] and by Cohen and Oosterbaan[52] for references and details concerning this subject and the next section.

There is excellent evidence for at least one anionic site on AChE which attracts the positively charged quaternary nitrogen of ACh; and evidence for one esteratic site at which hydrolysis occurs. These are shown below schematically.

Attraction of ACh to the anionic site is believed to be due to coulombic and Van der Waals forces, primarily the former. Comparisons of the inhibitory action of neostigmine, which is quaternary at all pH values, with that of physostigmine (pK amine of 8.1) demonstrate 16-fold greater binding of the charged than uncharged physostigmine. Similarly, dimethylaminoethanol, $(CH_3)_2\overset{+}{N}HCH_2CH_2OH$ is an inhibitor which binds 30 times as well as its uncharged analogue, isoamyl alcohol, $(CH_3)_2CHCH_2CH_2OH$; and the Michaelis constants of their respective acetyl esters, which are substrates, differ by a factor of 8. The inhibition of AChE by quaternary amines decreases with increasing ionic strength, and the Michaelis constant for ACh is raised by salt, in keeping with such coulombic attraction forces. The effects of the methyl groups of ACh have been deduced from inhibitor and substrate series of analogues with three, two, one, or no methyl groups; in brief, at least two methyl groups appear to form dispersion forces with the enzyme.

The attractive forces between ACh and the esteratic site of AChE appear to be due to a weak covalent bond between the carbonyl carbon atom of the acetyl group, and at least one basic group on the enzyme. The weakness of the attachment is apparent from the very high K_m values of substrates without cationic groups, e.g., 0.5 M for ethyl acetate. An increase in the electrophilicity of the carbonyl carbon atom decreases the K_m predictably, e.g., 10^{-4} M for acetic anhydride. The pH dependence of the velocity of hydrolysis is broad. Because ACh remains charged at all pH values, the pH-activity relationship has been interpreted as evidence for two charged groups at the esteratic site: one with a pK of 9–10.5 takes part in hydrolysis, and the other is considered a basic group (pK about 6.5) which binds the carbon atom. There has been a great deal of study and speculation as to whether the imidazole group of histidine, which has a pK in the right range, could serve as the basic group. Models incorporating this view have had to take into account the fact that the phosphorus atom of $DF^{32}P$, which presumably binds irreversibly to the same basic group as the acetate carbonyl carbon, is found attached to serine after enzyme degradation. Plausible models involving the basic atoms of both serine and histidine have therefore been proposed.[52]

Available evidence supports the view that AChE is acetylated by ACh and then reacts with water to free acetic acid. Schematically, the reaction between AChE and ACh (I) and the subsequent hydrolysis of the acetylated enzyme (II) are as follows, where dashed lines represent original bonds, and dotted lines represent new covalent bonds. B symbolizes one or more basic groups on the enzyme.

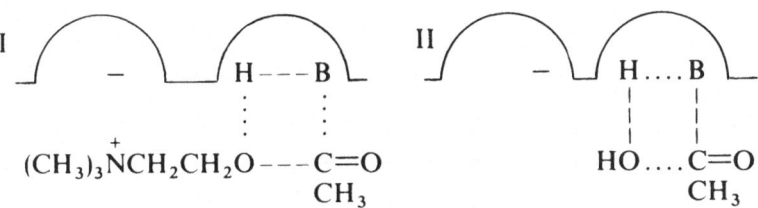

Formation of an enzyme intermediate is consistent with considerable evidence: by the backward reaction the enzyme catalyzes the formation of ACh from choline much more rapidly when the other substrate is ethyl acetate than free acetate; exchange of oxygen between water and acetate has been observed; and enzyme inactivation by organophosphorus compounds occurs at the esteratic site because the phosphorylated intermediate is not hydrolyzed significantly by water. With DFP (III), HF and a stable inactive enzyme are formed. Other organophosphates inhibit AChE in the same manner. A few, like echothiophate, $(CH_3)_3\overset{+}{N}CH_2CH_2SPO(OC_2H_5)_2$, react with both anionic and esteratic sites; this compound is used clinically because it is relatively stable in water. On the basis of such models, Wilson predicted that the phosphorylated enzyme could be reactivated by compounds more nucleophilic than water, particularly if quaternary groups on the molecule were attracted to the anionic site and fixed the nucleophilic atom near the phosphorus atom. Pyridine-2-aldoxime methiodide (IV) proved very effective; it and its *bis*-derivatives are used as antidotes for alkyl phosphate poisoning.

Reactions of AChE with reversible inhibitors can be depicted in a similar manner. Some like edrophonium (V) are not esters and are therefore active only as long as they are present in high enough concentration to compete successfully with ACh for anionic sites. Others are esters which inhibit because the intermediate enzyme complex is relatively stable; neostigmine (VI), for example, forms a carbamoyl enzyme which reacts with water at a millionth the rate of the acetyl enzyme.

5. Kinetics and Thermodynamics

The reaction catalyzed by AChE may be represented as

$$E + ACh \underset{k_2}{\overset{k_1}{\rightleftharpoons}} E(ACh) \underset{k_4}{\overset{k_3}{\rightleftharpoons}} \text{acetyl-E} + \text{choline} \underset{k_6}{\overset{k_5 H_2O}{\rightleftharpoons}} \text{acetic acid} + E.$$

When the rate of hydrolysis of ACh is plotted against the concentration of

ACh, a well-defined optimum substrate level between 1–10 mM is apparent, above which there is substrate inhibition. This inhibitory effect is explained as the result of formation of an inactive complex of two substrate molecules with one active site on the enzyme. It is not clear whether the two molecules are both ACh (one attached at the anionic and the other at the esteratic site), or whether ACh combines with the acetyl enzyme. Neither alternative clearly explains the lack of substrate inhibition seen with acetylhomocholine and acetyl-β-methylcholine. The substrate optimum is increased by salt, presumably because the salt ions compete with ACh for the anionic site, and by physostigmine (which forms a carbamoyl enzyme) at much lower concentrations. Plasma ChE does not show such substrate inhibition, and it binds the uncharged ACh analog 3,3'-dimethylbutylacetate, almost as well as ACh. These observations indicate that the anionic site of ChE is relatively unimportant; it is thought to be present because of the inhibitory effect of quaternary ions. The optimal substrate concentration for ChEs might therefore be expected to be high, and is 10–100 mM.

The Michaelis constant for ACh with AChE has been reported for various different preparations and conditions between 2.8×10^{-4} and 5×10^{-3} M. It is generally increased by cations and is independent of pH. A detailed discussion of the effects of inhibitors on apparent K_m values is given in Ref. 52. V_{max} for the crystalline eel enzyme is presumably higher than the peak assay value so far published.[54]

The pH optimum for eel AChE is approximately 8.25. At other values lower activity is explained as the effect of altering the ionization of essential groups at the esteratic site. However, the hydrolysis of acetylthiocholine and phenyl acetates is not decreased at high pH values.

The temperature dependence of overall ACh hydrolysis is complicated; Arrhenius plots of the logarithm of hydrolysis rate vs. $1/k^\circ$ are nonlinear. At high temperatures k_5 is limiting (at 35°C the eel enzyme $\log_{k_5} = 5.65$–5.56); and k_3 limits at low temperatures, especially for poor substrates. The activation energy for eel AChE is about 1700 cal/mole.

Low concentrations of divalent ions activate crude preparations of AChE, but their effect on highly purified preparations has not been carefully studied, and there is no reason to believe that AChE is a metaloenzyme. Isotonic NaCl and about 20 mM $MgCl_2$ are generally used for assays.

Relatively few data are available on the equilibrium positions of the intermediate and overall reactions. At 3°C $k_3 = k_5$. At low pH values Hestrin found the equilibrium constant for overall synthesis of ACh or its hydrolysis with eel AChE as $K = (choline)(H\ acetate)/(ACh)(H_2O) = 3.7$–$4.0$, which indicates a free energy of ACh hydrolysis of about 3150 cal/mole.

IV. REFERENCES

1. H. H. Dale, *Adventures in Physiology*, Pergamon Press, London (1953).
2. C. O. Hebb and K. Krnjević, *in Neurochemistry* (K. A. C. Elliott, I. H. Page, and J. H. Quastel, eds.), pp. 452–521, C. C. Thomas, Springfield, Illinois (1962).

3. R. Birks and F. C. MacIntosh, Acetylcholine metabolism of a sympathetic ganglion, *Can. J. Biochem. Physiol.* **39**:787–827 (1961).

4. B. Katz, The Sherrington Lectures: *The Release of Transmitter Substances*, C. C. Thomas, Springfield, Illinois (1969).

5. B. Katz, *Nerve, Muscle and Synapse*, McGraw-Hill, New York (1966).

6. D. Nachmansohn, in *Handbook of Experimental Pharmacology* (G. B. Koelle, ed.), Vol. 15, pp. 701–740, Springer-Verlag, Berlin (1963).

7. C. O. Hebb, Biochemical evidence for the neural function of acetylcholine, *Physiol. Rev.* **37**:196–220 (1957).

8. G. B. Koelle, in *Handbook of Experimental Pharmacology* (G. B. Koelle, ed.), Vol. 15, pp. 187–298, Springer-Verlag, Berlin (1963).

9. J. H. Burn and M. J. Rand, in *Advances in Pharmacology* (S. Garattini and P. A. Shore, eds.), Vol. 1, pp. 1–30, Academic Press, New York (1962).

10. C. B. Ferry, Cholinergic link hypothesis in adrenergic neuroeffector transmission, *Physiol. Rev.* **46**:420–456 (1966).

11. C. O. Hebb and D. Ratković, Choline acetylase in the placenta of man and other species, *J. Physiol.* **163**:307–313 (1962).

12. L. T. Potter and V. A. S. Glover, in *Methods in Enzymology* (H. Tabor and C. W. Tabor, eds.), Academic Press, New York (In press.)

13. A. J. D. Friesen, J. W. Kemp, and D. M. Woodbury, The chemical and physical identification of acetylcholine obtained from sympathetic ganglia, *J. Pharmacol.* **148**:312–319 (1965).

14. L. T. Potter, in *The Interaction of Drugs and Subcellular Components in Animal Cells* (P. N. Campbell, ed.), pp. 293–304, Churchill, London (1968).

15. V. P. Whittaker, in *Handbook of Experimental Pharmacology* (G. B. Koelle, ed.), Vol. 15, pp. 1–39, Springer-Verlag, Berlin (1963).

16. E. Frommel, A. Bischler, P. Gold, M. Faure, and F. Vallette, Le problème d'un sel d'acétylcholine d'action prolongée, *Schweiz Med. Woch.* **49**:1269–1276 (1947).

17. L-E. Tammelin, Substrates and inhibitors of cholinesterases, *Svensk. Kemisk Tidskrift.* **70**:4–181 (1958).

18. L. T. Potter, A radiometric microassay of acetylcholinesterase, *J. Pharmacol.* **156**:500–506 (1967).

19. S. Hestrin, The reaction of acetylcholine and carboxylic acid derivatives with hydroxylamine, and its analytical application, *J. Biol. Chem.* **180**:249–261 (1949).

20. J. R. Cooper, The fluorometric determination of acetylcholine, *Biochem. Pharmacol.* **13**:795–800 (1964).

21. L. T. Potter, V. A. S. Glover, and J. K. Saelens, Choline acetyltransferase from rat brain, *J. Biol. Chem.* **243**:3864–3870 (1968).

22. J. Crossland, in *Methods in Medical Research* (J. H. Quastel, ed.), Vol. 9, pp. 125–129, Year Book Publishers, Chicago (1961).

23. L. T. Potter and W. Murphy, Electrophoresis of acetylcholine, choline and related compounds, *Biochem. Pharmacol.* **16**:1386–1388 (1967).

24. J. Schuberth, Choline acetyltransferase, *Biochim. Biophys. Acta* **122**:470–481 (1966).

25. G. T. Girvin and J. W. Stevenson, Cell-free "choline acetylase" from *Lactobacillus plantarum*, *Can. J. Biochem. Physiol.* **32**:131–146 (1954).

26. C. O. Hebb, K. Krnjević, and A. Silver, Acetylcholine and choline acetyltransferase in the diaphragm of the rat, *J. Physiol.* **171**:504–513 (1964).

27. R. Birks, B. Katz, and R. Miledi, Physiological and structural changes at the amphibian myoneural junction, in the course of nerve degeneration, *J. Physiol.* **150**:145–168 (1960).

28. V. P. Whittaker, in *Progress in Biophysics and Molecular Biology*, Vol. 15, pp. 41–96, Pergamon Press, London (1965).

29. F. Fonnum, Choline acetyltransferase binding to and release from membranes, *Biochem. J.* **109**:389–398 (1968).

30. L. T. Potter, Synthesis, storage and release of acetylcholine-^{14}C in isolated rat diaphragm muscles, *J. Physiol.*, in press.
31. C. O. Hebb, in *Handbook of Experimental Pharmacology* (G. B. Koelle, ed.), Vol. 15, pp. 55–88, Springer-Verlag, Berlin (1963).
32. F. Fonnum, A radiochemical method for the estimation of choline acetyltransferase, *Biochem. J.* **100**:479–484 (1966).
33. D. Morris, The effect of sulphydryl and other disulphide reducing agents on choline acetyltransferase activity estimated with synthetic acetyl-CoA, *J. Neurochem.* **14**:19–27 (1967).
34. A. K. Prince, Properties of choline acetyltransferase isolated from squid ganglia, *Proc. Nat. Acad. Sci.* **57**:1117–1122 (1967).
35. G. L. Ellman, K. D. Courtney, V. Andres, and R. M. Featherstone, A new and rapid colorimetric determination of acetylcholinesterase activity, *Biochem. Pharmacol.* **7**:88–95 (1961).
36. R. B. Reisberg, Properties and biological significance of choline acetylase, *Yale J. Biol. Med.* **29**:403–435 (1957).
37. K. N. Mehrotra, Properties of choline acetylase from the house fly *Musca domestica* L., *J. Insect Physiol.* **6**:215–221 (1961).
38. H. Kumagai and S. Ebashi, Highly purified choline acetylase, *Nature* **173**:871–872 (1954).
39. D. Morris, The choline acetyltransferase of human placenta, *Biochem. J.* **98**:754–762 (1966).
40. A. S. V. Burgen, E. Burke, and M-L. Desbarets-Schonbaum, The specificity of brain choline acetylase, *Brit. J. Pharmacol.* **11**:308–312 (1956).
41. J. C. Smith, C. J. Cavallito, and F. F. Foldes, Choline acetyltransferase inhibitors: a group of styryl-pyridine analogs, *Biochem. Pharmacol.* **16**:2438–2441 (1967).
42. L. T. Kremzner and I. B. Wilson, A partial characterization of acetylcholinesterase, *Biochemistry* **3**:1902–1905 (1964).
43. L. T. Kremzner, R. J. Kitz, and S. Ginsburg, A partial purification and characterization of the acetylcholinesterase of human brain, *Fed. Proc.* (Abstract) **26**:296 (1967).
44. G. R. de Lores Arnaiz, M. Alberici, and E. De Robertis, Ultrastructural and enzymic studies of cholinergic and non-cholinergic synaptic membranes isolated from brain cortex, *J. Neurochem.* **14**:215–225 (1967).
45. E. Giacobini, The distribution and localization of cholinesterases in nerve cells, *Acta Physiol. Scand.* **45**:Suppl. **156**:1–45 (1959).
46. E. Koenig, Synthetic mechanisms in the axon. Local axonal synthesis of acetylcholinesterase, *J. Neurochem.* **12**:343–349 (1965).
47. L. Lubińska, S. Niemierko, B. Oderfeld, and L. Szwarc, The distribution of acetylcholinesterase in peripheral nerves, *J. Neurochem.* **10**:25–41 (1963).
48. L. Guth, W. C. Brown, and P. K. Watson, Studies on the role of nerve impulses and acetylcholine release in the regulation of the cholinesterase activity of muscle, *Exper. Neurol.* **18**:443–452 (1967).
49. B. C. R. Stromblad, Cholinesterase activity in skeletal muscle after botulinum toxin, *Experientia* **16**:458–460 (1960).
50. R. Miledi, *in* Ciba Symposium on *Enzymes and Drug Action* (J. L. Mongar and A. V. S. de Rueck, eds.), pp. 220–235, Little, Brown, Boston (1962).
51. K-B. Augustinsson, *in Handbook of Experimental Pharmacology* (G. B. Koelle, ed.), Vol. 15, pp. 89–128, Springer-Verlag, Berlin (1963).
52. J. A. Cohen and R. A. Oosterbaan, *in Handbook of Experimental Pharmacology* (G. B. Koelle, ed.), Vol. 15, pp. 299–373, Springer-Verlag, Berlin (1963).
53. O. Svensmark, Molecular properties of cholinesterases, *Acta Physiol. Scand.* **64**:Suppl. **245**:1–74 (1965).
54. W. Lenzinger and A. L. Baker, Acetylcholinesterase, I. Large-scale purification, homogeneity, and amino acid analysis, *Proc. Nat. Acad. Sci.* **57**:445–451 (1967).

55. J-P. Changeux, Responses of acetylcholinesterase from *Torpedo marmorata* to salts and curarizing drugs, *Mol. Pharmacol.* **2**:369–392 (1966).
56. I. B. Wilson, *in The Enzymes* (P. D. Boyer, H. Lardy, and K. Myrbäck, eds.), Vol. 4, pp. 501–520, Academic Press, New York (1960).

Chapter 13

AMINE OXIDASES

J. H. Quastel

Kinsmen Laboratories of Neurological Research
Faculty of Medicine, University of British Columbia, Vancouver, B.C., Canada

Much of the literature on monamine oxidase and diamine oxidase, since the very early observations on tyramine oxidation in the liver[1-4] and tissue histaminase,[5] has been reviewed in articles by Blaschko[6-9] and Zeller.[10] The present article will deal primarily with the distribution, properties, and functions of amine oxidases of the nervous system.

I. HISTORICAL

Hare,[2] in 1928, showed that tyramine is oxidized by a liver extract, the oxidation being insensitive to cyanide,[2,4] hydroxylamine, and iodoacetate[4] but strongly inhibited by methylene blue.[3] Hydrogen peroxide and ammonia are products of the reaction.[2,3,4] It also became clear that liver, as well as intestine, attacks epinephrine with an uptake of one atom of oxygen per molecule of epinephrine,[11] the amine being oxidized in these tissues in a similar manner to that of other primary, secondary, and tertiary amines.[4,12,13]

The first evidence of the presence of an amine oxidase in brain came from the work of Pugh and Quastel.[14] It had been shown[15] that tyramine, tryptamine, and isoamylamine, on being added to brain slices, incubated in a physiological saline-glucose medium, bring about a diminution of brain respiration. In seeking an explanation for this inhibition, it was found[14] that aliphatic amines are oxidized in brain to liberate ammonia and a dinitrophenylhydrazine-reacting substance, subsequently shown to be the corresponding aldehyde.[16] While the brain cortex slices (rat) are able to oxidize butylamine, amylamine, isoamylamine, and heptylamine, the rate of oxidation of ethylamine or methylamine is small or negligible.[14] It was then shown[16] that the amine oxidase in brain that attacks the aliphatic amines also attacks tyramine and tryptamine, liberating ammonia and the corresponding aldehyde. The brain amine oxidase was found to have similar properties to those of hepatic amine oxidase. The enzyme is not affected by cyanide and in its presence one atom of oxygen is taken up for each molecule of tryptamine or tyramine consumed and one molecule of ammonia is liberated. Hydrogen peroxide is formed, as shown by the coupled oxidation of ethanol.[16]

An important fact concerning tryptamine oxidation in the tissues was found to be the formation of a melaninlike pigment, brown or black, firmly adsorbed on the tissue slices. Brain cortex slices respiring in the presence of tryptamine take on a brown color. The presence of cyanide greatly reduces the pigment production, so that it became evident that two mechanisms are involved in tryptamine oxidation: (a) a cyanide-resistant mechanism resulting in the oxidation of the amino group and liberation of ammonia, and (b) a cyanide-sensitive mechanism resulting in the formation of a pigment either from tryptamine directly or from its product of oxidation by amine oxidase.[16]

The cerebral amine oxidase apparently has no ability to oxidize mescaline, histamine, or phenylisopropylamine (amphetamine). It was concluded that a common oxidase system is involved in the oxidation of aliphatic amines, tyramine, tryptamine, and phenylethylamine.

Independently of, but at the same time as, this work[16] Blaschko, Richter, and Schlossmann[13] came to the same conclusion that one enzyme system attacks epinephrine, tyramine, and aliphatic amines, that it exists in every vertebrate tested and in a number of invertebrates, and that it attacks only compounds containing a terminal amino group but not those with a substituent on the α-carbon atom such as amphetamine or ephedrine.

Analysis of the reaction products formed when isoamylamine and butylamine are oxidized by liver extracts showed[16] that the following transformations take place:

$$R \cdot CH_2NH_2 + O_2 + H_2O = R \cdot CHO + NH_3 + H_2O_2$$

$$2R \cdot CHO + H_2O = R \cdot CH_2OH + R \cdot COOH$$

Richter[12] (see also Kohn[4]), showed that the enzyme oxidizes phenylethylamine, tyramine, and arterenol to form ammonia; epinephrine, epinine, and sympatol to produce methylamine; and dl-alkamine and hordenine to give ethylamine and dimethylamine respectively. The corresponding aldehydes are also formed.

The general reaction was represented as follows:

$$R \cdot CH_2 \cdot \overset{+}{N}H \cdot R_2' + \tfrac{1}{2}O_2 = R \cdot CHO + \overset{+}{N}H_2 \cdot R_2'$$

where R = hydrogen or an alkyl group.

Competition for amine oxidase takes place between various amines. Thus, it was found by Pugh and Quastel[16] that competition exists between the following amines: tyramine, isoamylamine, tryptamine, and phenylethylamine. The latter amine, only feebly attacked, inhibits tyramine oxidation. Blaschko *et al.*[13] showed that competition occurs between tyramine, adrenaline, isoamylamine, and tryptamine. They also showed that *l*-ephedrine, triethylamine, triisoamylamine, hordenine, amines which are either slowly or not attacked by amine oxidase, have an affinity for the enzyme and inhibit the oxidation of amines which are attacked by the oxidase. Kohn[4] arrived at similar results. Amphetamine is either very feebly or not attacked by amine oxidase[13,16] but has affinity for the enzyme and inhibits amine oxidation.[17,18]

The fall in brain slice respiration brought about by the addition of tyramine or isoamylamine was shown to be due to the formation of the corresponding aldehyde, which exerts potent inhibitory effects on cerebral respiratory processes.[18] *p*-Hydroxybenzaldehyde, and isovaleric aldehyde, in common with other aldehydes[19] such as acetaldehyde or *n*-valeric aldehyde, inhibit brain respiration at relatively low concentrations (1–2 mM).

The presence of amphetamine, which is a competitive inhibitor of amine oxidase,[17,18] prevents the fall in brain respiration due to tyramine, tryptamine, or isoamylamine[18] by inhibiting the formation of the inhibitory aldehydes. L-Ephedrine[13,18] and certain phenylisopropylamine derivatives[18] also compete reversibly for amine oxidase and exert effects similar to amphetamine in retarding the fall in brain respiration, in glucose or succinate media, due to the presence of tyramine. It was suggested by Mann and Quastel[18] that the neurological effects of amphetamine and related substances, in their antidepressant action on the nervous system, may be related to their inhibitory effects on cerebral amine oxidase.

However, the brain is equipped with an aldehyde dehydrogenase[19,26] and presumably also with a mutase, or reducing system, that brings about the formation of the corresponding alcohol.[16,27,28] The quantitative significance of this system is at present unknown. It had been known for many years that tyramine is converted in the body to *p*-hydroxyphenylacetic acid[1] and, moreover, that the corresponding alcohols, as well as the acids, are formed from a number of amines such as tyramine and tryptamine during animal metabolism.[20] The presence in brain of active aldehyde-removing systems has turned attention away from the possible pharmacological implications of aldehyde formation from cerebral amines. Nevertheless, there are observations indicating that aldehydes derived from neuroamines may influence various metabolic changes and as such influence neuronal function.[64,73] Attention, today, is focused on the significance of the changes in cerebral amine levels and on a variety of other changes, brought about by administration of amine oxidase inhibitors, and their correlation with pharmacological, physiological, and clinical effects.[21]

What may be considered the modern era concerning the properties and function of amine oxidase commenced with the discovery of a variety of amines in the nervous system (e.g., serotonin, tryptamine, norepinephrine) and by the introduction of iproniazid (1-isonicotinyl-2-isopropylhydrazine), which, in contrast to amphetamine, is a highly active and irreversible inhibitor of amine oxidase *in vivo*.

Zeller[22] in 1938 showed that histaminase of pig kidney, which is inhibited by cyanide, also attacks putrescine (1,4-diaminobutane) or cadaverine (1,5-diaminopentane), which, like histamine, is not oxidized by amine oxidase.[13,15,16] He considered that the experimental data led to the conclusion that an enzyme, diamine oxidase (DAO), attacks both diamines and histamine[23,10] in much the same way as monoamine oxidase (MAO) acts on both alkylamines and arylalkylamines. The two enzyme systems are distinguished by their different sensitivities to carbonyl reagents, e.g., semicarbazide, which blocks the diamine oxidase but not the monoamine oxidase. This classification of amine oxidases (MAO and DAO) is current today, although it is known that certain diamines (e.g., 1,10-decamethylenediamine and 1,12-dodecamethylenediamine) are oxidized by MAO.[24] Moreover, an enzyme exists (benzylamine oxidase) which acts on monoamines and has the inhibitor characteristics of histaminase.[25]

The use, however, of the term MAO for the enzyme that attacks the pharmacological active monoamines, is so common today that it is proposed to continue with the use of this term until a more appropriate and exact term is considered necessary.

II. DETECTION AND ESTIMATION OF AMINE OXIDASE

A. Methods *in Vitro*

Histochemical studies of MAO make use of the fact, mentioned above, that in the presence of tryptamine or 5-hydroxytryptamine (serotonin), the enzyme produces a dark-colored pigment whose location gives an indication of its site in tissues.[29] According to Blaschko and Hellmann[29] the pigment, derived from tryptamine in presence of amine oxidase, is rapidly dissolved by ethanol, acetone, dioxane, and *n*-butanol. In experiments of Arioka and Tanimukai,[30] using serotonin as substrate, the pigment is relatively insoluble in ethanol, acetone, *n*-butanol, and formalin. These investigations have led to a histochemical technique for the detection of MAO in tissues. In this technique, brain slices (for example) are incubated for several hours at 37°C in oxygen in presence of serotonin and traces of heavy metals (Co, Fe, Cu, Mo, etc.) which accelerate pigment formation. The slices are washed and embedded in paraffin, after which sections are cut, deparaffinized, dehydrated in alcohol, and mounted in cedar oil. Sites of MAO are shown by the deposition of a dark brown pigment. The pigment is produced from 5-hydroxyindole-acetaldehyde which is also oxidized to, and excreted as, 5-hydroxyindole-acetic acid.[31,32,33] Pigment formation from serotonin in presence of MAO is inhibited by urea, octanol, or iproniazid.

The aldehyde formed from the amine reduces tetrazolium with the formation of formazan[34,35] which may, therefore, be used for the detection of the enzyme.[36]

Another method, using tryptamine as substrate, has been devised by Koelle and Valk.[37] This hydrazone-precipitation method gives a reliable direct localization of MAO but the procedure, with nervous tissue, is handicapped by experimental difficulties.[34,55]

Microfluorimetric estimations of MAO have been recently described. [38,39,40,201] Determination of MAO activity may be based on the rate of disappearance of the substrate (tyramine, serotonin) in presence of a tissue homogenate, incubated at 37°C in presence of oxygen, a fluorescence technique being used to estimate the amine.[39,40] Another method is based on the oxidation of kynuramine by a brain homogenate to 4-hydroxyquinoline, which fluoresces strongly in alkaline solution.[201]

A colorimetric method for the estimation of MAO, based on observations[14,16] of the red color, that turns to orange-yellow, formed by the solution in alkali of the dinitrophenylhydrazone of the aldehyde derived from the amine, has been described[41] and correlations of some of these histochemical and biochemical assays have recently been made.[42]

MAO activity is also estimated by measurement of the ammonia formed during incubation of the tissue with tyramine in a Conway diffusion chamber, the ammonia being absorbed by boric acid placed in the center well of the chamber.[58,59]

Another method involves the extraction and spectrofluorimetric assay of indoleacetic acid derived from tryptamine,[74] a method considered well suited for the study of MAO in rat tissues *in vitro*, so long as various critical conditions are observed.[49]

A radiometric method for the assay of MAO, utilizing labeled tryptamine, has been devised by Wurtman and Axelrod.[75]

B. Methods *in Vivo*

Tests needed to determine the efficacy *in vivo* of compounds inhibiting MAO have been described by Sjoerdsma.[43] The tests are based on the assumption that the inhibition of MAO will result in a proportionate decrease in excretion of the acid end products of amine metabolism. Such tests are: (a) conversion of serotonin to 5-hydroxyindoleacetic acid whose amounts in the urine are measured, inhibition of MAO being indicated by a decreased rate of conversion to the acid,[44] (b) urinary amine excretion (e.g., 45) (determined by column and two-dimensional chromatography, apart from tryptamine and tyramine for which fluorimetric methods are available), (c) pharmacological tests such as the potentiation of amine effects, in animals receiving the MAO inhibitor, and the tryptophan intoxication syndrome caused by the inhibitor. It is thought that the latter is the only test, so far discovered, for detecting MAO inhibition in the brain of man.

A method for studying MAO activity *in vivo*, based on the estimation of radioactivity in the urine following administration of N-methyl labeled epinephrine, has been described by Resnick[46] who showed that iproniazid administration leads to increased rates of radioactive excretion persisting for 2 wk after cessation of iproniazid treatment. Radioactive (^{14}C) serotonin has been used for detection of MAO *in vivo*.[64] The labeled serotonin is injected into mice by an endocranial-intraventricular route. The labeled aldehyde, formed by the action of MAO, is incorporated into acid-insoluble components of the brain. The incorporation is prevented by previous treatment of the animal with a MAO inhibitor.

III. DISTRIBUTION OF AMINE OXIDASE

A. Cellular

Although amine oxidase is found in all vertebrates tested, in mollusks, echinoderms, annelids, and in cockroaches,[8] in plants and in bacteria, the typical MAO is found only in animals. The enzyme occurs in a variety of vertebrate tissues, particularly liver. All glands contain the enzyme[8] and, in man, the parotid and submaxillary glands represent the richest source of amine oxidase.[47] It is present in the nervous system, in the gonads, in smooth muscle, and in smaller amounts in cardiac muscle. It is apparently absent from skeletal muscle and from blood.[8] It has high activity in the pineal glands of man, cow, rabbit, rat, but less so in those of pig and cat.[48,65]

Using the pigment-histochemical test, it is found that the dark-brown coloration indicative of MAO is greater in gray matter than in white matter, and that coloration in the hypothalamus is very intense.[30] The pigment granules are seen in the cytoplasm, especially the perinuclear region, and absent in the nuclei of nerve cells of the paraventricular structures.[30] No pigment formation is seen in the nerve fibers or in the glial cells.

Usually, relatively high amounts of MAO are found in structures richly endowed with sympathetic nerves as well as in the nerves themselves. The superior cervical ganglia of the rat are rich in MAO. In fact, the ganglionic MAO activity is greater than that of all other tissues investigated, in the rat, except the liver.[49] The brain differs in that it does not have a homogeneous distribution of the enzyme (see, e.g., Ref. 36), the greatest activities being localized in the midbrain. In dog brain, the highest activity of MAO is in the hypothalamus[51] but it is relatively evenly distributed throughout the brain[52] (see Table I).

MAO activity is found in the neurohypophysis,[65] oculomotor and optic nerves,[53] the myelin sheaths of rat sciatic and optic nerves,[54] the midbrain of the mouse[30] and rat,[55] and the pons and mesencephalon of the squirrel monkey.[56] Its distribution in various parts of the developing rat brain has been studied[57] by a histochemical technique.[34]

B. Subcellular

The greater part of liver MAO is to be found in the mitochondria,[60,61, 67,69] as is the case with adrenal medulla,[62,68] but the microsomal fraction of liver also contains amine oxidase activity.[61,67]

MAO is associated with the particulate fraction of brain.[66] Subcellular fractionation of whole rat brain shows that 70% of MAO activity is in the

TABLE I

Distribution of MAO in Dog Brain[52]

Tissue	MAO μg serotonin oxidized per g/hr
Amygdala	968 ± 23
Hypothalamus	1624 ± 510
Septal region	1212 ± 58
Midbrain	842 ± 122
Pyriform cortex	926 ± 183
Caudate nucleus	935 ± 200
Medulla	1117 ± 230
Thalamus	940 ± 164
Hippocampus	1176 ± 38
Pons	936 ± 110
Cortical grey matter	819 ± 300
Cerebellum	930 ± 87

mitochondrial fraction, 17.5% in the nuclear fraction, and 12.5% in the microsomal fraction.[63] There is a close parallelism between the distribution of MAO and succinic dehydrogenase in rat brain.[63] While both MAO and succinic dehydrogenase activities are absent in subfractions of myelin and membranes, very low in cholinergic nerve endings, they are high in the non-cholinergic nerve endings and highest in the mitochondria.[63] MAO activity also occurs in the microsomal fractions of nerve and sympathetically ennervated tissue.[50]

A soluble MAO, present in guinea pig liver, has been partially purified by use of ammonium sulfate fractionation and acid precipitation[70] and seems to have properties similar to those of mitochondrial MAO. It can be separated from the aldehyde dehydrogenase present, this being an NAD-linked enzyme. It oxidizes 5-hydroxyindoleacetaldehyde to 5-hydroxy-indoleacetic acid.[70] Blaschko and Jacobson[71] used lysolecithin to produce clear "solutions" of amine oxidase. The detergent, Cutscum, with iso-octylphenoxy-polyphenoxy-ethanol as the active principle, has been used for solubilization of the enzyme.[72] Studies of the distribution of MAO and other oxidizing enzymes by histochemical methods,[34] in the medulla oblongata and cerebellum of the squirrel monkey, showed that enzymes concerned with the glycolytic and citric acid cycle pathways are located in the perikarya and the neuropil, but MAO activity is more apparent in the nerve fibers, dendrites, and finer processes in the neuropil.[76] Some of the larger neurons (e.g., the Purkinje cells of the cerebellar cortex) show a fair amount of MAO as well as strong lactic dehydrogenase activity.[76] It was considered probable that the cells having MAO activity may be responsible for controlling the level of serotonin and norepinephrine in the brain as suggested by Brodie and Shore.[77]

C. In the Developing Brain

Judging MAO activity by the disappearance of tyramine from incubated homogenates of brains of rats of 1 to 40 days of age, it is found that the enzyme activity increases during the first 20 days by a factor of three, on a wet-weight basis, or twofold on a dry-weight basis.[36] The rate of increase is slow during the first 10 days. Serotonin is present in the brains of newborn rats at about one third the concentration in the adult rat brain. The brains of animals treated with reserpine during the first 20 days of life contain levels of MAO activity about 65%–75% of those of control animals examined at 20 days of age.[36] The brain levels of norepinephrine, serotonin, and MAO in the newborn guinea pig, whose functional activity is well developed, are almost as high as those in the adult brain, in contrast to the brain levels of the newborn rat whose functional development is poor.[78] Histochemical studies[57] of the cerebrum of developing rat brain show the presence of MAO in specific areas at 15 days of age but the rates of increase of MAO activity, with increasing age, apparently differ in different sites. There is a stronger activity at an earlier age in most brain stem nuclei than in various other centers.[79] Prominent MAO activity is present in the locus coeruleus and nucleus ambiguus at birth. At 5 days of age there is weak MAO activity localized mainly within the neuropil and glial cells, and at 10 days of age all nuclei possess MAO of various degrees of activity. The staining due to MAO in nerve fibers is negligible at 5 days of age but increases rapidly in some tracts at 15 days of age. No increase is apparent after 55 days.[79]

The development of MAO with maturation of the brain resembles that seen with a variety of other enzymes, e.g., succinic dehydrogenase, cytochrome oxidase, aldolase, phosphorylase, glutamic dehydrogenase, or Mg^{2+}-ATPase.

IV. EFFECTS OF INCREASED OXYGEN TENSION

Oxygen pressure is a rate-limiting factor for MAO in vitro.[3,4,80] Gey and Pletscher[81] point out that in spite of an increased rate of oxidation of tyramine by a rat brain particulate fraction on substituting oxygen for air in the gas phase, the degree of MAO inhibition by a given iproniazid dose is unaffected. Novick[82] reports a similar result.

V. SUBSTRATES OF MAO

MAO is responsible, in part, for the metabolism in the brain of many biogenic amines, e.g., tryptamine, 5-hydroxytryptamine, tyramine, catecholamines, O-methylcatecholamines. Primary and secondary amines are oxidized but the latter only if one of the substitutes is a methyl group. Tertiary amines are oxidized more slowly than the secondary amines, but the rate of oxidation of tertiary amines is higher in some species (e.g., cat) than in others (e.g., rabbit).[83]

Naturally occurring substrates of MAO are usually derivatives of β-phenylethylamine, tryptamine, or histamine. The catechol compounds are 3,4-dihydroxyphenylethylamine (dopamine), norepinephrine and epinephrine, to which should be added metanephrine and normetanephrine formed by methylation of the corresponding catecholamines. Tyramine and β-hydroxytyramine (octopamine) occur in octopods and in small amounts in mammals. The whole brain (rabbit) content of tyramine is less than 10 $\mu g/g$ wet-weight and it is considered not to be an important metabolic precursor of norepinephrine.[84] Serotonin (5-hydroxytryptamine) is the most important naturally occurring derivative of tryptamine. The N-methyl derivative has been found in amphibian brain stem. The N-dimethyl derivative (bufotenine) is slowly attacked by MAO.[31]

Tyramine and dopamine are among the amines most readily oxidized by MAO.[85] Metanephrine and normetanephrine are attacked at approximately similar rates to epinephrine and norepinephrine, indicating that methylation of the monohydroxy group of the catecholamines does not markedly affect the attack by MAO. A variety of ring methoxylated phenylethylamines (other than compounds substituted in the 2,6 position or containing more than three methoxyl groups) is attacked by liver amine oxidase, and among these only mescaline oxidation is inhibited by semicarbazide.[92] It would seem therefore that a number of methoxylated phenylethylamines is affected by MAO in a manner similar to tyramine or phenylethylamine. The ring-methoxylated phenylethylamines which are not deaminated do not compete with other amines for MAO.

While histamine is a weak substrate of MAO in the liver of ox or mouse,[84,87] the methyl derivative (1,4-methyl histamine), which exists in brain[105,106,107] is fairly well attacked by MAO.[88] It seems, therefore, possible that histamine, after ring methylation, may be metabolized with the help of MAO. In fact, the production of l-methylimidazoleacetic acid is retarded by MAO inhibitors.[89,90] Excretion of methyl histamine is increased in patients receiving a MAO inhibitor.[108]

Other cyclic amines attacked by MAO are m-xylylenediamine, 3-picolylamine, and 4-picolylamine.[91]

Oxidation of aliphatic amines by MAO has been the subject of a variety of reports and reviews.[14,16,6,24,8] In the homologous series, $CH_3 (CH_2)_n NH_2$, maximal rates are found with amylamine or hexylamine. Short-chain members of the series $NH_2 (CH_2)_n NH_2$, such as putrescine ($n = 4$) or cadaverine ($n = 5$), are not attacked by MAO and have little affinity for the enzyme. Longer chain members of the series are attacked.[24] For example, $H_2N (CH_2)_n, N^+(CH_3)_3$ where $n = 10$, 11, or 12 is oxidized by MAO. Also, $H_2N (CN_2)_{10} \cdot COOH$ is attacked although shorter chain ω-amino acids are not attacked. Presumably, either a positive or a negative charge may exert a disturbing influence on the interrelation between enzyme and substrate, an influence that diminishes with the increasing distance between the disturbing group and the amino group.[8]

An amine oxidase in the skin may bring about oxidation of specific epsilon amine groups of polypeptide-bound lysine to the corresponding aldehydes which then undergo aldol condensation and bring about cross-linking. It is not identical with liver MAO.[202]

VI. FOOD, AND SUBSTRATES OF MAO

Certain types of cheese contain relatively high quantities of tyramine,[93] and as a number of MAO inhibitors enhance the pressor response due to orally administered tyramine,[94] by preventing its breakdown by liver or intestinal MAO, it is possible that ingestion of certain cheeses may, in presence of a MAO inhibitor, give rise to hypertensive responses. This has already been reported[95] and a case with fatal outcome has been described.[96] It may be noted that isoamylamine, an effective substrate of MAO, is to be found in Roquefort cheese[97] and this amine has pharmacological activity.[97] Tyramine and β-phenylethylamine have been found in all cheeses investigated and tryptamine is present in most.[98] Broad beans (*Vicici faba* L.) may induce hypertension in patients on MAO inhibitors.[99] The beans contain 3,4-dihydroxyphenylalanine (DOPA) which exerts pressor effects in patients on MAO inhibitors.[100] This activity is ascribed to its conversion to dopamine.

Dietary tryptophan may play an important role in investigations on the effects of MAO inhibitors. Thus, it has been shown that, whereas there is very little tryptamine in the brain of the guinea pig even after administration of iproniazid, there is a considerable increase when a combination of iproniazid and tryptophan is given to the animal.[101] This occurs also with the brain of the rabbit, where convulsions can occur after administration. Udenfriend[102] points out that tryptophan administration (20 mg/kilo) to patients on an MAO inhibitor (phenylisopropylhydrazine) may produce marked mental effects.[103] Tryptophan administration to patients with depressive illness seems to potentiate the effects of an MAO inhibitor (e.g., tranylcypromamine).[104]

Methionine administration to schizophrenic patients on an MAO inhibitor results in the development of psychic symptoms regarded as either due to an intensification of schizophrenic symptoms or due to added toxic symptoms.[129–132] There are increased outputs of catecholamines, methionine metabolites, and amino acids,[133] indicating somewhat complex disturbances of metabolism.

VII. INHIBITORS OF MAO

A large literature now exists on the properties of MAO inhibitors and various reviews have been published.[21,8,109–111] Only broad aspects of this

subject will be considered in this article. Structural formulas of some inhibitors are given in Table II.

TABLE II

Pargyline	
Tranylcypromine	
Amphetamine	
Iproniazid	
Pheniprazine	
Isocarboxazid	
Nialamide	
Phenelzine	
Modaline	
Harmine	
Harmaline	

A. Reversible Inhibitors

Amines, in which one of the hydrogen atoms in the terminal carbon group is substituted by a methyl group, are usually inhibitors of MAO. Amphetamine (β-phenylisopropylamine), ephedrine and a number of isopropylamines were shown as long ago as 1939 to be competitive inhibitors of MAO, and their affinity constants to the enzyme have been measured.[17,18] Amphetamine and ephedrine, however, do not affect catecholamine metabolism *in vivo*.[114,115] The general rule that amines containing the side chain $- CH_2\text{-}CH(CH_3) \cdot NH_2$ or $- CH_2\text{-}CH(C_2H_5) \cdot NH_2$ are reversible inhibitors of MAO applies to the indole, and 5-hydroxyindole, class of compounds.[112] α-Ethyltryptamine inhibits MAO activity *in vitro* and *in vivo*.[112] α-Methyldopamine (α-methyl-dihydroxyphenyethylamine) and α-methyl-*m*-tyramine are inhibitors of MAO in rat brain.[113]

The harmala alkaloids (harmaline and harmine) are potent reversible amine oxidase inhibitors.[116,117] They are active at 10^{-7}M and ten times more active than the potent hydrazine inhibitors. Harmaline, applied *in vivo*, is able to protect an animal from some of the effects of long-acting amine oxidase inhibitors.[118]

B. Irreversible Hydrazine Inhibitors

The introduction, in 1952, of iproniazid (1-isonicotinyl-2-isopropylhydrazide) as an efficient inhibitor of MAO *in vitro*[119] and *in vivo*[120] focused attention on MAO as a controlling enzyme in amine metabolism both in the brain and other organs of the body.[110] Iproniazid gives maximal inhibition of enzyme activity only after exposure to oxygen[121] and it was postulated that the irreversible inhibition is due to dehydrogenation of iproniazid at the active center of the enzyme. As the isopropylhydrazine moiety of the molecule is responsible for the inhibitory action against MAO[122] it seemed possible that iproniazid is first cleaved to form free isopropylhydrazine which is the active inhibitor of MAO.[21,123] Using iproniazid, labeled with C^{14} in the isopropyl moiety, it was found that the substance is cleaved nonenzymatically, in presence of oxygen, to form a potent inhibitor of MAO that is either isopropylhydrazine or a product derived from it, which binds in an apparently irreversible manner to the enzyme.[124] The inhibition of MAO by iproniazid obeys first-order kinetics.[121]

Iproniazid, injected intraperitoneally into rats (100 mg/kg) causes an increase of the serotonin of the brain of about 100%. Isoniazid has but little effect, and isopropylhydrazine and N^1N^2-diisopropylhydrazine, injected at equivalent concentrations, increase the cerebral serotonin content by over 200%.[125] The inhibitory activity of these compounds on MAO of brain mitochondria parallels their activity *in vivo*.[125] The isopropylhydrazides of the amino acids, L-alanine and L-serine, are also inhibitors of MAO *in vitro* and *in vivo*.[125] The D-alanine derivative, however, has little effect. It seems possible that the stereospecificity shown by the alanyl-isopropylhydrazides may be due to differences in their rates of breakdown to isopropylhydrazine.

The properties of the two groups of hydrazides, viz, the alkylhydrazines and the acylhydrazides of the structure: R · CO · NH · NH · R have been reviewed.[21]

As iproniazid blocks MAO irreversibly, it is suggested [111] that the restoration of enzymatic activity in the course of several days (or weeks) may represent a feedback controlled synthesis rather than a reactivation of the inactivated enzyme molecule.

MAO seems to be present in some excess in rat brain, as a significant rise in brain amines does not take place after iproniazid administration until the MAO is inhibited by at least 85–90 %.[81] The presence of measurable quantities of amines, e.g., serotonin, in brain does not oppose this conclusion, for the amines may be stored in granules and protected from enzymatic attack.[126]

Out of the large number of hydrazine derivatives examined for anti-MAO activity, only a few have clinical value. These include iproniazid, pheniprazine, phenelzine, isocarboxazid, and nialamide. Phenelzine (β-phenylethylhydrazine) markedly reduces MAO activity in all tissues of the rat, leading to increased serotonin levels in the brain, liver, and kidney. Recovery occurs in the liver in 3 days, but not in the brain.[136] Pheniprazine injection into rats results in increased outputs of methylamine.[142] Recovery of MAO in mouse brain following a single dose of β-phenylethylhydrazine may take many days.[21]

In the intact animal, there is considerable variability in the qualitative nature of MAO inhibition by hydrazine compounds. Not only the rate, but the organ distribution of MAO inhibition varies according to the chemical structure of the inhibitors,[137,138] and tissue-drug interactions take place.[139]

The hydrazine derivatives not only affect MAO but also other enzymes (pyridoxal-dependent systems) and diamine oxidase. They often have undesirable side effects. A search made for nonhydrazine MAO inhibitors led to the finding of tranylcypromine (2-phenylcyclopylamine), pargyline, and modaline which are unrelated in structure but which inhibit MAO in an irreversible manner. They are long-acting.

C. Nonhydrazine Inhibitors

1. Phenylcyclopropylamines

Cis- and trans-phenylcyclopropylamines are among the most effective inhibitors of MAO in vitro.[127,128] The compound 2-phenylcyclopropylamine (tranylcypromine) has a structure resembling that of amphetamine and it has some of the pharmacological properties of this drug such as central stimulation and peripheral sympathomimetic action. It brings about increased serotonin[140,145] and norepinephrine[144,145] accumulation in rat brain.

2. Pargyline

Pargyline (N-benzyl-N-methyl-propynylamine) and modaline (2-methyl-3-piperidinopyrazine) exert irreversible inhibitions of MAO and act as antihypertensive and antidepressant agents. Pargyline administration to rats increases the brain accumulation of serotonin, norepinephrine, and dopamine.[145] It compares in activity with iproniazid and harmaline.[143]

D. Other MAO Inhibitors

Substances carrying the amidino group, $-C=(NH) NH_2$, may be strong inhibitors of MAO[134,135] but sensitivity of MAO to these compounds seems to vary with the animal species. Thus, rabbit liver MAO is far more sensitive than guinea pig liver MAO to pentamidine. The length of the polymethylene chain separating two amidino groups influences inhibitory activity but optimal chain length differs from species to species.[8,152] There are different degrees of reversibility according to the length of the carbon chain. Aminoguanidine, like other carboxyl reagents such as semicarbazide or hydroxylamine, is not an inhibitor of MAO.

Benactyzine (diethylaminoethyl ester of benzoic acid) inhibits MAO activity of various parts of the cat brain, the inhibition amounting to about 40% with a drug concentration of 4 mM, and about 11% at 0.4 mM. It is thus a relatively feeble MAO inhibitor. It has effects *in vivo* that vary with the part of the brain examined.[141]

1-Benzyl-2 methyl-5 methoxytryptamine (BAS), once tried as a serotonin antagonist, is an inhibitor of MAO in man.[146] It is of interest as it acts as a tranquilizer.

A number of thyroxine derivatives inhibit both liver and cerebral MAO, e.g., 3,5-diiodothyroacetic acid, 3,5,3'-triiodothyroacetic acid, and 3,5,DL-diiodothyroepinephrine.[147]

An extremely potent MAO inhibitor is a derivative of propylamine, viz, N-methyl-n-propargyl-3(2,4 dichlorophenoxy)-propylamine which is about 1000 times more active than iproniazid *in vitro*.[148] It possesses pharmacological properties associated with MAO inhibitors. Johnson[148] considers that the kinetics of inhibition of this drug point to the existence of more than one form of MAO.

A large group of styrylquinoliniums has been found to inhibit liver MAO, the 4-(-p-chlorostyryl)-1,methylquinolinium compound being five times more potent than harmaline and over 1000 times more potent than iproniazid.[149] The inhibition apparently is reversible.

Bretylium (O-brombenzyl-ethyl-dimethylammonium), an adrenergic neuronal blocking agent, is able to inhibit intraneural MAO as a typical MAO inhibitor.[150]

The oxidation of amylamine by MAO is inhibited by secondary and tertiary amines, the magnitude of the inhibition being determined by the size of the molecule.[153]

Thiol reagents (p-chloromercuribenzoate, iodoacetate, mercury salts, arsenicals, and salts of cadmium and silver) inhibit MAO,[151] pointing to the involvement of a thiol group in the activity of MAO. 2-Bromo-2-phenylacetaldehyde is a potent inhibitor.[203]

Octanol is a well-known inhibitor of MAO.

VIII. RIBOFLAVIN AND MAO

Riboflavin has been suspected of playing a role in the activity of MAO but there has been no direct evidence to support this view. Studies with riboflavin-deficient rats have not revealed any decrease in MAO activity, when this is examined *in vivo* by measurement of the excretion of 5-hydroxy-indoleacetic acid after serotonin administration.[154] After injection of iproniazid, however, return to control levels of MAO activity (when measured by dopamine oxidation) is slower in riboflavin-deficient animals and it would seem that riboflavin deficiency potentiates the effects of inhibitors.[155] Flavin as a possible constituent of MAO has been suggested by Belleau and Moran.[177]

IX. HORMONES AND MAO

Among physiological compounds so far investigated[81] only weak MAO inhibitors have been found. Cortisone[156] and thyroxine[157,158] in high doses depress enzyme activity by 30 and 50 % respectively and this degree of inhibition does not influence the brain amine content.[81] Thyroxine injection into the adult rat decreases liver MAO but not brain MAO. In female rats, thyroxine injection increases MAO activity in the hearts of young and adult animals.[169]

X. METABOLIC FUNCTIONS OF MAO

MAO seems to be primarily involved in the breakdown, in the animal body, of a variety of biologically active amines. It is evident that the enzyme is active *in vivo* as well as *in vitro*, and it is known that a number of amines, not attacked by the enzyme *in vitro*, are excreted from the body with the amine groups intact.[6] MAO brings about oxidative removal of terminal amino, or methylamino, groups, the aldehyde formed being converted, by enzymatic means, to the corresponding carboxylic acid, or to the corresponding alcohol, or undergoing further changes that have been described above. It is known that the attack by MAO on the amino group is not necessarily the first step in the metabolism of the biologically active amines.

It is generally believed that MAO inhibition results in the protection, and consequent accumulation in the brain, of serotonin and catecholamines. [21,159] Thus, the primary cause of the pharmacological response to MAO inhibitors would seem to be the increased levels in the body of these amines. However, many of the MAO inhibitors of the hydrazide type are inhibitors of 5-hydroxytryptophan and dihydroxyphenylalanine decarboxylases[160–162] and such inhibitors might, under certain circumstances, diminish the normal production of the relevant amines after MAO inhibition. It is, therefore, evident that the pharmacological behavior (due to changed amine levels) of

MAO inhibitors may be affected by the nonspecificity of some of these inhibitors. Moreover, as Horita points out,[163] the fact that a compound exerts MAO inhibition *in vitro* does not necessarily indicate that it will be effective in the animal. If the drug passes slowly through the blood–brain barrier, it may have only limited central activity. Three different methods are used to estimate the degree of MAO inhibition; (a) direct assay of brain MAO activity, (b) assay of brain levels of serotonin and for catecholamines, and (c) reversal of reserpine-induced depression.[163]

Gey and Pletscher[81] point out that MAO activity may have but little influence in controlling the total monoamine content of brain under physiological conditions, for the amines are partly stored in granules. The considerable excess of MAO in the brain suggests that the function of the enzyme lies in the rapid inactivation of amines after they have been released from storage depots.

That MAO activity is possibly linked with the functional activity of the brain seems to be indicated by the fact that levels of MAO are almost equal in the newborn and adult guinea pig brains while in the newborn rat brain the activity is only 30 % of that in the adult rat brain.[78] The guinea pig can walk and display organized patterns of behavior shortly after birth whereas this is not the case with the rat. Electrical activity typical of mature animals can be recorded from the cortex of the newborn guinea pig while similar activity does not appear in the rat until 2 wk after birth. The brain levels of norepinephrine and serotonin are lower in newborn rat than in the adult, whereas these levels are almost as high in the newborn guinea pig as in the adult. The low levels of amines in the newborn rat brain are attributed to a low capacity to store and synthesize these amines.[78]

Behavioral studies indicate that the lower the brain MAO level, the greater is the effect on behavior of administration of 5-hydroxytryptophan to pigeons. There is a lesser behavioral effect of this amino acid when the brain MAO levels recover after iproniazid treatment.[164] Injection of the amino acid leads to increased serotonin levels in four different brain areas of the pigeon but only in the telencephalon and midbrain are the changes related to the behavioral changes measured. There are also lowered dopamine levels in the midbrain and pons-medulla oblongata that require consideration.[165]

MAO inhibition (by tranylcypromine) has no effect on the resting temperature or the response to thermal loads in the conscious rabbit[166] and it seems therefore unlikely that normal temperature regulation depends on hypothalamic monoamine concentrations that are affected by MAO inhibition. This, however, is true only of the rabbit. Tranylcypromine causes a rise of body temperature in the unanesthetized cat when injected into the cerebral ventricle but intraperitoneal injection has little effect. The rat shows a temperature fall after intraventricular injection of tranylcypromine.[166,167]

Various MAO inhibitors block ganglionic transmission, as studied by perfusion through the sympathetic ganglion of the cat. Among such inhibitors are harmine (the most potent), iproniazid, and tranylcypromine.[168] The inhibition is slow in onset, gradual and progressive with time. It is possible that, in presence of an MAO inhibitor, an active amine accumulates at the ganglionic synapse, but there may also be interference with the normal functioning at the synapse.[168]

It is evident that MAO inhibitors represent important tools in physiological and pharmacological studies, but interpretations are still difficult in view of possible nonspecific effects of the inhibitors and of a variety of other factors.[170]

XI. ISOZYMES OF MAO

There is a possibility that brain contains at least two monoamine oxidases.[148] This is suggested by the fact that γ-morpholinobutyrophenone, which is an anticonvulsant and protects against the effects of reserpine, inhibits MAO *in vivo* in mouse brain by 40% using kynuramine as substrate and by 70% using tryptamine as substrate.[171] The concentrations required to give 50% inhibition depend on the inhibitors used. The drug produces a greater inhibition in preparations from brain and kidney than in those from liver. Squires[172] gives evidence that there are at least three types of mitochondrial MAO in the mouse capable of deaminating kynuramine, one type being harmine sensitive and another (containing two forms) being harmine resistant. Pigbrain mitochondria, however, seems to contain one MAO.[203]

Early observations on the properties of MAO of mouse tissues are most readily accounted for by assuming that there is more than one enzyme of this kind.[173–175,153]

There is a report of variable low levels of an enzyme in brain, insensitive to MAO inhibitors, which can catalyze the formation of 5-hydroxyindoleacetic acid from serotonin.[176]

XII. DOPAMINE-β-OXIDASE

Enzymatic conversion of 3,4-dihydroxyphenylethylamine to norepinephrine takes place in brain homogenates. Dopamine-β-oxidase is found in high concentrations in the hypothalamus and caudate nucleus, there being little activity in the cortex and cerebellum.[178] The enzyme is inhibited by MAO inhibitors containing a hydrazine structure.

XIII. OTHER AMINE OXIDASES

These have been described in some detail by Blaschko[8,9] and Zeller.[10, 111] They will be considered only briefly here as they are usually absent from the nervous system.

A. Diamine Oxidases

Diamine oxidases, which are lacking in the brain of man and other mammals, are more widely distributed in nature than MAO but are found in fewer mammalian organs. While the brains of mammals, reptiles, and amphibians show little or no diamine oxidase activity, the brains of birds show a low activity while those of fish show a relatively high activity.[179] The distribution in fish brain follows the order, in decreasing activity: diencephalon, prosencephalon = medulla oblongata, mesencephalon, metencephalon. The activity is about 100 times lower than that of MAO.[179] Putrescine or cadaverine is usually employed as a substrate for diamine oxidase.

The diamine oxidases attack histamine and a variety of diamines and are, in contrast to MAO, usually recovered in cell supernatants after removal of particulate matter from tissue homogenates by centrifugation. As diamines and polyamines occur widely in nature, the diamine oxidases may play important biological roles as controlling agencies. These enzymes do not remain constant during the life of an organism, and different enzyme levels are found during embryonal, postembryonal, and adult periods and during pregnancy.[10]

Diamine oxidases, in contrast to MAO, do not attack secondary amines and are not inhibited by 2-phenylcyclopropylamines, while MAO is not inhibited by hydrazine and unsubstituted acylhydrazides. Diamine oxidases are sensitive to hydrazine and semicarbazide. Although ethylenediamine and trimethylenediamine are poor substrates for diamine oxidase, they are relatively good inhibitors.[180] Certain guanidine derivatives, notably aminoguanidine, are potent inhibitors.[181] Metabolism of histamine in man is affected by aminoguanidine administration,[182] the excretion of labeled imidoleacetate from labeled histamine being diminished and that of methylhistamine increased. Diamidines, diguanidines, and diisothioureas are potent inhibitors of diamine oxidase,[183] while hydrazides such as iproniazid can block diamine oxidase, as well as MAO, *in vitro* and *in vivo*.[184]

Three groups of diamine oxidases seem to exist[10]: (a) bacterial, plant, and hog kidney DAO, (b) plasma amine oxidases, (c) mitochondrial rabbit liver amine oxidase (mescaline oxidase). Like MAO, they give rise to aldehydes corresponding to the amines, to ammonia, and to hydrogen peroxide.

B. Spermine Oxidase

This enzyme, which brings about the oxidation of spermine and spermidine, decamethylenediamine and dodecamethylenediamine, but not putrescine or cadaverine, exists in blood plasma.[9,185] It will also attack many aliphatic monoamines and benzylamines. The enzyme is strongly inhibited by cyanide and carbonyl reagents. Bovine spermine oxidase has been crystallized,[186] the molecular weight of the enzyme being 195,500.[9] It acts on benzylamine, mescaline, and histamine.

C. Benzylamine Oxidase

This enzyme has substrate and inhibitor specificities similar to those of spermidine oxidase, except that it cannot attack spermine and spermidine.[9] A benzylamine oxidase has been found in human blood plasma[187] and rabbit serum[188] from which it has been purified by fractionation procedures. Substrates most actively oxidized by it are substituted phenylethylamines (e.g., dopamine, tyramine, mescaline) and C-aryl methylamines. Less actively oxidized are tryptamine, serotonin, histamine, and simple aliphatic amines.[188]

Mescaline is also oxidized by a carbonyl-reagent sensitive amine oxidase present in rabbit liver mitochondria, whereas 3,4-dimethyoxyphenylethylamine is a substrate of MAO.[9,189]

D. Cofactors

The amine oxidases (attacking spermine, benzylamine, etc.) appear to be copper-proteins, first shown for plant (pea seedling) amine oxidase by Mann.[190] 2.7 to 3.7 g-atoms of copper per g-mole have been found in spermine oxidase and it has been shown that the copper content of pig kidney histaminase increases as purification progresses.[9]

Pyridoxal-5-phosphate is a coenzyme for these enzymes, though not for the plant amine oxidase[190] and it appears that there is present, in the enzyme, approximately one molecule of pyridoxal per gram-atom of copper.[191] It is likely that the characteristic sensitivity of the enzymes to the carbonyl reagents is due to their possession of pyridoxal. Inhibition of the enzyme by cysteamine is explained by its combination with pyridoxal.[192] MAO seems not to contain pyridoxal.[9]

There are large differences in MAO activities of blood platelets and plasma. The platelet enzyme is inhibited by antidepressant drugs but not by isoniazid or potassium cyanide. The plasma enzyme, on the other hand, is inhibited by isoniazid and cyanide and less by antidepressant drugs. When eighteen humans were given antidepressant drugs, or isoniazid, the inhibition of platelet amine oxidase correlated well with the urinary tryptamine excretion but the inhibition of plasma amine oxidase did not.[199]

XIV. SOME EFFECTS OF MAO INHIBITORS

Iproniazid can counteract the serotonin-removing ability of reserpine and this has been attributed to an action of the hydrazide on the tissue binding of the enzyme.[193,194] MAO inhibitors block reserpine-induced release, and the spontaneous slow release, of labeled norepinephrine previously taken up by the heart,[195] this being explained as due to a blockage of release of the amine from a tissue binding site. A similar explanation is given for the fact that serotonin levels in mouse brain, which increase in response to MAO inhibitors, continue to increase when excessive doses of the inhibitor are injected beyond the requirements for complete inhibition of MAO.[196] There is some dispute, however, as to whether MAO does inhibit release of biogenic amines from cells. This is founded on the observations that MAO inhibitors do not prevent release of serotonin from rabbit platelets *in vivo* and *in vitro*, that harmaline does not block release of brain amines by reserpine and that a variety of MAO inhibitors do not disturb the pharmacological action of reserpine.[197] There are species differences in the action of MAO inhibitors which may, however, be attributed to variations in the metabolism of the drugs, as well as to differences in their rates of penetration into tissues.[198]

XV. REFERENCES

1. A. J. Ewins and P. P. Laidlaw, The fate of *p*-hydroxyphenylethylamine in the organism, *J. Physiol.* **41**:78 (1910).
2. M. L. C. Hare, Tyramine oxidase—a new enzyme in liver, *Biochem. J.* **22**:968 (1928).
3. F. J. Philpot, Some observations on the oxidation of tyramine in the liver, *Biochem. J.* **31**:856 (1937).
4. H. I. Kohn, Tyramine oxidase, *Biochem. J.* **31**:1693 (1937).
5. C. H. Best and E. W. McHenry, The inactivation of histamine, *J. Physiol.* **70**:349 (1930).
6. H. Blaschko, Amine oxidase and amine metabolism, *Pharm. Rev.* **4**:415 (1952).
7. H. Blaschko, Enzymic oxidation of amines, *Brit. Med. Bull.* **9**:146 (1953).
8. H. Blaschko, in *The Enzymes* (P. D. Boyer, H. Lardy and K. Myrbäck, eds.), 2nd ed. Vol. 8. 337, Academic Press, New York (1963).
9. H. Blaschko, in *Molecular Basis of Some Aspects of Mental Activity* (O. Walaas, ed.), Vol. 1, p. 403, Academic Press, New York (1966).
10. E. A. Zeller, in *The Enzymes* (P. D. Boyer, H. Lardy and K. Myrbäck, eds.), 2nd ed. Vol. 8, p. 313, Academic Press, New York (1963).
11. H. Blaschko, D. Richter and H. Schlossmann, The inactivation of adrenaline, *J. Physiol.* **90**:1 (1937).
12. D. Richter, Adrenaline and amine oxidase, *Biochem. J.* **31**:2022 (1937).
13. H. Blaschko, D. Richter and H. Schlossmann, The oxidation of adrenaline and other amines, *Biochem. J.* **31**:2187 (1937).
14. C. E. M. Pugh and J. H. Quastel, Oxidation of aliphatic amines by brain and other tissues, *Biochem. J.* **31**:286 (1937).
15. J. H. Quastel and A. H. M. Wheatley, The effects of amines on oxidations by the brain, *Biochem. J.* **27**:1609 (1933).
16. C. E. M. Pugh and J. H. Quastel, Oxidation of amines by animal tissue, *Biochem. J.* **31**:2306 (1937).
17. P. J. G. Mann and J. H. Quastel, Benzedrine and brain metabolism, *Nature* **144**:943 (1939).
18. P. J. G. Mann and J. H. Quastel, Benzedrine (β-phenylisopropylamine) and brain metabolism, *Biochem. J.* **34**:414 (1940).
19. C. T. Beer and J. H. Quastel, Effects of aliphatic aldehydes on the respiration of rat brain cortex slices and rat brain mitochondria, *Canad. J. Biochem.* **36**:531 (1958).
20. M. Guggenheim and W. Loeffler, Das Schicksal proteinogener Amine in Tierkörpen, *Biochem. Zeit.* **72**:325 (1915).
21. Amine oxidase inhibitors, *Ann. N. Y. Acad. Sc.* **80**:551 (1959).
22. E. A. Zeller, Uber den enzymatischen Abbau von Histamin und Diaminen, *Helv. Chim. Acta* **21**:880 (1938).

23. E. A. Zeller, *in Advances in Enzymology* Vol. 2, p. 93, Academic Press, New York (1942).
24. H. Blaschko and J. Hawkins, Enzymic oxidation of aliphatic diamines, *Brit. J. Pharm. Chemotherap.* **5**:625 (1950).
25. H. Blaschko, P. J. Friedman, R. Hawes and K. Nilsson, The amine oxidases of mammalian plasma, *J. Physiol.* **145**:384 (1959).
26. J. H. Fellman, The oxidation of 3-4-dihydroxyphenylacetaldehyde, *Biochem. Biophys. Acta* **35**:530 (1959).
27. I. J. Kopin and J. Axelrod, 3-4-Dihydroxphenylglycol, a metabolite of epinephrine, *Arch. Biochem. Biophys.* **89**:148 (1960).
28. M. Goldstein, A. J. Friedhoff, S. Pomerantz and C. Simmons, The characterisation of a new metabolite of dopamine, *Biochim. Biophys. Acta* **39**:189 (1960).
29. H. Blaschko and I. Hellmann, Pigment formation from tryptamine and 5-hydroxytryptamine in tissues, *J. Physiol.* **122**:419 (1953).
30. I. Arioka and H. Tanimukai, Histochemical studies on monoamine oxidase in the midbrain of the mouse, *J. Neurochem.* **1**:311 (1957).
31. H. Blaschko and F. J. Philpot, Enzymic oxidation of tryptamine derivatives, *J. Physiol.* **122**:403 (1953).
32. E. D. Titus and S. Udenfriend, Metabolism of 5-hydroxytryptamine, *Fed. Proc.* **13**:411 (1954).
33. S. Udenfriend, E. D. Titus, H. Weissbach and R. E. Peterson, Biogenesis and metabolism of 5-hydroxyindole compounds, *J. Biol. Chem.* **219**:335 (1956).
34. G. G. Glenner, H. J. Burtner and G. W. Brown, Histochemical demonstration of monoamine oxidase activity by tetrazolium salts, *J. Histochem. Cytochem.* **5**:591 (1957).
35. G. G. Glenner, H. Weissbach and B. G. Redfield, Histochemical demonstration of enzymic activity by a nonenzymatic redox reaction. Reduction of tetrazolium salts by indoleacetaldehyde, *J. Histochem. Cytochem.* **8**:258 (1960).
36. V. T. Nachmias, Amine oxidase and 5-hydroxytryptamine in developing brain, *J. Neurochem.* **6**:99 (1960).
37. G. B. Koelle and A. T. Valk, Physiological implications of the histochemical localisation of monoamine oxidase, *J. Physiol.* **126**:434 (1954).
38. B. Century and K. L. Rupp, Comment on microfluorimetric determination of monoamine oxidase. *Biochem. Pharmacol.* **17**:2012 (1968).
39. S. Udenfriend and J. R. Cooper, The chemical estimation of tyrosine and tyramine, *J. Biol. Chem.* **196**:227 (1952).
40. S. Udenfriend, H. Weissbach and B. B. Brodie *in Methods of Biochemical Analysis* (D. Glick, ed.), Vol. 6, p. 95, Interscience Publ. Inc., New York (1958).
41. A. L. Green and T. M. Haughton, A colorimetric method for the estimation of monoamine oxidase, *Biochem. J.* **78**:172 (1961).
42. J. Southgate, E. C. G. Grant, W. Pollard, J. Pryse-Davies and M. Sandler, Cyclical variations in endometrial monoamine oxidase; correlation of histochemical and quantitative biochemical assays, *Biochem. Pharm.* **17**:721 (1968).
43. A. Sjoerdsma, Techniques for measuring monoamine oxidase inhibiting activity in man, *J. Neuropsychiat.* **2**: Supp. 1 ;159 (1961).
44. A. Sjoerdsma, L. Gillespie and S. Udenfriend, A method for the measurement of monoamine oxidase inhibition in man: application to studies on hypertension, *Ann. N. Y. Acad. Sci.* **80**:969 (1959).
45. J. A. Oates and P. Zaltzman, Amine oxidase inhibitors, *Ann. N.Y. Acad. Sci.* **80**:977 (1959).
46. O. Resnick, *Neuropharmacology* (H. A. Abramson, ed.), p. 81, Trans. 5th Conference. J. Macy Jr. Foundation (1960).
47. B. C. R. Strömblad, Observations on amine oxidase in human salivary glands, *J. Physiol.* **147**:639 (1959).

48. R. Håkanson and C. Owman, Pineal DOPA decarboxylase and monoamine oxidase activities as related to monoamine stores, *J. Neurochem.* **13**:597 (1966).
49. G. I. Klingman and J. D. Klingman, Monoamine oxidase activity in peripheral organs and adrenergic tissues of the rat, *Biochem. Pharmacol.* **15**:77 (1966).
50. L. Stjärne, R. H. Roth and N. J. Giarman, Microsomal monoamine oxidase in sympathetically innervated tissues, *Biochem. Pharmacol.* **17**:2008 (1968).
51. D. F. Bogdanski and S. Udenfriend, Serotonin and monoamine oxidase in brain, *J. Pharmacol.* **116**:7 (1956).
52. D. F. Bogdanski, H. Weissbach and S. Udenfriend, The distribution of serotonin, 5-hydroxytryptophan decarboxylase and monoamine oxidase in brain, *J. Neurochem.* **1**:272 (1957).
53. T. R. Shanthaveerapa and G. H. Bourne, Monoamine oxidase distribution in the rabbit eye, *J. Histochem. Cytochem.* **12**:281 (1964).
54. M. Härkonen, A. Mustakallio and M. Niemi, Monoamine oxidase activity in peripheral nerve myelin, *J. Neurochem.* **13**:269 (1966).
55. N. Shimizu, N. Morikawa and M. Okada, Histochemical studies of monoamine oxidase in the brains of rodents, *Z. Zellforsch.* **49**:389 (1959).
56. S. L. Manocha and G. H. Bourne, Histochemical mapping of monoamine oxidase and lactic dehydrogenase in the pons and mesencephalon of squirrel monkey, *J. Neurochem.* **13**:1047 (1966).
57. N. Robinson, Histochemistry of monoamine oxidase in the developing rat brain, *J. Neurochem.* **14**:1083 (1967).
58. G. C. Cotzias and V. P. Dole, Metabolism of amines I. Microdetermination of monoamine oxidase in tissues, *J. Biol. Chem.* **190**:665 (1951).
59. M. H. Aprison, R. Takahashi and T. L. Folkerth, Biochemistry of the avian central nervous system. I. Monoamine oxidase etc. in several discrete areas of the pigeon brain, *J. Neurochem.* **11**:341 (1964).
60. G. C. Cotzias and V. P. Dole, Metabolism of amines II: Mitochondrial localisation of monoamine oxidase, *Proc. Soc. Exp. Biol. Med.* **78**:157 (1951).
61. J. Hawkins, The localisation of amine oxidase in the liver cell, *Biochem. J.* **50**:577 (1952).
62. H. Blaschko, J. M. Hagen and P. Hagen, Mitochondrial enzymes and chromaffin granules, *J. Physiol.* **139**:316 (1957).
63. G. R. de Lores Arnaiz and E. D. P. De Robertis, Cholinergic and non-cholinergic nerve endings in the rat brain II. Subcellular localisation of monoamine oxidase and succinic dehydrogenase, *J. Neurochem.* **9**:503 (1962).
64. S. G. A. Alivisatos, F. Ungar, S. S. Parmar and P. K. Seth, Monoamine oxidase dependent labelling *in vivo* of mouse brain by ^{14}C-serotonin, *Biochem. Pharmacol.* **17**:1993 (1968).
65. B. Smith, Monoamine oxidase in the pineal gland, neurohypophysis and brain of the albino rat, *J. Anat. (London)* **97**:81 (1963).
66. A. N. Davison, Physiological role of monoamine oxidase, *Physiol. Rev.* **38**:729 (1958).
67. B. Jarrott and L. Iverson, Subcellular distribution of monoamine oxidase in rat liver and vas deferens, *Biochem. Pharmacol.* **17**:1619 (1968).
68. P. Laduron and F. Belpaire, Tissue fractionation and catecholamines II. Monoamine oxidase, tyrosine hydroxylase, dopa decarbosylase and dopamine hydroxylase in adrenal medulla, *Biochem. Pharmacol.* **17**:1127 (1968).
69. E. A. Zeller, J. Barsky and E. R. Berman, Amine oxidases XI. Inhibition of monoamine oxidase by 1-isonicotinyl-2-isopropylhydrazine, *J. Biol. Chem.* **214**:267 (1955).
70. H. Weissbach, B. G. Redfield and S. Udenfriend, Soluble monoamine oxidase; its properties and action on serotonin, *J. Biol. Chem.* **229**:953 (1957).
71. H. Blaschko and W. Jacobson, *in Cytology and Cell Physiology* (G. H. Bourne, ed.), p. 190, Clarendon, Oxford (1942).

72. G. C. Cotzias, I. Serlin and J. J. Greenough, Preparation of soluble monoamine oxidase, *Science* **120**:144 (1954).

73. S. H. Barondes, The influence of neuroamines on the oxidation of glucose by the anterior pituitary I. The role of monoamine oxidase, *J. Biol. Chem.* **237**:204 (1962).

74. W. Lovenberg, R. J. Levine and A. Sjoerdsma, A sensitive assay of monoamine oxidase *in vitro*. Application to heart and sympathetic ganglia, *J. Pharmacol. Exp. Therap.* **135**:7 (1962).

75. R. J. Wurtman and J. Axelrod, A sensitive and specific assay for the estimation of monoamine oxidase, *Biochem. Pharmacol.* **12**:1439 (1963).

76. S. L. Manocha and G. H. Bourne, Histochemical mapping of lactic dehydrogenase and monoamine oxidase in the medulla oblongata and cerebellum of the squirrel monkey, *J. Neurochem.* **15**:1033 (1968).

77. B. B. Brodie and P. A. Shore, *in Hormones, Brain Function and Behaviour* (H. Hoagland, ed.), p. 161, Academic Press, New York (1957).

78. N. Karki, R. Kuntzman and B. B. Brodie, Storage, synthesis and metabolism of monoamine oxidase in the developing rat brain, *J. Neurochem.* **9**:53 (1962).

79. N. Robinson, Histochemistry of rat brain stem monoamine oxidase during maturation, *J. Neurochem.* **15**:1151 (1968).

80. F. Bernheim and M. L. C. Bernheim, The oxidation of mescaline and certain other amines, *J. Biol. Chem.* **123**:317 (1938).

81. K. F. Gey and A. Pletscher, Activity of monoamine oxidase in relation to the 5-hydroxy-tryptamine and norepinephrine content of rat brain, *J. Neurochem.* **6**:239 (1961).

82. W. J. Novick, Effect of oxygen tension on monoamine oxidase activity, *Biochem. Pharmacol.* **15**:1009 (1966).

83. L. O. Randall, Oxidation of phenylethylamine derivatives by amine oxidase, *J. Pharmacol.* **88**:216 (1946).

84. L. M. Gunne and J. Jonsson, On the occurrence of tyramine in rabbit brain, *Acta Physiol. Scand.* **64**:434 (1965).

85. K. J. Beyer, The relation of molecular configuration to the rate of deamination of sympathomimetic amines, *J. Pharmacol & Exp. Therap.* **79**:85 (1943).

86. E. A. Zeller, P. Stern and L. A. Blanksma, Degradation of histamine by monoamine oxidase, *Naturwissenschaften* **43**:157 (1956).

87. Y. Kobayashi, A histamine metabolizing enzyme system of mouse liver, *Arch. Biochem. Biophys.* **71**:352 (1957).

88. S. E. Lindell and H. Westling, Enzymic oxidation of some substances related to histamine, *Acta Physiol. Scand.* **39**:370 (1957).

89. Y. Kobayashi and R. W. Schayer, *In vitro* demonstration of a new histamine metabolizing enzyme, *Arch. Biochem. Biophys.* **63**:480 (1956).

90. Z. Rothschild and R. W. Schayer, Synthesis and metabolism of a histamine metabolite, 1-methyl-4-(β aminoethyl)imidazole, *Biochim. Biophys. Acta* **30**:23 (1958).

91. H. Blaschko and T. L. Chrusciel, Observations on the substrate specificity of amine oxidase, *Brit. J. Pharmacol.* **14**:364 (1959).

92. L. C. Clark, F. Bennington and R. D. Morin, The effects of ring methoxy groups on biological deamination of phenylethylamines, *J. Med. Chem.* **8**:353 (1965).

93. A. C. Dahlberg and F. U. Kosikowsky, Influence of temperature of ripening on the tyramine content of American cheddar cheese, *J. Dairy Science* **32**:316 (1949).

94. D. H. Tedeschi and E. J. Fellows, Monoamine oxidase inhibitors: augmentation of pressor effects of peroral tyramine, *Science* **144**:1225 (1964).

95. B. Blackwell, Tranylcypromine, *Lancet* **I**:414 (1963). 95a.; B. Blackwell and E. Marley, Interaction between cheese and monoamine oxidase inhibitors in rats, *Lancet* **I**:530 (1964).

96. J. M. Cuthill, A. B. Griffiths and D. E. B. Powell, Death associated with tranylcypromine and cheese, *Lancet* **I**:1076 (1964).

97. M. Guggenheim, *Les Amine Biologiques*, p. 63, J. B. Bailliere et fils, Paris (1934).
98. A. M. Asatoor, A. J. Levi and M. D. Milne, Tranylcypromine and cheese, *Lancet* II:733 (1963).
99. J. V. Hodge, E. R. Nye and G. W. Emerson, Monoamine oxidase inhibitors, broad beans and hypertension, *Lancet* I:1108 (1964).
100. P. L. McGeer, J. E. Boulding, W. C. Gibson and J. G. Foulkes, Drug induced extrapyramidal reactions. Treatment with diphenylhydramine and dihydroxyphenylalanine, *J. Am. Med. Assn.* 177:665 (1961).
101. S. M. Hess, B. G. Redfield and S. Udenfriend, The effect of monoamine oxidase inhibitors and of tryptophan on the tryptamine content of animal tissues and urine, *J. Pharmacol. Exp. Therap.* 127:178 (1959).
102. S. Udenfriend, *in Neuropharmacology* (H. A. Abrahamson, ed.), p. 95, Trans. 5th Conf. J. Macy Jr. Foundation, (1960).
103. A. Sjoerdsma, J. A. Oates, P. Zaltzman and S. Udenfriend, Identification and assay of urinary tryptamine. Application as an index of monoamine oxidase inhibition in man, *J. Pharmacol. Exp. Therap.* 126:217 (1959).
104. A. Coppen, D. M. Shaw and J. P. Farrell, Potentiation of the antidepressive effect of a monoamine oxidase inhibitor by tryptophan, *Lancet* I:79 (1963).
105. D. H. Fram and J. P. Green, The presence of methylhistamine in the brain, *Pharmacologist* 5:253 (1963).
106. T. L. Perry, S. Hansen, J. G. Foulkes and G. M. Ling, Aliphatic and aromatic amines of cat brain, *J. Neurochem.* 12:397 (1965).
107. D. H. Fram and J. P. Green, Methylhistamine in guinea pig brain, *J. Neurochem.* 15:597 (1968).
108. J. P. Green and D. H. Fram, Methylhistamine in urine and brain, *Fed. Proc.* 25:559 (1966).
109. M. Ozaki, H. Weissbach, A. Ozaki, B. Witkop and S. Udenfriend, Monoamine oxidase inhibitors and procedure for their evaluation *in vivo* and *in vitro*, *J. Med. Pharmacol. Chem.* 2:591 (1960).
110. A. Pletscher, P. Zeller and K. F. Gey, *in Progress in Drug Research* (E. Jucker, ed.), Vol. 2, p. 417, Birkhauser, Basle (1960).
111. E. A. Zeller, *in Metabolic Inhibitors* (R. M. Hochster and J. H. Quastel, eds.), Vol. 2, p. 53, Academic Press, New York (1963).
112. M. E. Greig, R. A. Walk and A. J. Gibbons, The effect of three tryptamine derivatives on serotonin metabolism *in vitro* and *in vivo*, *J. Pharmacol.* 127:110 (1959).
113. D. F. Sharman and S. E. Smith, The effect of α-methyldopa on the metabolism of 5-hydroxytryptophan in rat brain, *J. Neurochem.* 9:403 (1962).
114. R. W. Schayer, K. Y. T. Wu, R. L. Smiley and Y. Kobayashi, Studies on monoamine oxidase in intact animals, *J. Biol. Chem.* 210:259 (1954).
115. U. S. von Euler and S. Hellner-Björkman, Effect of amine oxidase inhibitors on the noradrenaline and adrenaline content of cat organs, *Acta Physiol. Scand.* 33: Suppl. 118:21 (1955).
116. S. Udenfriend, B. Witkop, B. G. Redfield and H. Weissbach, Studies with reversible inhibitors of monoamine oxidase: harmaline and related compounds, *Biochem. Pharmacol.* 1:160 (1958).
117. K. Freter, H. Weissbach, B. G. Redfield, S. Udenfriend and B. Witkop, Oxindole analogs of 5-hydroxytryptamine and tryptophan as inhibitors of the biosynthesis and breakdown of serotonin, *J. Am. Chem. Soc.* 80:983 (1958).
118. A. Pletscher and H. Besendorf, Antagonism between harmaline and long acting monoamine oxidase inhibitors concerning their effects on 5-hydroxytryptamine and norepinephrine metabolism in brain, *Experientia* 15:25 (1959).
119. E. A. Zeller, J. Barsky, J. R. Fouts, W. F. Kirshheimer and L. S. Van Orden, Influence of

isonicotinic hydrazide and isonicotinylisopropylhydrazine on bacterial and mammalian enzymes, *Experientia* **8**:349 (1952).

120. E. A. Zeller and J. Barsky, *In vivo* inhibition of liver and brain monoamine oxidase by isonicotinylisopropylhydrazine, *Proc. Soc. Exp. Biol. Med.* **81**:459 (1952).

121. A. N. Davison, The mechanism of the irreversible inhibition of rat liver monoamine oxidase by iproniazid, *Biochem. J.* **67**:316 (1957).

122. J. Barsky, W. L. Pacha, S. Sarkar and E. A. Zeller, Amine oxidases XVII. Mode of action of 1-isonicotinyl-2-isopropylhydrazine on monoamine oxidase, *J. Biol. Chem.* **234**:389 (1959).

123. M. A. Schwartz, Monoamine oxidase inhibition by isocarboxazid, *J. Pharmacol. Exp. Therap.* **135**:1 (1962).

124. T. E. Smith, H. Weissbach and S. Udenfriend, Studies on monoamine oxidase. The mechanism of inhibition by iproniazid, *Biochemistry* **2**:746 (1963).

125. A. Pletscher, in *Proc. 4th Intern. Congress of Biochemistry, Vienna*, p. 124, Pergamon Press, Oxford (1959).

126. A. Pletscher, P. A. Shore and B. B. Brodie, Serotonin as a mediator of reserpine action in brain, *J. Pharmacol.* **116**:84 (1956).

127. A. R. Maass and M. J. Nimmo, A new inhibitor of serotonin metabolism, *Nature* **184**:547 (1959).

128. S. Sarkar, R. Banerjee, M. S. Ise and E. A. Zeller, Über die Wirkung von phenylcyclopropyl-aminen auf die monamin-oxydase und andere Enzyme Systeme, *Helv. Chim. Acta.* **43**:439 (1960).

129. W. Pollin, P. V. Cardon and S. S. Kety, Effects of aminoacid feedings in schizophrenic patients treated with iproniazid, *Science* **133**:104 (1961).

130. G. G. Brune and H. E. Himwich, Effects of methionine loading on the behaviour of schizophrenic patients, *J. Nerv. Ment. Dis.* **134**:447 (1962).

131. L. C. Park, R. J. Baldessarini and S. S. Kety, Methionine effects on chronic schizophrenics, *Arch. Gen. Psychiat.* **12**:346 (1965).

132. H. H. Berlet, K. Matsumoto, G. R. Pscheidt, J. Spaide, C. Bull and H. E. Himwich, Biochemical correlates of behaviour in schizophrenic patients receiving tryptophan or methionine with a monoamine oxidase inhibitor, *Arch. Gen. Psychiat.* **13**:521 (1965).

133. Y. Kakimoto, I. Sano, A. Kanazawa, T. Tsujio and Z. Kaneko, Metabolic effects of methionine in schizophrenic patients pretreated with a monoamine oxidase inhibitor, *Nature* **216**:1110 (1967).

134. F. Bernheim, The effect of propamidine on bacterial metabolism, *Science* **98**:223 (1943).

135. H. Blaschko and R. Duthie, The inhibition of amine oxidase by amidines, *Biochem. J.* **39**:347 (1945).

136. J. H. McNeill and B. E. Riedel, Effects of phenelzine on serotonin, noradrenaline and monoamine oxidase in the rat, *Canad. J. Physiol. Pharmacol.* **42**:33 (1964).

137. J. H. Gogerty and A. Horita, A comparison of the *in vivo* inhibitors of brain and liver monoamine oxidase by phenylisopropylhydrazine and iproniazid, *J. Pharmacol* **129**:357 (1960).

138. A. Horita and W. R. McGrath, Specific liver and brain monoamine oxidase inhibition by alkyl- and arylalkylhydrazines, *Proc. Soc. Exp. Biol. Med.* **103**:753 (1960).

139. A. Horita, Influence of drug-tissue interactions and the inhibition of monoamine oxidase by pheniprazine and iproniazid, *J. Pharmacol. Exp. Therap.* **142**:141 (1963).

140. H. Green, J. L. Sawyer, R. W. Erickson and L. Cook, Effect of repeated oral administration of monoamine inhibitors on rat brain amines, *Proc. Soc. Exp. Biol. Med.* **109**:347 (1962).

141. N. I. Lukshina, Effect of benactyzine on cat brain and liver monoamine oxidase activity, *Voprosy Meditsinskoi Khimii* **8**:256 (1962).

142. E. J. Davis and R. S. de Ropp, Metabolic origin of urinary methylamine in the rat, *Nature* **190**:636 (1961).

143. J. D. Taylor, A. A. Wykes, Y. C. Gladish and W. B. Martin, New inhibitor of monoamine oxidase, *Nature* **187**:941 (1960).

144. H. Green and R. W. Erickson, Effect of trans-2-phenylcyclopropylamine upon norepinephrine concentration and monoamine oxidase activity of rat brain, *J. Pharmacol. Exp. Therap.* **129**:237 (1960).

145. L. Valzelli and S. Garattini, Biogenic amines in discrete brain areas after treatment with monoamine oxidase inhibitors, *J. Neurochem.* **15**:259 (1968).

146. A. Feldstein, H. Hoagland and H. Freeman, Monoamine oxidase psychoenergizers and tranquilisers, *Science* **130**:500 (1959).

147. R. Michel, R. Truchot, N. Autissier and B. Rosner, Synthesis and biological activity of various thyroxine and adrenergic derivatives VII. Influence of various iodinated phenolic derivatives as monoamine oxidase, *Biochem. Pharmacol.* **15**:1127 (1966).

148. J. P. Johnston, Some observations upon a new inhibitor of monoamine oxidase in brain tissue, *Biochem. Pharmacol.* **17**:1285 (1968).

149. R. J. Taylor, E. Markley and L. Ellenbogen, The inhibition of monoamine oxidase by styrylquinoliniums, *Biochem. Pharmacol.* **16**:79 (1967).

150. A. Giachetti and P. A. Shore, Monoamine oxidase inhibition in the adrenergic neuron by bretylium and other adrenergic neuronal blocking agents, *Biochem. Pharmacol.* **16**:237 (1967).

151. T. L. Sourkes and A. D'Iorio, *in Metabolic Inhibitors* (R. M. Hochster and J. H. Quastel, eds.), Vol. 2, p. 89, Academic Press, New York (1963).

152. H. Blaschko and J. M. Himms, Inhibition of amine oxidase and spermine oxidase by amidines, *Brit. J. Pharmacol.* **10**:451 (1955).

153. E. V. Heegaard and G. A. Alles, Inhibitors specificity of amine oxidase, *J. Biol. Chem.* **147**:505 (1943).

154. M. H. Wiseman and T. L. Sourkes, The effect of riboflavin and mepacrine on the metabolism of 5-hydroxytryptamine, *Biochem. J.* **78**:123 (1961).

155. M. H. Wiseman-Distler and T. L. Sourkes, The role of riboflavin in monoamine oxidase activity, *Canad. J. Biochem. Physiol.* **41**:57 (1963).

156. J. S. Schweppe, E. A. Zeller and M. Higgins, Interrelationships between enzymes and hormones VI. Influence of age, sex, adrenalectomy and cortisone acetate upon the concentration of monoamine oxidase in the livers of white rats, *Proc. Mayo Clinic* **26**:371 (1951).

157. M. Zile and H. A. Lardy, Monoamine oxidase activity in liver of thyroid-fed rats, *Arch. Biochem. Biophys.* **82**:411 (1959).

158. M. H. Zile, Effects of thyroxine and related compounds on monoamine oxidase activity, *Endocrinology* **66**:311 (1960).

159. S. Spector, D. Prockop, P. A. Shore and B. B. Brodie, Effect of iproniazid on brain levels of norepinephrine and serotonin, *Science* **127**:704 (1958).

160. P. Hagen and A. D. Welch, *in Recent Progress in Hormone Research*, Vol. 12, p. 27, Academic Press, New York (1956).

161. B. B. Brodie, S. Spector and P. A. Shore, Interaction of drugs with norepinephrine in the brain, *Pharmacol. Rev.* **11**:548 (1959).

162. W. G. Clark, *in Metabolic Inhibitors* (R. M. Hochster and J. H. Quastel, eds.), Vol. I, p. 340, Academic Press, New York (1963).

163. A. Horita, Some pharmacological aspects of monoamine oxidase regulators, *J. Neuropsychiatry* Supp. I.S. 141 (1961).

164. M. H. Aprison and C. B. Ferster, Neurochemical correlates of behaviour. II. Correlation of brain monoamine oxidase activity with behavioural changes, *J. Neurochem.* **6**:350 (1961).

165. M. H. Aprison and J. N. Hingtgen, Neurochemical correlates of behaviour. IV. Norepinephrine and dopamine in pigeon brain following injection of 5-hydroxytryptophan, *J. Neurochem.* **12**:959 (1965).

166. W. I. Cranston and G. Rosendorff, Central temperature regulation in the conscious rabbit after monoamine oxidase inhibition, *J. Physiol.* **193**:359 (1967).

167. W. Feldberg and V. J. Lotti, Body temperature responses to the monoamine oxidase inhibitor tranylcypromine, *J. Physiol.* **190**:203 (1967); W. Feldberg and V. J. Lotti, 167a, Temperature changes in the rat by monoamines and tranylcypromine, *J. Physiol.* **191**:35P (1967).

168. S. B. Gertner, Effects of monoamine oxidase inhibitors on ganglionic transmission, *J. Pharmacol. Exp. Therap.* **131**:223 (1961).

169. A. Ho-Van-Hap, L. M. Babineau and L. Berlinguet, Hormonal action on monoamine oxidase activity in rats, *Canad. J. Biochem.* **45**:355 (1967).

170. R. J. Levine, Assessment of drug induced inhibition of monoamine oxidase activity, *Biochem. Pharmacol.* **15**:1645 (1966).

171. R. F. Squires and J. B. Lassen, Some pharmacological and biochemical properties of γ-morpholino-butyrophenone, a new monoamine oxidase inhibitor, *Biochem. Pharmacol.* **17**:369 (1968).

172. R. F. Squires, Additional evidence for the existence of several forms of mitochondrial monoamine oxidase in the mouse, *Biochem. Pharmacol.* **17**:1401 (1968).

173. P. Hagen and N. Weiner, Enzymic oxidation of pharmacologically active amines, *Fed. Proc.* **18**:1005 (1959).

174. D. B. Hope and A. D. Smith, Distribution and activity of monoamine oxidase in mouse tissues, *Biochem. J.* **74**:101 (1960).

175. V. Z. Gorkin, Monoamine oxidase, *Pharmacol. Rev.* **18**:115 (1966).

176. H. Weissbach, W. Lovenberg, B. G. Redfield and S. Udenfriend, *In vivo* metabolism of serotonin and tryptamine. Effect of monoamine oxidase inhibition, *J. Pharmacol.* **131**:26 (1961).

177. B. Belleau and J. Moran, Deuterium isotope effects in relation to the chemical mechanism of monoamine oxidase, *Ann. N.Y. Acad. Sci.* **107**:822 (1963).

178. S. Udenfriend and C. R. Creveling, Localisation of dopamine-β-oxidase in brain, *J. Neurochem.* **4**:350 (1959).

179. W. P. Burkard, K. F. Gey and A. Pletscher, Diamine oxidase in the brain of vertebrates, *J. Neurochem.* **10**:183 (1963).

180. E. A. Zeller, Diamine oxidase III. Enzymic degradation of polyamines, *Helv. Chim. Acta.* **21**:1645 (1938).

181. W. Schuler, Zur Hemmung den Diaminoxydase (Histaminase), *Experientia* **8**:230 (1952).

182. S. E. Lindell, K. Nilsson, B. E. Roos and H. Westling, The effect of enzyme inhibitors on histamine catabolism in man, *Brit. J. Pharmacol.* **15**:351 (1960).

183. H. Blaschko, F. N. Fastier and T. Wajda, The inhibition of histaminase by amidines, *Biochem. J.* **49**:250 (1951).

184. P. A. Shore and V. H. Cohn, Comparative effects of monoamine oxidase inhibitors on monoamine oxidase and diamine oxidase, *Biochem. Pharmacol.* **5**:91 (1960).

185. J. G. Hirsch, Spermine oxidase: an amine oxidase with specificity for spermine and spermidine, *J. Exp. Med.* **97**:345 (1953).

186. H. Yamada and K. T. Yasunobu, Monoamine oxidase I. Purification, crystallisation and properties of plasma monoamine oxidase, *J. Biol. Chem.* **237**:1511 (1962).

187. C. M. McEwen, Human plasma monoamine oxidase, *J. Biol. Chem.* **240**:2003, 2011 (1965).

188. C. M. McEwen, K. T. Cullen and A. J. Sober, Rabbit serum monoamine oxidase, *J. Biol. Chem.* **241**:4544 (1966).

189. Z. Huszti and J. Borsy, Differences between amine oxidases deaminating mescaline and the structurally related 3,4-dimethoxyphenylethylamine, *Biochem. Pharmacol.* **15**:475 (1966).

190. P. J. G. Mann, Further purification and properties of the amine oxidase of pea seedlings, *Biochem. J.* **79**:623 (1961).

191. H. Blaschko and F. Buffoni, Pyridoxal phosphate as a constituent of the histaminase (benzylamine oxidase) of pig plasma, *Proc. Roy. Soc. Ser. B.* **163**:45 (1965).
192. C. de Marco, M. Coletta and G. Bombardieri, Inhibition of plasma monoamine oxidase by cysteamine, *Nature* **205**:176 (1965).
193. A. Pletscher, Wirking von isonicotinsaurehydraziden auf den 5-hydroxytryptamine stoffwechsel *in vivo*, *Helv. Physiol. Acta* **14**:C76 (1956).
194. N. J. Giarman and S. Schanberg, The intracellular distribution of 5-hydroxytryptamine in rat brain, *Biochem. Pharmacol.* **1**:301 (1958).
195. J. Axelrod, G. Hertting and R. W. Patrick, Inhibition of H³-norepinephrine release by monoamine oxidase inhibitors, *J. Pharmacol.* **134**:325 (1961).
196. B. Dubnick, G. A. Leeson and G. E. Phillips, Effects of monoamine oxidase inhibitors on brain serotonin of mice, *J. Neurochem.* **9**:299 (1962).
197. S. Spector, R. Kuntzman, P. A. Shore and B. B. Brodie, Evidence for release of brain amines by reserpine in presence of monoamine oxidase inhibitors, *J. Pharmacol. Exp. Therap.* **130**: 256 (1960).
198. A. Pletscher, H. Göschke, K. F. Gey and H. Thölen, Species differences in the action of monoamine oxidase inhibitors, *Medicina Exper.* **4**:113 (1961).
199. D. S. Robinson, W. Lovenberg, H. Keiser and A. Sjoerdsma, Effects of drugs on human blood platelet and plasma amine oxidase activity *in vitro* and *in vivo*, *Biochem. Pharmacol.* **17**:109 (1968).
200. K. F. Gey, A. Pletscher and W. Burkard, Effects of inhibitors of monoamine oxidase on various enzymes and on the storage of monoamines, *Ann. N.Y. Acad. Sci.* **107**:1147 (1963).
201. M. Kraml, A rapid microfluorimetric determination of monoamine oxidase, *Biochem. Pharmacol.* **14**:1684 (1965).
202. W. Lovenberg, E. Dixon, H. R. Keiser and A. Sjoerdsma, A comparison of amine oxidase activity in human skin, rat skin and rat liver; relevance to collagen cross linking, *Biochem. Pharmacol.* **17**:1117 (1968).
203. K. F. Tipton and I. P. C. Spires, The homogeneity of pig brain monoamine oxidase, *Biochem. Pharmacol.* **17**:2137 (1968).

Chapter 14

ION MOVEMENT

Robert Katzman

The Saul R. Korey Department of Neurology
Albert Einstein College of Medicine, Bronx, New York

I. INTRODUCTION

The ionic environment of the nervous system of mammals (and many other vertebrates) is maintained within narrow limits by special transport mechanisms—the blood–brain barrier system.[1,2] The molecular nature of these transport mechanisms is still poorly understood. However, there has accumulated in recent years a large body of data covering the kinetics of ion movement and exchange between blood, brain, and CSF based upon the use of radioisotopes.

The use of radioisotopes to measure ion movement has been criticized by Nims,[3] who pointed out that the movement between compartments of such tracers does not necessarily reflect what is occurring with the parent species. In fact, however, it has been shown that in the usual biological system, tracer studies are valid means of monitoring ion flux, both on theoretical grounds[4] and on empirical grounds.[5] What is important to recognize is that in the normal physiological situation, a steady state occurs in which there is no *net* movement of ions between blood, brain, and CSF. However, substantial *exchange* of ions occurs, which can be measured directly by use of isotopes. It has been shown that coefficients obtained from measurement of ion exchange can be used to predict reliably the net movement of the ion that occurs if the system is perturbed.[5]

It is evident that the brain is not a single compartment but is exceedingly complex, that the blood–brain relationship varies from region to region simply on the basis of capillary density, and that extracellular space, glia, and neurons should perhaps be treated as separate compartments. Moreover, the interfaces between cerebrospinal fluid and brain are also heterogeneous, consisting of pia-glial as well as ependymal surfaces. Nevertheless, in comparing the existing data on isotope exchange, it is convenient to use a simple three-compartment system consisting of blood, brain, and cerebrospinal fluid. Ion flux will be denoted by J (isotope flux by J^*), the units being quantity/unit time.

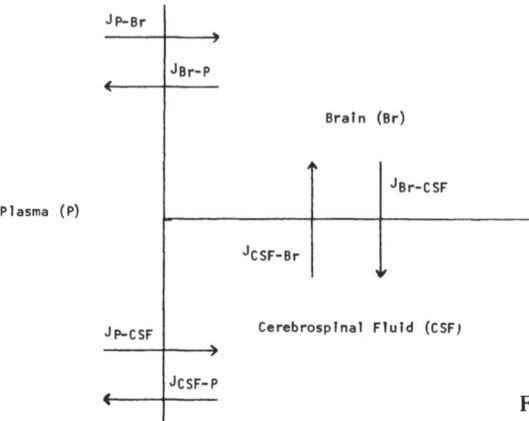

Fig. 1. Ion flux in a simple three-compartment system.

II. BLOOD–BRAIN BARRIER—RESTRICTION OF ION EXCHANGE

The terms "blood–brain barrier" and "blood-CSF barrier" have been applied to several different classes of observations. Most frequently these terms have been applied to the *restriction* in movement of an ion between blood and brain or CSF. Such restrictions may be a relatively low value of flux, that is, J_{P-Br} and J_{P-CSF}, as compared to flux of the ion across capillaries in other tissues. Moreover, in the case of the "blood–brain barrier" it must be shown that if the capillary wall is circumvented, e.g., by introducing the isotope into the CSF, then more rapid exchange occurs ($J_{P-Br} \ll J_{CSF-Br}$).

The data on ^{42}K exchange provides an excellent example of restricted exchange between blood and brain. When ^{42}K is injected intravenously, most of the injected dose is cleared from the blood stream within minutes, entering cells throughout the body.[6] Uptake by brain parenchyma is, however, extraordinarily slow, requiring 24 hr for half the brain K^+ to exchange (Fig. 2).[7] The low rate at which ^{42}K moves between blood and brain is in sharp contrast to the rate of its movement between an incubating media and a slice of brain tissue where exchange occurs in minutes.[8] The movement of ^{42}K into the brain from the capillary thus seems to be hindered, constituting a blood–brain barrier to K^+. That this more rapid movement into tissue slices is not entirely an artifact caused by tissue injury can be shown, if the blood–brain barrier is circumvented by introducing ^{42}K into CSF. Here, a rapid movement of the isotope into brain occurs.[5,7,9,10] This will be discussed later in the chapter.

The exchange of radioactive Na^+ and Cl^- between blood and brain is much faster than the exchange of ^{42}K, with the specific activity of the brain reaching half that of plasma in 2 or 3 hr.[11,12] However, even here the same phenomena are shown as with K^+, that is, the rate of uptake of both the Na^+ and Cl^- is slower than in other soft tissues. Moreover, the movement from

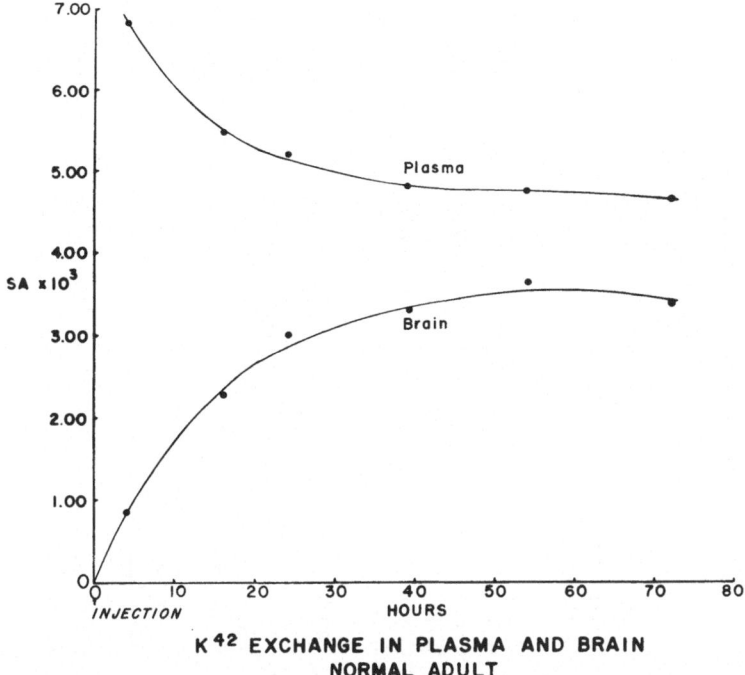

K⁴² EXCHANGE IN PLASMA AND BRAIN
NORMAL ADULT

Fig. 2. The specific activity of counts/min/meq is plotted as a function of hours after intraperitoneal injection of ^{42}K into normal adult rats. It can be seen that the specific activity of the brain does not reach half that of the plasma until 24 hr and that even at 72 hr exchange is not complete. From P. H. Leiderman and R. Katzman, *Am. J. Physiol.* **175**:271–275 (1953).

CSF into brain is more rapid than the movement from blood into brain.[13] The existence of a blood–brain barrier to Na^+ was shown more directly by Davson and Pollay[14] in experiments in which they demonstrated that during ventriculocisternal perfusions, movement of isotope from blood to brain was slower than from CSF to brain. The divalent ions differ in the time required for exchange, Ca^{2+} achieving half equilibrium in the brain 8 hr after parenteral injection,[15] whereas Mg^{2+} is almost as slow as K^+, requiring about 24 hr for half exchange.[16]

It should be noted that the rate of ion movement into brain is higher in immature animals. Thus, the time required for half equilibrium of brain Ca^{2+} with plasma in immature rats is $1\frac{1}{2}$–3 hr. In the case of K^+, the time of half exchange is 15 hr in 4-day-old rats as compared to the 24 hr in adult animals.

III. EQUATIONS FOR CALCULATION OF ION FLUX

It is possible to obtain quantitative estimates of the individual unidirectional fluxes; with such estimates the movement between plasma and

TABLE I Cation flux, plasma-brain J_{P-Br}

			Flux per unit weight meq/kilo/hr fresh weight	Cation content of brain meq/kilo/hr fresh weight	References
Potassium					
Rats					
Normal adult	plasma K = 5.48 meq/liter	Whole brain	2.89	100.9	44
4-day	10.84 meq/liter	Whole brain	3.90	84.94	44
Adrenalectomized	7.6 meq/liter	Whole brain	2.97	98.7	28
Rabbits					
Normal adult	4.15 meq/liter	Cortex	2.3[b]	99.6	29
	4.15 meq/liter	White[a]	1.7	93.3	29[b]
Low K diet	1.58 meq/liter	Cortex	1.4	96.7	29[b]
	1.58 meq/liter	White	1.2	94.7	29[b]
High K diet	7.09 meq/liter	Cortex	3.2	97.6	29[b]
	7.09 meq/liter	White	2.4	91.6	29[b]
Cats					
Normal adult		Cortex	1.9		
		White	1.5		
Sodium					
Rats					
Acidotic	Plasma pH 7.0–7.1	Whole brain	24		42[b]
Alkalotic	Plasma pH 7.5–7.6	Whole brain	28		42[b]
Chloride					
Rats					
Acidotic		Whole brain	17		
Normal		Whole brain	13		11[b]
Calcium					
Rats					
Adult		Whole brain	0.09		15[b]
Immature		Whole brain	0.40		15[b]
Magnesium					
Rats					
Adult		Whole brain	0.15		16[b]

[a] White matter from centrum semiovale.
[b] Data calculated from references.

brain and between CSF and brain can be compared. In addition, such information makes it easier to assess the effect on the ion transport of changes in ion concentration, presence of inhibitors, etc. If a more or less constant serum level can be maintained following intravenous infusion or intraperitoneal injection of isotope, then a simple two-compartmental analysis can give a good estimation of J_{P-Br} and J_{Br-P}. Here, the basic equation is

$$\frac{dC_{Br}^*}{dt} = k_{P-Br}\, C_P^* - k_{Br-P}\, C_{Br}^* \tag{1}$$

where C_P^* is the concentration of isotope in plasma and C_{Br}^* that in brain. This equation can be easily solved by graphical or analytic techniques.[8,17,18] The flux, J_{P-Br}, can be calculated as the product $k_{P-Br}\, C_P$, as discussed below. Examples of J_{P-Br} for the common cations are shown in Table I. However, if one is interested in transport into the CSF, because of the close relationship between CSF and brain, the system must be treated as a three-compartment one and the problem becomes mathematically more complex, although theoretically still tractable.[19] In such instances, interpreting the data in biological terms often becomes more difficult.[8,13] A major advance in such analysis was the application of the steady-state ventriculocisternal perfusion by Pappenheimer and his associates.[9,20,21]

In applying the ventriculocisternal perfusion technique to ion movement, two groups of animals are studied; in one, the isotope of interest is introduced into the perfusing fluid, and in the second, the isotope is maintained at a relatively constant level in the blood while the CSF system is perfused with isotope-free fluid. It is then possible to make reasonable estimates of each individual flux.[20,21]

When an isotope C* is added to the simulated spinal fluid, its efflux can be expressed as a clearance coefficient:[9,22,23]

$$k_{\text{efflux}}\, \overline{C}^* = (k_{CSF-P} + k_{CSF-Br})\, \overline{C}^* = CSF_i\, C_i^* - CSF_0 C_0^* - CSF_a C_0^* \tag{2}$$

where k_{efflux} is the milliliters of perfusate cleared of C*/min; $\overline{C}^*$ is the average concentration of C* in the CSF (taken arbitrarily as either the simple arithmetic average of C_i^* and C_0^* or as the exponential mean); CSF_i is the perfusion rate in ml/min; CSF_0 is the cisternal effluent in ml/min, and CSF_a is the measure of the perfusate absorbed elsewhere in the system as determined by the inulin clearance. This efflux coefficient can be separated into components k_{CSF-P} and k_{CSF-Br} representing the movement into plasma and brain, respectively, by determining the ratio of counts remaining in the brain and counts lost to the plasma during a recovery experiment following the isotopic perfusion.

Similarly, the movement of the isotope C* from plasma to CSF can be measured as a clearance coefficient of the plasma. Here, the ventriculocisternal system is perfused with a nonradioactive perfusate, while C* is perfused intravenously. Assuming that brain uptake of C^* from plasma is negligible, the following material balance holds

$$k_{P-CF}\, C_P^* = CSF_0\, C_0^* + CSF_a\, C_0^* + k_{efflux}\, \bar{C}^* \tag{3}$$

where k_{P-CSF} is the transport coefficient from plasma to CSF or clearance of $\bar{C}^*$ from plasma in ml/min. Cserr[9] in calculating the k_{P-CSF} separated this into two components, the first being due to the presence of the ion as part of the fluid secreted by the choroid plexus, and the second representing the molecular exchange of the ion across ependymal and pia-glial surfaces. It should be noted, however, that introducing acetazolamide into the perfusate may alter the rate of CSF without altering k_{P-CSF} proportionately.[9,7,22]

Fluxes can now be obtained from the transport coefficients by multiplying the transport coefficient by the concentration of isotope in the first compartment. Thus, the flux J_{P-CSF}^* is equal to the product $k_{P-CSF} \times C_P^*$. The unidirectional flux of the parent species is then determined by utilizing the general expression that $J_{12} = J_{12}^*/(S.A.)_1$, where J is the flux in moles/min, J_{12}^* is the flux of isotope in count rate/min, and $S.A.$ is the specific activity in count rate/mole (C^*/C). Then, $J_{P-CSF} = J_{P-CSF}^*/S.A._P$, assuming all of parent species are available for flux.

To determine J_{Br-CSF}^*, we use the fact that the brain is not appreciably labeled during the influx measurements. If our perfusate is initially free of both isotope and parent species, then the nonradioactive isotope appearing in the effluent can be treated as if it were a tracer coming from the brain. Here,

$$J_{Br-CSF}^* = J_{P-CSF}^* \left(1 - \frac{(S.A.)\ \text{effluent}}{(S.A.)\ \text{plasma}} \right) \Big/ \frac{S.A.\ \text{effluent}}{S.A.\ \text{plasma}} \tag{4}$$

Examples of cation flux into and out of CSF are calculated from these equations using the data obtained in perfused cats in Table II.

TABLE II
Exchange of Cations with Ventriculocisternal System of Perfused Cats[a]

	K^+	Mg^{2+}	Ca^{2+}
J_{P-Br}	1.25		
J_{P-CSF}	0.121	0.026	0.026
J_{Br-CSF}	0.178	0.013	0.015
J_{CSF-P}	0.039 $\bar{K}^+$	0.021 $\overline{Mg^{2+}}$	0.016 $\overline{Ca^{2+}}$
J_{CSF-Br}	0.057 $\bar{K}^+$	0.011 $\overline{Mg^{2+}}$	0.009 $\overline{Ca^{2+}}$

[a] Table 5 from R. Katzman, L. Graziani, and S. Ginsburg, in Brain-Barrier System (D. Ford and A. Lajtha, eds.) Elsevier, New York (in press).
J are in 10^{-6} mole/min.
The units of the coefficients are ml/min. $\bar{K}^+$, etc. average concentration of K^+ in CSF; Br, brain.

One aspect of the concept of the blood–brain barrier to ions can be tested using the data in Tables I and II. The term "blood–brain barrier" as we have used it implies that the movement across capillary epithelium is slower than the movement across the ependymal or pia-glial surfaces. To state this in another way, if the barrier from blood to brain is greater than the barrier from CSF to brain, then the flux J_{P-Br} per unit area of capillary membrane should be less than the flux J_{CSF-Br} per unit area of ependymal membrane. Estimates of capillary area and combined ependymal and pia-glial areas are necessarily crude, but conservative figures can be obtained by using those estimated for other species.[23-27] A low conservative estimate for the area of capillary epithelium in the cat would be 100 cm^2/g for combined gray and white matter or 2500 cm^2 for the cat brain, whereas a high estimate for the combined ependymal and pia-glial surfaces would be 50 cm^2. Hence, if K^+ flux is expressed in terms of these unit areas, we obtain a value of J_{P-Br} per square centimeter of capillary surface of 5×10^{-4} μmole/min/cm^2, whereas the flux from CSF into brain per square centimeter of surface is 35×10^{-4} μmole/min/cm^2. Thus, the movement from plasma into brain is only one-seventh as fast as from CSF into brain, and this is probably an under-estimate. Nevertheless, this ratio is less than that found between the uptake of whole brain and the uptake of tissue slices,[8] indicating that in the latter experiments injury plays a considerable role in the movement of ions.

The flux data included in Table II were derived from measurements of isotope exchange but were then applied to the prediction of movement of the parent species. The theoretical basis of such an application has already been discussed. A practical test of the validity of these flux data is now available from measurements of ion concentration changes occurring when the ventriculocisternal system is perfused with varying initial concentrations of K^+.[5,7] In Fig. 3, the concentration of K^+ in the cisternal effluent, K_0, is plotted against that of the perfusing fluid, K_{in}, and compared with the theoretical curves based upon the isotope flux data which predict a linear relationship between K_{in} and K_0. It is apparent that there is excellent agreement between the predicted and measured data. In addition, a good fit is obtained in animals in whom the perfusate contained 5×10^{-5} M ouabain. Here, there was an increase in J_{Br-CSF} and a decrease in the clearance coefficient, but the over-all linear relationship was not altered.

IV. MAINTENANCE OF CONSTANT IONIC ENVIRONMENT

The term "blood–brain barrier" has also been used to describe the observation that the concentration of metabolites in the brain (and in the CSF) is kept within narrow limits despite fluctuations in plasma levels.[1] The ions, K^+, Ca^{2+}, and Mg^{2+} certainly fit this concept. Elevation of serum K^+ to levels of 7 to 8 meq/liter either by bilateral adrenalectomy in rats[28] or by excessive feeding of K^+ to rabbits[29] does not alter the brain K^+ (Table I), while CSF K^+ was increased minimally but not significantly (from 2.83 $\pm$

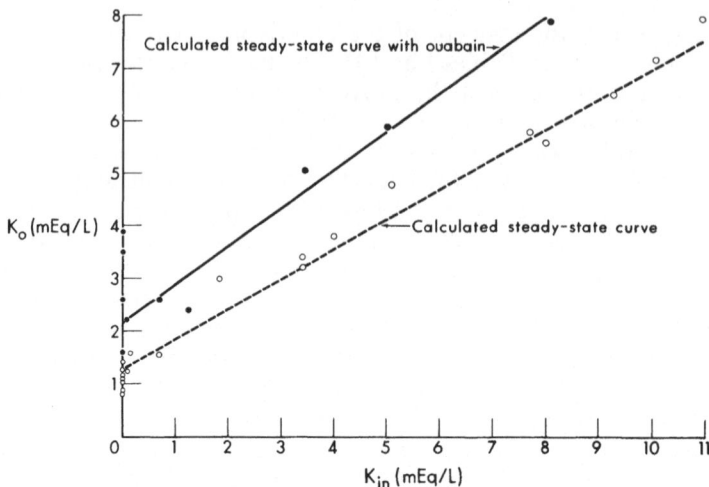

Fig. 3. The potassium concentration of the cisternal effluent K_0, as a function of the potassium concentration of the ventricular perfusate, K_{in}, in cats during steady-state ventriculocisternal perfusions. Open circles are the values for the control animals, the solid dots are the values for perfusates containing 5×10^{-5} M ouabain. The steady-state curves are calculated from Eq. (7) using coefficients obtained from experimental measurements of ^{42}K flux. [From R. Katzman, L. Graziani, and S. Ginsburg, *in Brain-Barrier Systems* (D. H. Ford and A. Lajtha, eds.) Elsevier, Amsterdam (1968).]

0.08 to 3.02 ± 0.12) in rabbits. Reduction of K^+ in the plasma by hypokalemic diets did not lead to any change in brain K^+ at a time when serum, muscle, and kidney levels were low (Table I),[30] and the drop in CSF K^+ in the rabbits was hardly perceptible (from 2.83 ± 0.08 to 2.72 ± 0.06). Similarly, the elevation of serum Ca^{2+} or Mg^{2+} by infusion, or in the case of Ca^{2+} by vitamin D intoxication or by administration of parathyroid hormone, produced either little or no change in brain and CSF levels.[31-39] Hypocalcemia has been produced by parathyroidectomy, and this in turn either leads to no change in CSF Ca^{2+} or minimal depression.[31]

In contrast, the brain concentrations of Na^+ and Cl^- when referred to the fresh weight of the brain, or the concentration of Na^+ and Cl^- in the CSF are both directly related to the serum concentration.[11,39] It is especially noteworthy that similar changes can be produced in the Na^+ and Cl^- levels of the brain following infusions of other hyperosmotic materials.[40] This close relationship appears to be due entirely to the osmotic movement of water, rather than to the movement of Na^+ and Cl^- per se.

V. MEDIATED FLUX

Of special interest are the mechanisms underlying the apparent constancy of brain cations. The restricted rate of influx described for these ions buffers the brain and CSF against transient changes in plasma levels. But how are brain and CSF concentrations maintained against longer standing alterations in plasma levels, especially those observed in experiments where the serum level is altered by adrenalectomy or parathyroidectomy, and also in patients with chronic electrolyte imbalance? One important element maintaining this homeostasis is the existence of mediated or concentration-independent flux. An excellent example of this is shown in the work on K^+ by Ames et al.[41] In these experiments, the choroid plexus fluid was sampled in animals in whom the lateral ventricle had been surgically exposed and filled with a transparent oil. Droplets of freshly formed fluid could then be sampled directly from the surface of the choroid plexus. Elevation of serum K^+ from 3 to 11 meq/liter in animals infused with added Ca^{2+} to prevent cardiac arrest showed no alteration in the choroid plexus fluid K^+. In animals made hypokalemic by infusion of glucose and insulin, there was some change in choroid plexus fluid K^+ with the decreased serum levels but again not a proportionate one. These results of Ames et al. are shown in Fig. 4. When the flux of a molecule is related to the concentration in this manner, it usually represents the presence of saturated carrier sites, implying carrier-mediated transport. Such results could also be obtained if there were restricted diffusion.

The demonstration of concentration-independent or mediated flux does not in itself prove the presence of active or energy-requiring transport. For the latter to be demonstrated, evidence of transport against an electrochemical gradient without counterflow must be adduced or other metabolic evidence must be found demonstrating the energy requirement of the transport. In the case of K^+ movement into CSF, this normally appears to be the gradient, whereas that of Mg^{2+} is against the gradient. However, the electrical

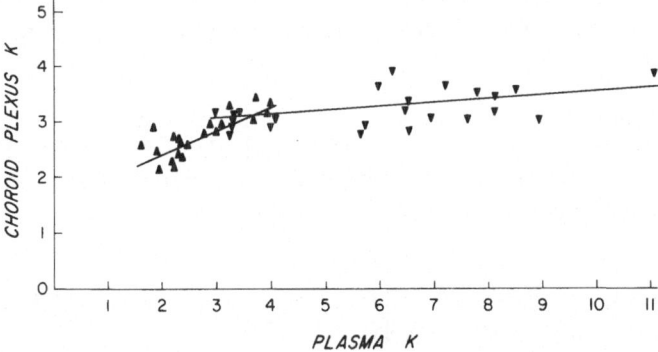

Fig. 4. Relation between choroid-plexus fluid K^+ and plasma K^+, before and after experimental induction of hypokalemia (seven cats) or hyperkalemia (eight cats). The two groups of experiments have been distinguished by different symbols and a separate regression line calculated for each. Concentrations in mmoles/kg H_2O. [From A. Ames, III, K. Higashi and F. B. Nesbett, J. Physiol. **181**:506–515 (1965).]

gradient between blood and CSF was found by Held *et al.*[42] to be determined by CSF K^+ concentration and by blood H-ion concentration. Altering these two factors to reverse the normally favorable electrochemical gradient does not change the rate of K^+ flux from plasma to CSF. Hence, it is quite probable that the transport mechanism involved has its own energy-requiring, driving force and is not dependent upon a favorable electrochemical gradient. However, the possibility of a compensating counterflow has not yet been ruled out.

The movement of Mg^{2+} from blood to CSF is almost entirely concentration independent.[16] Since this movement is also against an electrochemical gradient, it appears likely to involve carrier-mediated active transport. Calcium movement into CSF can be separated into two components: a concentration-independent and a concentration-dependent flux representing both carrier-mediated and diffusional components.[22]

The movement of Na ion may also be mediated by enzymatic processes as shown by the marked effect of acetazolamide on its flux.[43]

In a study of K^+ movement into brain in normal and adrenalectomized rats, Katzman and Leiderman[44] found no difference in K^+ flux in these two groups of animals, although the plasma K^+ concentration in the adrenalectomized animals was increased by almost 50% (Table I). However, Bradbury and Kleeman[29] in a study of rabbits in whom blood K^+ levels were markedly altered by feeding too much or too little K^+, found that the flux of K^+ into the brain varied with the plasma value, and two components could be discerned: a major flux component that was concentration independent, and a second concentration-dependent component having a regression coefficient of 0.09 to 0.28. It should be noted that whereas these two groups of investigators both used ^{42}K uptake to measure K^+ influx, Katzman and Leiderman[44] measured the uptake between 4 and 96 hr, whereas Bradbury and Kleeman[29] measured the uptake between 1 and 3 hr; moreover, the two groups calculated their flux data differently. Hence, it is uncertain whether the differences found between the adrenalectomized rats and K^+-fed rabbits represent a real difference or a methodological one.

Movement of Ca^{2+} and Mg^{2+} into brain tissue in the presence of varied serum-ion concentrations has been recently studied.[16,22,34] Again, the major portion of this flux appears to be independent of a serum concentration, but there is a small element that is proportional to the serum level.

VI. CARRIER-MEDIATED AND ACTIVE CLEARANCE OF ANIONS: BROMIDE AND IODIDE

Steady-state concentrations of Br^- in CSF have been studied for years in man, since Br^- was once a popular sedative. The fact that the CSF Br^- concentrations were lower than the plasma Br^- was described as a blood-CSF barrier to Br. It was found that the Br^-/Cl^- plasma concentration ratio changes in various pathological conditions (e.g., meningitis); this ratio was therefore considered an index of the integrity of the blood-CSF (and by analogy the blood–brain) barrier. Bromide levels are still occasionally used in this way. Wallace and Brodie[45] compared Br^-/Cl^- concentration ratios in brain, CSF, and various organs of animals given Br parenterally and found that this ratio was similar in brain and CSF but quite different from that in serum. The Br^-/Cl^- ratio in serum, however, was like that in other body organs such as lung, liver, etc. An identical phenomenon was found when I^- or CNS^- was administered; the I^-/Cl^- or CNS^-/Cl^- ratios were the same in CSF and brain but different from the rest of the body pool. In these three instances the ratios were less in the nervous system than elsewhere, as if Br^-, I^-, and CNS^- were systematically excluded.

In 1961, Becker[46] postulated that the low steady-state value of I^- in CSF was due to the active transport of I^- from CSF. He demonstrated that the choroid plexus of rabbit could accumulate ^{131}I in a concentration 30 times that of the media. This accumulation was temperature dependent and blocked by metabolic inhibitors. It is curious that the most potent of the metabolic inhibitors was ouabain, which lowered ^{131}I accumulation by 50% in a concentration of 10^{-7} M. In addition to the agents acting on cellular metabolism, anions which alter ^{131}I uptake by the thyroid were found effective. Thus, 50% inhibition of ^{131}I was produced by 2×20^{-6} M perchlorate, 5×10^{-6} M fluoroborate, or 3×10^{-5} M thiocyanate. Iodide in concentrations of 5×10^{-5} M inhibited ^{131}I uptake. Thus, transport mechanism showed both competitive inhibition and self-saturation. This active uptake by choroid plexus was confirmed by Welch.[27]

In addition, Becker[46] demonstrated *in vivo* that parenteral administration of perchlorate (3 mM/kg) or I (20 mM/kg) raised the steady-state ratio of CSF ^{131}I/plasma ^{131}I from 1.6% to 40%. Direct evidence that the effect of

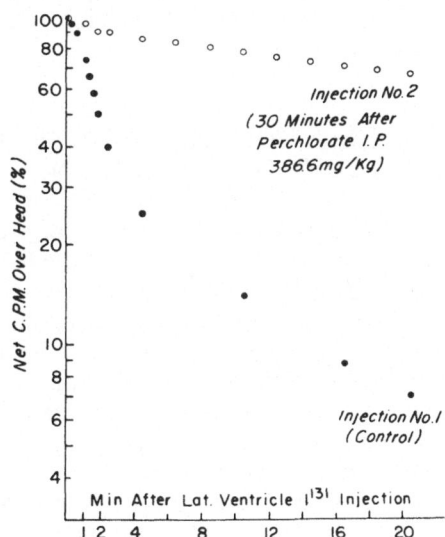

Fig. 5. Slowing of disappearance of ^{131}I from lateral ventricle with intraperitoneal perchlorate in the dog. [From L. A. Coben, J. D. Loeffler, and J. C. Elsasser, *Am. J. Physiol.* **206**:1373–1378 (1964).]

perchlorate was that of inhibiting ^{131}I clearance from CSF was provided by Coben,[47] who monitored the disappearance of ^{131}I from the head of dogs following intraventricular injection and demonstrated that parenteral injection of perchlorate inhibited this disappearance (Fig. 5). The active clearance of I^- from CSF has been confirmed by Davson and Pollay,[14] Reed and Woodbury,[48] and Reed et al.[49]

The evidence that Br^- is actively cleared by CSF and brain is not as straightforward as that obtained for I^-. Davson[50,51] found that ^{86}Br entered CSF as quickly as ^{24}Na, but then it reached a steady-state level half that of the plasma as if it were cleared. This problem was reinvestigated by Pollay and Curl,[26] who found the ratio of Br^- in brain as compared to plasma to be less than the CSF/plasma ratio. Moreover, this ratio gradually rose as the plasma Br^- was raised to levels of 20 mM. He interpreted these findings as suggesting that primary clearance occurred in the brain either directly or by metabolic conversion of the Br^- into its oxide which is then cleared by the blood. At high blood levels, the amount of Br^- entering the brain is sufficiently great to make the clearance mechanism less efficient.

VII. CATION EFFLUX

The mechanism of clearance of cations is not well understood. The efflux of Ca^{2+}, Mg^{2+}, and K^+ from CSF is directly proportional to the ion concentration in the CSF (Table II),[5,9,10,7,52,53] and this could be attributed to a simple diffusional mechanism. However, the marked reduction of K^+ clearance occurring with the addition of ouabain to the ventriculocisternal perfusates suggests the possibility that this clearance is dependent upon an underlying enzymatic process, but that here one is dealing with an enzyme that does not approach saturation within the wide limits of the ion concentrations studied.[9,7]

VIII. RELATIONSHIP OF CSF AND THE EXTRACELLULAR FLUID OF THE BRAIN

It is now widely assumed that a close equilibrium exists between the ions in the extracellular fluid of the brain and the CSF. This was originally suggested by the unique concentration ratios in brain and CSF of Br^-, and I^- to Cl^- all presumably extracellular.[45] The recent work on cations has been interpreted as supporting this hypothesis. The ionic composition in the CSF is quite different from that of the serum, particularly in respect to K^+ and Mg^{2+} concentrations. The rate of movement of ions such as ^{42}K from CSF to brain is greater than the rate of movement of the same ion from blood into brain or from blood into CSF. Two-thirds of ^{42}K cleared from the CSF can be found in brain parenchyma. Moreover, behavioral changes occur rapidly in unanesthetized animals when the ionic composition of the CSF is altered. Thus, during ventriculocisternal perfusion in rhesus monkeys with fluid containing an increased K^+ concentration, seizure activity occurred at a time when less than 1 μeq of K^+ had moved from the CSF into the adjoining tissue. Similarly, with perfusions of simulated CSF free of K^+, a sleeplike state is obtained in some primates, again at a time when 1 μeq of K^+ has been exchanged between the brain and the CSF.[54] However, such changes in excitability do not occur normally, since the CSF concentrations tend to be independent of fluctuations in the serum levels. Thus, it would seem to be considerably beneficial to the organism to have an extracellular ionic environment in the brain resembling CSF and maintained as constant as possible.

It must be emphasized that the supposition of an identity of CSF and extracellular fluid depends upon such indirect evidence. Moreover, there is evidence that indicates a close but dynamic equilibrium between these two fluids such that they must continually modify one another. For example, Ames et al.[41] found that the concentration of K^+ in the choroid plexus fluid, already different from that of plasma, underwent further modification reaching the cisterna magna, so that the concentration in the cisterna magna was consistently lower than that secreted by the choroid plexus. Bito and Davson[55] found the K^+ concentration of fluid drawn from the cortical subarachnoid space in cats to be 2.52 $\pm$ 0.06 meq/liter, lower than the value of 2.88 $\pm$ 0.08 in fluid drawn from cisterna magna. The dynamic equilibrium between CSF and brain extracellular fluid has received special emphasis by Davson and others who have suggested that the low levels of I^- found in brain are due to leakage into CSF, from whence it is cleared. The word "sink" has been used by Davson and his associates to describe this phenomenon.[10,27,55,56] This particular concept is still a controversial one.[57]

ACKNOWLEDGMENT

I wish to thank Drs. Ames and Coben for permission to reproduce their figures.

IX. REFERENCES

1. R. D. Tschirgi, in Handbook of Neurophysiology (J. Field, H. W. Magoun and V. E. Hall, eds.), Vol. III, pp. 1865–1890, American Physiological Society, Washington, D.C. (1960).
2. D. H. Ford and A. Lajtha, eds., in Brain-Barrier Systems, Elsevier, Amsterdam (1968).
3. L. F. Nims, in Head Injury (W. F. Caveness and A. E. Walker, eds.), Lippincott, London (1966).
4. O. Kedem and A. Essig, Isotope flows and flux ratios in biological membrane, J. Gen. Physiol. 48:1047–1070 (1965).
5. R. Katzman, L. Graziani and S. Ginsburg, in Brain-Barrier Systems (D. H. Ford and A. Lajtha, eds.), Elsevier, Amsterdam (1968).
6. G. G. Rudolph and N. S. Olsen, Transfer of potassium between blood, cerebrospinal fluid and other brain tissues, Am. J. Physiol. 186:157–160 (1956).

7. R. Katzman, L. Graziani, R. Kaplan and A. Escriva, Exchange of cerebrospinal fluid potassium with blood and brain, *Arch. Neurol.* **13**:513–524 (1965).
8. H. A. Krebs, L. V. Eggleston, and C. Terner, *In vitro* measurements of the turnover rate of K in brain and retina, *Biochem. J.* **48**:530–537 (1951).
9. H. Cserr, Potassium exchange between cerebrospinal fluid, plasma and brain, *Am. J. Physiol.* **209**:1219–1226 (1965).
10. M. W. B. Bradbury and H. Davson, The transport of potassium between blood, cerebrospinal fluid and brain, *J. Physiol.* **181**:151–174 (1965).
11. D. M. Woodbury, *in Biology of Neuroglia* (W. F. Windle, ed.), Charles C. Thomas, Springfield, Illinois, pp. 120–127 (1958).
12. R. Katzman and C. Wilson, Further observations on triethyltin edema, *A.M.A. Arch. Neurol.* **9**:178–187 (1963).
13. N. Olsen and G. G. Rudolph, Transfer of sodium and bromide ions between blood, cerebrospinal fluid and brain tissue, *Am. J. Physiol.* **183**:427–432 (1955).
14. H. Davson and M. Pollay, Influence of various drugs on the transport of I^{131} and PAH across the cerebrospinal fluid-blood barrier, *J. Physiol.* **167**:239–246 (1963).
15. L. Graziani and A. Escriva, in preparation.
16. S. Ginsburg, in preparation.
17. A. K. Solomon, Equations for tracer experiments, *J. Clin. Invest.* **28**:1297–1307 (1949).
18. J. S. Robertson, Theory and use of tracers in determining transfer rates in biological systems, *Physiol. Rev.* **37**:133–154 (1957).
19. H. E. Hart, Multi-compartment analysis of tracer experiments, *Ann. N.Y. Acad. Sci.* **188**:1–339 (1963).
20. J. R. Pappenheimer, S. R. Heisey, and E. F. Jordan, Active transport of diodrast and phenolsulfonphthalein from cerebrospinal fluid to blood, *Am. J. Physiol.* **200**:1–10 (1961).
21. S. R. Heisey, D. Held, and J. R. Pappenheimer, Bulk flow and diffusion in the cerebrospinal fluid system of the goat, *Am. J. Physiol.* **203**:775–781 (1962).
22. L. Graziani, A. Escriva, and R. Katzman, Exchange of calcium between blood, brain, and cerebrospinal fluid, *Am. J. Physiol.* **208**:1058–1064 (1965).
23. C. Crone, The permeability of capillaries in various organs as determined by use of the "indicator diffusion" method, *Acta Physiol. Scand.* **58**:292–305 (1963).
24. J. D. Fenstermacher and J. A. Johnson, Filtration and reflection coefficients of the rabbit blood–brain barrier, *Am. J. Physiol.* **211**:341–346 (1966).
25. E. A. Bering and O. Sato, Hydrocephalus: Changes in formation and absorption of cerebrospinal fluid within the cerebral ventricles, *J. Neurosurg.* **20**:1050–1063 (1963).
26. J. Pollay and F. Curl, Secretion of cerebrospinal fluid by the ventricular ependyma of the rabbit, *Am. J. Physiol.* **213**:1031–1038 (1967).
27. K. Welch, Secretion of cerebrospinal fluid by choroid plexus of the rabbit, *Am. J. Physiol.* **205**:617–624 (1963).
28. P. H. Leiderman and R. Katzman, Effect of adrenalectomy, desoxycorticosterone and cortisone on brain potassium exchange, *Am. J. Physiol.* **175**:271–275 (1953).
29. M. W. B. Bradbury and C. R. Kleeman, Stability of the potassium content of cerebrospinal fluid and brain, *Am. J. Physiol.* **213**:519–528 (1967).
30. R. Katzman, Effect of electrolyte disturbance on the central nervous system, *Ann. Rev. Med.* **17**:197–212 (1966).
31. H. H. Merritt and W. Bauer, The equilibrium between cerebrospinal fluid and blood plasma. IV. The calcium content of serum, cerebrospinal fluid, and aqueous humor at different levels of parathyroid activity, *J. Biol. Chem.* **90**:233–246 (1931).
32. F. K. Herbert, The total and diffusible calcium of serum and the calcium of cerebrospinal fluid in human cases of hypocalcemia and hypercalcemia, *Biochem. J.* **27**:1978–1991 (1933).
33. S. Morgulis and A. Perley, Studies on cerebrospinal fluid and serum calcium, with special reference to the parathyroid hormone, *J. Biol. Chem.* **88**:169–188 (1930).

34. M. Hilmy and G. G. Somjen, Distribution and tissue uptake of magnesium related to its pharmacological effects, *Am. J. Physiol.*, in press.

35. C. Pollis, I. McIntyre, and H. Anstall, Some observations on magnesium in cerebrospinal fluid, *J. Clin. Path.* **18**:762–784 (1965).

36. A. Kemeny, H. Goldizsar, and G. Pethes, The distribution of cations in plasma and cerebrospinal fluid following infusion of solutions of salts of sodium, potassium, magnesium and calcium, *J. Neurochem.* **7**:218–227 (1961).

37. R. J. Schain, Cerebrospinal fluid and serum cation levels, *Arch. Neurol.* **11**:330–333 (1964).

38. S. Wallach, J. B. Bellavia, J. Schorr, and P. J. Gamponia, Effect of vitamin D on tissue distribution and transport of electrolytes, ^{47}Ca and ^{28}Mg, *Endocrinology* **79**:773–782 (1966).

39. S. Wallach, J. V. Bellavia, J. Schorr, and D. L. Reizenstein, Tissue distribution of electrolytes Ca^{47} and Mg^{28} in acute hypercalcemia, *Am. J. Physiol.* **207**:553–560 (1964).

40. J. F. Sotos, P. Dodge, P. Meara, and N. Talbot, Studies in experimental hypertonicity. 1. Pathogenesis of the clinical syndrome, biochemical abnormalities and cause of death, *Pediatrics* **26**:925–938 (1960).

41. A. Ames, M. Sakanoue, and S. Endo, Na, K, Ca, Mg and Cl concentrations in choroid plexus fluid and cisternal fluid compared with plasma ultrafiltrate, *J. Neurophysiol.* **27**:672–681 (1964).

42. D. Held, V. Fencl, and J. R. Pappenheimer, Electrical potential of cerebrospinal fluid, *J. Neurospinal.* **27**:942–959 (1964).

43. H. Davson and C. P. Luck, The effect of acetazolamide on the chemical composition of the aqueous humour and cerebrospinal fluid of some mammalian species and on the rate of turnover of ^{24}Na in these fluids. *J. Physiol.* **137**:279–293 (1957).

44. R. Katzman and P. H. Leiderman, Brain potassium exchange in normal adult and immature rats, *Am. J. Physiol.* **175**:263–270 (1953).

45. G. B. Wallace and B. B. Brodie, The passage of bromide, iodide and thiocyanate into and out of the cerebrospinal fluid, *J. Pharmacol. Exp. Therap.* **68**:50–55 (1940).

46. B. Becker, Cerebrospinal fluid iodide, *Am. J. Physiol.* **201**:1149–1151 (1961).

47. L. A. Cohen, J. D. Loeffler, and J. C. Elsasser, Spinal fluid iodide transport in the dog, *Am. J. Physiol.* **206**:1373–1378 (1964).

48. D. J. Reed and D. M. Woodbury, Kinetics of movement of iodine, sucrose, inulin and radio-iodinated serum albumin in the central nervous system and cerebrospinal fluid of the rat, *J. Physiol.* **169**:816–850 (1963).

49. D. J. Reed, D. M. Woodbury, L. Jacobs, and R. Squires, Factors affecting distribution of iodine in brain and cerebrospinal fluid, *Am. J. Physiol.* **209**:757–764 (1965).

50. H. Davson, *Physiology of the Cerebrospinal Fluid*, p. 297, Little, Brown, Boston (1967).

51. H. Davson, *Physiology of the Ocular and Cerebrospinal Fluids*, Little, Brown, Boston (1956).

52. W. W. Oppelt, I. McIntyre, and D. P. Rall, Magnesium exchange between blood and cerebrospinal fluid, *Am. J. Physiol.* **205**:959–962 (1963).

53. W. W. Oppelt, E. D. Owens, and D. P. Rall, Calcium exchange between blood and cerebrospinal fluid, *Life Sci.* **2**:599–605 (1963).

54. R. Katzman, E. D. Weitzman, L. Graziani, and A. Escriva, Ventriculocisternal perfusion of Rhesus: Transport of Ca^{++} and K^+ with correlated behavioral changes, *Neurology* **14**:267 (1964).

55. L. Z. Bito and H. Davson, Local variations in cerebrospinal fluid composition and its relationship to the composition of the extracellular fluid of the cortex, *Exp. Neurol.* **14**:264–280 (1966).

56. W. H. Oldendorf and H. Davson, Brain extracellular space and the sink action of cerebrospinal fluid. Measurement of rabbit brain extracellular space using sucrose labeled with carbon 14, *Arch. Neurol.* **17**:196–205 (1967).

57. R. Katzman, H. Schimmel, and C. E. Wilson, Diffusion of inulin as a measure of extracellular fluid space in brain, *Leo M. Davidoff Festschrift*, Proc. Rudolf Virchow Medical Society, City of New York, Suppl. to Vol. 26, pp. 254–280, 1968.

Chapter 15

TRANSPORT OF MONOSACCHARIDES, AMINES, AND CERTAIN OTHER METABOLITES

K. D. Neame

Department of Physiology, University of Liverpool
Liverpool, England

I. INTRODUCTION

This chapter deals with the movement of certain metabolites into the brain and is only concerned with other processes such as binding to intracellular particles and metabolism insofar as these may affect or be misinterpreted as movement.

The entry of a substance into a tissue such as brain involves movement across diffusional barriers which may vary according to the route of administration. If injected into the blood, the substance has to pass through the capillary wall before entering the extracellular fluid or the cell. Movement may be complicated by variations in permeability of the membrane boundaries, and measured values can be affected by removal within the cell as a result of metabolism or binding.

When considering entry into the brain in particular, there is the added problem of a so-called blood–brain barrier, a term coined to indicate the apparently impeded entry of various substances into the brain. There has been considerable disagreement as to whether such a barrier is apparent rather than real,[1,2] and it has been claimed that the structures across which metabolites pass show permeability characteristics similar to those of other tissues.[3] A full discussion would be out of place here, but in view of its relevance to the problem of metabolite transport, a few comments are needed.

A blood–brain barrier must be thought of not as a structure affecting all substances similarly, but as a way of describing unexplained experimental phenomena related to the nature of the substance being investigated. It should be considered as having a definite entity only when adequate comparative studies between different tissues and between *in vivo* and *in vitro* techniques have been made. To illustrate this, two criteria which demonstrate the existence of a barrier mechanism are, first, inability to raise the concentration of a metabolite in brain above that in the blood when active transport of that metabolite by brain is known to exist, and, second, limitation of entry into the brain *in vivo* but no limitation *in vitro* when compared with entry into other tissues. It must of course be established that such differences are not due to other factors, such as metabolic breakdown. A difference in the rate of entry as between one tissue and another or between one metabolite and another is not, on its own, evidence of a barrier mechanism. For example, the presence of regional barriers within the brain itself has been inferred from local differences in entry of a metabolite, although such differences, which disappear on death,[4] could to a great extent be due to variations in local transport.[2]

The presence of a blood–brain barrier has been claimed for most of the metabolites to be discussed and in some instances such a claim appears to be valid. Energy-dependent transport systems may also be present at the same time, and it has been suggested in the case of amino acids[2] that an apparent barrier may to some extent be the manifestation of a balance between inward and outward transport.

II. MONOSACCHARIDES

Although the transport of sugars has been investigated in some detail in a number of tissues[5] such as intestine,[6,7] kidney[8] and erythrocyte,[9] there is less information available on their transport into brain tissue, particularly as regards detailed analysis of the kinetics involved.

There are two main groups of monosaccharides with respect to entry into the brain from the blood, one group (for which evidence of a carrier system has in most cases been demonstrated) entering the tissue considerably more readily than the other.[10,11] It is possible that the relationship between the structure of each sugar and its entry into brain may be similar to that in the human erythrocyte where there is also a comparable, though not so well-defined, division into two groups.[12] Here the affinity for a carrier system has been related by LeFevre and Marshall[12] to the three-dimensional conformation of the sugar molecule. They believe that "the 'carrier' system in the surface of the human erythrocyte, facilitating passage of monosaccharides through the cell membrane, reacts preferentially with those sugars in which the pyranose ring tends to assume the particular 'chair' shape designated by Reeves as the C_1 conformation. . . . Sugars predominantly stable in the other chair conformation ($1C$) showed extremely low affinity".[12] No affinity constants are available for the entry of sugars other than xylose[13] into the brain, but in most cases the ease of entry is related, as in the erythrocyte, to stability in the C_1 conformation (Table I). D-Ribose, however, does not at first sight appear to fit into the general pattern, but its conformation and affinity for the carrier system in the human erythrocyte are intermediate between the comparable properties for most of the other sugars. Its spatial conformation is also somewhat similar to that of L-arabinose, but these two sugars clearly enter brain differently.

When compared with movement across the intestine, on the other hand, there are certain differences. The only sugars listed in Table I which are also transported by intestine are D-glucose, 3-O-methylglucose, D-galactose, and D-xylose.[6,14] It must be emphasized that the nature of the transport process is not necessarily the same in different tissues. For instance, in the case of intestine but not the erythrocyte, certain sugars can move against a concentration gradient; in brain, only D-xylose has as yet been shown to do so.[13]

Some of the evidence upon which the transport data in Table I are based is shown in the upper half of Fig. 1 in which the amount of a sugar entering the brain from the blood is related to the height of the shaded column. (The left- and right-hand portions of the figure are derived from different types of experiment but can be indirectly related through the data given for L-arabinose and D-xylose.) It can be seen that the five sugars represented on the extreme

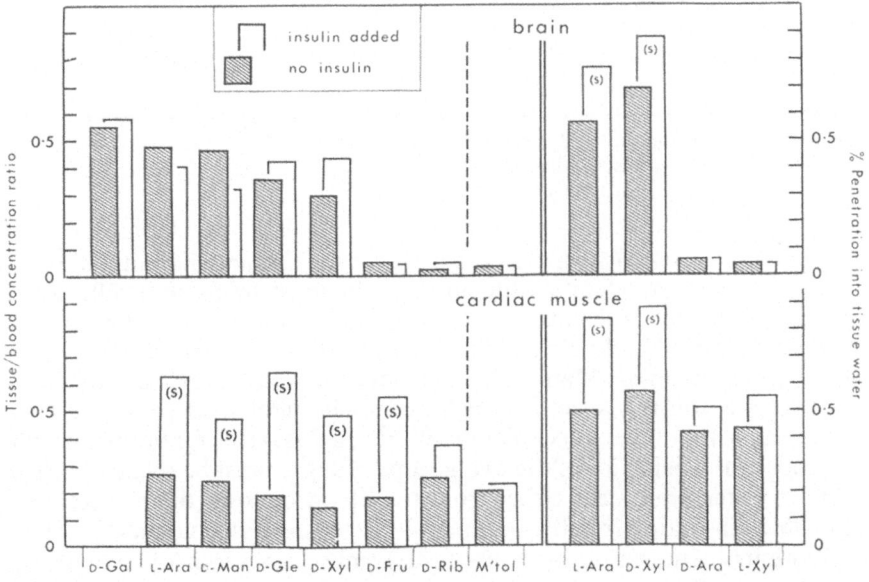

Fig. 1. Entry of monosaccharides into brain and cardiac muscle from the blood in the absence and presence of insulin. Data on the left are derived from different experiments from those on the right; the two groups cannot therefore be compared directly, but extrapolation may be made by means of L-arabinose and D-xylose, which are common to both. *ON LEFT.* Eviscerated rats infused intravenously with monosaccharides for 2 hr at constant rate; tissue samples collected 15 min later. Dose of monosaccharide adjusted to give similar final blood serum concentrations. Drawn from data published by Park *et al.*[10] *ON RIGHT.* Single injection of monosaccharide (1 g/kg) injected subcutaneously into cats; tissue samples collected 3 hr later. Drawn from data published by Sacks and Bakshy.[11] (S) = effect of insulin statistically significant ($P < 0.01$); (s) = significant effect claimed by authors, but statistical data not provided; M'tol = mannitol.

TABLE I

Comparison of Spatial Conformation of Certain Monosaccharides and of their Entry into Brain[a]

	Predominant "chair" conformation[12]	Entry into brain	References
Hexoses			
Fructose, D-**isomer**		· —	3, 10
Galactose, D-**isomer**	C_1	+	10, 27
Glucose, D-**isomer**	C_1	+	3, 10, 13, 27, 28, 32
2-Deoxyglucose	C_1	+	13
3-*O*-Methylglucose, D- isomer		+	32
Mannose, D-**isomer**	C_1	+	10, 27
Pentoses			
Arabinose { D-**isomer**	$1C$	—	11
{ L-isomer	C_1	+	10, 11
Ribose, D-**isomer**	C_1	—	10
Xylose, { D-**isomer**	C_1	+	10, 11, 13, 27
{ L-isomer	$1C$	—	11

[a] Sugars in bold type are commonly occurring natural isomers.[33]
+ = Entry by carrier transport demonstrated.
− = Entry from blood apparently impeded.

left entered brain readily, whereas D-fructose, D-ribose, and mannitol did not, a marked difference which is difficult to explain in terms of a diffusional barrier. The difference in entry between the stereoisomers of arabinose and xylose (in the right-hand portion of the figure) indicates that a stereopreferential carrier is involved. Mannitol, the alcohol derivative of mannose, was included as a control, since it is believed to be unable to penetrate cells although able to enter extracellular fluid[15,16]; if this is true for brain, then the difference in the behavior of the two groups of sugar must be related to their ability to enter brain cells either from the blood or from the extracellular fluid (which in brain probably occupies 4–5% of the tissue volume[17,18]).

The entry of the same sugars into cardiac muscle is shown for comparison in the lower half of Fig. 1 (galactose was omitted from this particular series, but it behaves similarly to the sugars nearby in the figure[19]). In this case there was no significant difference in behavior between any of the sugars on the left of the figure nor was there any between the stereoisomers of arabinose and xylose. The amount of each sugar entering the tissue was similar to that of mannitol, suggesting that all the sugars were able to enter the extracellular space (about 15% of the tissue volume in muscle[17,18]), but did not enter the cells.

The presence of insulin (Fig. 1, unshaded columns) resulted in stereopreferential stimulation of entry in muscle, but had no definite effect in brain.[10,11,20] Insulin thus appears able to activate a dormant transport system, as in muscle, but to have no effect where a carrier is already able to function, as found in a number of other tissues as well as in brain.[7,21,22]

The limited entry of fructose into brain has been to some extent attributed to the presence of a blood–brain barrier,[10] a view claimed to be supported by the inability of fructose (as opposed to glucose or mannose) in the blood to maintain cortical electrical activity.[23] However, fructose can, under suitable conditions, attain a concentration in the intact brain which approaches that of glucose[24] and it seems more likely that its inability to support normal function is related to the inability of the brain to oxidize fructose *in vivo*[25] in contrast to previous findings *in vitro*.[26]

Further evidence that a carrier system is involved has been provided by LeFevre and Peters[27] whose work merits consideration in some detail.

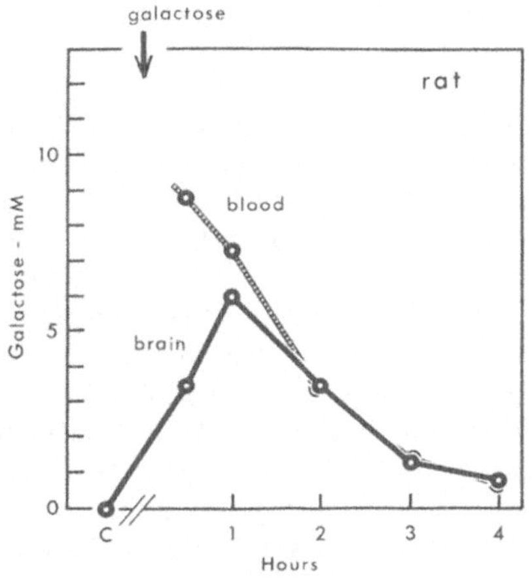

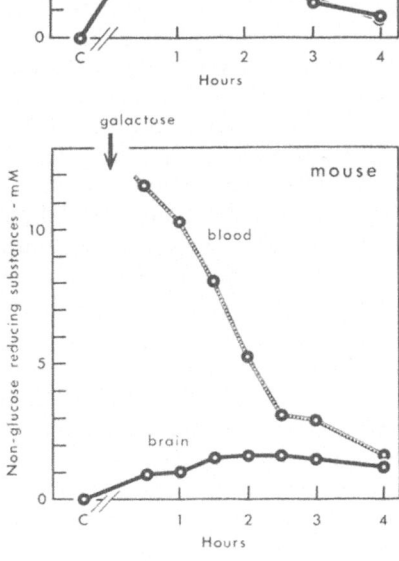

Fig. 2. Entry of D-galactose into rat brain from the blood after single intravenous injection (10 mmole/kg body weight). Each point represents the mean of results from at least three animals. Redrawn from LeFevre and Peters.[27] C = control; mM = millimolal.

Fig. 3. Entry of D-galactose into mouse brain from the blood after single intravenous injection (8.33 mmole/kg body weight). Each point represents the mean of results from at least three animals. Redrawn from LeFevre and Peters.[27] C = control; mM = millimolal.

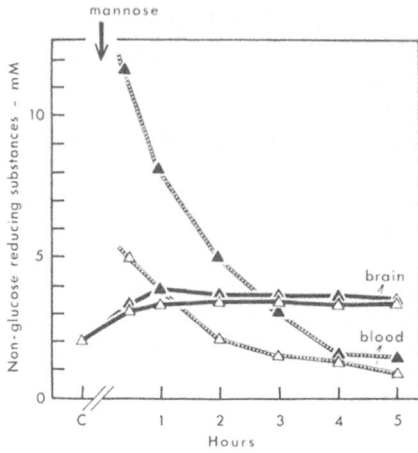

Fig. 4. Effect of size of dose of D-mannose on its entry into rat brain from the blood after single intravenous injection. Each point represents the mean of results from two animals. Redrawn from LeFevre and Peters.[27] C = control; mM = millimolal; △, dose of mannose 10 mmole/kg body weight; ▲, dose of mannose 20 mmole/kg body weight.

It was found that a difference in the species of animal can modify the entry of a sugar into the brain. In the rat, for instance, D-galactose (Fig. 2) and D-xylose penetrated the brain readily from the blood, the brain concentration rising to and then following that in the blood. In the mouse, however, D-galactose penetrated the brain much less readily (Fig. 3), while D-mannose entered slowly in both animals. As the authors point. out, metabolism of sugars in the brain may complicate interpretation. For instance, mannose and glucose, but not xylose or galactose, are stated to be metabolized by brain[27] and this would tend to lower the tissue concentration of the former. However, in the mouse, the pattern of entry of mannose into the brain was similar to that of galactose, suggesting that, with this sugar at least, the effect of metabolism may be small.

If a sugar enters a tissue by means of a carrier system, then with increasing concentration in the blood there should be evidence of saturation of the carrier, as shown by a limit to the amount transported in a given time. In Fig. 4 can be seen what is probably saturation of a carrier for D-mannose in the brain of the rat, as suggested by the fact that doubling the dose of the sugar made little difference to the amount which entered the brain. (It must be noted that the sugar content indicated in Fig. 4 represents also substances other than mannose). Similar saturation phenomena have also been found for mannose and glucose in other species,[20,27] but could not be shown for entry of xylose or galactose.[27]

Further evidence in favor of carrier-mediated transport was provided by LeFevre and Peters[27] in demonstrating the ability of certain sugars to compete with one another for entry into the tissue. If two substances are transported by a common carrier, each should inhibit the transport of the other by occupying a proportion of the carrier sites available. Thus D-mannose was found to inhibit the entry of D-glucose, as shown in Fig. 5. Glucose in the brain increased more after an intravenous injection of glucose alone than

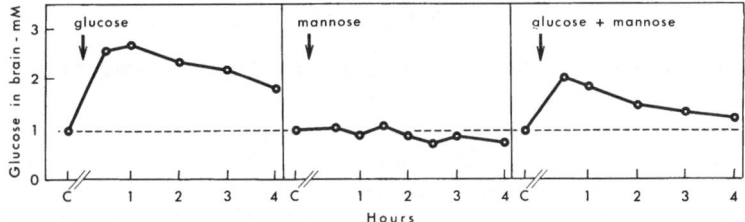

Fig. 5. Concentration of glucose in brain of mice after single intravenous injection of D-glucose and D-mannose either separately or as a mixture. Dose: 8.33 mmole of each sugar/kg body weight. Each point represents the mean of results from at least two animals. Redrawn from LeFevre and Peters.[27] mM = millimolal.

when the same amount of mannose was also present. (Mannose injected alone had no effect on brain glucose.) The effect of D-glucose on the entry of D-mannose was similar. This strongly suggests that the entry of glucose and mannose into the brain from the blood involves a common carrier system. In contrast, there was no such mutual inhibition between D-xylose and D-galactose.

The experimental results of LeFevre and Peters[27] are summarized in Table II, which includes also results of *in vitro* experiments by Gilbert[13] and indicates certain species differences.

Most experiments on the entry of monosaccharides into the brain have involved *in vivo* techniques in which sugars resemble amino acids[2] in their apparent inability to enter against a concentration gradient.[27] When brain slices (guinea pig) are used, however, xylose at least can be taken up against such a gradient.[13] The entry of this sugar into slices also shows evidence of saturation kinetics,[13] apparently absent *in vivo*[27] (Table II). Its entry appears to be inhibited competitively by the presence of glucose, but affinity for the transport carrier is low (K_m = 322 mM).[13]

Mention should also be made of the movement of monosaccharides between blood and cerebrospinal fluid, since in the dog at least there is evidence of a carrier system, probably situated in the choroid plexus, which transports glucose and 2-deoxyglucose but not fructose.[28]

It may be concluded, therefore, that the entry of certain monosaccharides into the brain is facilitated by a carrier system which may under certain circumstances be able to transport them against a concentration gradient. In addition to the properties already mentioned, the carrier, for glucose at least, shows evidence of a two-way flux, the net or overall movement into brain from the blood (influx less efflux) being less than the one-way movement (influx).[20] Thus entry into brain is apparently unlike entry into the erythrocyte where, although there is evidence of countertransport, sugars enter only down a concentration gradient[9,29,30]; in this respect brain more closely resembles intestine and kidney.[7,21]

It has been claimed that the brain shows permeability characteristics which, as far as can be judged, are similar to those of cell membranes in

TABLE II

Comparison of Various Aspects of the Entry of Certain Monosaccharides into brain

		Mouse in vivo[27]	Rat in vivo[27]	Guinea pig in vitro[13]
Concentration ratio developed[a]	D-Glucose			
	D-Mannose	0.2[b]	0.8[c]	
	D-Galactose	0.1[b]	0.9[c]	
	D-Xylose		1.0[c]	1.1, 1.8[d]
Saturation	D-Glucose	+		
	D-Mannose	+	+	
	D-Galactose		∘	
	D-Xylose		∘	+
Competition	D-Glucose	+⎫		+⎫
	D-Mannose	+⎭		⎬
	D-Galactose		∘⎫	⎭
	D-Xylose		∘⎭	+⎭

[a] Concentration ratio = $\dfrac{\text{concentration in tissue}}{\text{concentration in blood or suspending medium}}$

+ = phenomenon demonstrable.
∘ = phenomenon not demonstrable.
  Absence of symbol indicates no data available.
[b] Dose 8.3 mmole/kg. Time, 1 hr.
[c] Dose 10 mmole/kg. Time, 1 hr.
[d] Concentration in suspending medium 30 mM, 1 mM respectively. Time, 10 min.

general,[3] and it thus seems likely that the apparent limitation of entry of some sugars, such as fructose, is the result of confinement to a diffusional process only, in contrast to the entry of others where active transport is superimposed.[20]

Where a carrier system is involved, as with xylose, the intact brain appears to be able to take up the sugar less *in vivo* than *in vitro*.[13,27] Such limitation of entry could be due either to a structural or dynamic barrier, or to species difference, or possibly to differences in the nature of the fluid bathing the brain, since factors present in the blood are able to affect the entry of at least glucose.[25,31]

III. AMINES

A. Introduction

It is believed that the amines norepinephrine,[34] 5-hydroxytryptamine, [35] and histamine[36] may act as neurotransmitters in the central nervous system mainly by reason of their pharmacological properties and their presence (and that of related enzymes) in specific regions of the brain. For these substances to act at the appropriate site and no other, it is presumed that there must be either enzymes for their rapid inactivation, or mechanisms for their capture and localization, such as transport systems and binding sites. Binding sites are usually associated with intracellular particles, which themselves may possess membranes involved with the transport of metabolites into the particles.[37] Thus uptake by a cell may consist not only of entry into the cell from outside, but also of entry and binding involving subcellular particles, and data are usually insufficient to enable distinction to be made between the different mechanisms.

B. Catecholamines

In a number of peripheral tissues the catecholamines are taken up by a saturable mechanism located primarily in the postganglionic sympathetic nerves[38] and in many tissues they can be rapidly accumulated to a level considerably higher than that in the blood.[38-40] Brain, on the other hand, takes up catecholamines to a much smaller extent.[39-41] This can be seen in Fig. 6 which shows the uptake of norepinephrine-^{3}H and its O-methyl derivative normetanephrine (one of the principal metabolites in brain[42,43]) in a number of tissues after intravenous injection. In brain,[42,44] lung, and liver the breakdown of norepinephrine is rapid, and this would limit its rate of accumulation. In spleen and heart, on the other hand, and also in adrenal

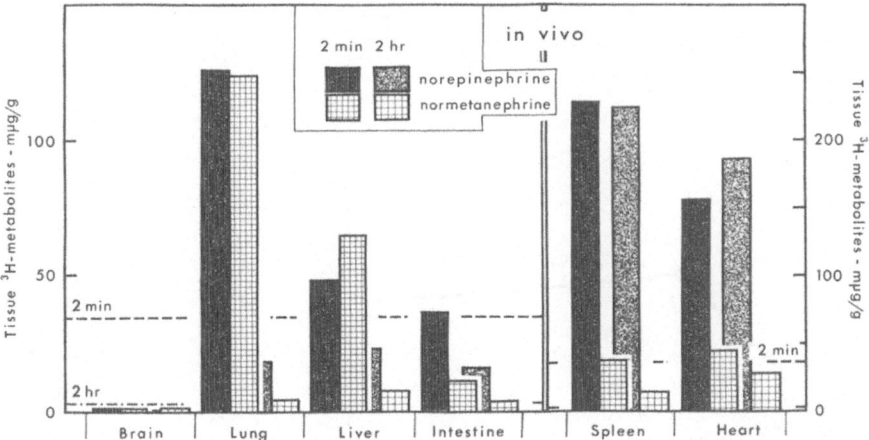

Fig. 6. Concentration of norepinephrine-^{3}H and normetanephrine-^{3}H in various tissues of the cat 2 min and 2 hr after the injection of DL-norepinephrine-^{3}H (25 μg/kg) injected intravenously within 1 min. Values represent the mean of results from three (2 min experiments) or four (2 hr experiments) cats. Horizontal interrupted lines represent plasma concentration of norepinephrine-^{3}H at the time indicated. No significance can be given to minor differences in concentration owing to large variations in individual values. Separation into left and right parts due only to differences in scale. Drawn from data published by Whitby et al.[40]

and salivary gland,[40] stability of the concentration over a period of time and the relative lack of catabolism suggests that in these tissues considerable binding of the amine occurs. Although not demonstrated in Fig. 6, brain can also bind some newly entered norepinephrine,[45,46] but in the brain as a whole the proportion bound *in vivo* is probably relatively small[47] (see below, and Fig. 7), rapid breakdown[42,44] probably predominating.

The relatively limited amount of norepinephrine-^{3}H which appears in brain as compared with other tissues after intravenous injection[40,47] (Fig. 6) suggests the existence of a blood–brain barrier for the amine, although the difference could also be explained if transport and binding in other tissues were considerably greater than in brain.

The distribution of norepinephrine-^{3}H in the various regions of the brain after entering from the blood is shown in Fig. 7; findings with epinephrine are in general similar.[48,49] In the pituitary, binding of the amine is suggested by the rapid rise and subsequent maintenance of the concentration, as also demonstrated elsewhere.[50] In the hypothalamus, too, there was a rise in concentration of norepinephrine-^{3}H, but the slower rate indicates that here binding, although suggested by the later stability of the concentration, was probably not a major factor in uptake. With certain other compounds,

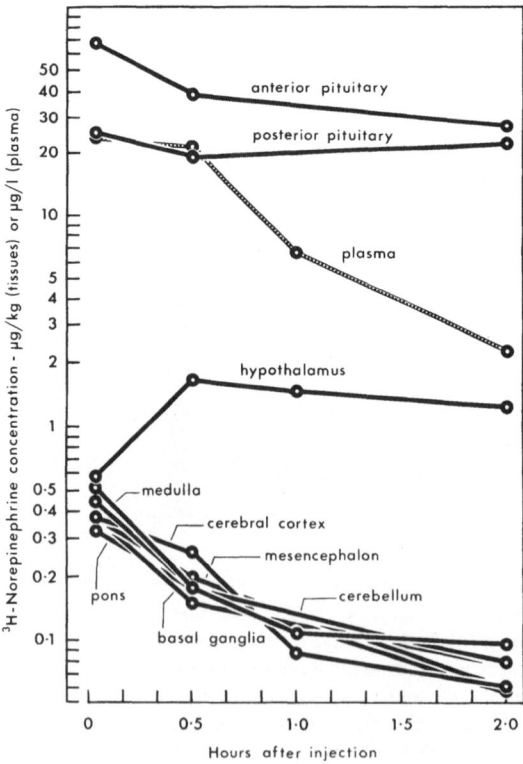

Fig. 7. Concentration of norepine-phrine-^{3}H in the pituitary and various areas of the brain of the cat after a single intravenous injection of DL-norepinephrine-^{3}H (25 µg/kg). Drawn from data published by Weil-Malherbe *et al.*[48] (semi-log scale).

greater uptake by the pituitary has been attributed to the local absence of a barrier which might be present in various regions of the brain[51]; the same might apply to norepinephrine but could also be due to a greater ability to transport and bind catecholamines.

Figure 7 also shows that in regions of the brain other than the hypothalamus the initial rise in concentration of norepinephrine-^{3}H was small and subsequently fell progressively, being maintained at 0.1–2.0% of that in the plasma. Since the amount of blood in the brain lies between 0.7% and 2.4% of the organ as a whole,[48] much of the norepinephrine appearing to enter these regions was probably located in trapped blood. However, about 1 hr after injection, concentrations declined more slowly than in plasma, and this may be attributed to retention of some norepinephrine-^{3}H by the tissue. (It should be pointed out that since the dose of norepinephrine-^{3}H was equivalent to about 0.2% of the concentration of endogenous amine in the hypothalamus,[48] the data should not necessarily be taken as evidence of a net rise in the concentration in the tissue, but could also be explained in terms of an exchange with norepinephrine already in the tissue.[48])

The selective entry of norepinephrine into the hypothalamus and the lack of binding in the cortex after peripheral injection has also been shown by radioautography.[47,52]

The route of entry of catecholamines into brain, as with other metabolites,[2] may affect uptake and distribution. For example, although entry from the blood appears to be impeded, that into brain slices does not,[45,53,54] as shown in Fig. 8. *In vivo* the concentration of norepinephrine-^{3}H in brain did not attain the concentration in the blood even after 2 hr, but *in vitro* it had risen considerably above that in the environment by the end of 1 hr, a difference not attributable to inability of the slices to break down the amine.[45] In the case of other tissues the opposite was found; uptake *in vitro* was considerably less than *in vivo*. The differences (also found with dopamine[44,45]) which apply to brain have been explained in terms of a blood–brain barrier which is supposed to be less effective in the region of the hypothalamus and ineffective *in vitro*. Why tissues other than brain, however, should show less uptake *in vitro* is not clear. Deterioration of the tissue, known to occur in liver,[55] might be held responsible, but should then also be expected to apply to brain tissue. That it does not is shown by the ability of brain slices to take up the amine against a concentration gradient.

When catecholamines are injected into the cerebral ventricles, the pattern of entry apparently differs from that found with the two types of technique already mentioned. Considerably more amine has been shown to be taken up by this route than after intravenous injection,[56] but the amount varies considerably from region to region[44] (Fig. 9). The interpretation of such variation is, however, uncertain, since accessibility of the different regions of the brain might be expected to be affected by distance from the site of administration and by thickness of tissue.[4] It is nevertheless valid to compare entry of one catecholamine with that of another, and it appears that the relationship between the entry of norepinephrine and of dopamine into

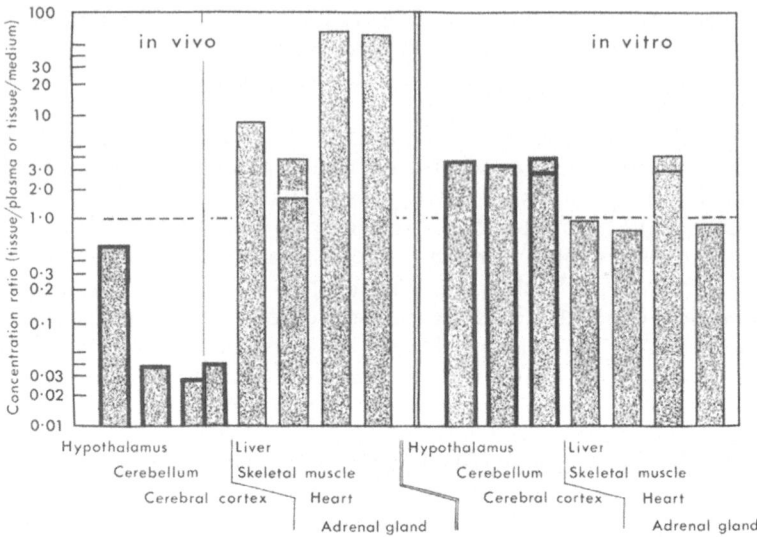

Fig. 8. Uptake of DL-norepinephrine-³H by brain regions and other tissues *in vivo* and *in vitro*. *IN VIVO. On left:* mean concentration ratio for DL-norepinephrine-³H developed between brain regions and plasma of cat 2 hr after single intravenous injection (25 mμg/g) n = 7–8. Concentration in plasma 2.7 mμg/g. Drawn from data published by Weil-Malherbe *et al.*[48] *On right:* mean concentration ratio for DL-norepinephrine-³H developed between various tissues and plasma of cat 2 hr after single intravenous injection (25 mμg/g), n = 3–4. Concentration in plasma 2.25 mμg/g. Column representing skeletal muscle signifies diaphragm (above) and abdominal wall (below). Drawn from data published by Whitby *et al.*[40] *IN VITRO.* Mean concentration ratio for total radioactivity developed between isolated tissues of cat and suspending medium which initially contained 5 mμg/ml DL-norepinephrine-³H. In cerebral cortex and heart about 70% of the radioactivity in the tissue was derived from norepinephrine, represented by horizontal black line of appropriate columns; related values for other tissues not available. Incubation 1 hr, 37°C, n = 4–15. Drawn from data published by Dengler *et al.*[45] (log scale).

the various regions was similar except that proportionately more dopamine than norepinephrine entered the corpus striatum and proportionately less entered the cerebellum (Fig. 9). It is possible that the relatively high content of endogenous dopamine which is present in most of the corpus striatum[57] and the relatively large amount of endogenous norepinephrine which is present in the hypothalamus[57,58] may be related to uptake ability,[44] a relationship which may apply more generally.[44,54] However, when comparing anything but gross differences this may not apply, since the nature of the preparation can affect relative uptake in the different regions. For example, when brain slices are used, uptake of norepinephrine by the corpus striatum is greater than by the hypothalamus, and by the cerebral cortex than by the medulla.[162]

At the subcellular level, fractionation techniques have demonstrated that part of the norepinephrine-³H entering the brain is found in the super-

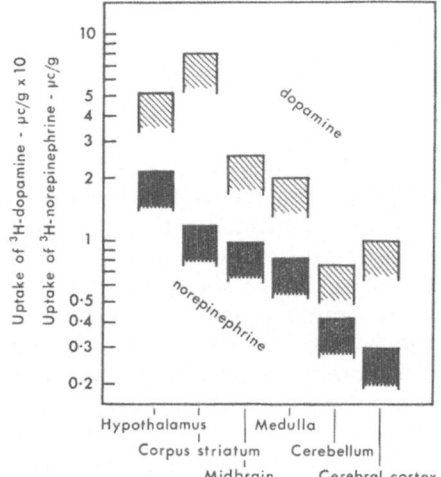

Fig. 9. Comparison of the uptake of nore-
pinephrine-^{3}H and dopamine-^{3}H by various
regions of the brain after injection of each into
cerebral ventricle. Tissues analyzed 1 hr after
injection. Dose: norepinephrine-^{3}H 6.3 μC
(specific activity 5C/mmole), dopamine-^{3}H
17 μC (specific activity 1.5 C/mmole). Drawn
from data published by Glowinski and Iversen[44] (log scale).

natant fraction, while part is in the nerve endings[59,60] (normally associated
with endogenous amines[61]). The proportions are subject to regional variation[59,62] and alter with time owing to more rapid disappearance from the
supernatant fraction.[59] The nerve endings themselves appear to be able to
take up norepinephrine by a transport mechanism and to retain it on binding
sites.[163]

The existence of a carrier system for norepinephrine and other catecholamines has been demonstrated by experiments using brain slices.[45,46,
54,63–65] Not only can norepinephrine be accumulated against a concentration gradient, as shown in Fig. 8, but its entry also shows evidence of saturation kinetics and is inhibited by the presence of ouabain (which interferes
with local energy processes involved in active transport[5]) or of metabolic
inhibitors,[45,64] suggesting that active transport is involved.

A considerable amount of attention has been devoted to the action of
reserpine on the catecholamine content of brain.[38] This drug, which has
hypotensive and tranquilizing effects, is known to lower the concentration of
amines in the brain.[54,66–69] In the platelet it appears to act by interfering
with transport,[70] while in adrenergic nerves, such as those of the heart,[71] it
primarily affects binding sites and in some instances has no effect on transport.[72,73] Although causing depletion of catecholamines which are already
in the brain, reserpine does not reduce their entry from the blood,[48] but this
is not surprising in view of the apparent barrier already present. It does,
however, reduce uptake from the cerebral ventricles,[74,75] and it also reduces
uptake by brain slices[45,53,76] although not apparently if incubation is kept
short.[77] Its effect can be comparable with that found in heart, suggesting, in
agreement with other evidence,[76,78,79] an action mainly on amine storage
sites,[74] although in brain these may not all show uniform sensitivity to the
drug.[75] On the other hand, reserpine reduces the content of endogenous and

exogenous norepinephrine in the cytoplasmic or soluble fraction of the tissue more readily than in the particulate fraction,[41,59,80] which suggests possibly an increase in catabolism[68] or an inhibition of transport, although some energy-dependent transport is unaffected by reserpine.[79] Such diversity makes it seem likely that reserpine has a multiple action involving mainly binding[34] but possibly also transport and metabolism to a limited extent.

The uptake of norepinephrine (and probably dopamine) by brain thus seems to be effected at least in part by an active transport system which can show saturation properties and can be altered by certain substances known to inhibit active transport in general. Some of the norepinephrine which enters brain is bound and some is rapidly altered metabolically, two processes which make interpretation of experimental results difficult. Entry into the intact brain is greatly impeded in some way by a blood–brain barrier mechanism; the nature of such a mechanism is unknown, but it is possibly located in the walls of the cerebral blood vessels,[79] or in structures adjacent.[2]

A final comment must here be added concerning the transport of norepinephrine by the choroid plexus. This contains a carrier system which in vitro actively transports amines such as norepinephrine and 5-hydroxytryptamine.[50,81] The direction of movement is probably from the cerebrospinal fluid to the blood[50] although there may be some movement in the opposite direction.[82]

C. 5-Hydroxytryptamine

It is generally believed that there also exists a blood–brain barrier for 5-hydroxytryptamine, a view based originally on the limitation of its entry into the brain from the blood as compared with that of the related amino acid 5-hydroxytryptophan.[83,84] This relative limitation, however, might be accounted for to some extent by the existence of an active transport mechanism which is considerably more effective for the amino acid than for the amine.[85–87] In fact, active transport of the amine in vitro could not initially be demonstrated,[86] but nevertheless, under suitable conditions, it can be shown to exist[160] (see below).

Although there is no doubt that 5-hydroxytryptamine can enter the brain from the blood,[88–91] it is taken up far more readily by a number of other tissues[92,93] (Fig. 10), reinforcing the idea of a barrier in the brain. However, this difference does not always apply, since accumulation of the amine by intestine is comparable with that by brain[90,92] (Fig. 10), and this could well be due to a relative deficiency of active transport of the amine in both tissues.[86,94] In addition, there is probably considerable binding of the amine in some of the tissues, as indicated by a low rate of disappearance, and this may well contribute toward a relatively greater accumulation in those tissues despite the occurrence in brain of some binding by structures such as nerve-ending particles and microsomes.[61,95–97]

Part of the apparent limitation of entry of 5-hydroxytryptamine into the brain might be accounted for, as with the catecholamines, by disappear-

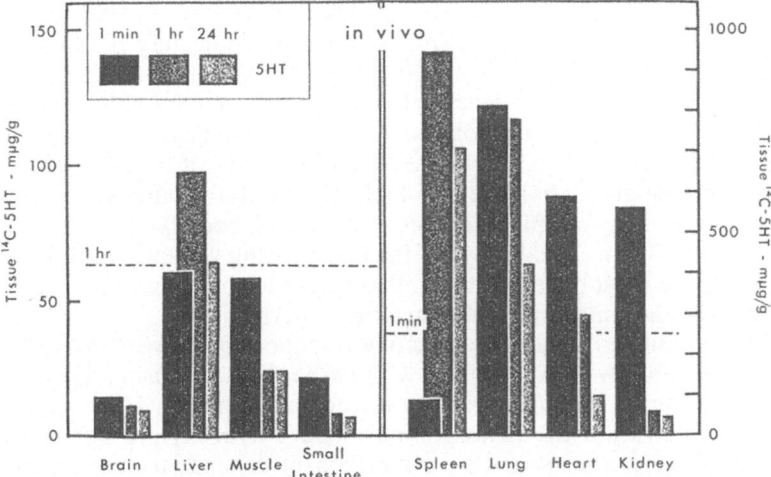

Fig. 10. Uptake of 5-hydroxytryptamine-^{14}C by various tissues of the mouse after injection of 25 μg 5-hydroxytryptamine-3-^{14}C-creatinine sulfate intravenously. Height of each column represents mean tissue content of 5-hydroxytryptamine-^{14}C after time interval indicated; $n = 2$ (1 min) or 3 (1 hr and 24 hr). (In some cases there was wide variation between individual values.) Horizontal interrupted lines represent plasma concentration at time indicated; $n = 1$ (1 min) or 3 (1 hr). Separation into right and left parts is due only to difference in scale. Drawn from data published by Axelrod and Inscoe.[92]

ance of the amine as a result of rapid breakdown,[90,98,164] some indication of which is shown in Fig. 11. After the intravenous injection of various amounts of 5-hydroxytryptamine, at least as much of the derivative 5-hydroxyindoleacetic acid (produced by the action of the enzyme monoamine oxidase[99]) as of the amine itself appeared in brain. (Although it is possible that some of the derivative may have come from peripheral metabolism, its entry into the brain appears to be limited.[100]) When breakdown of the

Fig. 11. Increase in concentration of 5-hydroxytryptamine and its derivative, 5-hydroxyindoleacetic acid (5-HIAA), in brain 10 min after intravenous injection of various doses of 5-hydroxytryptamine. Values on ordinate represent difference between concentration after injection of 5-hydroxytryptamine and concentration after injection of saline. Each point represents the mean of two experiments in which three brains were pooled. Drawn from data published by Bulat and Supek.[90]

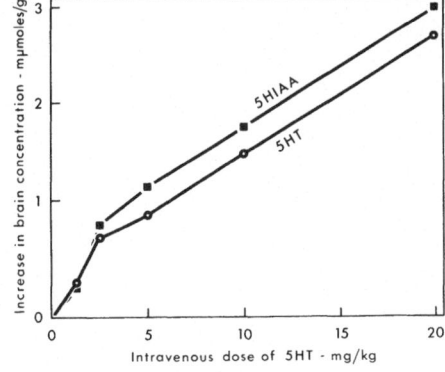

amine is inhibited by the monoamine oxidase inhibitor iproniazid, the rate of accumulation in the brain after intravenous injection may be doubled, as shown in Fig. 12. (In the figure absolute concentrations of the amine should not be compared, control levels also being affected by the presence of iproniazid. It must also be pointed out that iproniazid does not always produce the effect shown.[88]) Some idea of the rate of breakdown can be given by the metabolic half-life of 5-hydroxytryptamine after entry into the brain. After an intravenous injection of 10 mg/kg body weight, it has been found to be 5–10 min, breakdown of the amine being virtually complete after 80 min,[2] but it must be remembered that values for processes of this sort can be affected by the concentration of metabolite in the tissue.

The inability to demonstrate active transport of 5-hydroxytryptamine into the brain involved experiments with tissue slices and using the amine at a concentration of 10 μg/ml or more in the environment. Although at this sort of concentration the related amino acid 5-hydroxytryptophan shows evidence of being transported by a mechanism dependent upon metabolic energy, 5-hydroxytryptamine does not.[86,90] However, if much lower concentrations are used (approximately 5 mμg/ml, as in comparable experiments with norepinephrine), evidence of active transport can be demonstrated, showing some features similar to the findings with catecholamines, such as localization subcellularly, mainly in nerve endings.[62,160,163] However, in view of the lack of mutual competition for receptor sites during entry,[160,161] the mechanism of uptake involved appears to be different from that which transports catecholamines. (The choroid plexus, in contrast, is able actively to transport both types of amines by means of a common carrier.[50])

The much lower concentration which has to be used in order to demonstrate active transport of 5-hydroxytryptamine as compared with 5-hydroxytryptophan suggests that transport of the amine is either less effective or has a more limited distribution in the brain. As with the catecholamines,

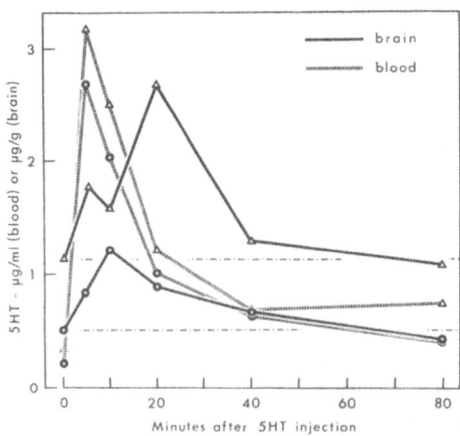

Fig. 12. Concentration of 5-hydroxytryptamine in brain and blood after intravenous injection (10 mg/kg) into rats. Each point represents the mean of results from at least two experiments. The horizontal interrupted lines represent the levels of 5-hydroxytryptamine in brain at zero time in each type of experiment. Redrawn from diagrams published by Kärki and Paasonen.[89] △, rats pretreated with iproniazid phosphate subcutaneously (100 mg/kg). ○, rats not pretreated with iproniazid.

5-hydroxytryptamine is able to enter from the blood certain regions in the base of the brain more readily than elsewhere in that organ.[91] Such local differences have been considered as evidence in favor of the existence of a blood–brain barrier elsewhere in the brain, but could well be due to the limitation of active transport to those regions only.

Reserpine has an action on 5-hydroxytryptamine in brain which is similar to that on catecholamines; it depletes the stores of the amine in the brain,[101,102] reduces its entry from the cerebral ventricles,[98] and also reduces its accumulation by tissue slices.[160,161]

It would seem, then, that some of the entry of 5-hydroxytryptamine into the brain involves an active transport mechanism which, although in many respects resembling that which transports catecholamines, is not identical with it. Presumably a considerable proportion of the amine may also enter by diffusion. To what extent the entry of 5-hydroxytryptamine into the brain is impeded by a barrier in the intact animal is, however, uncertain, particularly in view of the similarity of the amount which can be accumulated in brain and in intestine. Although this relationship between the two tissues (Fig. 10) is different from that with respect to norepinephrine (Fig. 6), any interpretation derived from such comparisons should be made with caution in view of the differences in the concentration of amine in each group of experiments involved.

D. Histamine

Although histamine can be taken up readily by many tissues,[103–106] it also resembles the catecholamines and 5-hydroxytryptamine in that its entry into the brain from the blood is relatively limited as compared with that into other tissues, whether considered in terms of histamine alone or in terms of histamine together with its derivatives.[105–107] This can be seen from Fig. 13 which also shows that in many tissues administered histamine appears to be rapidly metabolized[104,107–111] rather than bound, since most of that present 1 hr after intraventricular injection had disappeared within 24 hr. It is interesting to compare Fig. 14 with Figs. 6 and 8, although it must be noted that different time intervals were involved.

There are two metabolic pathways for the breakdown of histamine, that which predominates depending upon the animal species; in man, dog, cat and rabbit it is primarily by way of ring-N-methylation to produce 1,4-methylhistamine,[109,110,112] in the rat it is primarily by oxidation to imidazoleacetic acid,[112] while in the mouse and guinea pig the amine can be metabolized by either route.[107,112]

The rapidity with which the product imidazoleacetic acid-^{3}H appears in the brain of the mouse during entry of histamine-^{3}H (primarily into the gray matter[4,113]) is shown in Fig. 14. It is clear that, over the range of dosage indicated, about as much histamine was converted to imidazoleacetic acid as appeared as histamine. (Although appearing to be similar, Fig. 14 cannot validly be compared with Fig. 11 owing to differences in dosage and route of

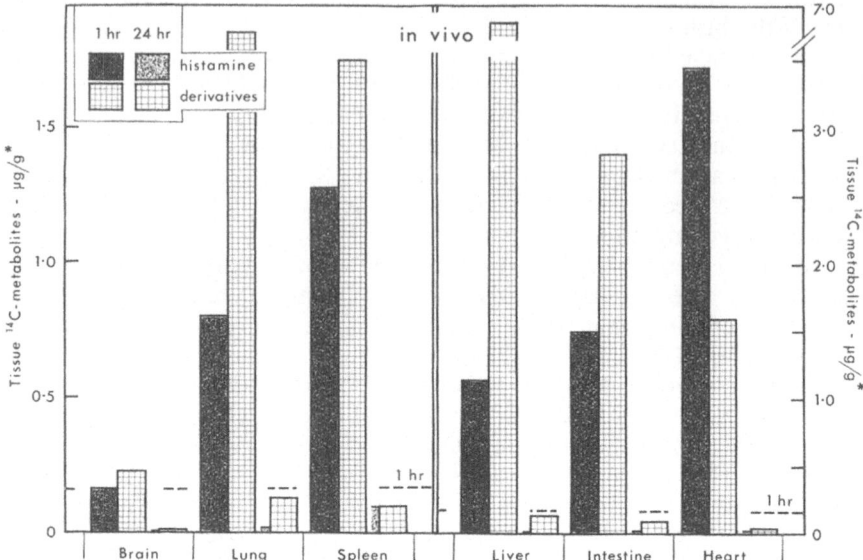

Fig. 13. Concentration of histamine-^{14}C and derivative metabolites in various tissues of the rat, 1 hr and 24 hr after a single subcutaneous injection of 520 μg histamine-^{14}C. Each value represents the mean of results from two animals; it must be noted that there was considerable variation between individual results. Horizontal interrupted lines represent serum levels of histamine-^{14}C at time indicated. Separation into left and right parts is due only to difference in scale. Drawn from data published by Snyder et al.[107]
*Values for ^{14}C-derivatives have been calculated by subtracting data given for histamine-^{14}C from those for total radioactivity; concentration given in terms of the equivalent amount of histamine-^{14}C.

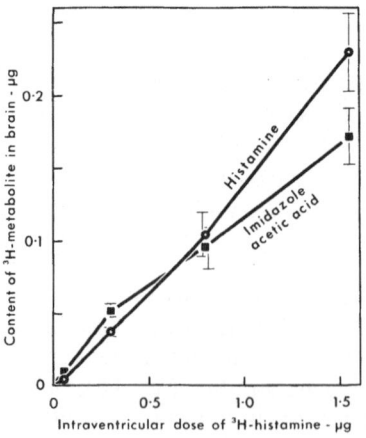

Fig. 14. Mean concentration of histamine-^{3}H and its derivative imidazoleacetic acid-^{3}H in the brain of rats 2 hr after intraventricular injection of various amounts of histamine-^{3}H; $n = 4$–6. Values, with standard error of the mean, calculated from data published by Snyder et al.[111]

administration.) With increase in the concentration of injected amine there was a significant[111] falling off in the rate of increase in the amount of the derivative, indicating saturation of the oxidative enzyme; entry of the amine, on the other hand, does not show evidence of saturation kinetics, but at the low dosage used this cannot necessarily be considered as evidence of the absence of active transport.

As with the amines already discussed, histamine can be bound in certain tissues,[106] and in brain the localization of uptake and binding by nerve ending particles and microsomes resembles that for other amines.[61,97,114] Binding also occurs in the pituitary gland (the region of the greatest concentration of endogenous amine[115]), where mast cells, which are particularly involved with storage[116,117] probably contain much of the local histamine.[118]

Owing to the relatively small amount of histamine which enters the brain in comparison with other tissues, it has been claimed that there is a blood–brain barrier also for this amine.[111] In support of such a claim is the finding that histamine does not enter the hypothalamus or the area postrema to a significant extent whereas it does enter the anterior lobe of the pituitary,[119] suggesting that there might be a barrier in the first two regions but not in the latter. However, preferential uptake of histamine by the pituitary could well be explained in terms of local transport or binding, while the similarity in concentration in whole brain and blood 1 hr after peripheral injection (Fig. 13) shows that it enters the brain more rapidly than do the other two amines discussed.

There is no evidence that histamine enters brain to any appreciable extent by active transport. The concentration developed in tissue slices can rise slightly above that in the environment, while the effect of metabolic inhibitors on uptake is considerably less than on the uptake of the related amino acid, histidine, and the related carboxylic acid, dihydrourocanic acid, both of which are actively transported[120] (Fig. 15). This suggests that histamine enters brain primarily by diffusional-type movement, although the existence of a small active component, perhaps localized, is possible, and may yet be revealed by the use of concentrations of the amine lower than those used heretofore.

The ready entry of histamine *in vitro* together with the absence of active transport indicates relatively little involvement with a barrier mechanism.

E. Conclusions

Various factors, such as diffusion, active transport, binding and metabolism, which will affect the movement of both ionized and un-ionized metabolites into a tissue, have already been discussed in association with the appropriate metabolites. The amines, however, are mainly in the form of positively charged ions and hence their movement will be influenced in addition by the factors tending to establish a Gibbs-Donnan equilibrium. These would tend to raise their intracellular concentration in relation to their

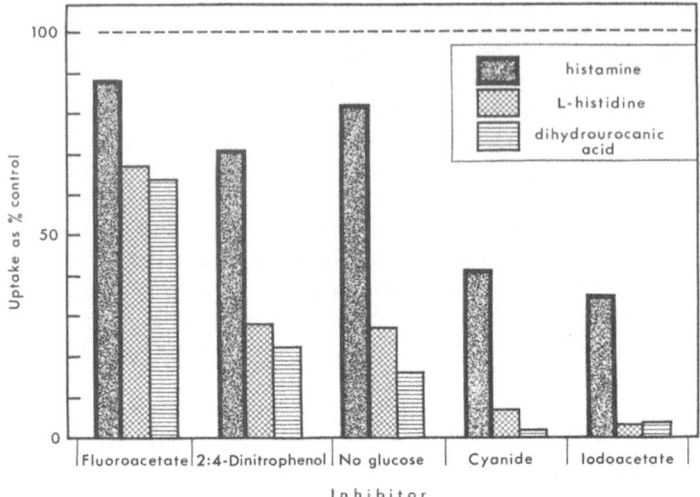

Fig. 15. Effect of metabolic inhibitors, etc. on uptake of imidazole derivatives by rat brain slices. (Uptake = concentration in tissue less concentration in suspending medium.) Control uptake (uptake in absence of inhibitor) for histamine = 1.7 mM, for L-histidine = 18.0 mM and for dihydrourocanic acid = 5.0 mM. Initial concentration of each imidazole
derivative in suspending medium = 2 mM; apparent blank value (primarily
attributable to histidine[121]) of each in incubated tissue = 0.3 mM. Incubation 1 hr, 37°C. Drawn from data published by Neame.[120]

extracellular concentration (the reverse being the case with negatively
charged ions). Information as to the magnitude of this effect in brain is not,
however, available, and so the component which it contributes toward total
entry of amines cannot be assessed. However, the distribution ratio achieved
by some comparable organic cations and anions between brain slices and a
suspending medium under various conditions does not diverge from unity
by a factor of more than about 2.[86,120] It has therefore been assumed in this
review that such a component is relatively insignificant and may be dis-
regarded. Future studies, however, may well show it to be of some significance
in particular cases.

The entry of catecholamines, 5-hydroxytryptamine, and histamine into
the brain from the blood is nevertheless in most instances clearly limited
when compared with their entry into other tissues (Figs. 6, 12, and 14). This in
itself does not prove the existence of a barrier in brain, but could equally well
indicate the presence elsewhere of more effective transport and binding. In
brain, norepinephrine and 5-hydroxytryptamine, but not histamine, appear
to be actively transported, but the entry of the catecholamines and perhaps
5-hydroxytryptamine is limited by a barrier mechanism. The ability of
histamine to achieve fairly rapidly a concentration in whole brain similar to
that in the blood (Fig. 14) suggests that its entry is not obviously affected by

a barrier mechanism, while the existence of a barrier to 5-hydroxytryptamine is open to some doubt. A direct comparison of the movement of the different amines would go far to remove many of the uncertainties which at present exist. For the results of such investigations to be of value, movement of the amines would have to be compared under identical conditions, both in *vivo* and in *vitro*, and using a range of concentrations, time periods, and tissue samples. Such comparative data are not at present available.

IV. MISCELLANEOUS COMPOUNDS

A. Quaternary Ammonium Compounds

A number of quaternary ammonium compounds, such as acetylcholine, appear to be able to pass between the blood and the neurons of the brain or spinal cord, as shown by the production of neuropharmacological and electrophysiological responses after peripheral injection.[122-124] It has, nevertheless, been claimed that between the blood and the central nervous system there is a barrier to certain of these compounds,[123,125] but this belief appears to be based only on the finding that central responses to peripherally injected prostigmine (neostigmine) are often negative.[123] On the other hand, the fact that responses *can* be positive[123] suggests that variations in technique or in neuronal anatomy may be responsible for differences found in experimental results. Since the quantitative movement of such compounds into the central nervous system does not appear to have been measured, a definite answer cannot be given as to whether a barrier to these compounds does in fact exist in the brain.

Nevertheless, there is in brain an active transport system for quaternary ammonium compounds, as investigations into the uptake of choline,[126] acetylcholine,[127,128] and carbachol[129-131] by tissue slices have shown. As well as an active component, predominating at lower concentrations and saturable at higher concentrations, there is also, as shown with choline and acetylcholine, a passive component which predominates at higher concentrations and presumably represents diffusion. The active component demonstrates the usual properties associated with active transport, since both compounds can be taken up against a concentration gradient, and uptake is reduced by factors which inhibit metabolism. Uptake is also inhibited competitively by the presence of HC-3 (a hemicholinium[132]), physostigmine, atropine, and morphine[126-128] which have certain structural similarities. The transport system in brain may be related to that in choroid plexus which transports not only hexamethonium and related compounds but also the amines norepinephrine and 5-hydroxytryptamine.[50,81,133]

Data which indicate the accumulation of these compounds in brain, must, however, be treated with caution. Interpretations tend to be based on the total amount in the tissue on the assumption that it is all in the free form, but it has been found that, of acetylcholine accumulated by brain slices, two-thirds may be in an inactive bound form.[134,135] Nevertheless, the obvious effect of metabolic inhibition on total uptake does suggest that the presence of much of the choline or acetylcholine accumulated by brain slices is related mainly to its passage across the cell membrane.

In possessing a carrier system actively transporting acetylcholine, brain appears to differ from muscle, in which the acetylcholine entering the tissue is found in the extracellular space rather than in the cells.[136]

B. Urea

Urea is generally considered to be able to diffuse across most cell membranes, and to distribute itself freely throughout the extracellular and intracellular water,[137,138] but in some species its entry into the brain from the blood appears to be restricted. For instance, high blood levels can result initially in a lowering of the hydrostatic pressure in the central nervous system with accompanying shrinkage of tissue.[139-141] This has been attributed to an osmotic effect believed to result from limited entry of urea into the brain.[142] That such limitation exists is shown by a comparison with entry into muscle. After 1 hr of intravenous infusion of urea into cats or rabbits, the tissue/plasma concentration ratio may only be 0.3–0.6 for brain at a time when it approaches 1.0 for muscle.[141,142] Equilibration in muscle can be approached within 1–4 hr, but for brain it can take more than 9 hr.[142,143] (After infusion into the cerebrospinal fluid, equilibration with brain can take even longer, probably owing to loss from the brain into the blood, since entry per unit surface area appears to be greater from the cerebrospinal fluid than from the blood.[142]) Thus it has been postulated that there is a blood–brain barrier for urea (believed by some[142] to lie in the capillary walls). Although these interpretations have been criticized by Dobbing,[17] the difference between the rate of entry into brain and into muscle is far greater than can be accounted for by differences in other factors such as extracellular fluid volume. In addition, the rate of entry into different regions and into different compartments of the brain is not the same,[140,142-144] which suggests the presence of regional barriers (or possibly regional transport systems, although there has been no evidence of these with respect to urea).

Conclusions reached from experiments on one animal cannot, however, necessarily be applied to another in view of species differences. In the dog the concentration of urea in most regions of the brain can reach that in the plasma after 6 hr of continuous intravenous infusion, but in the cat and rabbit the concentration in the brain after a similar time may only be about 70 % of that in the plasma.[143] In the chicken, on the other hand, equality of concentration can be reached by the end of 1 hr.[141]

The hypothesis of a blood–brain barrier for urea may possibly be strengthened when the data mentioned above from experiments on mammals *in vivo* are compared with experiments *in vitro*. When brain slices from the guinea pig are incubated in a medium containing urea, the concentration of urea in the tissue is found to have risen to a level comparable with that in the suspending medium by the end of 1 hr.[145] Thus the impediment to entry that is found *in vivo* appears to be absent *in vitro*, but, in view of the different species involved in the two types of technique, a definite conclusion cannot be reached. No comparable data on a single species appear to be available.

C. Cholesterol

It has been reported that when a small amount of cholesterol is added to the blood it is able to enter the brain of the immature animal more readily than that of the adult.[146–149] This difference was originally interpreted as an increase in the effectiveness of a blood–brain barrier with age, similar to that originally claimed, but later disproved, for the dye trypan blue.[150] However, alterations in the amount of cholesterol entering the brain during development have been found to follow rather closely alterations in the amount of cholesterol laid down, both phenomena having a peak of activity in the rat when the body weight has reached about 20 g.[149] Thus the greater limitation of entry of cholesterol with age may simply result from the decreased rate of deposition and the greater metabolic inertness which is found in the brain of the adult as opposed to that of the young animal.[151–155] There is no evidence of a mechanism which transports cholesterol into the brain, and it seems probable that diffusion alone is responsible for its entry.

D. Other Metabolites

The following compounds are apparently able to enter the brain from the blood in small amounts, as shown by evidence of an increase in their concentration in the brain or of their involvement in brain metabolism: lactate,[156] pyruvate,[156,157] acetate,[157,158] butyrate,[158] propionate[158] and cortisol.[159] Succinate, on the other hand, has not been shown to pass from the blood into the brain.[156] It must be remembered that those compounds which are readily metabolized by brain will rapidly disappear from the tissue and hence evidence of their entry may remain undetected.

V. CONCLUSIONS

The extent to which a metabolite appears to enter a tissue such as brain can be affected by a number of factors apart from the actual rate of movement itself. This chapter has illustrated how experimental findings may be influenced by the route of entry, the rate of binding, the rate of breakdown, and the type of preparation. It must be remembered that the proportion of a compound which is broken down or bound can be affected by its concentration as a result of saturation of receptor sites. Nevertheless, it is possible to draw general conclusions by surveying the findings which have been derived from different experimental techniques. For most of the compounds discussed, the concept of a blood–brain barrier has been postulated as an explanation for relative limitation of entry into the brain or into certain regions of the brain. The concept is valid in the case of actively transported metabolites (amino acids,[2] certain monosaccharides, norepinephrine, and possibly 5-hydroxytryptamine and quaternary ammonium compounds) which can enter brain more readily *in vitro* than *in vivo*, or which can be accumulated *in vitro* at least as readily by brain as by other tissues. There is, however, little evidence of a barrier mechanism for some metabolites which do not conform to these criteria (certain monosaccharides and histamine); for these, more detailed studies would be needed before a valid conclusion could be reached.

The nature of the barrier mechanism, where demonstrated, is unknown. It is possible that it may be a concomitant of active transport, and, as suggested for amino acids,[2] could possibly be one facet of a balance between inward and outward transport.

ACKNOWLEDGMENTS

I would like to thank Dr. E. C. Pickering for helpful and constructive criticism. I would also like to thank the publishers of the *Journal of Neurochemistry*, of *Acta Pharmacologica et Toxicologica* and of the *American Journal of Physiology* and the respective authors for permission to publish Figs. 2, 3, 4, 5 and 12, and the quotation on page 330.

VI. REFERENCES

1. J. Dobbing, The blood–brain barrier, *Physiol. Rev.* **41**:130–188 (1961).
2. K. D. Neame, *in Applied Neurochemistry* (A. N. Davidson and J. Dobbing, eds.), pp. 119–177, Blackwell Scientific Publications, Oxford (1968).
3. C. Crone, The permeability of brain capillaries to non-electrolytes, *Acta Physiol. Scand.* **64**:407–417 (1965).
4. W. Feldberg, *in A Pharmacological Approach to the Brain from its Inner and Outer Surface*, pp. 18–29, Edward Arnold, London (1963).
5. J. H. Quastel, Molecular transport at cell membranes, *Proc. Roy. Soc. Ser. B* **163**:169–196 (1965).
6. R. K. Crane, Intestinal absorption of sugars, *Physiol. Rev.* **40**:789–825 (1960).
7. G. Wiseman, *Absorption from the Intestine*, p. 43, Academic Press, London (1964).
8. A. Kleinzeller, J. Kolínská, and I. Beneš, Transport of monosaccharides in kidney-cortex cells, *Biochem. J.* **104**:852–860 (1967).
9. P. G. LeFevre, Sugar transport in the red blood cell: structure-activity relationships in substrates and antagonists, *Pharmacol. Rev.* **13**:39–70 (1961).
10. C. R. Park, L. H. Johnson, J. H. Wright, and H. Batsel, Effect of insulin on transport of several hexoses and pentoses into cells of muscle and brain, *Am. J. Physiol.* **191**:13–18 (1957).
11. J. Sacks and S. Bakshy, Insulin and tissue distribution of pentose in nephrectomized cats, *Am. J. Physiol.* **189**:339–342 (1957).
12. P. G. LeFevre and J. K. Marshall, Conformational specificity in a biological sugar transport system, *Am. J. Physiol.* **194**:333–337 (1958).
13. J. C. Gilbert, Mechanism of sugar transport in brain slices, *Nature* **205**:87–88 (1965).

14. F. Alvarado, D-Xylose active transport in the hamster small intestine, *Biochim. Biophys. Acta* **112**:292–306 (1966).

15. J. R. Elkington, The volume of distribution of mannitol as a measure of the volume of extracellular fluid, with a study of the mannitol method, *J. Clin Invest.* **26**:1088–1097 (1947).

16. J. R. Robinson, Metabolism of intracellular water, *Physiol. Rev.* **40**:112–149 (1960).

17. J. Dobbing, The blood–brain barrier: some recent developments, *Guy's Hosp. Rep.* **112**:267–286 (1963).

18. J. Dobbing, Changing views of the blood–brain barrier, *London Hospital Gazette* **67**:ii–xi (1964).

19. C. R. Park and L. H. Johnson, Effect of insulin on transport of glucose and galactose into cells of rat muscle and brain, *Am. J. Physiol.* **182**:17–23 (1955).

20. C. Crone, Facilitated transfer of glucose from blood into brain tissue, *J. Physiol. (London)* **181**:103–113 (1965).

21. C. R. Park, R. L. Post, C. F. Kalman, J. H. Wright, L. H. Johnson, and H. E. Morgan, The transport of glucose and other sugars across cell membranes and the effect of insulin, *Ciba Found. Colloq. Endocrinol.* **9**:240–260 (1956).

22. W. Wilbrandt and T. Rosenberg, The concept of carrier transport and its corollaries in pharmacology, *Pharmacol. Rev.* **13**:109–183 (1961).

23. S. Maddock, J. E. Hawkins, and E. Holmes, The inadequacy of substances of the "glucose cycle" for maintenance of normal cortical potentials during hypoglycemia produced by hepatectomy with abdominal evisceration, *Am. J. Physiol.* **125**:551–565 (1939).

24. J. R. Klein, R. Hurwitz, and N. S. Olsen, Distribution of intravenously injected fructose and glucose between blood and brain, *J. Biol. Chem.* **164**:509–512 (1946).

25. A. Geiger, J. Magnes, R. M. Taylor, and M. Veralli, Effect of blood constituents on uptake of glucose and on metabolic rate of the brain in perfusion experiments, *Am. J. Physiol.* **177**:138–149 (1954).

26. J. R. Klein, Oxidation of fructose by brain *in vitro*, *J. Biol. Chem.* **153**:295–300 (1944).

27. P. G. LeFevre and A. A. Peters, Evidence of mediated transfer of monosaccharides from blood to brain in rodents, *J. Neurochem.* **13**:35–46 (1966).

28. R. A. Fishman, Carrier transport of glucose between blood and cerebrospinal fluid, *Am. J. Physiol.* **206**:836–844 (1964).

29. W. Wilbrandt, in *Membrane Transport and Metabolism* (A. Kleinzeller and A. Kotyk, eds.), pp. 388–397, Academic Press, London (1961).

30. D. M. Regen and H. E. Morgan, Studies of the glucose-transport system in the rabbit erythrocyte, *Biochim. Biophys. Acta* **79**:151–166 (1964).

31. A. Geiger and S. Yamasaki, Cytidine and uridine requirement of the brain, *J. Neurochem.* **1**:93–100 (1956).

32. W. F. Agnew and C. Crone, Permeability of brain capillaries to hexoses and pentoses in the rabbit, *Acta Physiol. Scand.* **70**:168–175 (1967).

33. W. W. Pigman and R. M. Goepp, *Chemistry of the Carbohydrates*, Chapter III, Academic Press, New York (1948).

34. J. Glowinski and R. J. Baldessarini, Metabolism of norepinephrine in the central nervous system, *Pharmacol. Rev.* **18**:1201–1238 (1966).

35. I. H. Page, Serotonin (5-hydroxytryptamine); the last four years, *Physiol. Rev.* **38**:277–335 (1958).

36. J. P. Green, Histamine and the nervous system, *Fed. Proc.* **23**:1095–1102 (1964).

37. S. Varon and W. Wilbrandt, in *Intracellular Transport* (K. B. Warren, ed.) pp. 119–139, Academic Press, London (1966).

38. L. L. Iversen, *The Uptake and Storage of Adrenaline in Sympathetic Nerves*, Chapters 7 and 10, University Press, Cambridge (1967).

39. J. Axelrod, H. Weil-Malherbe, and R. Tomchick, The physiological disposition of H[3]-epinephrine and its metabolic metanephrine, *J. Pharmacol. Exp. Therap.* **127**:251–256 (1959).

40. L. G. Whitby, J. Axelrod, and H. Weil-Malherbe, The fate of H[3]-norepinephrine in animals, *J. Pharmacol. Exp. Therap.* **132**:193–201 (1961).

41. H. Weil-Malherbe and A. D. Bone, The effect of reserpine on the intracellular distribution of catecholamines in the brain stem of the rabbit, *J. Neurochem.* **4**:251–263 (1959).

42. J. Glowinski, I. J. Kopin, and J. Axelrod, Metabolism of [³H]norepinephrine in the rat brain, *J. Neurochem.* **12**:25–30 (1965).

43. C. O. Rutledge and J. Jonason, Metabolic pathways of dopamine and norepinephrine in rabbit brain *in vitro*, *J. Pharmacol. Exp. Therap.* **157**:493–502 (1967).

44. J. Glowinski and L. L. Iversen, Regional studies of catecholamines in the rat brain—I. The disposition of [³H]norepinephrine, [³H]dopamine and [³H]dopa in various regions of the brain, *J. Neurochem.* **13**:655–669 (1966).

45. H. J. Dengler, I. A. Michaelson, H. E. Spiegel, and E. Titus, The uptake of labeled norepinephrine by isolated brain and other tissues of the cat, *Intern. J. Neuropharmacol.* **1**:23–38 (1962).

46. S. B. Rose and A. L. Renyi, Uptake of some tritiated sympathomimetic amines by mouse brain cortex slices *in vitro*, *Acta Pharmacol. Toxicol.* **24**:297–309 (1966).

47. B. H. Marks, T. Samorajski, and E. J. Webster, Radioautographic localization of norepinephrine-H[3] in the tissues of mice, *J. Pharmacol. Exp. Therap.* **138**:376–381 (1962).

48. H. Weil-Malherbe, L. G. Whitby, and J. Axelrod, The uptake of circulating [³H]norepinephrine by the pituitary gland and various areas of the brain, *J. Neurochem.* **8**:55–64 (1961).

49. H. Weil-Malherbe, J. Axelrod, and R. Tomchick, Blood–brain barrier for adrenaline, *Science* **129**:1226–1227 (1959).

50. Y. Tochino and L. S. Schanker, Transport of serotonin and norepinephrine by the rabbit choroid plexus *in vitro*, *Biochem. Pharmacol.* **14**:1557–1566 (1965).

51. C. W. M. Wilson and B. B. Brodie, The absence of blood–brain barrier from certain areas of the central nervous system, *J. Pharmacol. Exp. Therap.* **133**:332–334 (1961).

52. T. Samorajski and B. H. Marks, Localization of tritiated norepinephrine in mouse brain, *J. Histochem. Cytochem.* **10**:392–399 (1962).

53. C. W. M. Wilson, A. W. Murray, and E. Titus, The effects of reserpine on uptake of epinephrine in brain and certain areas outside the blood–brain barrier, *J. Pharm. Exp. Therap.* **135**:11–16 (1962).

54. B. Hamberger and D. Masuoka, Localization of catecholamine uptake in rat brain slices, *Acta Pharmacol. Toxicol.* **22**:363–368 (1965).

55. K. D. Neame, Uptake of L-histidine, L-proline, L-tyrosine and L-ornithine by brain, intestinal mucosa, testis, kidney, spleen, liver, heart muscle, skeletal muscle and erythrocytes of the rat *in vitro*, *J. Physiol.* **162**:1–12 (1962).

56. G. Milhaud and J. Glowinski, Métabolisme de la dopamine-¹⁴C dans le cerveau du rat. Étude du mode d'administration, *Compt. Rend. Acad. Sci. Paris* **255**:203–205 (1962).

57. E. P. Kindwall and N. Weiner, The distribution and rates of formation of catecholamines in several regions of bovine brain, *J. Neurochem.* **13**:1523–1531 (1966).

58. M. Vogt, The concentration of sympathin in different parts of the central nervous system under normal conditions and after the administration of drugs, *J. Physiol. (London)* **123**:451–481 (1954).

59. J. Glowinski, S. H. Snyder and J. Axelrod, Subcellular localization of H[3]-norepinephrine in the rat brain and the effect of drugs, *J. Pharmacol. Exp. Therap.* **152**:282–292 (1966).

60. L. T. Potter and J. Axelrod, Subcellular localization of catecholamines in tissues of the rat, *J. Pharmacol. Exp. Therap.* **142**:291–298 (1963).

61. V. P. Whittaker, *in Progress in Brain Research*, Vol. 8, pp. 90–117 (1964).

62. J. Glowinski and L. Iversen, Regional studies of catecholamines in the rat brain—III: Subcellular distribution of endogenous and exogenous catecholamines in various brain regions, *Biochem. Pharmacol.* **15**:977–987 (1966).

63. J. Häggendal and B. Hamberger, Quantitative *in vitro* studies on noradrenaline uptake and its inhibition by amphetamine, desipramine, and chlorpromazine, *Acta Physiol. Scand.* **70**:277–280 (1967).

64. B. Hamberger and T. Malmfors, Uptake and release of α-methylnoradrenaline *in vitro* after reserpine treatment, *Acta Physiol. Scand.* **70**:412–418 (1967).

65. S. B. Ross and A. L. Renyi, Blocking action of sympathomimetic amines on the uptake of tritiated noradrenaline by mouse cerebral cortex tissues *in vitro*, *Acta Pharmacol. Toxicol.* **21**:226–239 (1964).

66. M. Holzbauer and M. Vogt, Depression by reserpine of the noradrenaline concentration in the hypothalamus of the cat, *J. Neurochem.* **1**:8–11 (1956).

67. A. Pletscher, P. A. Shore and B. B. Brodie, Serotonin as a mediator of reserpine action in brain, *J. Pharmacol. Exp. Therap.* **116**:84–89 (1956).

68. J. Glowinski, J. Axelrod, and L. L. Iversen, Regional studies of catecholamines in the rat brain. IV. Effects of drugs on the disposition and metabolism of H^3-norepinephrine and H^3-dopamine, *J. Pharmacol. Exp. Therap.* **153**:30–41 (1966).

69. B. B. Brodie, J. S. Olin, R. G. Kuntzman, and P. A. Shore, Possible interrelationship between release of brain norepinephrine and serotonin by reserpine, *Science* **125**:1293–1294 (1957).

70. F. B. Hughes and B. B. Brodie, The mechanism of serotonin and catecholamine uptake by platelets, *J. Pharmacol. Exp. Therap.* **127**:96–102 (1959).

71. L. L. Iversen, J. Glowinski, and J. Axelrod, The uptake and storage of H^3-norepinephrine in the reserpine-pretreated rat heart, *J. Pharmacol. Exp. Therap.* **150**:173–183 (1965).

72. H. Corrodi, T. Malmfors, and Ch. Sachs, Differences in the uptake of secondary catecholamines by the adrenergic nerves, *Acta Physiol. Scand.* **67**:358–362 (1966).

73. T. Malmfors, Studies on adrenergic nerves. The use of rat and mouse iris for direct observations on their physiology and pharmacology at cellular and subcellular levels, *Acta Physiol. Scand.* **64**:Suppl. **248**:1–93 (1965).

74. J. Glowinski, L. L. Iversen, and J. Axelrod, Storage and synthesis of norepinephrine in the reserpine-treated rat brain, *J. Pharmacol. Exp. Therap.* **151**:385–399 (1966).

75. J. Glowinski and J. Axelrod, Effect of drugs on the uptake, release, and metabolism of H^3-norepinephrine in the rat brain, *J. Pharmacol. Exp. Therap.* **149**:43–49 (1965).

76. S. B. Ross and A. L. Reyni, *In vitro* inhibition of noredrenaline-^{3}H uptake by reserpine and tetrabenazine in mouse cerebral cortex tissues, *Acta Pharmacol. Toxicol.* **24**:73–88 (1966).

77. S. B. Ross and A. L. Renyi, *In vivo* inhibitions of ^{3}H-noradrenaline uptake by mouse brain slices *in vitro*, *J. Pharmacol. Pharmacol.* **18**:322–323 (1966).

78. A. Dahlström, K. Fuxe, and N.-Å. Hillarp, Site of action of reserpine, *Acta Pharmacol. Toxicol.* **22**:277–292 (1965).

79. B. Hamberger, Reserpine-resistant uptake of catecholamines in isolated tissues of the rat, *Acta Physiol. Scand.* **71**:Suppl. **295**:1–56 (1967).

80. J. Glowinski and J. Axelrod, Effects of drugs on the disposition of H^3-norepinephrine in the rat brain, *Pharmacol. Rev.* **18**:775–785 (1966).

81. Y. Tochino and L. S. Schanker, Serum and tissue factors that inhibit amine transport by the choroid plexus *in vitro*, *Am. J. Physiol.* **210**:1229–1233 (1966).

82. M. Draškoci, W. Feldberg, and P. S. R. K. Haranath, Passage of circulating adrenaline into perfused cerebral ventricles and subarachnoidal space, *J. Physiol.* (*London*) **150**:34–49 (1960).

83. S. Udenfriend, H. Weissbach, and D. F. Bogdanski, Biochemical findings relating to the action of serotonin, *Ann. N.Y. Acad. Sci.* **66**:602–608 (1957).

84. D. W. Woolley, *The Biochemical Bases of Psychoses*, p. 78, John Wiley, New York (1962).

85. S. Schanberg and N. J. Giarman, Uptake of 5-hydroxytryptophan by rat brain, *Biochim. Biophys. Acta* **41**:556–558 (1960).

86. S. M. Schanberg, A study of the transport of 5-hydroxytryptophan and 5-hydroxytryptamine (serotonin) into brain, *J. Pharmacol. Exp. Therap.* **139**:191–200 (1963).

87. S. E. Smith, Uptake of 5-hydroxy[^{14}C]tryptophan by rat and dog brain slices, *Brit. J. Pharmacol.* **20**:178–189 (1963).

88. E. Costa and M. H. Aprison, Distribution of intracarotidly injected serotonin in the brain, *Am. J. Physiol.* **192**:95–100 (1958).

89. N. T. Kärki and M. K. Paasonen, The influence of iproniazid and raunescine on the penetration of 5-hydroxytryptamine into brain tissue from the circulation, *Acta Pharmacol. Toxicol.* **16**:20–27 (1959).

90. M. Bulat and Z. Supek, The penetration of 5-hydroxytryptamine through the blood–brain barrier, *J. Neurochem.* **14**:265–271 (1967).

91. W. Lichtensteiger, U. Mutzner, and H. Langemann, Uptake of 5-hydroxytryptamine and 5-hydroxytryptophan by neurons of the central nervous system normally containing catecholamines, *J. Neurochem.* **14**:489–497 (1967).

92. J. Axelrod and J. K. Inscoe, The uptake and binding of circulating serotonin and the effect of drugs, *J. Pharmacol. Exp. Therap.* **141**:161–165 (1963).

93. M. Ritzén, L. Hammarström and S. Ullberg, Autoradiographic distribution of 5-hydroxytryptamine and 5-hydroxytryptophan in the mouse, *Biochem. Pharmacol.* **14**:313–321 (1965).

94. L. L. Cohen and K. C. Huang, Intestinal transport of tryptophan and its derivatives, *Am. J. Physiol.* **206**:647–652 (1964).

95. A. Inouye, K. Kataoka, and J. Shinagawa, Uptake of 5-hydroxytryptamine by sub-cellular particles of rabbit brain, *Nature* **198**:291–293 (1963).

96. C. N. Gillis, N. J. Giarman, and D. X. Freedman, Retention of 5-hydroxytryptamine by subcellular fractions of rat brain homogenates, *Biochem. Pharmacol.* **13**:1457–1474 (1964).

97. J. D. Robinson, J. H. Anderson, and J. P. Green, The uptake of 5-hydroxytryptamine and histamine by particulate fractions of brain, *J. Pharmacol. Exp. Therap.* **147**:236–243 (1965).

98. D. Palaić, I. H. Page, and P. A. Khairallah, Uptake and metabolism of [^{14}C]serotonin in rat brain, *J. Neurochem.* **14**:63–69 (1967).

99. A. N. Davison, Physiological role of monoamine oxidase, *Physiol. Rev.* **38**:729–747 (1958).

100. B.-E. Roos, On the occurrence and distribution of 5-hydroxyindoleacetic acid in brain, *Life Sci.* **1**:25–27 (1962).

101. P. A. Shore, A. Pletscher, E. G. Tomich, A. Carlsson, R. Kuntzmann and B. B. Brodie, Role of brain serotonin in reserpine action, *Ann. N.Y. Acad. Sci.* **66**:609–615 (1957).

102. M. K. Paasonen and N. J. Giarman, Brain levels of 5-hydroxytryptamine after various agents, *Arch Intern. Pharmacodyn.* **114**:189–200 (1958).

103. B. Rose and J. S. L. Browne, The distribution and rate of disappearance of intravenously injected histamine in the rat, *Am. J. Physiol.* **124**:412–420 (1938).

104. B. N. Halpern, Th. Neveu, and C. W. M. Wilson, The distribution and fate of radioactive histamine in the rat, *J. Physiol. (London)* **147**:437–449 (1959).

105. J. D. Robinson and J. P. Green, Presence of imidazoleacetic acid riboside and ribotide in rat tissues, *Nature* **203**:1178–1179 (1964).

106. J. P. Green, Uptake and binding of histamine, *Fed. Proc.* **26**:211–218 (1967).

107. S. H. Snyder, J. Axelrod, and H. Bauer, The rate of C^{14}-histamine in animal tissues, *J. Pharmacol. Exp. Therap.* **144**:373–379 (1964).

108. R. W. Schayer, Biogenesis of histamine, *J. Biol. Chem.* **199**:245–250 (1952).

109. T. White, Formation and catabolism of histamine in brain tissue *in vitro*, *J. Physiol. (London)* **149**:34–42 (1959).

110. T. White, Formation and catabolism of histamine in cat brain *in vivo*, *J. Physiol.* (*London*) **152**:299–308 (1960).

111. S. H. Snyder (*sic*), J. Glowinski, and J. Axelrod, The physiologic disposition of H^3-histamine in the rat brain, *J. Pharmacol. Exp. Therap.* **153**:8–14 (1966).

112. R. W. Schayer, Catabolism of physiological quantities of histamine *in vivo*, *Physiol. Rev.* **39**:116–126 (1959).

113. M. Draškoci, W. Feldberg, K. Fleischhauer, and P. S. R. K. Haranath, Absorption of histamine into the blood stream on perfusion of the cerebral ventricles, and its uptake by brain tissue, *J. Physiol.* (*London*) **150**:50–66 (1960).

114. K. Kataoka and E. de Robertis, Histamine in isolated small nerve endings and synaptic vesicles of rat brain cortex, *J. Pharmacol. Exp. Therap.* **156**:114–125 (1967).

115. H. M. Adam, *in Regional Neurochemistry* (S. S. Kety and J. Elkes, eds.) pp. 293–305, Pergamon Press, London (1961).

116. I. L. Thon and B. Uvnäs, Mode of storage of histamine in mast cells, *Acta Physiol. Scand.* **67**:455–470 (1966).

117. C.-H. Åborg, J. Novotny, and B. Uvnäs, Ionic binding of histamine in mast cell granules, *Acta Physiol. Scand.* **69**:276–283 (1967).

118. H. M. Adam and H. K. A. Hye, Concentration of histamine in different parts of brain and hypophysis of cat and its modification by drugs, *Brit. J. Pharmacol.* **28**:137–152 (1966).

119. H. M. Adam, H. K. A. Hye, and N. G. Waton, Studies on uptake and formation of histamine by hypophysis and hypothalamus in the cat, *J. Physiol.* (*London*) **175**:70–71 P (1964).

120. K. D. Neame, Uptake of histidine, histamine and other imidazole derivatives by brain slices, *J. Neurochem.* **11**:655–662 (1964).

121. R. Nakamura, The transport of histidine and methionine in rat brain slices, *J. Biochem.* (*Tokyo*) **53**:314–322 (1963).

122. A. Schweitzer, E. Stedman, and S. Wright, Central action of anticholinesterases, *J. Physiol.* (*London*) **96**:302–336 (1939).

123. J. C. Eccles, R. M. Eccles, and P. Fatt, Pharmacological investigations on a central synapse operated by acetylcholine, *J. Physiol.* (*London*) **131**:154–169 (1956).

124. D. R. Curtis, J. C. Eccles, and R. M. Eccles, Pharmacological studies on spinal reflexes, *J. Physiol.* (*London*) **136**:420–434 (1957).

125. D. R. Curtis and R. M. Eccles, The effect of diffusional barriers upon the pharmacology of cells within the central nervous system, *J. Physiol.* (*London*) **141**:446–463 (1958).

126. J. Schuberth, A. Sundwall, B. Sörbo, and J.-O. Lindell, Uptake of choline by mouse brain slices, *J. Neurochem.* **13**:347–352 (1966).

127. R. L. Polak and M. M. Meeuws, The influence of atropine on the release and uptake of acetylcholine by the isolated cerebral cortex of the rat, *Biochem. Pharmacol.* **15**:989–992 (1966).

128. J. Schuberth and A. Sundwall, Effects of some drugs on the uptake of acetylcholine in cortex slices of mouse brain, *J. Neurochem.* **14**:807–812 (1967).

129. R. Creese and D. B. Taylor, Effect of *d*-tubocurarine on uptake of labeled carbaminoyl choline in brain slices of the rat, *Nature* **206**:310 (1965).

130. R. Creese and D. B. Taylor, Effect of atropine on the uptake of labeled carbachol by rat brain slices, *Life Sci.* **4**:1545–1547 (1965).

131. R. Creese and D. B. Taylor, Entry of labeled carbachol in brain slices of the rat and the action of *d*-tubocurarine and strychnine, *J. Pharmacol. Exp. Therap.* **157**:406–419 (1967).

132. F. W. Schueler, A new group of respiratory paralyzants. I. The "hemicholiniums", *J. Pharmacol. Exp. Therap.* **115**:127–143 (1955).

133. Y. Tochino and L. S. Schanker, Active transport of quaternary ammonium compounds by the choroid plexus *in vitro*, *Am J. Physiol.* **208**:666–673 (1965).

134. P. J. G. Mann, M. Tennenbaum, and J. H. Quastel, On the mechanism of acetylcholine formation in brain *in vitro*, *Biochem. J.* **32**:243–261 (1938).
135. K. A. C. Elliott and N. Henderson, Factors affecting acetylcholine found in excised rat brain, *Am. J. Physiol.* **165**:365–374 (1951).
136. K. Krnjević and J. F. Mitchell, Diffusion of acetylcholine in agar gels and in the isolated rat diaphragm, *J. Physiol.* (*London*) **153**:562–572 (1960).
137. R. A. McCance and E. M. Widdowson, A method of breaking down the body weights of living persons into terms of extracellular fluid, cell mass and fat, and some applications of it to physiology and medicine, *Proc. Roy. Soc. Lond. Ser. B.* **138**:115–130 (1951).
138. C. A. Keele and E. Neil, *Samson-Wright's Applied Physiology*, pp. 13–14, Oxford University Press, London (1965).
139. J. Stubbs and J. Pennybacker, Reduction of intracranial pressure with hypertonic urea, *Lancet i*:1094–1097 (1960).
140. D. J. Reed and D. M. Woodbury, Effect of hypertonic urea on cerebrospinal fluid pressure and brain volume, *J. Physiol.* (*London*) **164**:252–264 (1962).
141. R. V. Coxon, *in Comparative Neurochemistry* (D. Richter, ed.) pp. 261–274, Pergamon Press, London (1964).
142. C. R. Kleeman, H. Davson, and E. Levin, Urea transport in the central nervous system, *Am. J. Physiol.* **203**:739–747 (1962).
143. M. W. B. Bradbury and R. V. Coxon, The penetration of urea into the central nervous system at high blood levels, *J. Physiol.* (*London*) **163**:423–435 (1962).
144. J. C. Schoolar, C. F. Barlow, and L. J. Roth, The penetration of carbon-14 urea into cerebrospinal fluid and various areas of the cat brain, *J. Neuropath. Exp. Neurol.* **19**:216–227 (1960).
145. R. V. Coxon, *in Metabolism of the Nervous System* (D. Richter, ed.) pp. 153–157, Pergamon Press, London (1957).
146. K. Bloch, B. N. Berg, and D. Rittenberg, The biological conversion of cholesterol to cholic acid, *J. Biol. Chem.* **149**:511–517 (1943).
147. M. D. Morris and I. L. Chaikoff, Concerning incorporation of labeled cholesterol, fed to the mothers, into brain cholesterol of 20-day-old suckling rats, *J. Neurochem.* **8**:226–229 (1961).
148. R. Clarenburg, I. L. Chaikoff, and M. D. Morris, Incorporation of injected cholesterol into the myelinating brain of the 17-day-old rabbit, *J. Neurochem.* **10**:135–143 (1963).
149. J. Dobbing, The entry of cholesterol into rat brain during development, *J. Neurochem.* **10**:739–742 (1963).
150. J. W. Millan and A. Hess, The blood–brain barrier: an experimental study with vital dyes, *Brain* **81**:248–257 (1958).
151. A. N. Davison, J. Dobbing, R. S. Morgan, and G. P. Wright, The deposition and disposal of (4-^{14}C)cholesterol in the brain of growing chickens, *J. Neurochem.* **3**:89–94 (1958).
152. A. N. Davison, J. Dobbing, R. S. Morgan, and G. P. Wright, Metabolism of myelin: the persistence of (4-^{14}C)cholesterol in the mammalian nervous system, *Lancet i*:658–662 (1959).
153. A. N. Davison and M. Wajda, Persistence of cholesterol-4-^{14}C in the central nervous system, *Nature* **183**:1606–1607 (1959).
154. A. A. Khan and J. Folch-Pi, Cholesterol turnover in brain subcellular particles, *J. Neurochem.* **14**:1099–1105 (1967).
155. P. A. Srere, I. L. Chaikoff, S. S. Treitman, and L. S. Burstein, The extrahepatic synthesis of cholesterol, *J. Biol. Chem.* **182**:629–634 (1950).
156. J. R. Klein and N. S. Olsen, Distribution of intravenously injected glutamate, lactate, pyruvate and succinate between blood and brain, *J. Biol. Chem.* **167**:1–5 (1947).
157. P. J. McMillan, G. W. Douglas, and R. A. Mortensen, Incorporation of C^{14} of acetate-1-C^{14} and pyruvate-2-C^{14} into brain cholesterol in the intact rat, *Proc. Soc. Exp. Biol. Med.* **96**:738–740 (1957).

158. R. M. O'Neal, R. E. Koeppe, and E. I. Williams, Utilization *in vivo* of glucose and volatile fatty acids by sheep brain for the synthesis of acidic amino acids, *Biochem. J.* **101**:591–597 (1966).

159. N. A. Peterson and I. L. Chaikoff, Uptake of intravenously-injected [4-^{14}C]cortisol by adult rat brain, *J. Neurochem.* **10**:17–23 (1963).

160. K. J. Blackburn, P. C. French, and R. J. Merrills, 5-Hydroxytryptamine uptake by rat brain *in vitro*, *Life Sci.* **6**:1653–1663 (1967).

161. S. B. Ross and A. L. Renyi, Accumulation of tritiated 5-hydroxytryptamine in brain slices, *Life Sci.* **6**:1407–1415 (1967).

162. S. H. Snyder, A. Green, E. D. Hendley, and E. Gfeller, Noradrenaline: kinetics of accumulation into slices from different regions of rat brain, *Nature* **218**:174–176 (1968).

163. D. F. Bogdanski, A. Tissari, and B. B. Brodie, Role of sodium, potassium, ouabain and reserpine in uptake, storage and metabolism of biogenic amines in synaptosomes, *Life Sci.* **7**:Part I, 419–428 (1968).

164. M. Bulat and Z. Supek, Passage of 5-hydroxytryptamine through the blood–brain barrier, its metabolism in the brain and elimination of 5-hydroxyindoleacetic acid from the brain tissue, *J. Neurochem.* **15**:383–389 (1968).

Chapter 16

INSULIN ACTION

Ole J. Rafaelsen

Head of Psychochemical Laboratory
University Clinic of Psychiatry
Rigshospitalet, Copenhagen

and

Erling Mellerup

Research Assistant, Psychochemical Laboratory
University Clinic of Psychiatry
Rigshospitalet, Copenhagen

I. GENERAL ACTION OF INSULIN

Insulin has been known since 1922, but in spite of intense research activity, the biochemical function of this hormone is not fully understood. Much is known of the physiological changes which follow administration of insulin to an organism, but the basic biochemical effect (or effects) remains to be clarified.

With our present knowledge insulin seems to act in at least two different ways: by increasing transport of some sugars and other metabolites and by influencing the rate of synthesis of some enzymes.

The most pronounced effect of an insulin injection is a rapid fall of glucose and free fatty acid concentration of blood. Insulin increases glucose uptake of several body tissues, especially muscle and adipose tissue, and decreases the release of free fatty acid from adipose tissue. In muscle and fat cells the glucose concentration is low under all conditions, and the transport of glucose into the cells is therefore rate limiting for glucose metabolism. Besides the increase of glucose uptake, insulin also increases the uptake of some other sugars. More important, however, is that the uptake of amino acids is increased.

Liver cells are permeable for glucose, and thus the glucose concentration in these cells is approximately the same as in blood. Insulin does not increase glucose uptake in the liver, but the hormone exerts an influence on glucose release from the liver.

The other type of insulin effect is the influence on the rate of synthesis of some enzymes.

These enzymes are involved in the synthesis and degradation of glycogen. The influence of insulin on enzymatic synthesis cannot be demonstrated until some hours after insulin administration, whereas the facilitation of transport can be seen after some minutes. The mode of action of insulin on enzyme synthesis is not known in details. It is, for example, unknown whether insulin penetrates the cell or performs its effect from the cell membrane. It has been shown, however, that besides increasing amino acid uptake, insulin also stimulates RNA synthesis.[1]

The effect of insulin on the formation of enzymes is not only caused by changed concentrations of certain metabolites deriving from permeability changes. This can be seen from the fact that insulin increases the amount of some enzymes, especially in the liver.

The two main lines of insulin action—independently and together—cause an increase in glucose oxidation and glycogen synthesis in muscle and liver, an increased protein synthesis, and in liver and adipose tissue an increased lipid synthesis.

Besides the *direct* insulin effects on various tissues, there are *indirect* effects caused by the fall in blood sugar. This means that it may be difficult to decide whether the local effects seen after insulin administration are due to the fact that insulin has reached the tissue in question and has acted thereon, or whether the changes observed are conditioned by the hypoglycemia. These indirect effects may naturally include other metabolic changes provoked by insulin per se or by the hypoglycemic state, e.g., the release of other hormones.

These problems are of special importance for the central nervous system because this organ system is the first to signal important consequences of a general hypoglycemia: clouding of consciousness and eventually coma. These cerebral effects of reduced glucose supply to the brain are so evident that it has been difficult to ascertain whether insulin also exerts direct effects on brain tissue. During later years there has, however, appeared decisive evidence for a direct action of insulin on central and peripheral nervous tissue.

II. *IN VITRO* STUDIES

When tissues are examined *in vitro*, the risk of indirect insulin effects is eliminated. Such techniques have yielded good results with adipose and muscle tissues, and methods have been developed to estimate insulin activity on rat diaphragm[2] and rat adipose tissue and isolated cells,[3] these tissues being sensitive to very small amounts of insulin.

As the central nervous system is very sensitive to a decrease in glucose supply, *in vitro* experiments should be very suitable, but a major difficulty is the inhomogeneity of the central nervous system, the results therefore being very dependent on the type of preparation used.

Rafaelsen[4] found that insulin in small concentrations increased glucose

TABLE I

In Vitro Studies

Preparation	Parameter studied	Amounts of insulin	Comments	Reference
Brain slice (rat)	Glucose uptake: 10% increase	10^{-2}-10^{-1} U/ml	Only in "first" brain slice	(5)
Brain slice (rat)	Glucose incorporation into lipid: no effect. Acetate incorporation into lipid: 100% increase	3 U/ml		(7)
Brain slice (mouse)	Anaerobic glycolysis: 30% increase	0.15 U/ml	The slices were preincubated with insulin	(8)
Anterior pituitary (calf and rat)	Gluxose uptake: 25% increase Alloxan diabetes: reduced glucose uptake normalized by insulin *in vivo*	1 U/ml		(6)
Spinal cord (rat)	Glucose uptake: 25–100% increase Alloxan diabetes: reduced glucose uptake normalized by insulin *in vitro*	10^{-4}-10^{-1} U/ml		(4)
Spinal cord (rat)	Glucose conversion to carbon dioxide and glycogen formation: no effect	10^{-4}-10^{-1} U/ml		(9)
Spinal cord (rat)	Glucose incorporation into lipid: no effect. Acetate incorporation into lipid: 150% increase	10^{-4}-10^{-1} U/ml		(10)
Isolated sciatic nerve (rabbit)	Glucose uptake: a release was turned to an uptake	0.3 U/ml		(11)

uptake of isolated rat spinal cord by 40–80%. Parallel experiments with rat brain slices showed only a slight increase in glucose uptake, and only in "first" brain slices (with only one cut surface), whereas no increase was seen in "second" brain slices[5]; much higher insulin concentrations were necessary in the brain slice than in the spinal cord experiments.

When slices of calf and rat anterior pituitary gland were incubated, insulin increased glucose uptake by 25% both when the hormone was added to the medium and when insulin was injected into alloxan diabetic rats before sacrifice.[6]

Both with brain slices and with spinal cord pieces the results obtained have depended on the parameters studied. Brain slices from growing and adult rats incorporated acetate into cholesterol and fatty acid to a greater extent when insulin was added to the medium, but the incorporation of glucose was not influenced by insulin.[7] It has also been shown that anaerobic glycolysis rises up to 30 % after preincubation of brain slices with insulin.[8]

Also, in spinal cord experiments no effect of insulin was seen when incorporation of glucose into glycogen or lipids was studied[9]; on the other hand, an increase was seen when acetate incorporation into rat spinal cord lipid was studied.[10]

Nervous tissue outside the central nervous system has also been proved to be insulin-sensitive. In rabbit sciatic nerve, insulin caused an increase of glucose uptake and of glucose incorporation into lipid.[11] In embryonic sensory and sympathetic nerve cells Levi-Montalcini[12] compared nerve growth factor and insulin. Insulin in these experiments stimulated a number of biochemical processes such as glucose utilization, uridine incorporation into RNA, and acetate incorporation into lipids.

In cat superior cervical ganglion, insulin inhibited transmission and enhanced the effect of several ganglion blocking agents.[13]

III. ANIMAL EXPERIMENTS *IN VIVO*

A. Direct Action

The first question that arises is whether insulin reaches the central nervous system or not. Elucidation of this problem has been tried by insulin-^{131}I studies or by measuring insulinlike activity in spinal fluid. Earlier investigations were negative,[14,15] but criticism can be raised from a theoretical point of view against using radioactively labeled insulin with a short biological half-life for a study involving passage of the blood–brain barrier and against measuring insulinlike activity in spinal fluid with not very sensitive bioassays. Recently Deckert et al.[16] using immunoassay techniques found human spinal fluid insulin values about 10 % of serum levels.

Margolis and Altszuler,[17] using radioimmunoassay, found in dogs an average insulin content of spinal fluid of 3 μU/ml or some 25 % of plasma insulin. Insulin was also infused intravenously, and when plasma insulin was increased to 150–300 μU/ml, CSF insulin increased up to tenfold. It was found that it took 3–4 hr for insulin levels to stabilize in the CSF.

When the *direct* effect of insulin is studied in *in vivo* experiments, it is advisable to take measures to avoid hypoglycemia. This can be done by giving glucose together with insulin, or by using such minute amounts of the hormone that no major changes in blood glucose levels ensue. Few workers have realized the importance of this consideration.

The route of administration of insulin is of similar importance for the results obtained. The appropriate route of administration has, for example, been decisive for the demonstration of a direct action of insulin on liver.[18] Different routes of administration (and different species) may be the reason why Park et al.[19,20] found no effect of intravenously administered insulin on the transport of several hexoses and pentoses into rat brain, whereas Sachs and Bakshy[21] did find an increase in the distribution volume of xylose and arabinose of cat brain after subcutaneous administration of insulin.

De Ropp and Snedeker[22] measured several amino acids in rat brain after treatment with glucose or insulin or glucose plus insulin. In the latter experiment (where hypoglycemia was avoided) they found an increase in γ-aminobutyric acid and alanine which was greater than the increase seen when glucose was given alone.

TABLE IIA

Direct Action
Animal Experiments *In Vivo*

Preparation	Parameter studied	Amounts of insulin	Comments	Reference
Infusion of hexoses and pentoses (rat)	Distribution volume in brain: no effect	About 25 U/kg i.v. 2 hr	20 U were given at the start of infusion. 5 U inf. together with sugar	(19, 20)
Infusion of hexoses and pentoses (cat)	Distribution volume in brain: increased	5 U/kg s.c. 3 hr		(21)
Inj. of 2 g glucose per rat	Amino acids: increase in GABA and Ala	20 U/kg i.p. 5 hr	With glucose alone, there was also an increase, but smaller	(22)
Dogs	Blood glucose, CSF-glucose: decrease of both	5–40 U/dog suboccipitally 0–4 hr		(24)
Vagotomized dogs	Blood glucose: no change. CSF-glucose: decrease	5–40 U/dog suboccipitally 0–4 hr		(25)
Rats	Glycogen content of the brain: increase	0.1 U/rat suboccipitally 4 hr		(26)
Cerebral arteriovenous difference (dogs)	Glucose uptake in brain: increase	2 U/kg s.c. 0.5–3 hr		(23)

TABLE IIB

Indirect Action
Animal Experiments *In Vivo*

Preparation	Parameter studied	Amounts of insulin	Comments	Reference
Brain (cat)	Content of pyruvic acid, glucose, glycogen, phosphocreatine: decreased	5 U/kg i.m. 2.5–7 hr		(27)
Brain (rat)	Free amino acids. Decrease in Glu, Gln and Ala. Increase in Asp.	50–100 U/kg i.p. 2 hr		(28, 29)
Brain (rat)	Free amino acids. Decrease in Glu, Gln and GABA. Increase in Asp.	350 U/kg i.p. 4 hr		(30)
Brain (rat)	Free amino acids. Decrease in Glu, Gln and GABA. Increase in ethanolamine and phospho-ethanolamine	100 U/kg s.c. 5 hr		(31)
Brain (rat)	Free amino acids. Decrease in Glu, Gln, GABA and Gly. Increase in Asp and ethanolamine	100 U/kg i.p. 3 hr		(22)
Brain (rat)	Incorporation of ^{32}P-phosphate into phosphorylcholine and phosphoethanolamine: uptake into the latter depressed	40–120 U/kg i.p. 3 hr		(33)

A Russian group[23] has studied glucose uptake in dog brain using arteriovenous differences, but without measuring cerebral blood flow. After subcutaneous administration of insulin, cerebral glucose uptake rose by 50% although blood glucose decreased to 50%.

Some very important experiments were reported by Chowers, Lavy and Halpern in 1961 and 1966.[24,25] Five to 40 units of insulin were injected suboccipitally (intraventricularly) to dogs, and glucose levels were estimated every 30 min in blood and cerebrospinal fluid. A decrease in glucose level of both was noted during the first 30 min and reached a minimum after 90–120 min. The reduction was both absolutely and relatively more pronounced in cerebrospinal fluid (50 mg/100 ml) than in blood (36 mg/100 ml), and it also lasted longer before preinjection values were regained in CSF glucose than in blood glucose. Insulin labeled with ^{131}I was injected intraventricularly in some of the experiments, and no radioactivity was detected in blood, thyroid, or urine during the first 24 hr. In more recent investigations the dogs were bilaterally vagotomized before insulin injections; in these experiments there was only a slight decrease in blood glucose, while the decline in CSF glucose was of the same magnitude as reported earlier. The authors concluded that insulin exerts a direct effect on the central nervous system by increasing glucose uptake, and they suggested an indirect effect on insulin secretion by the pancreas via stimulation of brain parasympathetic centers.

Also, in rats, suboccipital, intraventricular administration of insulin influenced brain carbohydrate metabolism. Insulin in a dose of 0.1 U/rat caused in 4 hr a 10–20 % increase in brain glycogen.[26]

B. Indirect Action

As mentioned before, insulin administration leads to fall in blood glucose, and as the brain is dependent on a steady supply of glucose, this organ is the first to show disturbance of function with clouding of consciousness and coma due to the decrease in glucose uptake.

This effect of insulin on the central nervous system is primarily indirect, but it cannot be excluded that direct effects are also involved. The biochemical changes which follow diminished cerebral glucose uptake have been studied by several workers.

Decrease in pyruvate, glucose, glycogen, phosphocreatinine, and ATP in brain after intramuscular injection of insulin was reported by Olsen and Klein.[27] Amino acids are also changed after insulin hypoglycemia. Brain glutamic acid, glutamine, alanine, and γ-amino buturic acid are reduced, whereas aspartic acid is increased.[28–30] Similar results have been reported by De Ropp and Snedeker,[22] who also found glycine decreased and ethanolamine increased. Knauff and Böck[31] found the same pattern, except that they saw no decrease in aspartic acid. In total there was a decrease in brain amino acids after diminished glucose uptake, which is understandable as glucose very quickly enters the brain amino acid pool.[32]

The incorporation of orthophosphate-^{32}P into phosphorylcholine and phosphoethanolamine has been studied[33]; here insulin hypoglycemia caused a significant depression of the uptake into phosphoethanolamine, phosphatidylcholine, and phosphatidylethanolamine, but had no influence on the uptake into phosphorylcholine.

IV. HUMAN STUDIES

Measurements of human brain blood arteriovenous differences have also been applied to determine changes during insulin hypoglycemia. Himwich[34] and later Kety and co-workers[35] found that both cerebral glucose

uptake and oxygen uptake were reduced during insulin hypoglycemia, the
fall in glucose uptake being greater than that of oxygen consumption. In
experiments where the fall in blood glucose was less pronounced, no change
of cerebral metabolic rate for oxygen was seen, although the glucose uptake
was reduced.[36,37]

Cerebral arteriovenous glutamate differences have been studied,[38] and
it was found that under normal conditions no glutamate was taken up by the

TABLE III

Human Studies

Preparation	Parameter studied	Amounts of insulin Duration of experiment	Comments	Reference
Arteriovenous differences	Glucose uptake: decrease Oxygen uptake: decrease Blood flow: no change	110–500 U total about 4 hr		(34)
Arteriovenous differences	Glucose uptake: decrease Oxygen uptake: lesser decrease Blood flow: no change			(35)
Arteriovenous differences	Glucose uptake: decrease Oxygen uptake: increase Blood flow: no change	0.15–0.25 U/kg 20 min		(36)
Arteriovenous differences	Glucose uptake: decrease Oxygen uptake: no change Blood flow: no change	0.1 U/kg 20 min		(37)
Arterio venous differences	Glucose uptake: decrease Oxygen uptake: lesser decrease Glutamic acid uptake: increase	100 U total 10 min–4.5 hr		(38)
Cerebral blood flow. Infusion of glucose and/or insulin	Glucose uptake: slight increase after glucose alone. 50% increase after insulin + glucose	24 U total 20-35 min	Insulin was infused together with glucose or saline	(39)
Cerebral blood flow. Inj. of glucose and insulin	Demonstration of glucose threshold in brain which is insulin-sensitive	0.1 U/kg i.v. 20–60 min		(40)

brain (but perhaps some was released); during insulin hypoglycemia, on the other hand, glutamate, lactate, and pyruvate were taken up.

Recently two groups have tried to elucidate the direct action of insulin on human brain.

Gottstein and co-workers[37,39] determined cerebral blood flow by the nitrous oxide method in all their experiments, in addition, they determined cerebral arteriovenous differences of glucose, oxygen, carbon dioxide, lactic acid, and pyruvic acid. After basal values were obtained, glucose or insulin or glucose plus insulin was given by intravenous infusion at a fixed speed. Thirty minutes later new determinations were made. Glucose infusion alone caused a slight but insignificant increase in cerebral glucose uptake in spite of a considerable increase in blood glucose. Insulin infused alone caused a considerable decrease in cerebral glucose uptake—perhaps secondary to fall in blood glucose. Combined infusion of glucose and insulin caused a significant increase in cerebral glucose uptake which rose from 5.62 to 8.29 mg/100 g/min; this finding was even more remarkable as blood glucose levels did not rise as high as when glucose was infused alone. These findings were obtained in persons without symptoms or signs of cerebral or endocrine disorder. Results obtained in diabetic patients confirmed the finding of an increase of cerebral glucose uptake caused by insulin. Among other remarkable results obtained by Gottstein and his group is the constancy of oxygen uptake both when glucose uptake was reduced, and when it was increased. This may indicate that other metabolites than glucose were utilized when glucose uptake was low, and that the extra glucose taken up under the opposite circumstance might be converted to amino acids, for example. It is difficult to reach a final conclusion on glucose–oxygen ratios as these experiments have been of relatively short duration, and a state of equilibrium may not have been obtained; the problem needs further investigations.

Butterfield and co-workers[40] have previously demonstrated that there exists a blood glucose threshold below which peripheral tissues do not take up glucose, and they have shown that this threshold is lowered by insulin. By calculating brain glucose uptake under varying blood glucose values, both with and without insulin, they could demonstrate the existence of a similar and insulin-sensitive threshold for brain. According to these workers, brain glucose threshold was lower than that of peripheral tissues, and it responded more slowly to insulin stimulation.

V. CONCLUSION

Insulin exerts pronounced effects on central and peripheral nervous tissue in hypoglycemic conditions. In addition it can under certain circumstances be demonstrated that insulin increases glucose uptake by brain, spinal cord, and peripheral nerve. Direct action of insulin on pituitary tissue has been demonstrated and action on cerebral parasympathetic centers has been postulated. Insulin interaction with other hormones may play a prominent role in various aspects of nerve metabolism.

NOTE ADDED IN PROOF

The effect of insulin has been studied in the early phase, before hypoglycemia has developed, and it has been found that insulin treated rats during this phase accumulated 3 times as much ^{14}C-glucose in the free amino acids in brain.[41]

On the other hand, in a study with an isolated, perfused rat brain preparation, electrical activity, efflux of potassium, and glucose consumption were measured without and with insulin the perfusate. No effect of insulin was found for any of the parameters.[42]

VI. REFERENCES

1. I. G. Wool, in Action of Hormones on Molecular Processes (G. Litwack and D. Kritchevsky, eds.) pp. 422–469, Wiley, New York (1964).
2. J. Vallance-Owen and B. Hurlock, Estimation of plasma-insulin, Lancet L:68–70 (1954).
3. M. Rodbell, Metabolism of isolated fat cells, J. Biol. Chem. 239:375–380 (1964).
4. O. J. Rafaelsen, Action of insulin on carbohydrate uptake of isolated rat spinal cord, J. Neurochem. 7:33–44 (1961).
5. O. J. Rafaelsen, Action of insulin on glucose uptake of rat brain slices and isolated rat cerebellum, J. Neurochem. 7:45–51 (1961).
6. C. J. Goodner and N. Freinkel, Studies of anterior pituitary tissue in vitro; effects of insulin and experimental diabetes mellitus upon carbohydrate metabolism, J. Clin. Invest. 402:261–272 (1961).
7. E. Grossi, P. Paoletti, and M. Poggi, The effect of insulin on brain cholesterol and fatty acid biosynthesis, Wld. Neurol. 3:209–215 (1962).
8. M. Woods, J. Hunter, and D. Burk, Insulin: anti-insulin regulation of glucose metabolism in mouse brain, Fed. Proc. 171:339 (1958).
9. O. J. Rafaelsen and T. Clausen, Fate of glucose in isolated rat spinal cord and diaphragm incubated in the absence and presence of insulin, J. Neurochem. 7:52–59 (1961).
10. O. J. Rafaelsen, L. C. Adams, and R. A. Field, Insulin effect in vitro on acetate incorporation into lipids of isolated rat spinal cord, (To be published).
11. R. A. Field and L. C. Adams, Insulin response of peripheral nerve, Medicine 43:275–279 (1964).
12. R. Levi-Montalcini, in The Molecular Basis of Some Aspects of Mental Activity (O. Walaas, ed.) Vol. 1, pp. 385–388, Academic Press, London and New York (1966).
13. E. Minker and M. Koltai, Effect of insulin on synaptic transmission in cat's superior cervical ganglion, Naturwissenschaften 52:189–190 (1965).
14. N. Haugaard, M. Vaughan, E. S. Haugaard, and W. C. Stadie, Studies of radioactive injected labeled insulin, J. Biol. Chem. 208:549–563 (1954).
15. W. A. Mahon, J. Steinke, G. M. McKhann, and M. L. Mitchell, Measurement of I^{131}-insulin and of insulinlike activity in cerebrospinal fluid of man, Metabolism 11:416–420 (1962).
16. J. Deckert, J. Lyngsøe, and O. J. Rafaelsen, Insulin in human cerebrospinal fluid, Paper read at the 2. meeting of the European Diabetes Association, Aarhus/Denmark (1966).
17. R. U. Margolis and N. Altszuler, Insulin in the cerebrospinal fluid, Nature 215:1375–1376 (1967).
18. L. L. Madison and R. H. Unger, The physiologic significance of the secretion of endogenous insulin into the portal circulation, J. Clin. Invest. 37:631–639 (1958).
19. C. R. Park and L. H. Johnson, Effect of insulin on transport of glucose and galactose into cells of rat muscle and brain, Am. J. Physiol. 182:17–23 (1955).

20. C. R. Park, L. H. Johnson, J. H. Wright Jr., and H. Batsel, Effect of insulin on transport of several hexoses and pentoses into cells of muscle and brain, *Am. J. Physiol.* **191**:13–18 (1957).

21. J. Sacks and S. Bakshy, Insulin and Tissue Distribution of Pentose in Nephrectomized Cats, *Am. J. Physiol.* **189**:339–342 (1957).

22. R. S. De Ropp and E. H. Snedeker, Effect of drugs on amino acid levels in the rat brain: Hypoglycemic agents, *J. Neurochem.* **7**:128–134 (1961).

23. S. G. Genes, Effect of insulin on the brain, *Usp. sovrem. Biol.* **51**:188–202 (1961).

24. I. Chowers, S. Lavy, and L. Halpern, Effect of insulin administered intracisternally in dogs on the glucose level of the blood and the cerebrospinal fluid, *Exp. Neurol.* **3**:197–205 (1961).

25. I. Chowers, S. Lavy, and L. Halpern, Effect of insulin administered intracisternally on the glucose level of the blood and the cerebrospinal fluid in vagotomized dogs, *Exp. Neurol.* **14**:383–389 (1966).

26. E. T. Mellerup and O. J. Rafaelsen, Brain glycogen after intracisternal insulin injection, *J. Neurochem.* **16**:777–781 (1969).

27. N. S. Olsen and J. R. Klein, Effect of insulin hypoglycemia on brain glucose, glycogen, lactate, and phosphates, *Arch. Biochem.* **13**:343–347 (1947).

28. R. M. C. Dawson, Studies on the glutamine and glutamic acid content of the rat brain during insulin hypoglycemia, *Biochem. J.* **47**:386–391 (1950).

29. R. M. C. Dawson, Cerebral amino acids in fluoroacetate-poisoned, anaesthetised, and hypoglycaemic rats, *Biochim. Biophys. Acta* **11**:548–552 (1953).

30. R. O. Cravioto, G. Massieu, and J. J. Izquierdo, Free amino-acids in rat brain during insulin shock, *Proc. Soc. Exp. Biol. (N.Y.)* **78**:856–858 (1951).

31. H. G. Knauff and F. Böck, Über die freien Gehirnaminosäuren und das Äthanolamin der normalen Ratte, sowie über das Verhalten dieser Stoffe nach experimenteller Insulinhypoglykämie, *J. Neurochem.* **6**:171–182 (1961).

32. R. Vrba, H. S. Bachelard, and J. Krawczynski, Interrelationship between glucose utilization of brain and heart, *Nature* **197**:869–870 (1963).

33. G. B. Ansell and S. Spanner, The effect of insulin on the formation of phosphorylcholine and phosphorylethanolamine in the brain, *J. Neurochem.* **4**:325–331 (1959).

34. H. E. Himwich, K. M. Bowman, C. Daly, J. F. Fazekas, J. Wortis, and W. Goldfarb, Cerebral blood flow and brain metabolism during insulin hypoglycemia, *Am. J. Physiol.* **132**:640–647 (1941).

35. S. S. Kety, R. B. Woodford, M. H. Harmel, F. A. Freyhan, K. E. Appel, and C. F. Schmidt, Cerebral blood flow and metabolism in schizophrenia, *Am. J. Psychiat.* **104**:765–770 (1948).

36. S. Eisenberg and H. S. Seltzer, The cerebral metabolic effects of acutely induced hypoglycemia in human subjects, *Metabolism* **11**:1162–1168 (1962).

37. U. Gottstein and K. Held, Insulinwirkung auf den menschlichen Hirnmetabolismus von Stoffwechselgesunden und Diabetikern, *Klin. Wschr.* **45**:18–23 (1967).

38. M. Gruss, Veränderungen des energieliefernden Hirnstoffwechsels in der Insulinhypoglykämie des Menschen, Dissertation Philipps-Universität zu Marburg/Lahn (1962).

39. U. Gottstein, K. Held, H. Sebening, and G. Walpurger, Der Glucoseverbrauch des menschlichen Gehirns unter dem Einfluss intravenöser Infusionen von Glucose, Glucagon und Glucose-Insulin, *Klin. Wschr.* **43**:965–975 (1965).

40. W. J. H. Butterfield, M. E. Abrams, R. A. Sells, G. Sterky, and M. J. Whichelow, Insulin sensitivity of the human brain, *Lancet* **I**:557–560 (1966).

41. K. Konitzer and S. Voigt, The effect of insulin on the incorporation of blood glucose into free amino acids of rat brain, Second International Meeting of the International Society for Neurochemistry (Milan 1969) p. 253.

42. H. Yamada and H. A. Sloviter, Does insulin have a direct effect on brain metabolism? *Fed. Proc.* **28**:574 (1969).

Chapter 17

NEUROSECRETION

Howard Sachs

Roche Institute of Molecular Biology
Nutley, New Jersey

I. INTRODUCTION

The structure and function of a class of neurons that have been termed "neurosecretory" form the basis for the subject of this chapter. Although Speidel[1] in 1919 described secretory-appearing neurons in the caudal region of the spinal cord of elasmobranchs, it remained for the pioneering work of E. Scharrer during the 1930's to focus attention on the widespread occurrence of hypothalamic "glandular neurons" in vertebrates. A voluminous literature has accumulated on the occurrence and structure of neurosecretory systems in both vertebrate and invertebrate forms, and within the past several years the results and conclusions of these studies have been admirably documented and reviewed by several of the leading workers in the field.[2-8] The studies of neurosecretion have consisted largely of morphological and histochemical investigations at the levels of light and electron microscopy interspersed with physiological studies directed to the delineation of the function of those neurons designated as "neurosecretory." Of such groups of neurosecretory cells, the hypothalamo–neurohypophyseal complex (HNC) of higher vertebrates has been studied most extensively. This chapter attempts primarily to bring together in a selective manner those studies which have contributed to our knowledge of the molecular structures and biochemistry of these hypothalamic neurons. It is useful, however, first to examine briefly and in a general way, the concept of neurosecretion, the distribution, significance, and some properties of the neurosecretory cell.

II. CHARACTERISTICS OF THE NEUROSECRETORY CELL

A. Classical Definition

The classical formulation on the concept of neurosecretion has stemmed largely from studies on the vertebrate hypothalamo-neurohypophyseal complex (HNC). On the basis of these studies the neurosecretory cell was described as a neuron with the morphological and functional characteristics of both nerve and endocrine cells. Although neurosecretory cells may receive inputs from other neurons, and are capable of propagating action potentials down their axons, their axons do not form synapses with other neurons or effector organs. Instead, the axons of neurosecretory cells end at blood spaces into which they release the secretory material. In addition to containing the cytoplasmic inclusions characteristic of most neurons (i.e., Nissl bodies, Golgi complexes, neurofibrillae, etc.), the cytoplasm of the neurosecretory cell contains "neurosecretions" (presumably of polypeptide nature) or neurosecretory material (NSM) whose presence may be demonstrated by means of standard cytological methods at the level of the light microscope. The stains most commonly used are those developed by Gomori[9] in the study of pancreatic islet cytology as well as those reactive toward S—S bonds. It is clear, however (as pointed out by Bern[2-4]), that there are no universal, selective stains for NSM and, in fact, a variety of different stains have now been used to demonstrate secretionlike inclusions in neurons of many species. The cytological criteria for the identification of a neurosecretory neuron were further refined with the introduction of the electron microscope. The high resolving power of the electron miscroscope has revealed the presence of electron-dense granules (termed neurosecretory granules or NSG) with diameters in the 1000–3000 Å range, and apparently the NSM consists of aggregates of such NSG. Observations on the proximodistal movement of the NSM stimulated the formulation of the concept of the "neurosecretory process." This process, as formulated by the Scharrers[6] and others,[10,11] ascribed an anatomical division of labor to the neurosecretory cell. The cell body synthesizes and packages its neurohormones within NSG which then move as aggregates (NSM) in a protoplasmic flow along the axon to the region of the nerve terminals where release into the bloodstream occurs. The term neurohormone as used here conforms to the conventional definition of a hormone, i.e., a unique chemical mediator (or messenger), produced by a specific group of cells and carried via the bloodstream to target sites. In contrast, neurons influence other neurons or effector organs by the release of transmitter substances at specific synaptic junctions. These transmitter substances must diffuse over a distance of only a few hundred angstrom units in order to reach postsynaptic receptor sites. The time scale involved for the onset of response at these sites is of the order of milliseconds; furthermore, the response is of relatively short duration following the cessation of the release of transmitter. Many diverse types of neurons produce the same transmitter substances, which are typically low molecular weight amines stored in quantal units within synaptic vesicles with diameters from 200 to 650 Å. These synaptic vesicles are largely localized within the presynaptic nerve terminals, and although the structure and function of the terminals depend upon the perikaryon, the actual formation of synaptic vesicles and synthesis of the transmitter probably takes place within the nerve ending proper at the site of synaptic activity.[12] In contrast to synaptic transmission, the onset and duration of the response following excitation of the neurosecretory neurons can usually be measured on a time scale ranging from minutes to hours or even days. By definition, the diameter of the NSG, in contrast to the synaptic vesicle, is greater than 1000 Å and the "neurosecretions" sequestered within the granule are thought to be polypeptide in nature. Presumably the synthesis of both components takes place solely within the perikaryon.

B. Inadequacies of the Classical Formulation of Neurosecretion

More recent investigations have made it clear that the criteria enumerated above which attempt to define the neurosecretory cell are much too narrow in scope and no longer tenable. For example, De Robertis[12] has pointed out that some adrenergic neurons secrete into the bloodstream. In the tetrapod median eminence, there are nerve fibers, unquestionably neurosecretory, which end on capillaries, but whose axon terminals display small, catecholamine-containing vesicles.[13] Neurons which by most criteria would be designated as neurosecretory have, furthermore, been shown to make direct contact with endocrine cells of the corpus allatum of insects,[14] and with the pars intermedia cells of the pituitary of elasmobranchs,[15] amphibians,[16] and mammals.[17] In teleosts,[18] neurosecretory fibers may invade the proximal pars distalis and make contact with endocrine cells believed to be gonadotrophs. Such contacts have been referred to as "neurosecretomotor junctions."[19] The studies of Knowles and Vollrath[20] on the pars intermedia and adenohypophysis of the eel indicated yet another pathway whereby neurosecretory products may reach target cells. These endocrine glands are innervated by at least two different kinds of neurosecretory fibers, termed by Knowles as type A and type B fibers. Type A fibers generally produce peptide hormones and contain spherical electron-dense vesicles with a diameter greater than 1000 Å. Type B fibers appear to contain monoamines and irregular vesicles which are usually smaller than 1000 Å in diameter. In the hypophysis of the eel and other lower vertebrates, type A and type B neurons may discharge into a narrow extravascular channel (perivascular space) through which the neurohormones diffuse in order to reach their target cells.

From the material presented, it is evident that there exist several distinctive types of neurosecretory cells (summarized in Fig. 1) and upon closer scrutiny, it is equally apparent that the boundary between the neurosecretory neurons and the "neurons banaux" has become less distinct. Koelle,[21] De Robertis,[12] and others have in fact emphasized the basic similarities between nerve cells and neurosecretory elements. De Robertis[12] and Welsh[22] have pointed out that essentially all neurons engage in secretory activity by which active substances are synthesized and released. Is it useful then to maintain a distinction between the neurosecretory and other neurons? This question has received an affirmative answer from Knowles and Bern[23] who have proposed a functional definition of a neurosecretory

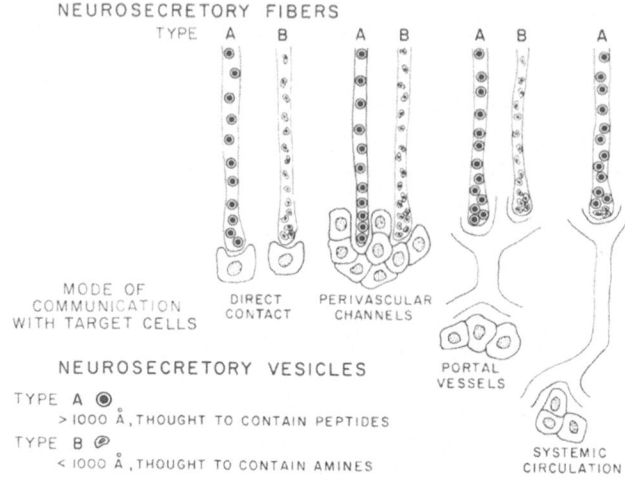

Fig. 1. Types of neurosecretory cells and modes of communication
with target cells. (After Knowles[30] and Bern.[4])

neuron as one which is "engaged directly or indirectly in endocrine control
and may form all or part of what may be regarded as a traditional endocrine
organ," and it is this activity which constitutes the fundamental criterion for
designating a neuron as "neurosecretory." This definition captures the
essence of the neurosecretory neuron, but the language is so general that a
major portion of the nervous system would have to be included (e.g., many
sensory, internuncial, and cortical neurons indirectly influence endocrine
function). Although greater precision could be achieved by specifying the
last neuron in a neuroendocrine system as neurosecretory (see below), this
raises further complications by the necessity to include such neurons as the
preganglionic fibers innervating the adrenal medulla or the neurons of the
intrinsic nerve plexus controlling the release of the gastrointestinal hormones. The latter two types of nerve fibers would ordinarily be excluded from
the category of "neurosecretory" on the grounds that they produce conventional transmitter substances as opposed to neurohormones. It should not be
too distressing that what at first sight appeared as a simple well-defined concept has become complex and problematical. An alternative and perhaps
fruitful way of viewing the problem is to use some of the simple concepts of set
theory.[24] The structural and functional characteristics of neurosecretory
cells may then be examined with respect to the intersection (shared characteristics) and nonintersection of the individual sets, i.e., neurons, neurosecretory neurons, and endocrine cells. In the Venn diagram of Fig. 2, the
total area of the figure represents the union of three sets, (1) all neurons (N)
except neurosecretory neurons (represented by the area ANIL), (2) neurosecretory cells (Ns, area CDHK), and (3) endocrine cells (E, area MFGJ). The
intersection of the sets N and Ns (N ∩ Ns) is represented by the area BNIK,

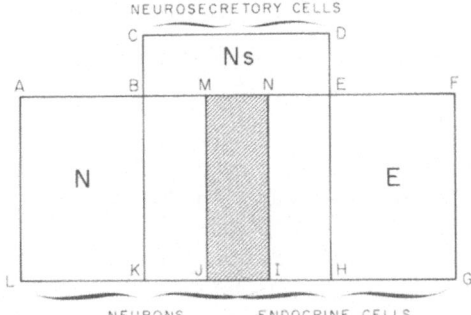

Fig. 2. Venn diagram of the union and intersection of three sets of cells; E, endocrine cells, area MFGJ; Ns, neurosecretory cells, CDHK; N, all neurons except neurosecretory cells, ANIL.

Ns ∩ E would be MEHJ and the intersection of all three sets (i.e., properties common to N, Ns, and E) is represented by the central area MNIJ. It is now possible to tabulate* some of the structural and functional properties common to N (all neurons except neurosecretory cells) and Ns (i.e., N ∩ Ns), as well as those for Ns ∩ E, Ns − N (i.e., the set of elements which belong to Ns and not N denoted by the area of the figure bounded by BCDHIN) and N − Ns (area ABKL). It is apparent that this type of formulation leaves room for all manner of new data irrespective of whether or not it conforms to preconceived ideas of the neurosecretory neuron.

III. SIGNIFICANCE OF THE NEUROSECRETORY CELL

One of the most distinguishing features of the course of evolution from lower to higher forms is the progressive increase in the complexity and capacity of the nervous system. Most conspicuously, there has been a phylogenetic development of an elaborate neural apparatus of correlation and integration as well as an increase of the sensory receptor elements, and the neuromotor mechanisms of response. The other significant feature of this course of evolution is that the reproductive processes have assumed a relatively minor role and instead, there has been the steady development of organ systems which function to maintain a relatively stable internal milieu and to ensure the survival of the genetic line during intermittent feeding or during adverse environmental periods connected with low food supply, drought, the presence of predators, etc. It is clear, however, that the harmonious functioning of these organ systems and the coordinated somatic, visceral, and genetic behavior of the entire organism depend upon direction from the two great control systems of the body, the nervous system and endocrine system. One of the most striking developments during the past 15–20 years has been the growing realization that these two control systems function in a closely coordinated way. A large part of the endocrine system may, in fact, be considered as an effector arm of the nervous system, and alterations in hormone secretion may be in a sense an integrative function of

* In preparation.

the nervous system. Endocrine cells involved in such diverse processes as homeostasis, adaptation, reproduction, and normal growth and development have been shown to be under the influence of the endogenous and rhythmic (cyclic) activity of central neurons, as well as neuronal activity arising from sensory information due to interoceptive and exteroceptive stimuli. The analogy with the somatic motor organization is readily apparent and in this sense the last neuron in the neuroendocrine system may be compared to Sherrington's "final common path." The fundamental significance of the neurosecretory cell here emerges, for it is a neurosecretory cell which is the last neuron in the neuroendocrine system. It is thus apparent that the neurosecretory neuron is the "final common path" over which a vast array of neural information may be channeled to endocrine organs and by which neural inputs are translated into endocrine function. The neurosecretory cell and the phenomenon of neurosecretion are ubiquitous among metazoans and almost without exception, wherever a central nervous system exists along with a degree of cephalization, some nerve cells with secretory activity can be found.[2] Although in many instances the precise function of "neurosecretory-appearing" cells has not been delineated, there is a reasonable expectation that the cells will be implicated in some phase of endocrine activity.

IV. DISTRIBUTION AND FUNCTIONS OF NEUROSECRETORY CELLS

No effort will be made here to document the voluminous literature that exists on this subject for both vertebrate and invertebrate forms. The interested reader is, instead, referred to a number of reviews which have attempted to tabulate the neurosecretory systems that have been found among the animal phyla.[2,5,7] The discussion here is confined to the neurosecretory cells of the vertebrate hypothalamus with brief mention of the pineal and the caudal urophysis of fishes.

The most important neurosecretory systems in vertebrates are found in the hypothalamus, which contains both a diffuse distribution and specific groupings of neurosecretory cells. These hypothalamic neurosecretory fibers are presumed to be involved in the control of hypophyseal function; this is accomplished in a variety of ways. In the case of the neurohypophysis, this structure is largely composed of neurosecretory fibers arising from specific nuclei in the anterior hypothalamus. These fibers are intermingled with neuroglial cells (pituicytes) and their axonal bulbs mostly terminate near blood vessels in the distal portion (infundibular process) of the neurohypophysis. The infundibular process thus falls within the category of what Knowles and Carlisle[25] have defined as a "neurohemal organ" (i.e., a storage-release area for the neurosecretory products). The neurosecretory products of this particular neurohemal organ have been characterized chemically for many species and they consist of a family of octapeptide hormones with closely related structures. These polypeptide neurohormones are released into the bloodstream and act directly on effector organs such as the kidney, the bladder and skin of lower vertebrates, the smooth musculature of blood vessels, the myoepithelial cells of the mammary gland, and the uterine smooth muscles. The functions most clearly defined for these neurohormones are intimately associated with those processes involved in the maintenance of the osmotic composition of the extracellular fluid (ECF), the volume of the vascular compartment, and milk ejection and parturition in higher vertebrates. In the subsequent section, the hypothalamo–neurohypophyseal neurosecretory cells are examined in greater detail.

The other group of neurosecretory fibers that originate in the hypothalamus are those which have been shown to be essential for the proper functioning of the endocrine cells of the pars distalis. Their relationship to adenohypophyseal function has been studied most extensively in lower mammals (for reviews see Ref. 26) where they are found diffusely distributed throughout the medial basal hypothalamus. This region of the hypothalamus has been termed by Halasz and co-workers[27] as the "hypophysiotropic area" since this area contains the neurosecretory cells which produce the substances responsible for the maintenance of normal pituitary structure and function. These hypophysiotropic neurons (in higher vertebrates) do not directly innervate hypophyseal cells but instead their axons terminate in the region of the median eminence where they discharge their products into a network of capillary tufts which drain into a hypothalamo-hypophyseal portal vessel system. It is now well established that there is a distinct neurohormonal factor concerned with the secretion of each of the six established anterior pituitary hormones. Although the chemistry of these factors has been studied intensively over the past several years, no precise structures have as yet been established; this subject is discussed in Chapter 20 of this volume. Whether these substances which affect the release of the respective pituitary hormones are also responsible for maintaining the structural integrity of the related pituitary cell remains to be determined.

The hypophysiotropic neurosecretory fibers that terminate in the median eminence (another neurohemal organ) show a considerable diversity with regard to stainability and content of neurosecretory granules and/or vesicles which may range in size from that characteristic of synaptic vesicles to that found for the classical neurosecretory granule. The products of neurosecretion are presumably both polypeptide and amine[13,28,29] in nature.

In lower vertebrates hypothalamic neurosecretory fibers invade the pars distalis and either impinge directly on epithelial cells or discharge their products into a system of perivascular channels by which the neurohormones make contact with target cells. Knowles[30] has suggested "that direct contact between neurosecretory fibers and intrinsic endocrine cells may be a primitive feature which later was replaced during phylogeny by an aggregation of fibers to form a kind of neuropil from which substances reach by diffusion the endocrine cells which they affect." He points out, furthermore, that by this means relatively few neurosecretory fibers could control a larger number of pars distalis cells. In this regard, still greater efficiency has been achieved with the further development of the median eminence neurohemal organ and the hypophyseal portal vessel system. It is apparent, however, that this latter type of arrangement present in higher vertebrates must depend upon what Horridge[31] has termed a "chemical addressing system" rather that the usual point-to-point contact for the transmission of information. As already indicated (see also Chapter 20), there is at hand experimental evidence for the existence of a separate chemical mediator (i.e., hypothalamic releasing factor) which is only recognized and acted upon by a specific type of endocrine-producing cell of the pars distalis.

The rate of secretion of the melanocyte-stimulating hormones by the pars intermedia cells has also been shown to be under hypothalamic control, which in this case is of an inhibitory nature (see Etkin[32]). Hypothalamic control appears to be mediated by means of penetrating neurosecretory fibers derived from the hypothalamus; this type of arrangement is present not only in fishes and amphibians but also in those lower mammals which have been studied.[32]

Two other neurosecretory systems found in the CNS of vertebrates are the caudal neurosecretory system of fishes and those fibers innervating the pineal gland. The caudal portion of the spinal cord of most fishes contains neurosecretory cells whose axons terminate in a neurohemal area known as the urophysis; this organ is thought to be involved in osmoregulation.[33] Electron microscope studies of the pineal gland by De Robertis[12] have indicated the presence of neurosecretory fibers that apparently produce catecholamines. These fibers presumably control the secretory activity of the pinealocytes which have been implicated in the control of skin coloration and gonadal function.[34]

In addition to the neurosecretory systems described above, various investigators have reported "neurosecretion-bearing" fibers elsewhere in the brain of vertebrates. Some of these reports have been reviewed by Bern[2] and others.[5-7] The functional significance of these additional "neurosecretory" pathways remains obscure.

V. THE HYPOTHALAMO–NEUROHYPOPHYSEAL COMPLEX, A MODEL SYSTEM

Of all the neurosecretory systems in the CNS, those fibers which comprise the hypothalamo-neurohypophyseal complex of higher vertebrates are most clearly defined with respect to their molecular organization and functions. This can be attributed to a combination of fortunate circumstances. These neurosecretory fibers form the bulk of the elements of a relatively simple neurohemal organ (i.e., the neurohypophysis) which is readily accessible and amenable to investigation in a wide variety of experimental animals. The most important physiological function of the neurohypophysis was delineated very early in the game as a result of studies concerned with the clinical syndrome of diabetes insipidus and with the biological actions of the neurohypophyseal peptide hormones isolated in bulk quantities from the pituitaries of slaughterhouse animals. It is also particularly evident that as a result of the individual and combined efforts of investigators from varied disciplines, our understanding of these neurosecretory cells has progressed far beyond the speculations of the morphologists and/or the information obtained by the analytical approaches of the biochemist.

Over the past several years there have been a considerable number of reviews and monographs concerned with various facets of neurohypophyseal structure and function. These publications have dealt principally with the morphological and ultrastructural aspects of the mammalian HNC as well as the chemistry, release, bioassay, peripheral action, and metabolism of its associated polypeptide hormones. The more general reviews have for the most part concentrated heavily on studies with the light and electron microscope. The remainder of this chapter, in contrast, will place particular emphasis on the biochemical and physiological studies which have taken place at the molecular and cellular level. An analysis of these latter investigations within the context of the cytochemical and ultrastructural data may suggest and/or begin to unfold an integral picture of the structure and function of these specialized hypothalamic neurons. Hopefully, this may also provide a model for the study of other neurosecretory systems.

A. Morphological Aspects

The nerve and blood supply of the neurohypophysis as well as the gross anatomical features of the mammalian hypothalamo–neurohypophyseal system have been described in great detail in several reviews.[8,35–37] A schematic representation of the dog hypothalamo–neurohypophyseal complex is shown in Fig. 3, the main features of which may be found in slightly modified form in most mammals. According to the standard nomenclature of Rioch et al.,[38] the neurohypophysis consists of three parts: the infundibular process (neural lobe), the infundibular stem, and the median eminence of the tuber cinereum (the latter two components comprising the infundibulum or neural stalk). The anatomy of the nerve supply to the neurohypophysis was first studied more than a half century ago by Ramon y Cajal[39] who described the rich nerve fiber connections which united the hypothalamus with the posterior pituitary, forming the hypothalamo–hypophyseal tract. These nerve fibers are practically all unmyelinated and derive principally from the supraoptico–hypophyseal tract and secondarily from the tubero–hypophyseal tract. The total number of nerve fibers in the neural stalk varies in different species; according to Rasmussen,[40] the estimates are: 10,000 in the rat, 40,000 in the Macaque monkey, 60,000 in the dog, and 100,000 in man. The majority of the tubero–hypophyseal fibers end in the median eminence on or near the primary plexus of the hypophyseal portal systems. The sites of origin of these fibers have not been precisely defined but they are believed to originate chiefly from the basal medial region of the hypothalamus. In the

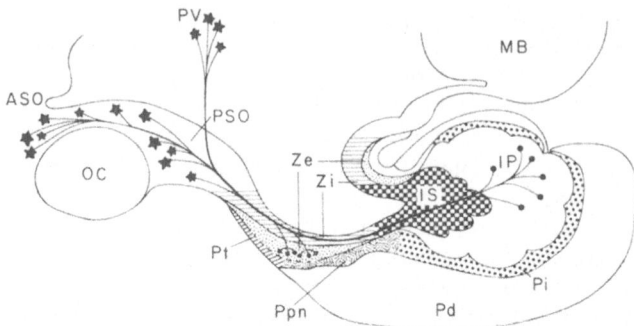

Fig. 3. Schematic representation of the dog hypothalamic neurohypophyseal complex of the dog, sagittal view; modified after
Zambrano and DeRobertis[41] and Rothballer and Skoryna.[42]
ASO, anterior division of the supraoptic nucleus; IP, infundibular
process; IS, infundibular stem; MB, mammilary body; OC, optic
chiasm; Pd, pars distalis; Ppn, pars paranuclearis; PSO, posterior
division of the supraoptic nucleus; Pt, pars tuberalis; PV, paraventricular nucleus; Ze, zona externa; ZI, zona interna.

preceding section these neurons have been discussed briefly in relation to the
control of anterior pituitary function. The schematic representation in Fig. 3
shows only the supraoptico–hypophyseal tract whose fibers originate in the
anterior hypothalamus, chiefly in the supraoptic nucleus and to a lesser extent
in the paraventricular nucleus. In the dog[41] as in other species, the supraoptic nucleus may be divided into an anterior and posterior region. After
coursing through the inner zone of the infundibulum, the supraoptico–
hypophyseal fibers enter the most distal division of the neurohypophysis
(infundibular process). In most mammals the infundibular process is composed of tiny lobules which receive the neurosecretory fibers whose endings
cluster in the peripheral (or "palisade") regions of these lobules; the nerve
terminals abut on a basement membrane which is separated by a perivascular space from the basement membrane of the capillary. The network of
perivascular channels found in the neural lobe is a characteristic of secretory
organs in general and contrasts sharply with the very small extravascular
spaces found throughout most of the CNS. Also found in the neurohypophysis, intermingled with fibers and nerve terminals, are "neuroglial" cells or
pituicytes.

B. Hormonal Activities, Chemistry, Functions, and Evolutionary Aspects

Neurohypophyseal extracts of vertebrates have yielded seven octapeptide hormones whose structures have been defined and whose biological
activities have been extensively investigated. There also appears to be pharmacological evidence[43] for the presence of several other hormonal peptides.

The chemical structures of these neurohypophyseal hormones as well as their major biological activities are presented in Table I; the data for this table have been selected from the very comprehensive reviews by Walter et al.[44] and Sawyer.[45] These peptides fall into two categories: vasopressor-antidiuretic and oxytocic-milk ejecting. In the first category there are the three dibasic peptides: arginine vasopressin, lysine vasopressin, and arginine vasotocin; the monobasic oxytocinlike peptides include oxytocin, mesotocin, isotocin, and glumitocin. The pituitaries of mammals contain only oxytocin and either arginine or lysine vasopressin. Lysine vasopressin has been found only in pituitaries from the suborder *Suina*, some of whose members can apparently synthesize both lysine and arginine vasopressin.[45,46]

Clearly, the most important physiological function of the vasopressin peptides stems from their ability to act as an antidiuretic agent in fantastically minute quantities (e.g., 2.0×10^{-13} M/kg); and it has been amply demonstrated that the neurons of the supraoptico-hypophyseal tract are essential components of the interlocking systems controlling the volume and composition of the body fluids. Oxytocins role in parturition and milk ejection in the female is reasonably well established. Whether this hormone has a function in the male is at present unknown. It is beyond the scope of this chapter to discuss the physiology of the mammalian neurohypophysis and/or those studies concerned with the mechanisms of action and metabolism of vasopressin and oxytocin; these subjects have been reviewed in a number of existing publications.[47–52] In the submammalian orders the precise functions of the various vasopressin- and oxytocinlike pituitary peptides have not been defined.

The study of the phyletic distribution of the neurohypophyseal hormones has provided some fascinating insights in the areas of comparative endocrinology and biochemical evolution.[45,46] Examination of the structures of these peptide hormones (Table I) reveals a class of closely similar octapeptides which have the following features in common: an N-terminal CyS, a C-terminal Gly amide, a positive charge at neutral pH, and 1,6,CyS residues enclosing a 20-atom disulfide ring followed by a tripeptide side chain. Variations in amino acid residues occur only in the 3, 4, and 8 position. The finding that only arginine vasotocin (8-Arg oxytocin) is present in the most primitive vertebrates, the cyclostomata, led to the suggestions that this hormone is ancestral to the other vertebrate neurohypophyseal hormones[45] and that two peptide series probably evolved after doubling of the vasotocin controlling gene.[53] In one series a variety of oxytocin principles derived from vasotocin by consecutive mutations, whereas in the other series vasotocin remained unchanged until the appearance of the primitive mammalia. In the most primitive mammalia arginine vasopressin appears due to the substitution of phenylalanine for isoleucine in position three. This mutation can be readily explained[53] by the single base change A → U in the triplets AUC or AUU coding for isoleucine, resulting in the triplets UUC or UUU for phenylalanine. The persistence, however, of arginine vasotocin throughout the vertebrate phyla indicates that this hormone has survived for nearly

TABLE I
Neurohyphophyseal Hormones[a]

Hormone	Phyletic occurrence	Biological activity, IU/mg			
		Rat antidiuresis	Rat blood pressure	Rat uterus in vitro (no Mg^{2+})	Frog bladder
Arginine-vasopressin Cys-Tyr-Phe-Glu(NH_2)-Asp(NH_2)-Cys-Pro-Arg-Gly-NH_2	All mammals except domestic pig, hippopotamus	400	400	20	70–90
Lysine-vasopressin Cys-Tyr-Phe-Glu(NH_2)-Asp(NH_2)-Cys-Pro-Lys-Gly-NH_2	Suina	250	270–300	4–7.5	5–12.5
Oxytocin Cys-Tyr-Ile-Glu(NH_2)-Asp(NH_2)-Cys-Pro-Leu-Gly-NH_2	Birds, mammals, reptiles, amphibia, probably lungfish	1.1–5.0	3–7	500	360–500
Arginine-vasotocin Cys-Tyr-Ile-Glu(NH_2)-Asp(NH_2)-Cys-Pro-Arg-Gly-NH_2	All vertebrates other than mammals	74–250	245	75–160	10,000–34,000
Isotocin Cys-Tyr-Ile-Ser-Asp(NH_2)-Cys-Pro-Ile-Gly-NH_2	Ray-finned fishes	0.18	0.06	150	—
Mesotocin Cys-Tyr-Ile-Glu(NH_2)-Asp(NH_2)-Cys-Pro-Ile-Gly-NH_2	Reptiles, amphibians, lungfish	1.1	6	289	800
Glumitocin Cys-Tyr-Ile-Ser-Asp(NH_2)-Cys-Pto-Glu(NH_2)-Gly-NH_2	Some elasmobranchs	—	—	8	—

[a] Modified after Walter et al.[44] and Sawyer.[43]

400 million years,[45] which is indeed a remarkable example of evolutionary stability of a peptide. This contrasts strongly with the common occurrence of mutations of other, larger peptide hormones in the same vertebrate class. The implication that any mutation changing an amino acid moiety within vasotocin must have been deleterious finds confirmation from the results of studies on structure-activity relations[44]; these studies have for the most part shown that minor alterations in the structures of either the vasopressin or oxytocic molecules lead to impairment of their respective biological activities.

C. The Neurosecretory Process

1. Physiological and Morphological Correlates

According to the Scharrers[6] and others,[10] the neurosecretory process consists of the following events: (a) synthesis and packaging of the hormones within NSG in the perikarya of the neurons of the supraoptico–hypophyseal tract; (b) movement of aggregates of NSG in a protoplasmic flow along the axon to the region of the nerve terminals (i.e., infundibular process); and (c) the release of the hormones from the nerve terminals into the bloodstream upon appropriate stimulation. A number of physiological and morphological correlates provided by many independent investigations support the concept of neurosecretion in the HNC as formulated above. Historically, some of the more pertinent findings may be summarized as follows: (1) Studies on extracts of neural lobe tissue established the principal biological activities present in the gland.[54–57] (2) The elegant physiological experiments of Verney[58] and Geiling and Robbins[59] demonstrated that the antidiuretic hormone was a true secretion of the neural lobe. (3) Ranson and his colleagues[60] showed that experimental diabetes insipidus could be produced by either bilateral destruction of the supraoptic nuclei or lesions in the neural stalk, procedures which demonstrated that the proper functioning of the neural lobe required the integrity of the supraoptico–hypophyseal tract. (4) The cytochemical observations of the Scharrers,[6] Bargmann,[5] and others[10,11] on the secretory-appearing inclusions within the nerve fibers of the supraoptico–hypophyseal tract suggested that the neuronal elements (rather than the pituicytes) were responsible for the synthesis and secretion of the neuro–hypophyseal hormones. Support for this suggestion came from studies which precluded the pituicytes as the source of hormones[61] and established a correlation between the stainable inclusions and assayable hormones[62] (with regard to both amounts and anatomical distribution). (5) After surgical interruption of the nerve fibers, the NSM as well as the hormones disappear distal to the lesion and accumulate in large masses in the swollen stumps of the fibers proximal to the lesion,[11,63] an observation consistent with the concept of axonal transport of secretory material from cell body to nerve terminal. This cellulofugal movement of NSG is analogous to the phenomenon of "axoplasmic flow" observed and studied most extensively in peripheral nerve.[64]

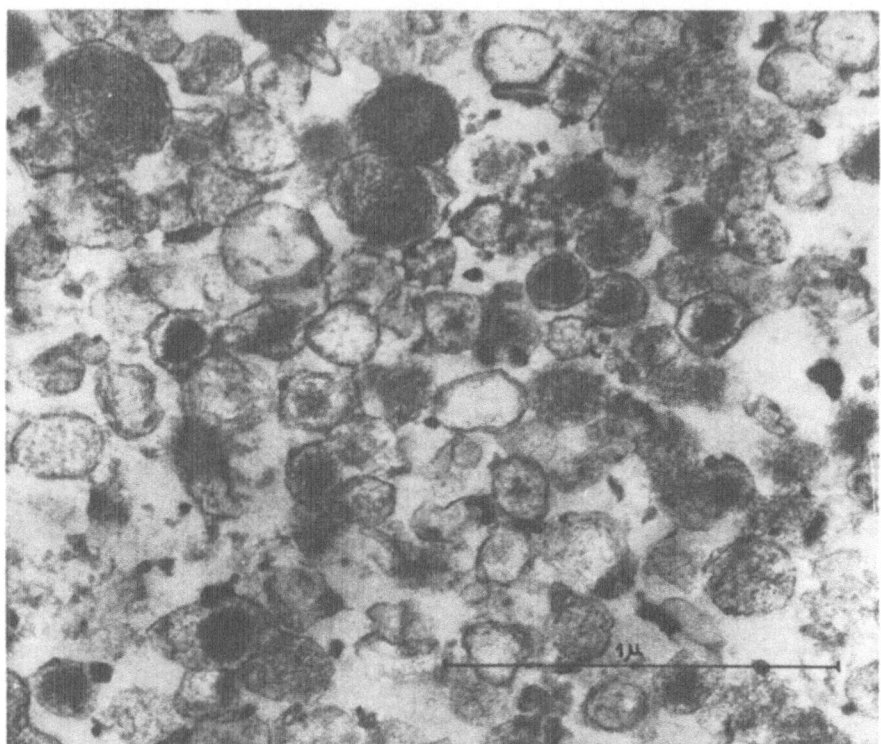

Fig. 4. Electron micrograph of neurosecretory vesicles isolated from the dog neurohypophysis.
× 50,000.

2. Cytoarchitecture and Biochemical Correlates

a. NSG and the Stainable Substance. The cytochemistry and fine structure of the supraoptico–hypophyseal neurons of a wide variety of species have
been studied with the light and electron microscopes; and this subject has
also been reviewed by several previously cited authors. In brief, these neurosecretory cells show many structural characteristics common to other
neurons of the CNS. Examination of the neurons in the supraoptic nuclei
reveals axodendritic and axosomatic synapses, Nissl substance, Golgi apparatus, mitochondria, large inclusion bodies, and neural filaments or
canaliculi extending down the length of the nerve fibers. The neurosecretory
neurons are, however, readily distinguished by the presence within their
axoplasm of dense granules, about 0.1 to 0.3 μ in diameter, numerous in the
nerve terminals, but also found throughout the length of the neuron. Aggregates of these granules may be observed with the light microscope using
staining techniques commonly employed to detect "neurosecretory material."
These granules, termed "neurosecretory granules" (or NSG), were presumed
to contain oxytocin and vasopressin and as such could be considered as the

cytochemical representation of the hormones of the neurohypophysis. Ultimate verification of this supposition, however, required the isolation and characterization of these "NSG." The initial efforts[65,66] to isolate NSG from sucrose homogenates of neural lobe tissues by means of differential

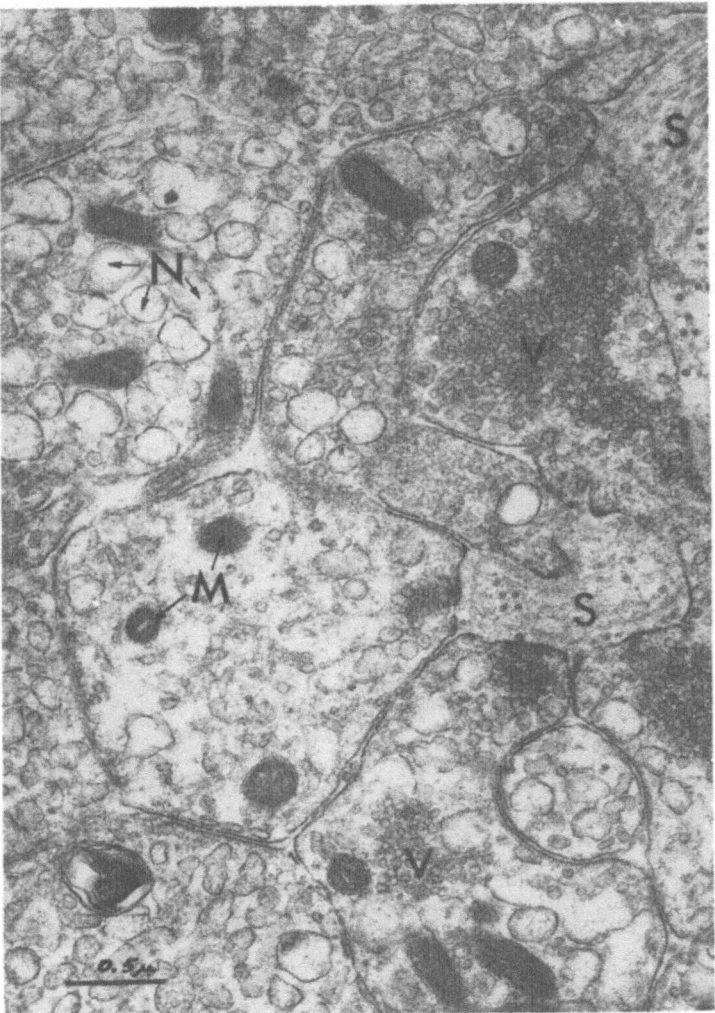

Fig. 5. Neural lobe from a dog, fixed *in situ* by perfusion with glutaraldehyde followed by postfixation in osmium tetroxide. Axon terminals near collagen-containing perivascular space (S) enclose swarms of the small vesicles (V) and a few mitochondria (M). Neurosecretory vesicles (N) do not display an inner dense matrix despite "double" staining in uranyl acetate and lead citrate. × 31,800. (From Osinchak, Haller, and Sachs, unpublished; cf. Ref. 72.)

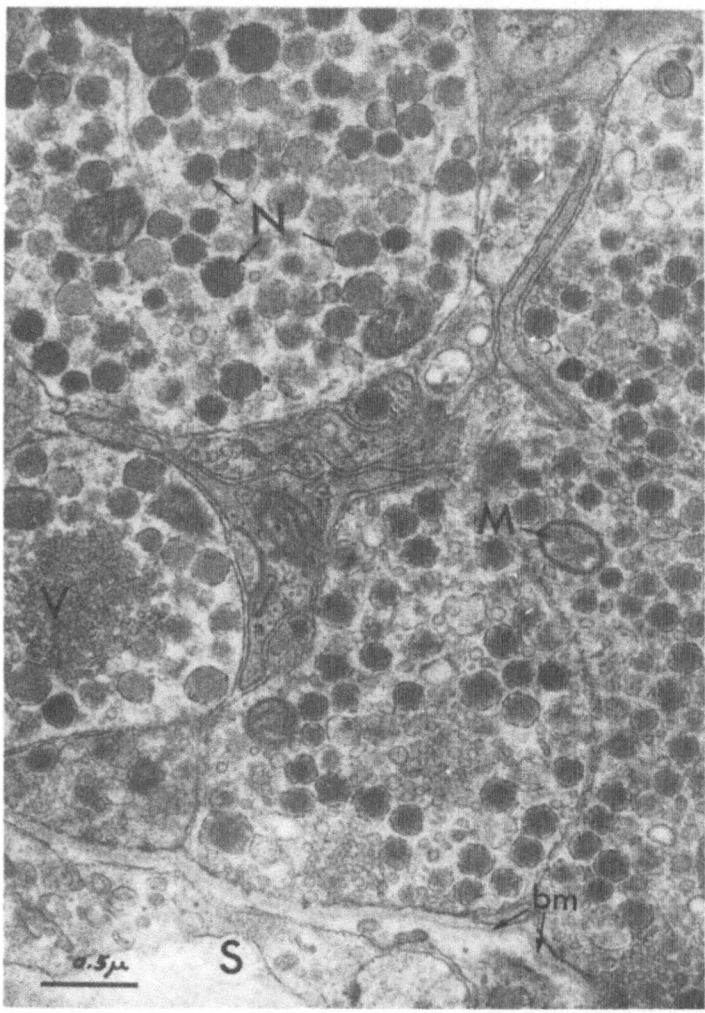

Fig. 6. Neural lobe of the rat fixed by immersion in glutaraldehyde followed by postfixation in osmium tetroxide. Several axon profiles are seen near basement membrane (bm) of perivascular space (S). The axons contain a full complement of dense neurosecretory vesicles (N), occasional mitochondria (M), and varying amounts of small "synaptic-like" vesicles (V). × 30,250. (Reproduced through the courtesy of J. Osinchak.)

centrifugation succeeded in verifying that the hormonal activities were bound to particulate material. The hormone-rich fractions were quite crude, and these experiments failed to resolve the NSG from mitochondrial and other cellular structures. In our laboratory, by the use of differential and density-gradient centrifugation techniques, NSG were isolated in relatively pure form from sucrose homogenates of dog neural lobes.[67] Similar feats were accomplished almost simultaneously by several investigators working with sucrose homogenates of the posterior pituitaries from a number of different species.[68–70] Electron micrographs of the isolated NSG from the dog (Fig. 4) showed a population of vesicles, approximately circular in profile, each bounded by a single membrane, and with diameters mostly in the 0.1–0.2 μ range. There was no morphological or biochemical evidence for the presence of significant quantities of mitochondria, endoplasmic reticulum, or lysosomes. Furthermore, the isolated granules were rich in vasopressin; the full biological activity was expressed only after the particles were subjected to treatments designed to disrupt the limiting membrane (e.g., incubation with detergents, hypotonic media, low pH, or boiling). Morphologically the NSG obtained from homogenates of the dog neural lobe appearing identical in all respects to the NSG or the dog HNC described *in situ* by Sloper,[71] Sachs and co-workers,[72] (Fig. 5) and others.[41] In the dog, in contrast to the findings in most other species (cf. Fig. 6), the majority of the NSG (*in vivo* or *in vitro*) did not display an inner matrix of high density.

The astute conjecture of the morphologists that the NSG were the major storage forms of the neurohypophyseal hormones has thus received convincing experimental support from the combined biochemical, electron microscope studies. These studies, however, were confined to neural lobe tissues (i.e., the distal segments of the neurosecretory fibers) and the intracellular distribution of the hormones in the perikarya remained to be established. The dog was chosen for such studies because of the unusually large amount of vasopressin which is present in the hypothalamus relative to that found in other species.[73] Differential centrifugation of homogenates of dog hypothalamic–median eminence (HME) tissues in dense sucrose media again yielded an NSG fraction containing about 50% of the total vasopressin and only a very small part of the cellular nucleic acid, cytochrome oxidase, and β-glucuronidase activities (Table II).[74] Centrifugation of this material, through either a discontinuous or continuous sucrose gradient, led to further purification (Fig. 7). Although the NSG isolated from the dog HME complex have not been examined in the electron microscope, it is clear that the NSG found within the cell body and upper portion of the neuron (HME) have chemical and sedimentation properties similar to the NSG found in the more distal portions of the axon (infundibular stem and process).

The fractionation studies on homogenates of the dog HME complex indicated that small, but relatively constant amounts of vasopressin are found in association with a number of heterogeneous particulate cell fractions other than the NSG (Table II). Aside from the NSG, however, there is no information as to the precise nature of the particulate materials that are associated with the hormone and as to what extent this association is artifactual. Nevertheless, a considerable number of isotope experiments have clearly indicated the existence and obligatory role of such cell structures in the intermediate stages of hormone biosynthesis (see Section VII).

TABLE II

Subcellular Distribution of Vasopressin, Neurophysin-^{35}S and Other Cellular Constituents of Dog Hypothalamic-Median Eminence Tissues

Cell fraction	Percent of original homogenate ± s.d.[a]					
	Vasopressin	Neurophysin-^{35}S[b]	Protein	Nucleic acid	Cytochrome oxidase	β-Glucuronidase
Nuclei, cell debris	7.4 ± 2.6	11.8	17.2 ± 6.1	43.4 ± 6.95	6.8 ± 2.7	11.4
70,000 × g pellet (containing NSG)	46.3 ± 12.9	48.0	1.82 ± 0.6	6.56 ± 3.6	2.63 ± 1.8	4.1
Large granules	13.3 ± 2.71	7.7	24.1 ± 6.0	12.7 ± 4.7	31.8 ± 4.7	22.2
Microsomes	8.5 ± 2.6	2.1	11.5 ± 3.5	16.5 ± 8.1	8.5 ± 5.8	12.4
Cell sap	29.6 ± 8.0	30.6	24.0 ± 1.8	8.0 ± 2.1	—	19.2

[a] From 6 to 8 experiments, see Sachs.[74]
[b] After the infusion of cysteine-^{35}S into the third ventricle of a dog's brain for 4 hr (Fawcett and Sachs, unpublished); the HME was removed 30 min later.

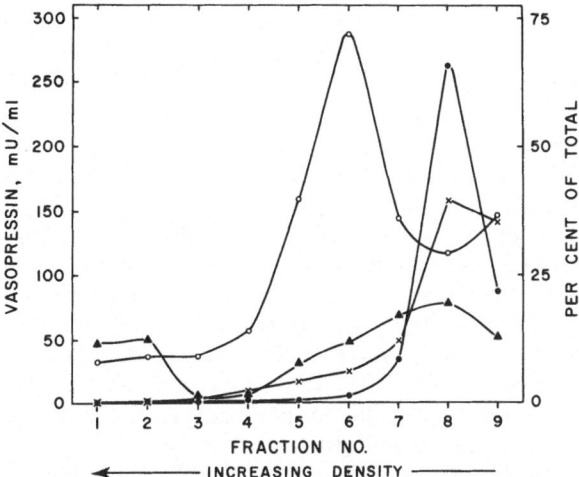

Fig. 7. Isolation of neurosecretory granules of the dog HME by means of centrifugation of the 70,000 × g pellet (Table II) through a continuous sucrose gradient (1.5–2.5 M); O—O vasopressin; mU/ml; ●—● cytochrome oxidase; ×—× protein, ▲—▲ nucleic acid; the latter three components expressed as percent of total material applied to the gradient. (From Sachs.[74])

An important advance in the cytochemical study of the neurosecretory cell was due to the success with which the NSM could be intensely stained by means of techniques devised by Bargmann[5] and Adams and Sloper.[75]

These methods depend upon the presence of high concentrations of protein-bound disulfide groups (for a review of this subject see Sloper[8]). Information as to the origin of such disulfide groups of the NSM has been provided by biochemical studies on NSG from neurohypophyseal and hypothalamic sources. La Bella et al.[76] subjected purified preparations of NSG from bovine neural lobes to amino acid analysis and found the values to be remarkably similar to that of the protein first isolated by Van Dyke and co-workers[77] from acid extracts of beef posterior pituitaries. This protein (Van Dyke protein) is characterized by an unusually high content of cystine disulfide bonds and was reported to have oxytocin and vasopressor activities in a mole ratio of 1:1 per 30,000 molecular weight. It has now been shown that the Van Dyke protein actually consists of a number of closely related proteins (known collectively as neurophysin) in loose association with the octapeptide hormones.[78,79] The fractionation studies of Ginsburg and Ireland[80] and Dean and Hope,[81] and their co-workers indicated that the bulk of the soluble proteins of the beef posterior pituitary is neurophysin and that the NSG are essentially packets of hormone–neurophysin complexes. These complexes would thus constitute the stainable neurosecretory substance. Further support for this postulate derives from the findings of several independent investigators[82–84] demonstrating that prolonged dehydration (or an osmotic stimulus) leads to the depletion of neurophysinlike proteins; repletion of neurophysin occurs after the dehydrated animals are permitted free access to water. Under these conditions of chronic stimulation, the decline in the amount of neurophysin in the posterior pituitary is paralleled by decreases in the amount of assayable hormones and stainable neurosecretory material. However, the synchronous loss of NSM, and hormonal and neurophysin peptides is not observed, following the application of an acute stimulus. Although painful stimuli, ether anesthesia, and acute hemorrhage unquestionably lead to the release of vasopressin[47] and neurophysin,[85] the amounts discharged are not commensurate with the dramatic disappearance of NSM[86,87] and loss of osmiophilicity of the NSG.[88] In this regard it is noteworthy that the NSG isolated from homogenates of dog neural lobes[67] appear optically "empty" despite the fact that they contain 14–24 units vasopressin per milligram protein (a value equivalent to that of the Van Dyke protein) as well as demonstrable quantities of neurophysin (Fawcett, Haller and Sachs, unpublished). This phenomenon of dissociation between intensity of staining (or osmiophilicity) of the NSG and their content of hormones and neurophysin will be examined in Section VIII.

D. Separate Neurosecretory Granules for Oxytocin and Vasopressin

The available evidence is consistent with the view that oxytocin and vasopressin are produced, stored, and released by separate neurons and therefore separate NSG. Studies on the distribution of the neurohypophyseal hormones within the hypothalamus of a number of species[89] suggest that the neurons of the paraventricular nuclei may be especially concerned with the elaboration of oxytocin. Support for this postulate has been obtained from studies on the effects of hypothalamic lesions in the rat[90] and in the cat[91]; bilateral destruction of the paraventricular nuclei causes a preferential depletion of oxytocin in the pituitary. Recent studies by Sokol and Valtin[92] using a strain of rats with hereditary diabetes insipidus also support the separate neuron hypothesis. Ultracentrifugation of sucrose homogenates of posterior lobes has indicated the possibility of separating distinct oxytocin- and vasopressin-containing NSG.[70,93] The concept of separate neurons for each hormone provides the anatomical basis for the differential release of oxytocin and vasopressin observed under a variety of experimental conditions.[94–96]

VI. NEUROPHYSIN, CHEMISTRY AND FUNCTIONS

Acher and co-workers[97] showed that the Van Dyke protein could be dissociated into its component parts by relatively mild procedures, and they suggested the name "neurophysin" for the protein moiety. Recent investigations have shown that neurophysin is heterogeneous and consists of several closely related, acidic, low molecular weight proteins.[78,79] Neurophysin isolated from acetone-dried posterior lobes of bovine pituitary glands has been separated into seven distinct protein fractions, one of which had no hormone-binding activity.[78] Dean, Hollenberg, and Hope[98] compared the soluble proteins of the NSG with the constituents of neurophysin; the granule neurophysin consisted of only two components and these authors concluded that those components of neurophysin not present in the NSG arise as a result of the degradation of the two granular proteins. This conclusion was supported by the observation that extracts of neural lobes exhibited catheptic activity under precisely those conditions employed for the isolation of the protein–hormone complex (i.e., pH 3–4); isolation of the complex at pH value incompatible with catheptic activity (i.e., pH 7.4 or pH 1.5) yielded neurophysin with only the two constituents present in the NSG. The most definitive work in this area has been performed by Hope and co-workers who recently succeeded in obtaining crystalline complexes of neurophysin with arginine vasopressin and oxytocin.[99,100] The neurophysin moieties consist of two constituents (termed neurophysin I and neurophysin II) separable by electrophoresis at pH 8.1. A molecular weight of about 19,000 and 21,000 has been determined for neurophysin I and II respectively; like the Van Dyke protein, these materials contain a relatively high content of cystine, glutamic acid, and glycine residues. Neurophysin II is distinguished from neurophysin I by the absence of histidine. The binding capacity of neurophysin I was estimated as 3 moles of either vasopressin or oxytocin per mole of protein whereas neurophysin II binds only 2 molecules of each hormone per molecule of protein. These results are in essential agreement with the data of Breslow and Abrash[79] but differ from that of Ginsburg and Ireland.[101] The latter authors reported that as many as 7 moles of oxytocin and 4 of vasopressin could be bound by a molecule of neurophysin. In a separate communication, Ginsburg et al.[102] reported the isolation of a neurophysin constituent which could bind oxytocin but not lysine vasopressin, an observation which so far has not been confirmed. The binding is maximal in the range pH 5.2–5.8 and is inhibited by cystine and low concentrations of Ca^{2+}[103,104]; oxytocin and vasopressin can apparently compete for the same binding sites.[79] A free primary amino group is an essential requirement for binding since neither deamino oxytocin[105] nor deamino (8-arginine) vasopressin[99] are bound by neurophysin. Breslow and Abrash[80] have shown that the binding also depends upon the nature of the amino acid residues in positions 2 and 3, and apparently the total binding energy represents the sum of the small energies of interaction (electrostatic, hydrogen binding, hydrophobic, and $\pi - \pi$) between neurophysin and the peptides. Breslow and Abrash[80] calculated an intrinsic binding constant of 1.8×10^5 for the binding of oxytocin to one of their neurophysin fractions, signifying that the interaction is very strong. Their experiments carried out in pyridine-acetate buffer, pH 5.8, 0.10 M KCl, make it unclear as to what extent the calculated binding constants are relevant to the in situ situation. It would also appear that the binding of the hormones (octapeptides with relatively rigid structures) to neurophysin leads to marked conformational changes in the protein structure.[100,101] This postulate is supported by the unique ability of (8-arginine) vasopressin to induce crystallization of neurophysin-hormone complexes and of both hormones to render the protein insoluble.

What is the cellular function of neurophysin? Is it also present in the perikaryon (i.e., hypothalamus) and does it reach the neural lobe via axoplasmic flow? Recent isotope experiments initiated in our laboratory[85,106] on the biosynthesis and release of neurophysin have shed some light on the latter question. The labeling of neurophysin was readily accomplished by the infusion of cysteine-^{35}S or tyrosine-^{3}H into the third ventricle of the dog's brain for periods of 4–6 hr. Whereas neurophysin and the proteins of the hypothalamus were extensively labeled immediately following such infusions, the proteins of the posterior pituitary contained only minimal quantities of radioactivity. About 10 days later however, the neural lobe had large amounts of highly labeled soluble protein that was almost entirely neurophysin, a result consistent with axoplasmic flow. The labeled neurophysin isolated from dog hypothalamic and neural lobe tissues had solubility, chromatographic, gel filtration, electrophoretic, and hormone-binding properties similar to beef neurophysin. Furthermore, the labeled material from the dog showed immunological reactivity with rabbit antisera to a beef neurophysin preparation, forming both insoluble and soluble complexes with the beef neurophysin antibody (Fig. 8). The observation that almost all of the label in the soluble protein fraction of the dog neural lobe resides in neurophysin is consistent with the analytic data[77,107] which indicates that most of the soluble protein of the beef posterior pituitary is neurophysin and that the NSG are principally packets of hormone–neurophysin complexes.

In the hypothalamus, shortly after the infusion of isotope, the subcellular distribution of radioactive neurophysin was remarkably parallel to that of vasopressin (Table II). The labeling pattern in the hypothalamus was complex, and soluble radioactive proteins other than neurophysin were detectable. Apparently these proteins either belong to cells other than the neurons of the supraoptico-hypophyseal tract or they do not move into the nerve terminals.

The question as to whether the role of neurophysin is confined to a storage function is at present unknown. Certainly, the analytic and isotope studies have indicated that the hormone–protein complex is a discrete entity of the neuron and probably constitutes the major storage form of the hormonal peptides of the NSG. Such a storage mechanism would provide for the neutralization of the positively charged hormones and for a decreased number of osmotically active molecules which need to be contained within the NSG. The presence of the specific binding protein(s) might also serve to prevent the diffusion of the small polypeptide hormones from the NSG and/or the neuron under quiescent or steady-state conditions. The possibility that the hormone–protein complex is involved in the secretory process is discussed below in Section VIII.

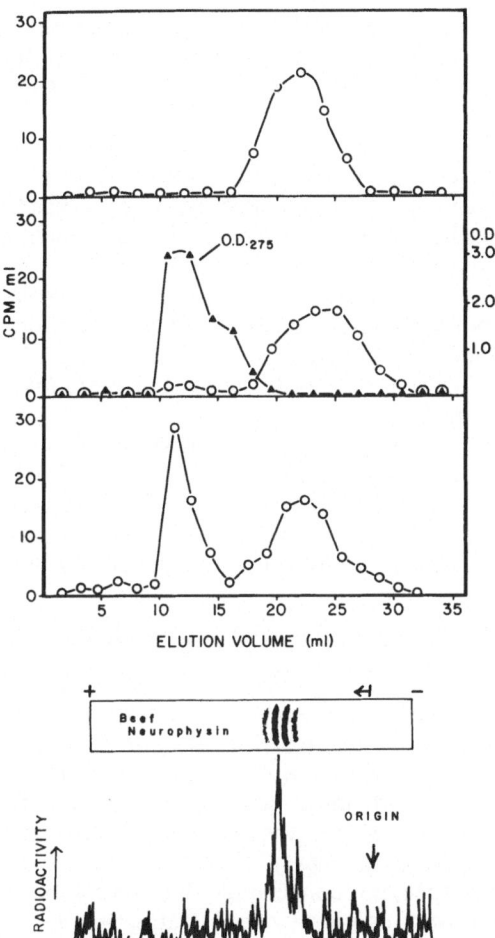

Fig. 8. Fractionation of neurophysin-^{35}S from the dog hypothalamus (upper portion) on Sephadex G-100 in the presence of nonimmune rabbit sera (upper curve) or rabbit antisera to beef neurophysin (lower curve); experiments performed in the region of antigen excess; ○—○ cpm/ml; ▲—▲ OD_{275} shown only for the upper curve; lower curve, electrophoretic patterns of beef neurophysin and neurophysin-^{35}S from a dog neural lobe. (From Fawcett, Powell, and Sachs.[85])

Examination of the neural lobes of several species[84,85,107] has indicated the presence of at least two neurophysinlike constituents. Although this suggests the possibility of a separate neurophysin for vasopressin and for oxytocin, this postulate must await experimental verification.

VII. BIOSYNTHETIC PATHWAYS

The function of the neurosecretory cells of the HNC is to synthesize and secrete vasopressin and oxytocin according to the needs of the body and in response to a variety of physiological stimuli. The biosynthesis of vasopressin and oxytocin may be examined within the context of several pertinent and related questions which, to a large extent, emerge from the physiological and morphological data. A number of such questions are:

(1) Which anatomical portions of the neuron engage in hormone biosynthesis?

(2) Are the NSG self-contained units in that they possess the enzymatic machinery for the *de novo* biosynthesis of hormones and other major constituents of the neurosecretory substance (i.e., neurophysin); or, are hormone and/or neurophysin molecules assembled at other intracellular loci and subsequently stored within the NSG?

(3) What is the nature of the enzymatic pathways involved in the biosynthesis of the packets of neurosecretory substance (i.e., NSG, hormones, neurophysin, etc.)?

(4) Are separate neurons or NSG involved in the formation of each hormone?

(5) Does the biosynthesis of hormone require the biosynthesis of its respective carrier protein?

(6) What factors regulate the rate at which hormone biosynthesis occurs? To what extent are the biosynthetic events responsive to neural inputs? How do neural inputs modulate the cellular biosynthetic processes?

A. Anatomical and Intracellular Loci of Hormone Biosynthesis

The lesion experiments described above in Section V, D, showed that cutting the nerve fibers led to the disappearance of the hormones and NSM distal to the lesion. While these experiments demonstrate that the integrity of the hypothalamic neurons is required for the sustained activity of their axonal elements, they do not preclude the possibility that these latter nerve segments can carry out the biosynthesis of oxytocin and vasopressin. For example, it is conceivable that the cell body merely furnishes a necessary and continuous supply of substrates, or cofactors, or must replenish essential but short-lived components of the biosynthetic apparatus. Recent experiments on peripheral nerve[108] and isolated synaptosomes[109] have indicated that the axon possesses a limited capacity to carry on both protein and RNA synthesis. Furthermore, the results of Vogt[110] and De Robertis[111] suggested that the distal portions of the HNC may play a significant role in the synthesis of one or both of the neurohypophyseal hormones. For example, the electron microscope studies of De Robertis[111] have shown that in at least two species, toad and rat, the size of the NSG increases along the supraoptico–hypophyseal tract (contrary to the findings of Sloper and Bateson[71]). Vogt[110] observed marked differences between hypothalamic and posterior pituitary oxytocin/vasopressin ratios, and postulated that either the posterior lobe must have an important modifying influence on whatever has been formed in the hypothalamus, or that the distal portion of the neuron is responsible for oxytocin biosynthesis. While the results of De Robertis and Vogt may be marshaled in favor of the idea that the axonal elements play a significant role in hormone biosynthesis, the nature of the evidence is at best indirect and is subject to a number of alternative interpretations. An often-used example to indicate that the perikaryon cannot be the sole source of hormone is taken from Green and Maxwell.[36] They postulated that if the cell body was the sole source of hormone, the rate of movement of NSG would have to be impossibly high in order to account for the observed rate of repletion of vasopressin in a dog after a period of dehydration. They calculated an average rate of movement down the stalk of 130 NSG/min per nerve fiber. Using Green and Maxwell's figure of 3×10^{11} granules per neural lobe (half of which may be oxytocin granules) and a value of about 6.3×10^{-12} U vasopressin in each NSG,[80] the rate of repletion of hormone is then about 56 mU/day, a figure entirely consistent with the available data. Estimates of the rate of axoplasmic flow of hormone have also been made by Sloper[8] who makes the unsubstantiated and unlikely assumption that acute secretion rates under a variety of circumstances are synonymous with the rates of axoplasmic flow. He has also attempted to recalculate the isotope data of Sachs[112] but unfortunately, Sloper's calculations and formulas[8] are incorrect. It may be relevant that recent investigations on axoplasmic flow[113] have suggested that this is a heterogeneous process; some cellular constituents can apparently move down the axon at relatively high rates compared to the estimates of the rate of movement of the bulk of the axoplasm.[64] It is not inconceivable that the movement of the NSG is a directed process and their rate of flow is many times greater than that of the surrounding axoplasm.

Over the past several years it has become possible to study the biosynthesis of vasopressin directly by means of isotope techniques.[113] The results of our isotope experiments on the incorporation of radioactive amino acids into vasopressin are in accord with the concept that the hypothalamus is the major, if not obligatory site for the synthesis of the peptide bonds in vasopressin. The experimental findings are as follows: (1) After the continuous infusion of cysteine-^{35}S into dogs for periods of from 8 to 36 hr, the specific activity of the hormone isolated from the hypothalamus was 2–3 times greater than that of neurohypophyseal vasopressin.[113] (2) In one experiment in which a stalk section was performed prior to infusion of cysteine-^{35}S into the third ventricle of a dog, labeled hormone was found in only the hypothalamus at the termination of the experiment. (3) Hypothalamic–median eminence tissue of the guinea pig was capable of de novo hormone biosynthesis in vitro. Under identical incubation conditions, isolated portions of the neurohypophysis (infundibular stem and process) could not carry out the incorporation of labeled amino acids into vasopressin.[114]. (4) Puromycin inhibits vasopressin biosynthesis in vivo or in vitro; these results suggest that the peptide bonds in vasopressin are constructed at sites on or resembling ribosomes. Except for mitochondrial structures,[115] the axonal portions of the neurosecretory cells in the posterior pituitary do not appear to contain other protein-forming units requiring the participation of nucleic acids. Although considerable evidence has been accumulated in favor of the essential role of the perikaryon in the initial steps in hormone biosynthesis, it should be stressed that none of the data preclude the possibility that the later stages of hormone biosynthesis (e.g., release of vasopressin from a bound form) can occur within portions of the axonal elements.

Isotope experiments performed on the intact dog have ruled out the possibility that the NSG represent the primary cellular loci for hormone biosynthesis.[116] After the continuous infusion of cysteine-^{35}S into the third ventricle of anesthetized dogs over relatively short time intervals, the vasopressin molecules associated with the NSG were always found to have the lowest specific activity compared to vasopressin associated with a number of other particulate cell structures (Table III).

TABLE III

Comparison of the Specific Activities of Vasopressin-^{35}S and Protein Associated with Cell Fractions of the Dog Hypothalamus

Cell fraction	Vasopressin		Protein counts/min/μg
	μg	counts/min/μg	
NSG	0.58	96	26.5
Cell sap	0.85	76	10.7
Nuclei, cell debris	0.36	229	11.9
Large granules	0.38	319	11.1
Microsomes	0.36	155	15.1

Cysteine-^{35}S, 1.7×10^9 counts/min infused into the third ventricle of a dog's brain for 3 hr. From Sachs.[116]

Although it was reasonable to assume that nucleic acid structures (i.e., ribosomes) would constitute the primary and obligatory site for the synthesis of the polypeptide hormone, the experimental data were not in accord with such an assumption. It has in fact been a consistent finding that after the infusion of cysteine-^{35}S into the third ventricle of dogs for 3–6 hr, the most highly labeled vasopressin molecules did not follow the ribosome-rich fractions (see Table III and Sachs[116]). Instead, vasopressin-^{35}S with the highest specific activity was found in association with particulate material which sedimented in a relatively low centrifugal field. While it cannot be excluded that these results were in part artifactual (i.e., due to distortion of pool sizes by disruption of cell structures, nonspecific binding, etc.), other interpretations of the isotope data are possible. Two of these are that: (a) nucleic acid structures are not involved in vasopressin biosynthesis in analogy with the biosynthesis of the pentapeptide nucleotide of *Staphylococcus aureus*,[117] or gramicidin S[118,119] or the tyrocidines[120]; or (b) that the biosynthesis of vasopressin occurs on ribosomes but in a bound, biologically inactive form (i.e., as part of a macromolecule), and labeled vasopressin first appears at a site removed from the ribosome. The results of a subsequent series of isotope experiments have provided information in support of the latter interpretation.

In the event that a biologically inactive precursor exists in significant quantities and that liberation of active hormone is relatively slow, then labeling experiments might be expected to show a lag period followed by the appearance of radioactive vasopressin at loci removed from and independent of the initial biosynthetic events. Should the formation of the precursor require the participation of nucleic acid structures, then it might be expected that puromycin would be an effective inhibitor of vasopressin biosynthesis. It is also apparent, however, that the efficacy of the puromycin inhibition would depend upon the time of addition of the drug; e.g., puromycin added after the lag period (or after the formation of precursor had taken place) would not be as effective as puromycin added at the start of the experiment. These expectations have indeed been borne out by both *in vivo* and *in vitro* studies on the incorporation of labeled amino acids into vasopressin.

Infusion of a radioactive amino acid into the third ventricle of a dog for 1.5 hr did not give rise to significant quantities of labeled hormone, whereas if the dog was permitted to survive for an additional 4.5 hr after the initial isotope infusion, considerable quantities of labeled vasopressin appeared. Furthermore, when puromycin was present from the start of the experiment, labeled hormone did not appear; if, however, puromycin was administered after the initial isotope infusion, then considerable quantities of labeled hormone were found at the end of the experiment.[121] These findings suggest that there is a lag period in vasopressin biosynthesis and that only the initial biosynthetic events in the production of hormone are puromycin sensitive. This postulate received confirmation from a series of experiments in which the release of radioactive vasopressin from some "precursor" (labeled during the initial 90-min infusion period *in vivo*) was permitted to take place *in vitro* (Table IV). Either cysteine-^{35}S or tyrosine-^{3}H was infused into the third ventricle of a dog for 1.5 hr. The HME was then excised and divided in the medial plane and one half was used for the immediate isolation of vasopressin (left-hand side, Table IV). The other half of the HME tissue was sliced and incubated at 37°C for 4.5 hr in a modified Krebs–Ringer buffer containing puromycin and unlabeled cysteine (or tyrosine). It can be seen (Table IV) that in each case incubation of the HME slices *in vitro* gave rise to labeled hormone under conditions which precluded *de novo* peptide bond synthesis. Furthermore, the production of radioactive hormone *in vitro* occurred only with dog HME slices which had previously been exposed to cysteine-^{35}S or tyrosine-^{3}H for 1.5 hr *in vivo*. Apparently, the formation of some labeled precursor took place *in vivo* during the 1.5 hr infusion period. It is also clear that whereas the initial biosynthetic events (i.e., synthesis of the precursor molecule) were inhibited by puromycin, the subsequent release (or production) of vasopressin from precursor was puromycin insensitive.

More recently Takabatake and Sachs[114] have been able to achieve the *de novo* synthesis of vasopressin by guinea pig HME tissues *in vitro*. The experimental findings *in vitro* are consistent with the *in vivo* isotope experiments. The incorporation of cysteine-^{35}S into vasopressin *in vitro* with HME

TABLE IV

Production of Labeled Vasopressin *In Vitro* by HME Slices Taken from Dogs Which Were Infused with a Radioactive Amino Acid for 1.5 hr Prior to Sacrifice

| | | $\frac{1}{2}$ HME immediately after 1.5 hr infusion *in vivo*[a] | | | $\frac{1}{2}$ HME after 4.5 hr incubation *in vitro*[a] | | |
| | | Vasopressin | | Protein | Vasopressin | | Protein |
Exp.	Amino acid infused 1.5 hr *in vivo*	μg	cpm/μg	cpm/μg	μg	cpm/μg	cpm/μg
1	Cys-^{35}S	1.8	80	28	1.9	500	32
2	Cys-^{35}S	3.0	116	123	2.3	1057	124
3	Cys-^{35}S	2.9	9	44	2.8	61	25
4	Tyr-^{3}H	2.4	79	33	2.7	216	27

[a] HME taken from dogs which were infused with either cysteine-^{35}S (1.4×10^9 cpm, Exp. 1 and 3; 2.4×10^9 cpm, Exp. 2) or tyrosine-^{3}H (1.4×10^9 cpm) for 1.5 hr; $\frac{1}{2}$ HME homogenized in 10% (w/v TCA immediately after 1.5 hr infusion *in vivo*, $\frac{1}{2}$ HME sliced and incubated in a modified Krebs-Ringer bicarbonate medium containing 2×10^{-4} M puromycin and either 5×10^{-3} M labeled cysteine (Exp. 1–3) or tyrosine (Exp. 4), respectively. From Sachs and Takabatake.[121]

tissue taken from the guinea pig has consistently shown a lag period of about 1 hour's duration followed by the appearance of labeled hormone. Furthermore, the biosynthesis of vasopressin was completely inhibited by puromycin if the drug was present from the start of the experiment. If, however, puromycin was added after the lag period (or after the synthesis of precursor had already taken place), then it did not prevent the subsequent appearance of labeled hormone.[122]

B. Enzymatic Pathways of Hormone Biosynthesis

The presently available data regarding the nature of the enzymatic steps involved in ADH and oxytocin biosynthesis are at most indirect and circumstantial. On the basis of the results of isotope studies (described in part above), Sachs and Takabatake[121] proposed a "precursor model" for vasopressin biosynthesis (depicted schematically in Fig. 9). According to this model, the biosynthesis of the peptide bonds in vasopressin would occur solely in the perikaryon, on ribosomes, via pathways common to the biosynthesis of other peptide chains (i.e., involving transfer RNA, messenger RNA, etc.). It was further proposed that the biosynthesis of vasopressin proceeds via a bound, biologically inactive form (i.e., as part of a precursor molecule) and that the appearance of the biologically active octapeptide occurs at a time and place removed from the initial biosynthetic events. Conceivably, the release of the octapeptide from the precursor molecule would take place during the formation and maturation of the NSG. The choice of the "Golgi region" for the formation of the NSG is based on the electron microscopic

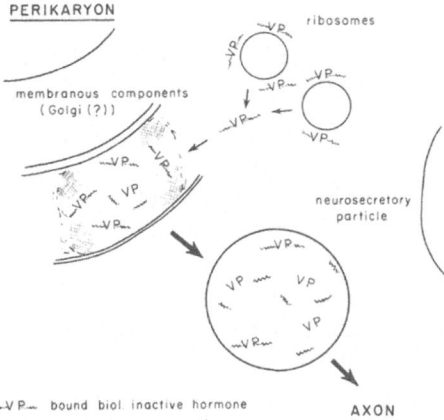

Fig. 9. A precursor model for vasopressin biosynthesis; ($\sim$ VP $\sim$) precursor molecule containing bound, biologically inactive vasopressin; VP, biologically active octapeptide hormone. [Reproduced through the courtesy of *Endocrinology*, H. Sachs and Y. Takabatake, **75**:943–948 (1964).]

studies of Bern *et al.*[3] and Scharrer and Brown.[123] The notion that ribosomes or similar nucleic acid structures play an essential role in the initial stages of peptide bond synthesis derives from the observations that puromycin inhibits the early phase of vasopressin biosynthesis. Further support comes from the results of electron microscopy and chemical analysis of supraoptic neurons of rats subjected to prolonged dehydration or osmotic stimulation. Under these circumstances there are increased numbers of ribosomes and polysomes,[124,125] increased amounts of RNA,[126] and an enhanced rate of ADH biosynthesis.[114] While the overall process of vasopressin biosynthesis (*in vitro*) has been shown to require an energy source such as glucose, it is not known whether the energy requirement is confined entirely to the formation of the "precursor molecule." At present, there is little information regarding either the nature of the "precursor molecule" or the enzymatic steps involved in the release (or production) of vasopressin from this "precursor." Although it is possible that the "precursor" is of small molecular weight (e.g., the —SH open chain form of vasopressin), we are inclined to the view that a more likely candidate is a macromolecule (i.e., a protein). If indeed the precursor molecule is a protein, then it should be possible to explain the evolution of the structures of the neurohypophyseal hormones according to current theories regarding protein evolution (i.e., substitution of one amino acid for another in polypeptide chains as a result of single base changes in codons of m-RNA). Vliegenthart and Versteeg[53] have recently presented such a scheme on the phyletic distribution of the hormones compatible with the available information on the genetic code. The formation of the protein precursor containing bound vasopressin might require a polycistronic messenger; the release of the free, biologically active hormone would conceivably involve one or more hydrolytic or ammonolytic steps. This hypothesis has a number of well-known analogies, such as the formation of angiotensin II, as well as that of enzymes involved in digestion and blood clotting. Recent studies on the biosynthesis of insulin[127] have indicated that this polypeptide hormone is also made via the intermediate formation of a larger protein unit. It is obvious, however, that ultimate proof of the existence of a "precursor molecule" in vasopressin biosynthesis must await its isolation and chemical characterization. Unfortunately, parallel investigations have not been carried out on oxytocin biosynthesis and virtually nothing is known about its mode of synthesis. Nevertheless, by virtue of the close similarities in chemical structure and cellular storage and release of oxytocin and vasopressin, it would not be unreasonable to assume similar biosynthetic pathways.

C. Synthesis of Oxytocin and Vasopressin in Separate Neurons

As discussed previously, the available evidence is consistent with the view that oxytocin and vasopressin are produced, stored, and released by separate neurons.

D. Simultaneous Biosynthesis of Hormones and Their Carrier Proteins

While the studies of Sachs and co-workers[85] have demonstrated the parallel release of labeled neurophysin and vasopressin, there is still a paucity of experimental data on the question as to whether there is a synchronous and/or mutually dependent synthesis of the neurohypophyseal hormones and their respective carrier proteins. The results of several investigations suggest an affirmative answer to this question. Rennels[83] and Frieden and Astwood[84] observed that prolonged dehydration (a treatment leading to hormone depletion) led to depletion of neurophysin from the neural lobes of rats and a number of other species. If dehydrated rats were permitted free access to water, there occurred a repletion of neurophysin during a time period in which the repletion of vasopressin and oxytocin is also taking place. Furthermore, in the Brattleboro strain of rats, which is apparently incapable of synthesizing vasopressin,[128] the concentrations of neurophysin,[83] NSM,[129] and NSG[130] are greatly reduced. Presumably, the reduced quantities of these constituents of the neural lobe are confined to oxytocin-producing neurons.[93] In the light of the foregoing, two interesting prospects emerge; these are: (1) neurophysin and vasopressin are made as part of the same precursor molecule; or (2) there is a common genetic unit. The development of isotope methods for the study of neurophysin biosynthesis[85] offers exciting prospects for future studies on this question.

E. The Rates of Hormone Biosynthesis

For many endocrine organs (e.g., adrenal cortex, gonads, parathyroids) *de novo* hormone biosynthesis is intimately linked to gland stimulation and hormone secretion; i.e., acute secretory stimuli bring about the rapid release of hormone in amounts far greater than that initially present in the gland. In the case of the HNC, the biosynthetic and secretory events often appear temporally unrelated. It is noteworthy that these neurosecretory neurons are not only a factory for the synthesis of oxytocin and vasopressin, but they are also a commodious storehouse. From the available estimates of "basal" vasopressin secretion rates, it is apparent that the neurohypophysis contains enough hormone to maintain a "basal" rate of secretion of vasopressin for several weeks. Furthermore, the large stores of pituitary hormones are more than sufficient to meet the needs of acute stimuli. The question nevertheless arises as to what extent the biosynthesis of hormone is attuned to the secretory activities of the cell. What factors are involved in the control of the rate of hormone biosynthesis?

Actual measurements (e.g., turnover studies) of the rate of either vasopressin or oxytocin biosynthesis have not been performed. Although under steady-state conditions the biosynthetic rate could be equated with the rate of secretion, precise estimates of the latter are not available. The rate of vasopressin biosynthesis in the dog under "resting" conditions is assumed to be on the order of 1–5 mU ($2.5–12.5 \times 10^{-3}$ μg/hr; this is based on the work of Shannon[131] who observed that an infusion rate of 1–5 mU vasopressin/hr in dogs with diabetes insipidus led to maximal antidiuresis. Estimates based on half-life time measurements[50] give a value for the secretion rate in man of about 50 mU/hr. In the guinea pig, a crude estimate of the capacity of hypothalamic–median eminence tissues to synthesize vasopressin *in vitro* was obtained from isotope studies[114]; and this was on the order of a few tenths of a milliunit per hour per hypothalamus. The availability of an *in vitro* system capable of hormone biosynthesis afforded the opportunity to examine the question of whether a prolonged secretory stimulus such as dehydration leads to an enhanced synthesis of vasopressin. It has been a consistent finding that the HME slices taken from guinea pigs deprived of water for 4 days incorporated 2 to 5 times more radioactivity into vasopressin than similar slices from guinea pigs with free access to water (Table V). Analogous results have also been obtained with HME tissues of guinea pigs allowed to drink 2.0 % sodium chloride for a period of 10 days (Exp. 4, Table V).

What cellular mechanisms are involved in the activation of hormone biosynthesis under these circumstances? Assuming that the formation of the peptide chain constitutes the rate-limiting step and that this takes place on ribosomes, then control may be exercised at the level of either transcription, or translation, or both. Relevant to this discussion is the finding that the RNA content of the supraoptic neurons of rats subjected to a prolonged osmotic stimulus or dehydration is greater than that of normal rats.[126] Furthermore, electron micrographs of the supraoptic neurons of rats subjected to chronic dehydration have shown increased numbers of ribosomes and a morphologic picture indicative of enhanced activity of the protein and RNA synthesizing machinery (Fig. 10).[124,125] Hence, the RNA studies, the morphologic observations, and the isotope experiments indicate that the chronic reception of nerve impulses effective in the release of vasopressin may also be translated

TABLE V
Effect of Dehydration on the Labeling of Vasopressin *In Vitro*

		Vasopressin			
		Control		Dehydrated	
Exp.[a]	Incubation time, (n)	cpm/μg	total cpm	cpm/μg	total cpm
1	6	17,760	3,700	56,600	19,800
2	6	8,200	12,500	37,000	26,400
3	4	2,380	450	3,400	640
4	4	4,160	3,255	8,000	4,220

[a] Exp. 1–3: deprived of water for 4 days; Exp. 4., "dehydrated group," consisted of guinea pigs in which 2.0% NaCl was substituted for water 10 days prior to sacrifice; all "controls" were permitted free access to water.
Exp. 1 and 2: each flask contained hypothalamic-median eminence slices from 10 animals (either dehydrated or control guinea pigs and cysteine-^{35}S, 5.1×10^8 cpm).
Exp. 3 and 4: tyrosine-^{3}H and HME slices from 8 and 10 animals per flask respectively.
Exp. 3: 3.6×10^8 cpm per flask; Exp. 4, 1.0×10^9 cpm per flask. From Takabatake and Sachs.[114]

into an enhanced synthesis of both specific RNA molecules, and entire biosynthetic units involved in polypeptide hormone biosynthesis.

What can be said about the relationship between acute stimuli and hormone biosynthesis? Will an intense stimulus applied over a short time interval lead to a burst of biosynthetic activity? For example, the disappearance of NSM after the application of an acute, painful stimulus,[86] and the rapid reappearance of NSM after removal of the stimulus, were taken to indicate that the neurohypophysis had discharged and resynthesized its hormonal content at a rate several orders of magnitude greater than the unstimulated rate of hormone synthesis. The studies of Moses *et al.*[87] and Daniel and Lederis,[88] however, have clearly shown that while acute stimuli lead to the loss of NSM (i.e., stainable aggregates of NSG) and loss of osmiophilicity of the NSG, this is not synonymous with hormone depletion; and in all probability these phenomena are related to the process of secretion rather than to the biosynthetic mechanisms. Furthermore, labeling experiments carried out in the dog in our own laboratory have thus far failed to show an enhanced uptake of cysteine-^{35}S into vasopressin, either during or shortly after hemorrhage. Neither did electrical stimulation of HME tissues of the guinea pig enhance the incorporation of labeled amino acids into vasopressin *in vitro*. To date, there is no convincing evidence that an acute secretory stimulus leads to an immediate increase in the rate of hormone biosynthesis as is the case with other secretory organs (e.g., the adrenal, gonads, etc.). The enhanced synthesis of vasopressin after dehydration or prolonged osmotic stress may

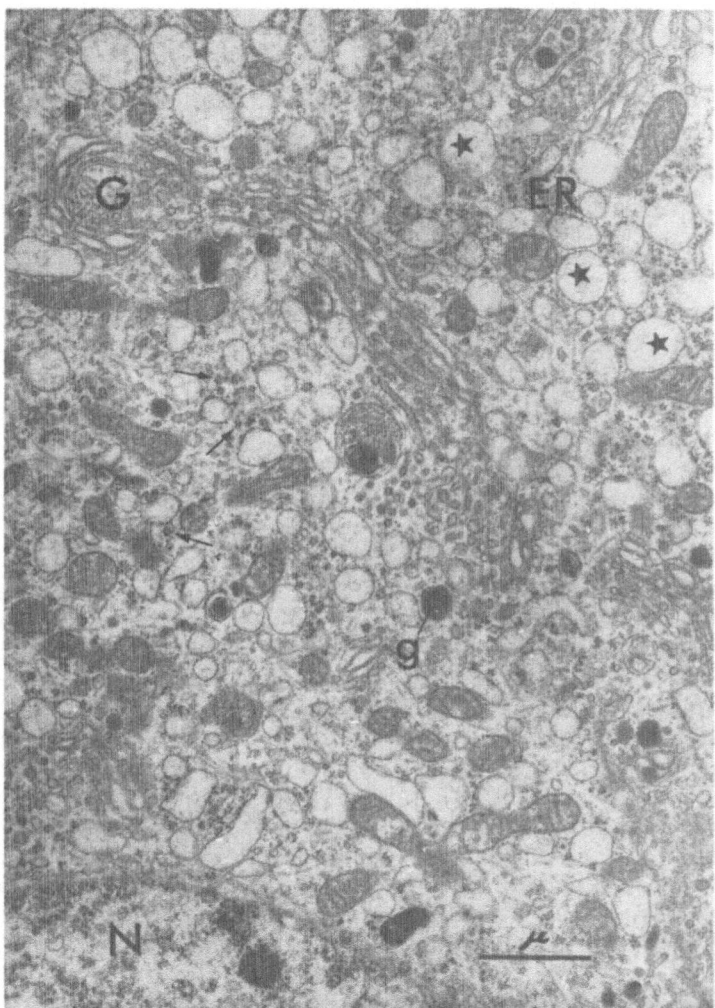

Fig. 10. Electron micrograph of rat supraoptic nucleus after a period of water deprivation for 24 hr. Cisternae of rough endoplasmic reticulum (ER) are extensively dilated and contain a filamentous matrix (*). In addition many polysomelike clusters (arrows) of ribosomes may be seen in the intercisternal spaces. N, nucleus; G, Golgi apparatus; g, neurosecretory vesicle. × 17,600. (Reproduced through the courtesy of J. Osinchak.)

thus be construed as an adaptive response to a set of changing environmental conditions. The nature of the cellular events which intervene between the reception of nerve impulses and enhanced RNA and hormone biosynthesis, remains, of course, a central problem in neurosecretion and neurochemistry.

In this regard, it should be stressed that these neurosecretory neurons, like many other peripheral and central neurons, have a very high capacity for the general synthesis of protein and its protoplasmic constituents. It is evident that extensive stimulation results in the activation of many biosynthetic pathways and in some instances there may be a marked hypertrophy[128] of the neurosecretory perikarya. As compared to other neurons of the CNS, the neurons of the HNC show an unusual pattern of reaction to injury and apparently the neurosecretory neurons can call upon its biosynthetic apparatus in order to bring about a limited regeneration of their terminals.[8]

VIII. SECRETORY MECHANISMS

The physiological stimuli and afferent pathways involved in the reflex secretion of the posterior pituitary hormones in the intact animal have been reviewed by several authors.[47–49] In this section I would like to examine some of the current concepts and investigations concerned with the neural mechanisms of vasopressin and oxytocin secretion. Although most of the work has been done with vasopressin, probably because of the greater ease and availability of methods of quantitation, it is tacitly assumed that both hormones are released from neurons via similar pathways. The problem of hormone secretion at the cellular level involves several pertinent and related questions. These are: (a) What factors trigger the extrusion of hormone from the neuron? (b) Do the other constituents of the NSG (i.e., neurophysins) accompany the release of the hormones? (c) What are the anatomical and ultrastructural sites for the release of these polypeptides? (d) What cellular and enzymatic events are crucial to the secretory process? (e) What can be said about the control of hormone secretion at the level of the neuron?

It will become evident that there are large gaps in our understanding of the secretory mechanisms and in reality, some of the questions that have been posed above merely delineate important areas for future research. It is also apparent that the problems pertaining to the secretion of the neurohypophyseal hormones apply to a wide variety of cell types engaged in the secretion of chemical substances (e.g., enzymes, protein and peptide hormones, transmitter substances) initially packaged and stored within membrane-bound structures. In many instances these diverse cell types display some remarkably common features associated with their secretory processes.

A. Nerve Excitation and Hormone Release

Electrophysiological studies (recently reviewed by Cross and Silver[132] and Brooks et al[133] support the view that nerve impulses traveling down the supraoptico–hypophyseal tract trigger the liberation of hormone. It has been shown repeatedly[132,133] that stimuli known to evoke the release of vasopressin increase the rate of firing of supraoptic neurons; in the case of the paraventricular nuclei, Brooks and co-workers[134] have shown what appears to be a clear correspondence between unit activity and the secretion of oxytocin. In response to single stimuli of the supraoptic nucleus, the pituitary stalk exhibits action potentials with a conduction velocity characteristic of C fibers.[135] The activity of these hypothalamic neurons is apparently influenced by impulses originating in many parts of the central nervous system,[136] some of which are of an inhibitory nature.[133] Histochemical studies have indicated that both adrenergic and cholinergic fibers impinge on the cell bodies of the neurons found in the supraoptic and paraventricular nuclei. Whereas the injection of acetylcholine will excite the supraoptic and paraventricular neurons[134,136] and cause antidiuresis,[137] there are data which suggest that the catecholamines and serotonin inhibit the discharge of vasopressin[136] and oxytocin[138] respectively. The isolated neurosecretory neurons (i.e., the deafferented hypothalamus), fire at a faster than normal rate [133] and are functionally capable of responding to osmotic stimuli and maintaining a degree of water balance in the animal.[139] In contrast to the studies of conventional neurons on the release of transmitters at synaptic junctions, it has not been possible to measure quantitatively the relationship between the amount of hormone released and various facets of the electrical activity of the neurosecretory neuron (e.g., rate and patterns of firing, shape and duration of action potentials, changes in membrane potential, etc.). The supraoptico–hypophyseal axons are only about $1\,\mu$ in diameter, making microelectrode penetration virtually impossible and, furthermore, whereas the unique structure of the synaptic junction as well as presently available instrumentation make it possible to measure the quantal release of transmitters (recorded as postsynaptic potentials), presently available methods are incapable of estimating the amount of hormone that could be released from a single neurosecretory nerve fiber.

B. Excitation–Secretion Coupling

The cellular and enzymatic events which intervene between the arrival of action potentials and the secretion of hormone from the neuron constitute a process that has been termed[140] "excitation–secretion coupling." Two major hypotheses regarding this process have evolved from two separate lines of experimental work. The first of these was formulated entirely on the basis of histochemical and morphological observations. By means of histochemical methods, Abrahams et al.[141] and Koelle and Geesey[142] showed that true cholinesterase is present in the cell bodies and throughout the length of the neurons of the hypothalamo–neurohypophyseal tract. Abrahams et al.[141] speculated that if the hypothalamo–neurohypophyseal nerve fibers were cholinergic, "this would produce the unique situation of a neuron's own transmitter substance providing the stimulus for the release of its own endocrine product." Gerschenfeld and his colleagues[143] reached a similar conclusion on the basis of their studies on the ultrastructure of the neurohypophysis of the toad. These workers and Palay[144] described a small vesicular component in the nerve terminals of the neurohypophysis (toad and rat) that has the morphologic characteristics of the synaptic vesicles seen in many nerve endings. Both groups of investigators observed that a chronic dehydration of several days led to marked diminution in the density of the NSG initially and subsequently in their number; these changes, however, did not occur in the "synaptic" or microvesicles, and during this period they may actually have increased in number. Similar increases in the number of microvesicles have also been observed after acute stimuli both *in vivo* and *in vitro*.[144,72] Gerschenfeld *et al.*[143] postulated that the arrival of action potentials at the nerve terminals (in response to some secretory stimulus) could elicit the release of a transmitter substance from the "synaptic vesicles" and that this transmitter substance in turn would cause the release of vasopressin from the nerve. Acetylcholine was picked as the most reasonable choice for the identity of the transmitter substance. The actual site of action of acetylcholine was never specified and presumably this mediator could exert its action on either or both of the nerve cell or NSG membranes or at some unspecified locus. The general outline of the hypothesis finds some analogy in the situations described by Bern and Rand.[145]

The evidence in favor of this hypothesis is entirely circumstantial, namely, the demonstration of the presence within the hypothalamo–neurohypophyseal complex of (1) cholinesterase,[141,142] (2) choline acetylase,[146] (3) acetylcholine[147] and (4) "synaptic vesicles".[144] Whether acetylcholine (recently demonstrated in the neural lobe of the rabbit[147]) is actually contained within the microvesicles remains a central question. Although Bindler *et al.*[148] have recently succeeded in isolating nerve endings (neurosecretosomes) and a fraction rich in microvesicles from bovine posterior pituitaries, it is still unclear whether or not these latter axonal structures contain transmitter substances or hormones. Furthermore, experiments with isolated posterior pituitary preparations have so far failed to provide data which support the view that acetylcholine plays a key role in the release of hormone at the nerve terminal. Acetylcholine, atropine, or eserine did not effect the release of vasopressin by neural lobe tissue *in vitro* under conditions of "rest" or "stimulation."[140] A list of other known transmitter substances (e.g., norepinephrine, serotonin, epinephrine, histamine) were likewise shown to be ineffective.[149] These experiments, however, suffer from the drawback that little information is available regarding the extent to which such transmitters may penetrate the axon. Further doubt regarding the validity of the Gerschenfeld hypothesis is raised by the suggestion of Bern[150] and of Lederis[151] *et al.* that the microvesicles are merely degradation products of the NSG.

The second hypothesis implicates Ca^{2+} as a major factor in bridging the gap between nerve excitation and hormone secretion. Douglas and coworkers[140] were able to show that Ca^{2+} is an obligatory requirement for the release of vasopressin from rat posterior pituitaries incubated *in vitro* and stimulated either electrically or with 56 mM K^+. On this basis they postulated that the arrival of action potentials at the nerve terminals led to membrane depolarization which permitted the entrance of Ca^{2+} into the neuron, and that it was the presence of free Ca^{2+} in the axon terminals which in some way brought about the release of hormone. Ginsburg and Ireland[80] attempted to define the role of Ca^{2+} in the light of their findings on the subcellular distribution of neurophysin–hormone complexes in homogenates of bovine neurohypophyses and in view of their observation that Ca^{2+} inhibits the association of vasopressin to neurophysin. They suggested that hormone–neurophysin complexes are present in both the NSG and in the axoplasm. The postulated role of Ca^{2+} is to dissociate extragranular vasopressin–neurophysin complexes; the free octapeptide hormone would then be free to migrate to the exterior of the neuron. As intracellular free Ca^{2+} is restored to resting levels, extragranular complexes would be re-formed by drawing on hormone from the granules. This hypothesis specifies that the release of vasopressin is unaccompanied by neurophysin proteins and that the cellular path of hormone secretion consists of the movement of vasopressin from neurophysin complexes within the NSG to similar complexes with extragranular neurophysin, and finally, to the exterior of the neuron. Although this hypothesis has many attractive features, recent investigations in our laboratory[85] have demonstrated convincingly that neurophysin accompanies the release of vasopressin. It is thus unlikely that neurophysin serves merely as a way station in the transport of vasopressin from NSG to the cell exterior and that the site of action of Ca^{2+} is at extragranular hormone-neurophysin complexes. Whereas the precise role of Ca^{2+} in the secretion of the neurohypophyseal hormones remains unsolved, it is noteworthy that the requirement for Ca^{2+} in "stimulus–secretion coupling" is common to a wide variety of cell types engaged in the secretion of catecholamines,[152] hormones,[153] enzymes,[154] and transmitter substances.[155] The secretion of vasopressin appears to be more analogous to the release of catecholamine from chromaffin cells[156] than the release of transmitter from presynaptic nerve terminals.[155] In contrast to transmitter release at the mammalian neuromuscular junction in response to increased K^+ or lowered Na^+, the secretion of catecholamine in response to K^+ or acetylcholine does not appear to be a function of the initial membrane potential.[156] Furthermore, in the latter case secretion does not appear to be tightly coupled to depolarization and Douglas has proposed[156] that depolarization of the chromaffin cell in response to acetylcholine may be no more than the electrical sign of increased permeability to Ca^{2+} whose movement into the cell sets the stage for secretion. The studies of Douglas et al.[140] with isolated rat posterior pituitaries suggest that a similar mechanism may be operative with respect to vasopressin secretion.

Studies with isolated neural lobes in vitro have indicated that the secretion of vasopressin requires an energy source. Either inhibition of electron transport or oxidative phosphorylation,[157] or the absence of glucose and O_2[122,158] depressed the K^+-stimulated release of hormone from rat and dog posterior pituitaries respectively. These observations appear analogous to the findings of several investigators[154] on the energy requirement for the secretion of protein and enzymes from either pancreas or salivary gland slices. Experiments with parotid gland slices demonstrated that energy was required not only initially for the induction of secretion, but continuously for its maintenance. Schramm and co-workers[154] concluded that the secretion of amylase is an active, tightly coupled energy-requiring process. Recently, these authors and others have provided evidence that 3',5'-cyclic AMP is involved in the secretory processes of several cell types.[154] In the case of the isolated neural lobe it has been shown only that vasopressin secretion in response to K^+ is sharply reduced after an hour or more or deprivation of glucose and O_2, or inhibition of electron transport or oxidative phosphorylation. A number of pertinent questions remain unanswered; these are: (a) Is hormone secretion tightly coupled to a supply of energy? (b) Is ATP or some other high energy derivative involved? (c) What stages of the secretory process require energy? Krass and La Bella[159] have demonstrated a highly active hexosemonophosphate shunt in the posterior pituitary as well as in several other endocrine glands and the question arises as to the significance of this pathway of glucose metabolism to the secretory process. (d) Does 3',5'-cyclic AMP play a role in the secretion of the neurohypophyseal hormones? With regard to the latter question, preliminary experiments in our laboratory have so far failed to show a consistent stimulation of hormone release from neural lobes incubated in the presence of either 3',5'-cyclic AMP or the ^{6}N-2-O-dibutyryl derivative. Incubation of dog neural lobes in medium containing 56 mM K^+ leads to a decrease in the tissue concentration of 3',5'-cyclic AMP (Haller, Rall, and Sachs, unpublished), an observation whose meaning is at present unclear.

In view of the energy requirement for the secretion of hormone, it is pertinent that dog neural lobes, stimulated to secrete vasopressin, take up O_2 at a higher rate than unstimulated glands.[158] From Table VI it can be seen that neural lobes of bled dogs took up O_2 in vitro at about twice the rate of glands taken from unbled animals; and, furthermore, 56 mM K^+ stimulated the rate of O_2 uptake in both cases.

TABLE VI

Oxygen Uptake of Isolated Dog Neural Lobes

	Oxygen uptake[a] μM O_2/min/mg wet tissue $\times 10^3$	
	Unbled	Bled
Number of experiments	7	7
Control medium (8.0 mM K^+)	1.31 ± 0.17	2.48 ± 0.37
56 mM K^+ in medium	2.34 ± 0.12	3.76 ± 0.61

[a] Mean $\pm$ SEM calculated at a time when constant rates of O_2 uptake were maintained (see Sachs and Haller[158]); differences are significant at $P < 0.001$.

C. Secretion of the Total Neurosecretory Substance

As indicated in a previous section, the bulk of the neurosecretory substance within the NSG appears to consist of hormone–neurophysin complexes. The observed loss of soluble protein[82] and/or neurophysin[84,85] from the neural lobe following a period of dehydration suggests that the release of hormones is accompanied by the simultaneous release or degradation of neurophysin. The latter possibility seemed more reasonable in view of the failure of Thorn[160] to demonstrate the simultaneous secretion of neurophysin and vasopressin from rat posterior pituitaries stimulated *in vitro* with 56 mM K$^+$. Thorn's test for the presence of neurophysin in the incubation medium consisted of adjustment of the pH to 3.9 and of the NaCl concentration to 10%, followed by centrifugation of the medium at 2000 × g for 30 min in order to sediment a vasopressin–neurophysin complex. Although these conditions of pH and salt concentration are routinely employed for the precipitation of the vasopressin–neurophysin complex, it is questionable whether the complex at a concentration of a few micrograms per milliliter would form sedimentable aggregates. It was therefore imperative to apply more sensitive techniques to test whether neurophysin leaves the neuron under circumstances which lead to the release of vasopressin. The sensitive tool required for such studies was provided by the adaptation of isotope procedures previously employed for the study of hormone biosynthesis. Ten

TABLE VII

Secretion of Labeled Neurophysin, Protein, and Vasopressin from Dog Neural Lobes Incubated *In Vitro*[a]

			Secreted into the medium	
Exp.	Incubation Media, conditions	Vasopressin mU	Radioactive protein, counts/min	Radioactivity coincident with neurophysin, counts/min
1	Control (5.6 mM K$^+$)	63	6,320	1,760
	Stimulation (56 mM K$^+$)	209	14,740	8,600
2	Control	70	13,700	8,500
	Stimulation	354	45,600	44,000
3	Control	83	236	—
	Stimulation	507	457	—

[a] The labeling regimen prior to death of the animals was as follows: Dogs, dehydrated for a period of 5 days were used for the intraventricular infusion of either Cys-^{35}S (Exp. 1, 2) or Tyr-^{3}H (Exp. 3). After the infusion of isotope, the dogs had free access to water until they were killed 10–19 days later. From Fawcett, Powell, and Sachs.[85]

to 20 days after the infusion of Cys-^{35}S or Tyr-^{3}H into the third ventricle of the dog's brain, the neural lobe contained considerable quantities of highly labeled proteins and vasopressin. Incubation of such posterior pituitaries in media containing either 5.6 mM or 56 mM K$^+$ led to the release of labeled protein and vasopressin (Table VII). Under the latter conditions (56mM K$^+$) the amounts of both radioactive protein and vasopressin secreted into the incubation medium were severalfold greater than that observed with media containing the low K$^+$ concentration (5.6 mM). Further fractionation of the labeled protein demonstrated that the preponderance of counts resided in material which had gel filtration and electrophoretic properties similar to both bovine neurophysin and the labeled neurophysinlike material isolated from dog posterior pituitary glands (Fig. 8).

Two experiments analogous to those described above were also carried out *in vivo*. In confirmation of the *in vitro* studies it was shown that vasopressin and neurophysin were released into the blood in response to the stimulus of bleeding.[85] These results suggest that as in the case of the catecholamines[161–164] and the pancreatic zymogens,[165] a secretory stimulus leads to the extrusion of the entire contents of the storage granule to the exterior of the cell. The cellular mechanisms whereby this is accomplished are at present unknown; some possibilities are discussed below.

1. Cytology of Secretion

The cytology of the supraoptico–hypophyseal neurons has been described at the levels of light and electron microscopy by several investigators.[5–11] As already indicated, the main cellular structures implicated in the storage and secretion of the neurohypophyseal hormones are the NSG, the NSM (aggregates of NSG), and the microvesicles. In studying the cytological aspects of hormone secretion, the strategy of the morphologists has been to examine the ultrastructure and cytochemistry of the neurohypophyses before and after the application of either an acute or prolonged stimulus. The results of some of these studies have been discussed in preceding sections of this chapter. In the light of these observations as well as the demonstration that the bulk of the contents of the NSG leave the neuron, it is reasonable that hormone secretion probably occurs via a process of micropinocytosis (also termed exocytosis[166]). The prevailing view is that a secretory stimulus leads to the breakdown of the elementary granule (or NSG) into microvesicles which then fuse with the neuronal membrane and empty their contents of neurosecretory substance into the extracellular space. The choice of the microvesicle rather than the NSG for the final stage of secretion stems largely from the observation that following stimulation, the number of microvesicles increases and may become oriented toward the axonal membrane. The finding that such orientations are found in other than the terminal portions of the axon suggests that secretion may also occur from nonterminal segments of the nerve fiber. While this cytological description of the secretory process is in accord with the available evidence, it obviously requires experimental verification that the microvesicles contain the hormones and neurophysin. There remain, of course, many other unanswered questions such as: (a) What chemical or enzymatic events are involved in the vesiculation of the elementary granule following a secretory stimulus? (b) Is the formation of microvesicles from NSG the terminal event in a sequence which begins with the disappearance of NSM (i.e., disaggregation of NSG) and loss of electron density from the NSG? (c) Is the population of NSG metabolically homogeneous? (d) What chemical or enzymatic reactions participate in the fusion of microvesicle to nerve membrane? (e) What is the fate of the microvesicle membrane once the contents of the vesicle have been discharged? Although precise answers to these questions must await further investigation, recent work in our laboratory has demonstrated that the pool of neurohypophyseal vasopressin is heterogeneous and that the release of hormone from this pool is a selective process. These results, which pertain to question (c), are discussed below.

2. Capacity of the Neurohypophysis to Release Vasopressin; Evidence for a "Readily Releasable Pool"

At the present time there is relatively little information regarding the neurophysiological parameters that control the rate of secretion of hormone from the supraoptico–hypophyseal neurons (see Cross and Silver[132] and Brooks et al.[133]). It is also unknown as to what extent those neurophysiological factors implicated in the control of transmitter release from conventional neurons are operative in the case of the neurosecretory nerves. A discussion of these neurophysiological parameters is beyond the competence of this author and has recently been reviewed by Kandel[167] and others.[168,169]

The existence of one or more regulatory mechanisms limiting the vasopressin secretion rate was first indicated by the experimental results on the release of vasopressin during hemorrhage.[72,158,170] Within a few minutes following reduction of the dog's blood pressure to 50 mm Hg by bleeding, the rate of hormone release rises to a value several orders of magnitude greater than the estimated rate of vasopressin secretion in unbled animals (Fig. 11). However, the initial and rapid secretory response is not sustained beyond the first few minutes of hemorrhage; instead, the rate of hormone release declines rapidly although the animal remains under conditions of hypovolemia at a blood pressure of 50 mm Hg and despite a well-maintained content of pituitary hormone. Unfortunately, neurophysiological studies on this preparation have not been carried out and it is unknown whether the supraoptico–hypophyseal neurons show habituation or whether the firing of inhibitory fibers impinging on the neurosecretory neurons becomes dominant. Another line of investigation, however, provided convincing evidence that the attenuated release of vasopressin which occurs shortly after hemorrhage is due in part to factors which reside at the level of the pituitary itself. For example, it was found that: (a) Pituitaries taken from bled dogs release much less vasopressin in response to electrical or K^+ stimulation in vitro than pituitaries from nonbled animals. (b) Isolated pituitaries (from nonbled dogs

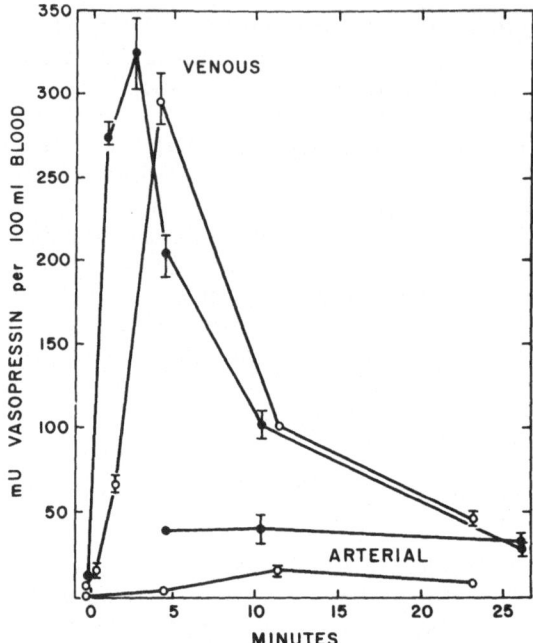

Fig. 11. Time course of discharge of vasopressin into the blood of dogs which have been rapidly bled and maintained at a blood pressure of 50 mm Hg. ●—●, dog, No. 1; ○—○, dog, No. 2. Venous blood obtained from the superior vena cava; arterial blood from a femoral artery. [Reproduced through the courtesy of *Endocrinology*, H. Weinstein, R. H. Berne, and H. Sachs, **66**:712 (1960).]

or guinea pigs) stimulated continuously *in vitro* by 56 mM K$^+$ or by electric pulses respectively, also show a progressive decline in vasopressin secretion; the fractional secretion rates decrease more rapidly than does the content of pituitary hormone. Furthermore, morphological and functional criteria (e.g., electron microscopy, EEG recording, O$_2$ uptake and respiratory response to K$^+$) indicated that the neurosecretory fibers had not suffered tissue damage during hemorrhage.

  In vitro experiments failed to provide evidence for a feedback inhibition at the level of the neural lobe. Instead, the results of *in vivo* and *in vitro* studies seemed to indicate that the decline in hormone secretion depends upon the prior release of about 10–20 % of the total hormone in the gland. On the basis of these findings, we proposed that the huge pool of pituitary vasopressin is metabolically heterogeneous and that there is a "readily releasable" pool which comprises about 10 to 20 % of the total hormone content of the gland. After this "readily releasable" pool of hormone has been discharged, the neurohypophysis is still capable of releasing vasopressin in response to appropriate stimuli, but at a greatly reduced rate. This postulate is consistent with the available evidence and is analogous to the activation of a perfused ganglion which shows a progressive decline in the output of acetylcholine to a low level despite a well-maintained content of transmitter in the ganglion.[171] Direct confirmation of the heterogeneity of the pituitary vasopressin pool was sought by an experimental approach that combined the techniques previously devised for biosynthetic studies, and for measuring the secretion of vasopressin *in vivo* and *in vitro*. It was possible to label the pool of neuro-hypophyseal vasopressin and to induce such glands to secrete radioactive hormone *in vitro* and *in vivo*.[158] If the radioactive vasopressin molecules were released in a random manner from the total hormonal pool, or if the pool was metabolically homogeneous, then it might be expected that the specific activity of vasopressin secreted would be identical with the specific activity of vasopressin remaining in the gland. This was clearly not the case. The specific activity of the hormone secreted in response to excess K^+ *in vitro* or to bleeding *in vivo* was several times greater than that of the hormone in the gland (Table VIII). With continued stimulation, the specific activity of vasopressin which was released at later time intervals declined and thus approached the specific activity of the labeled hormone remaining in the neural lobe. While the results of these experiments clearly established that the

TABLE VIII

Release of Vasopressin-^{35}S *In Vivo* in Response to Hemorrhage and *In Vitro* in Response to 56 mM K^+

| Exp. | Vasopressin-^{35}S | | Spec. act. ratio Vasopressin in blood or medium |
	Source	Spec. activity, counts/min/mμ	Vasopressin in gland
1a[a]	Blood,[b] 10 sec after bleeding	17.0	3.6
	Blood, 30 min after bleeding	6.9	1.5
1b[c]	Medium 5.6 mM K^+	3.7	0.8
	Medium 56 mM K^+	7.3	1.5
2[d]	Medium, 5.6 mM K^+	4.9	0.6
	First stimulation 56 mM K^+	15.6	1.8
	Second stimulation 56 mM K^+	13.5	1.6

[a] The dog was dehydrated for 5 days, and infused with cysteine-^{35}S into the third ventricle over 6 hr (2.3×10^9 counts/min) and then permitted to survive with free access to water for 10 days. From Sachs and Haller.[158]

[b] Cavernous sinus blood, collection of samples required about 2 min.

[c] After collection of the 30-min blood sample, the dog (from 1a) was killed, the posterior pituitary excised, cut in half and incubated at 37°C as described.[158]

[d] The neural lobe was taken from a dog which had undergone the labeling regimen described above.

pool of neurohypophyseal vasopressin is heterogeneous and hormone release a selective process, the data provide no information as to the precise anatomical or cellular location of the "readily releasable" pool. There are a number of possibilities which include the following: (a) the releasable hormone is a function of its spatial distribution within the axon; e.g., only those NSG in close apposition to the neuronal membrane in the nerve terminals can discharge their contents into the perivascular space; (b) only a limited number of neurons are "actively" producing, transporting and secreting hormone [125]; or (c) a combination of the above factors are operative. Which one, if any, of these alternatives is correct must await further experimentation.

IX. THE NEUROSECRETORY NEURON AND THE INTERSTITIAL NEUROGLIAL CELL

The various types of neuroglial cells found throughout the HNC have been described elsewhere.[8] In the hypothalamus, typical astrocytes and oligodendroglia are found in close association with the perikarya of neurosecretory cells. The neural lobe, on the other hand, contains more specialized interstitial cells (i.e., pituicytes) which resemble astrocytes and which comprise the only nucleated elements to be found among the nerve fibers. Despite the intimate morphological contacts seen between the neurosecretory and neuroglial elements, their functional interrelationships remain obscure. The various roles that have been ascribed to the neuroglial cells of the HNC include: (a) playing a supportive role with respect to the metabolic requirements of the neuronal elements (cf. Hyden[172]); (b) providing a means of propelling material along the nerve fiber; or (c) are essential in the final stages of the secretory process (i.e., the discharge of the hormones into the circulation). At present, there is no evidence in favor of any of the above proposals. Nevertheless, there are a number of observations that do tend to support the notion of a functional relationship between the neuroglial and neurosecretory cells of the HNC. Rats and guinea pigs, either deprived of water for 7 to 10 days or treated with nicotine sulfate, showed parallel cytological alterations in both the neurons and neuroglia located in the SON and PVN of the anterior hypothalamus.[173] It would appear that a number of stimuli that provoke the secretion and/or synthesis of hormonal peptides also induce "activation" of the glial cells surrounding the neurosecretory perikarya.[173] The pituicytes in the neural lobe are also affected by either water deprivation or prolonged osmotic stimuli. Under these conditions, there is an apparent increase in the mitotic activity of the pituicytes in the rat neural lobe.[174,175] The peak increase in mitotic activity occurs at about the fourth day after substitution of 2.0–2.5% NaCl for drinking water. It is, however, unknown whether or not hyperplasia and hypertrophy of the pituicytes can account for the increase in the size of the neural lobe of rats subjected to chronic dehydration or osmotic stimuli.[84]

We have begun to examine at a biochemical level the effect of substitution of a 2% NaCl solution for drinking water on the RNA metabolism of pituicytes of the rat neural lobe.[176] In nine experiments, the neural lobes of rats subjected to these conditions for a period of 6–18 days incorporated 1.5 to 3 times as much uridine-^{3}H into RHA *in vitro* as did the neural lobes of rats with free access to water. Actinomycin D at a concentration of 1.5 μg/ml of incubation medium inhibited the incorporation of label into RNA. Radioautography showed that the label was localized primarily in the nuclear regions of the pituicytes (Fig. 12). The nature of these newly synesized RNA molecules, the cellular control mechanisms involved, as well as the significance of these results are presently under investigation. It is noteworthy that the posterior lobe of the pituitary represents a model system for the study of cellular and metabolic interactions between neuroglial cells (i.e., pituicytes) and specialized neurons (i.e., neurosecretory cells). This tissue is virtually devoid of neuronal perikarya and is largely composed of well-defined nerve fibers intermingled with pituicytes. Furthermore, these nerve fibers can be stimulated in a controlled and reproducible manner *in vivo* and *in vitro* and many metabolic processes of the pituicyte can be readily distinguished from that occurring in the axonal elements.

X. SUMMARY

According to classical concepts, the neurosecretory cell is described as a neuron with the morphological and functional characteristics of nerve and endocrine cells. The two criteria most commonly used to distinguish neurosecretory cells from conventional neurons are: (1) the axons of neurosecretory cells do not form synapses with other neurons but instead end at blood spaces into which they release their "neurosecretions"; and (2) their "neurosecretions" are neurohormones, polypeptide in character and are sequestered within membrane-bound vesicles of diameters greater than 1000 Å. Aggregates of the vesicles are demonstrable by standard cytochemical procedures employed for the detection of high concentrations of protein-bound S—S groups. The concept of the "neurosecretory process" ascribes an anatomical division of labor to the neurosecretory cell; the cell body synthesizes and packages its neurohormones within NSG which then move as aggregates in a protoplasmic flow along the axon to the region of the nerve terminals, where release into the bloodstream occurs. This process has been compared and contrasted with the biosynthesis and release of conventional transmitter substances. Recent investigations, however, have uncovered a variety of "neurosecretory neurons" which may innervate directly endocrine cells and/or contain aminelike neurosecretions within granules of diameter less than 1000 Å. As such, the properties of these neurons do not conform to the classical formulation of the neurosecretory cell described above. The discovery of a variety of types of "neurosecretory cells" has blurred the boundary between the neurosecretory and conventional neuron. The description of the

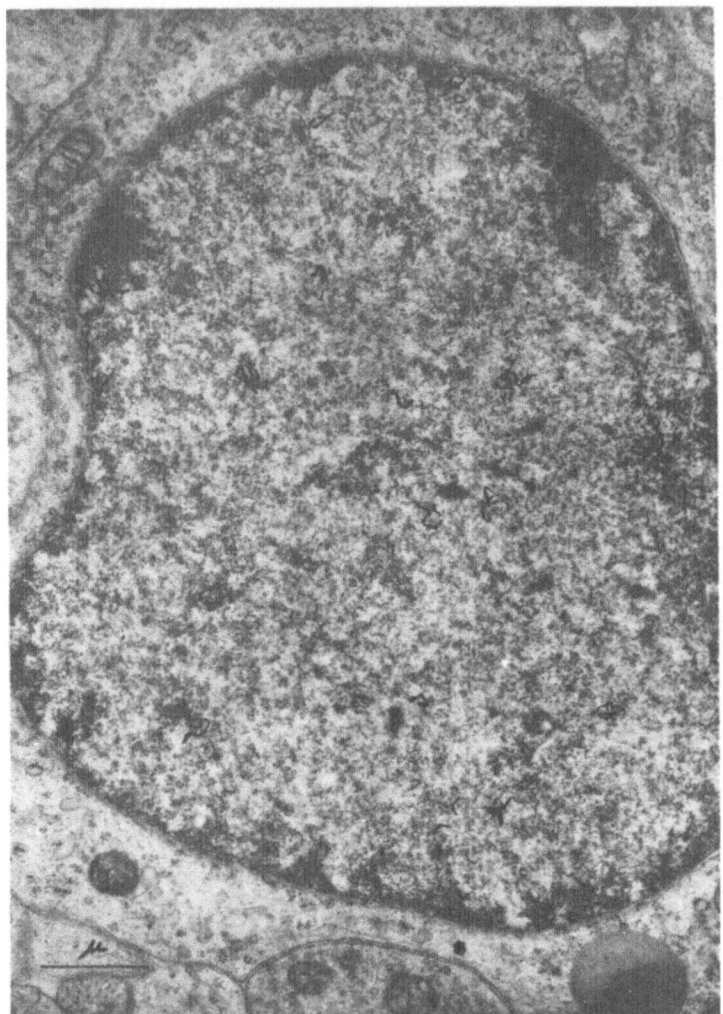

Fig. 12. Radioautography of a rat neural lobe after incubation *in vitro* with uridine-³H for 40 min. The rat had 2% NaCl substituted for drinking water for 11 days prior to sacrifice. Numerous silver grains are evident over nucleus of a pituicyte. × 17,100. (From Sachs and Osinchak, unpublished.)

neurosecretory cell as a neuron which translates neural inputs into endocrine activity defines its principal function and captures the essence of the neuro-secretory neuron. However, other neurons fulfilling similar functions would be excluded from the category of "neurosecretory" on the grounds that they produce conventional transmitter substances as opposed to neurohormones. An alternative way of viewing the problem is presented which attempts to use some of the simple concepts of set theory. This type of formulation leaves room for all manner of new data irrespective of whether or not they conform to preconceived ideas of the neurosecretory neuron.

The distribution and functions of the major neurosecretory cells of the vertebrate CNS are briefly described. Of all the neurosecretory neurons in the CNS, those which comprise the hypothalamo–neurohypophyseal complex of higher vertebrates are most clearly defined with respect to their molecular organization and function and may be regarded as a model for the study of other neurosecretory systems. The remainder of the chapter is devoted to a detailed discussion of these neurons of the HNC whose characteristics actually fulfill the classical criteria of the neurosecretory cell. Neurohypophyseal extracts of vertebrates have yielded at least seven octapeptide hormones whose structures have been defined. These peptides fall into two categories: vasopressor-antidiuretic and oxytocic-milk ejecting. The available evidence is consistent with the view that oxytocin and vasopressin are produced, stored, and released by separate neurons. These peptide hormones are contained largely within NSG (of diameters between 1000–3000 Å) in association with a carrier protein (neurophysin) of about 20,000 molecular weight. Although there are at least two neurophysin proteins in the neural lobes of a variety of species, it is unknown whether oxytocin and vasopressin are each associated with a distinct neurophysin molecule. Analytic and isotope studies have indicated that hormone and neurophysin biosynthesis are intimately related events. On the basis of results of isotope studies, a "precursor model" has been proposed for vasopressin biosynthesis. According to this model, the biosynthesis of the peptide bonds in vasopressin would occur solely in the perikaryon, on ribosomes, via pathways common to the biosynthesis of other peptide chains; initially, however, the hormone would be constructed as part of a macromolecule (probably a protein). The release of the octapeptide from the precursor molecule presumably takes place during the formation and maturation of the NSG. It is unknown whether or not neurophysin biosynthesis occurs via a similar mechanism or whether the hormones and neurophysin share a common precursor.

In contrast to many endocrine glands, hormone biosynthesis and secretion in the HNC appear as temporally unrelated events. So far there is little, if any, evidence to indicate that an acute secretory stimulus also leads to a burst of biosynthetic activity. On the other hand, dehydration or a prolonged osmotic stimulus leads to an enhanced rate of vasopressin biosynthesis as well as a general activation of the RNA and protein-synthesizing machinery of the neuron. The nature of the cellular events which intervene between the reception of nerve impulses and enhanced RNA and hormone biosynthesis remains a central problem on neurosecretion and neurochemistry.

Despite many experimental difficulties, some progress has been made regarding the cellular processes involved in hormone secretion. Electrophysiological studies support the view that nerve impulses traveling down the supraoptico–hypophyseal tract trigger the release of hormone. The activity of these hypothalamic neurons is influenced by impulses originating in many parts of the CNS, some of which are cholinergic, or adrenergic, or of an inhibitory nature. *In vitro* studies with isolated posterior pituitaries have indicated that hormone secretion occurs following a chain of events initiated by membrane depolarization; the presence of Ca^{2+} in the incubation medium is an obligatory requirement. Although the secretion of hormone in response to elevated K^+ does not occur in the absence of oxidative phosphorylation, the data do not permit any conclusions as to whether the secretory process is tightly coupled to a supply of energy. Experiments carried out both *in vivo* and *in vitro* have demonstrated that the release of vasopressin is accompanied by the release of neurophysin. These results and electron microscope studies suggest that secretory stimuli lead to the breakdown of the elementary granule (or NSG) into microvesicles which then fuse with the neuronal membrane and empty their contents of hormones and carrier proteins into the extracellular space. Evidence has been presented that the huge pool of pituitary vasopressin is metabolically heterogeneous and that there is a "readily releasable" pool which comprises about 10 to 20 % of the total hormone content of the gland. At present there is little information regarding the interrelationships between the neurosecretory neuron and the interstitial neuroglial cell. Recent biochemical studies have shown that the pituicytes of neural lobes taken from osmotically stressed rats incorporate more uridine-^{3}H into RNA than those of unstressed animals. It is noteworthy that the posterior lobe of the pituitary represents a model system for the study of cellular and metabolic interactions between neuroglial cells (i.e., pituicytes) and neurosecretory cells.

ACKNOWLEDGMENT

I am especially grateful to my students, Mr. D. Sunde and Mr. R. Portanova for their help and criticisms of this paper.

XI. REFERENCES

1. C. G. Speidel, Gland-cells of internal secretion in the spinal cord of the skates, *Carnegie Inst., Wash.* **13**(281):1–31 (1919).
2. H. A. Bern and I. R. Hagadorn, *in Structure and Function in the Nervous Systems of Invertebrates* (T. H. Bullock and G. A. Horridge, eds.) Vol. 1, pp. 353–429, W. H. Freeman, San Francisco and London (1965).
3. H. A. Bern and F. G. W. Knowles, *in Neuroendocrinology* (L. Martini and W. F. Ganong, eds.) Vol. 1, pp. 139–186, Academic Press, New York and London (1966).
4. H. Bern, On the production of hormones by neurons and the role of neurosecretion in neuroendocrine mechanisms, *Society for Exper. Biol. Symp. No. 20*, "Nervous and Hormonal Mechanisms of Integration," (1966).

5. W. Bargmann, in *International Review of Cytology* (C. H. Bourne and J. F. Danielli, eds.) Vol. 19, pp. 183–201, Academic Press, New York (1966).

6. E. Scharrer and B. Scharrer, Hormones produced by neurosecretory cells, *Rec. Prog. Hormone Res.* **10**:183–240 (1954).

7. B. Scharrer, Recent progress in the study of neuro-endocrine mechanisms in insects, *Archs Anat. Micros.* **54**:331–342 (1965).

8. J. C. Sloper, in *The Pituitary Gland* (G. W. Harris and B. T. Donovan, eds.) Vol. 3, pp. 131–239, University of California Press, Berkeley and Los Angeles (1966).

9. G. Gomori and R. D. Chessick, Esterases and phosphatases of the brain. Histochemical study, *J. Neuropath.* **12**:387–396 (1953).

10. W. Bargmann and E. Scharrer, The site of the origin of the hormones of the posterior pituitary, *Am. Scientist* **39**:255–259 (1951).

11. W. Hild, Experimentell-morphologische Untersuchungen über das Verhalten der "neurosekretorischen Bahn" nach Hypophysenstieldurchtrennungen, Eingriffen in den Wasserhaushalt und Belastung der Osmoregulation, *Virchows Arch.* **319**:526–546 (1951).

12. E. D. P. De Robertis, *Histophysiology of Synapses and Neurosecretion*, Pergamon Press, Oxford (1964).

13. K. Fuxe and T. Hökfelt, in *Neurosecretion* (F. Stutinsky, ed.) Vol. 4, pp. 165–177, Springer-Verlag, Berlin, Heidelberg, New York (1967).

14. R. L. Schultz, Electron microscopic observations of the corpora allata and associated nerves in the moth, *Celerio lineata, J. Ultrastruct. Res.* **3**:320–327 (1960).

15. J. Mellinger, Thesis, University of Strasbourg, *Imprimerie Alsatia Colmar*, 1963.

16. A. B. Dawson, Evidence for the termination of neurosecretory fibers within the pars intermedia of the hypophysis of the frog, *Anat. Rec.* **115**:63–69 (1953).

17. K. Kurosumi, T. Matsuzawa, and S. Shibasaki, Electron microscope studies on the fine structures of the pars nervosa and pars intermedia and their morphological interrelation in the normal rat hypophysis, *Gen. Comp. Endocr.* **1**:433–452 (1961).

18. J. M. Dodd, in *Techniques in Endocrine Research* (F. Knowles and P. Eckstein, eds.) pp. 161–185, Academic Press, London (1963).

19. Sir Frances Knowles, Evidence for a dual control, by neurosecretion, of hormone synthesis and hormone release in the pituitary of the dogfish *Scylliorhinus Stellaris, Phil. Trans. of the Roy. Soc. of London* **249**:435–456 (1965).

20. F. G. W. Knowles and L. Vollrath, A functional relationship between neurosecretory fibers and pituicytes in the eel, *Nature (London)* **208**:1343 (1965).

21. G. B. Koelle, A proposed dual neurohumoral role of acetylcholine: its functions at the pre- and post-synaptic sites, *Nature* **190**:208–211 (1961).

22. J. H. Welsh, in *The Physiology of Crustacea* (T. H. Waterman, ed.) Vol. 2, Academic Press, New York (1961).

23. F. G. W. Knowles and H. Bern, The function of neurosecretion in endocrine regulation, *Nature* **210**:271–272 (1966).

24. G. Stephenson, *Matrices, Sets and Groups: An Introduction for Students of Science and Engineering*, New York (1965).

25. F. G. W. Knowles and D. B. Carlisle, Endocrine control in the crustacea, *Biol. Rev.* **31**:396–473 (1956).

26. L. Martini and W. F. Ganong, eds., *Neuroendocrinology*, Vols. 1 and 2, Academic Press, New York and London (1967).

27. B. Halasz, L. Pupp, and S. Uhlarik, Hypophysiotropic area in the hypothalamus, *J. Endocrinol.* **25**:147–154 (1962).

28. S. Kawashima and N. Takasugi, Neurosecretory material in the hypothalamus and the median eminence—Infundibular stem in posterior-lobectomized rats in relation to the anterior hypophysial function, *J. Fac. Sci. Tokyo Univ.* **11**:29–42 (1966).

28a. S. Kawashima, T. Hirano, T. Matsui, and D. Takewaki, Neurosecretion and neuro-hypophysial hormones in the rat median eminence after total hypophysectomy, with reference to histological and fine structure of the region, *J. Fac. Sci. Tokyo Univ.* **11**:11–28 (1966).

29. S. Kawashima and N. Takasugi, Neurosecretory material in the hypothalamus and the median eminence–infundibular stem in posterior-lobectomized rats in relation to the anterior hypophysial function, *J. Fac. Sci. Tokyo Univ.* **11**:29–42 (1966).

29a. S. Kawashima, T. Hirano, T. Matsui, and D. Takewaki, Neurosecretion and neurohypo-physial hormones in the rat median eminence after total hypophysectomy, with reference to histological and fine structure of the region, *J. Fac. Sci. Tokyo Univ.* **11**:11–28 (1966).

30. F. Knowles, in *Neurosecretion* (F. Stutinsky, ed.) pp. 8–19, Springer-Verlag, Berlin, Heidelberg, New York (1967).

31. G. A. Horridge, in *Nervous Inhibition* (E. Florey, ed.) pp. 395–409, Pergamon Press, New York (1961).

32. W. Etkin, in *Neuroendocrinology* (L. Martini and W. F. Ganong, eds.) Vol. 2, pp. 261–277, Academic Press, New York and London (1967).

33. G. Fridberg, H. Bern, and R. Nishioka, The caudal neurosecretory system of the iso-pondylous teleost, *Albula Vulpes*, from different habitats, *Gen. Comp. Endocrinol.* **6**:195–212 (1966).

34. J. Kitay, in *Neuroendocrinology* (L. Martini and W. Ganong, eds.) Vol. 2, pp. 641–657, Academic Press, New York and London (1967).

35. G. W. Harris, in *Handbook of Physiology: Neurophysiology* (J. Field and H. W. Magoun, eds.) Vol. 2, pp. 1007–1038, American Physiological Society, Washington, D.C. (1960).

36. J. D. Green and D. S. Maxwell, in *Comparative Endocrinology* (A. Gorbman, ed.) pp. 368–392, Wiley, New York (1959).

37. J. F. Christ, in *The Pituitary Gland* (G. W. Harris and B. T. Donovan, eds.) Vol. 3, pp. 62–130, University of California Press, Berkeley and Los Angeles (1966).

38. D. Rioch, G. Wislocki, and J. O'Leary, A precis of preoptic, hypothalamic and hypophysial terminology with atlas, *Assoc. for Res. in Nervous and Mental Dis. Proc.* **20**:3–30 (1940).

39. S. Ramón y Cajal, *Histologie du Système Nerveux de l'Homme et des Vertébrés*, Maloine, Paris (1911).

40. A. T. Rasmussen, Innervation of the hypophysis, *Endocrinology* **23**:263–278 (1938).

41. D. Zambrano and E. de Robertis, Ultrastructure of the hypothalamic neurosecretory system of the dog, *Z. Zellforsch.* **81**:264–282 (1967).

42. A. Rothballer and S. Skoryna, Morphological effects of pituitary stalk section in the dog, with particular reference to neurosecretory material, *Anat. Rec.* **136**:5–25 (1960).

43. W. H. Sawyer, in *Neurohypophysial Hormones and Related Polypeptides. Handbuch der experimentellen Pharmakologie* (B. Berde, ed.) Vol. 23, Springer-Verlag, Berlin, Heidelberg, New York (1967).

44. R. Walter, J. Rudinger, I. L. Schwartz, Chemistry and structure-activity relations of the antidiuretic hormones, *Am. J. Med.* **42**:653–677 (1967).

45. W. H. Sawyer, Evolution of antidiuretic hormones and their functions, *Am. J. Med.* **42**:678–686 (1967).

46. D. R. Ferguson and H. Heller, Distribution of neurohypophysial hormones in mammals, *J. Physiol. (Lond.)* **180**:846–863 (1965).

47. H. Heller and M. Ginsburg, in *The Pituitary Gland* (G. W. Harris and B. T. Donovan, eds.) Vol. 3, pp. 330–373, University of California Press, Berkeley and Los Angeles (1966).

48. L. Share, Vasopressin, its bioassay and the physiological control of its release, *Am. J. Med.* **42**:701–712 (1967).

49. R. J. Fitzpatrick, in *The Pituitary Gland* (G. W. Harris and B. T. Donovan, eds.) Vol. 3, pp. 453–504, University of California Press, Berkeley and Los Angeles (1966).

50. H. D. Lauson, Metabolism of antidiuretic hormones, *Am. J. Med.* **42**:713–744 (1967).
51. A. Leaf, Membrane effects of antidiuretic hormone, *Am. J. Med.* **42**:745–756 (1967).
52. I. L. Schwartz and R. Walter, Factors influencing the reactivity of the toad bladder to the hydro-osmotic action of vasopressin, *Am. J. Med.* **42**:769–776 (1967).
53. J. F. G. Vliegenthart and D. H. G. Versteeg, The evolution of the vertebrate neurohypophysial hormones in relation to the Genetic code, *J. Endocrinol.* **38**:3–12 (1967).
54. W. H. Howell, The physiological effects of extracts of the hypophysis cerebri and infundibular body, *J. Exp. Med.* **3**:245–258 (1898).
55. H. H. Dale, The Action of extracts of the pituitary body, *Biochem. J.* **4**:427–447 (1909).
56. I. Ott and J. C. Scott, The action of infundibulin upon the mammary secretion, *Proc. Soc. Exp. Biol.* **8**:48–49 (1910).
57. E. Frank, Über Beziehungen der Hypophyse sum Diabetes insipidus, *Berliner klin. Wschr.* **49**:393–397 (1912).
58. E. B. Verney, The secretion of pituitrin in mammals as shown by perfusion of the isolated kidney of the dog, *Proc. Roy. Soc. Ser. B* **99**:487–517 (1926).
59. E. M. K. Geiling and L. I. Robbins, The posterior lobe of the pituitary gland of the whale; pituitrin and its fractions pitocin and pitressin, *Res. Publ., Assoc. Res. Nerv. Mental Disease* **17**.
60. C. Fisher, W. R. Ingram, and S. W. Ranson, *Diabetes Insipidus and the Neurohormonal Control of Water Balance: A Contribution to the Structure and Function of the Hypothalamico-Hypophyseal System*, Edwards Bros., Ann Arbor (1938).
61. W. Hild, Das morphologische, kinetische und endokrinologische Verhalten von hypothalamischem und neurohypophysärem Gewebe *in vitro*, *Z. Zellforsch.* **40**:257–312 (1954).
62. W. Hild and G. Zettler, Experimenteller Beweis fur die enstejing der sog, Hypophysenhinterlappenwirkstoffe im Hypothalamus, *Pfluegers Arch. Ges. Physiol.* **257**:169–201 (1953).
63. J. C. Sloper and C. W. M. Adams, The hypothalamic elaboration of the posterior pituitary principle in man; evidence derived from hypophysectomy, *J. Pathol. Bacteriol.* **72**:587–602 (1956).
64. P. Weiss, Evidence by isotope tracers of perpetual replacement of mature nerve fibers from their cell bodies, *Science* **129**:1290 (1959).
65. T. H. Schiebler, Cytochemische und elektronen mikrokopische Untersuchungen an granulären Fraktionen der Neurohypophyse des Rindes, *Z. Zellforsch.* **36**:563–576 (1952).
66. A. V. Pardoe and M. Weatherall, The intracellular localization of oxytocic and vasopressor substances in the pituitary glands of rats, *J. Physiol. (Lond.)* **127**:201–212 (1955).
67. H. Weinstein, S. Malamed, and H. Sachs, Isolation of vasopressin containing granules from the neurohypophysis of the dog, *Biochim. Biophys. Acta (Amst.)* **50**:386–389 (1961).
68. H. Heller and K. Lederis, Characteristics of isolated neurosecretory vesicles from mammalian neural lobes, *Mem. Soc. Endocrinol.* **12**:35–50 (1962).
69. S. Ishii, I. Yasumasu, H. Kobayashi, Y. Oota, T. Hirano and A. Tanaka, Isolation of neurosecretory granules and nerve endings from bovine posterior lobe, *Annotationes Zool. Japan* **35**:121–127 (1962).
70. F. LaBella, G. Beaulieu, and R. Reifenstein, Evidence for the existence of separate vasopressin and oxytocin-containing granules in the neurohypophysis, *Nature* **193**:173–174 (1962).
71. J. C. Sloper and R. G. Bateson, Ultrastructure and neurosecretory cells in the supraoptic nucleus of the dog and rat, *J. Endocrinol.* **31**:139–150 (1965).
72. H. Sachs, L. Share, J. Osinchak, and A. Carpi, Capacity of the neurohypohysis to release vasopressin, *Endocrinology* **81**:755–770 (1967).
73. H. B. Van Dyke, K. Adamsons, Jr., and S. L. Engel, *in The Neurohypophysis* (H. Heller, ed.) pp. 65–76, Butterworths, London (1957).

74. H. Sachs, Studies on the intracellular distribution of vasopressin, *J. Neurochem.* **10**:289–297 (1963).

75. C. W. M. Adams and J. C. Sloper, The hypothalamic elaboration of the posterior pituitary principles in man, the rat and dog. Histochemical evidence derived from performic acid-alcian blue reaction for cystine, *J. Endocrinol.* **13**:221–228 (1956).

76. F. S. LaBella, S. Vivian, and E. Bindler, Amino acid composition of neurohypophysial secretory granules and van dyke protein, *Biochem. Pharmacol.* **16**:1126–1130 (1967).

77. H. B. Van Dyke, B. F. Chow, R. O. Greep, and A. J. Rothen, The isolation of a protein from the pars neuralis of the ox pituitary with constant oxytocic, pressor and diuresis inhibiting effects, *J. Pharm. (Lond.)* **74**:190–209 (1942).

78. M. O. Hollenberg and D. B. Hope, Fractionation of neurophysin by molecular-sieve and ion-exchange chromatography, *Biochem. J.* **104**:122–127 (1967).

79. E. Breslow and L. Abrash, The Binding of oxytocin and oxytocin analogues by purified bovine neurophysins, *Proc. Nat. Acad. Sci., U.S.A.* **56**:640–646 (1966).

80. M. Ginsburg and M. Ireland, The role of neurophysin in the transport and release of neurohypophysial hormones, *J. Endocrinol.* **35**:289–298 (1966).

81. C. R. Dean and D. B. Hope, The isolation of purified neurosecretory granules from bovine pituitary posterior lobes, *Biochem. J.* **104**:1082–1088 (1967).

82. R. Albers and M. Brightman, A major component of neurohypophysial tissue associated with antidiuretic activity, *J. Neurochem.* **3**:269–276 (1959).

83. M. Rennels, A study of proteins of the posterior pituitary of normal and dehydrated rats using disc electrophoresis, *Endocrinology* **78**:659–660 (1966).

84. H. G. Friessen and E. B. Astwood, Changes in neurohypophysial proteins induced by dehydration and ingestion of saline, *Endocrinology* **80**:278–287 (1967).

85. P. Fawcett, A. E. Powell, and H. Sachs, Biosynthesis and release of neurophysin, *Endocrinology* **83**:1299–1310 (1968).

86. A. B. Rothballer, Changes in rat neurohypophysis induced by painful stimuli with particular reference to neurosecretory material, *Anat. Rec.* **115**:21–41 (1953).

87. A. M. Moses, T. F. Leveque, M. Giambattista, and C. W. Lloyd, Dissociation between the content of vasopressin and neurosecretory material in the rat neurohypophysis, *J. Endocrinol.* **26**:273–278 (1963).

88. A. R. Daniel and K. Lederis, Effects of ether anaesthesia and haemorrhage on hormone storage and ultrastructure of the rat neurohypophysis, *J. Endocrinol.* **34**:91–104 (1966).

89. K. Lederis, *in Neurosecretion* (H. Heller and R. B. Clark, eds.) pp. 227–239, Academic Press, London and New York (1962).

90. H. Olivecrona, Paraventricular nucleus and pituitary gland, *Acta Physiol. Scand.* Suppl. **40**(136):1–178 (1957).

91. D. W. Nibbelink, Paraventricular nuclei, neurohypophysis and parturition, *Am. J. Physiol.* **200**:1229–1232 (1961).

92. H. W. Sokol and H. Valtin, Evidence for the synthesis of oxytocin and vasopressin in separate neurons, *Nature* **214**:314–316 (1967).

93. R. Barer, H. Heller, and K. Lederis, The isolation, identification and properties of the hormonal granules of the neurohypophysis, *Proc. Roy. Soc. Ser. B.* **158**:388–416 (1963).

94. C. W. Bisset, S. M. Hilton, and A. M. Poisner, Hypothalamic pathways for independent release of vasopressin and oxytocin, *Proc. Roy. Soc. Ser. B.* **166**:422–442 (1967).

95. J. Roberts and L. Share, Release of oxytocin and vasopressin from guinea pig posterior pituitaries. *The Endocrine Society, 48th Meeting* p. 36 (1966).

96. B. J. Clark and M. Rocha e Silva, Jr., An afferent pathway for the selective release of vasopressin in response to carotid occlusion and haemorrhage in the cat, *J. Physiol. (Lond.)* **191**:529–542 (1967).

97. R. Acher and C. Fromageot, *in The Neurohypophysis* (H. Heller, ed.) pp. 39–50, Butterworths, London (1957).

98. C. R. Dean, M. C. Hollenberg, and D. B. Hope, The relationship between neurophysin and the soluble proteins of pituitary neurosecretory granules, *Biochem. J.* **104**:8c–10c (1967).

99. M. D. Hollenberg and D. B. Hope, The composition of crystallin complexes of neurophysin-M with [8-Arginine]-vasopressin and oxytocin, *Biochem. J.* **105**:921–926 (1967).

100. M. D. Hollenberg and D. B. Hope, The isolation of the native hormone-binding proteins from bovine pituitary posterior lobes, *Biochem. J.* **106**:557–564 (1968).

101. M. Ginsburg and M. Ireland, The preparation of bovine neurophysin and the estimation of its maximum capacity to bind oxytocin and arginine vasopressin, *J. Endocrinol.* **32**:187–198 (1965).

102. M. Ginsburg, K. Jayasena and P. J. Thomas, The preparation and properties of porcine neurophysin and the influence of calcium on the hormone-neurophysin complex, *J. Physiol. (Lond.)* **183**:4p (1965).

103. P. J. Thomas and M. Ginsburg, Inhibition by L-cystine of the binding of 8-lysine vasopressin by porcine neurophysin, *Biochem. J.* **100**:9c–10c (1966).

104. M. Ginsburg, K. Jayasena, and P. J. Thomas, The preparation and properties of porcine neurophysin and the influence of calcium on the hormone-neurophysin complex, *J. Physiol. (Lond.)* **184**:387–401 (1966).

105. J. E. Stouffer, D. B. Hope, and V. du Vigneaud, in *Perspectives in Biology* (C. F. Corci, V. G. Foglia, L. F. Leloir, and S. Ochoa, eds.) Elsevier, Amsterdam (1963).

106. H. Sachs, Biosynthesis and release of vasopressin, *Am. J. Med.* **42**:687–700 (1967).

107. C. R. Dean and D. B. Hope, The isolation of neurophysin-I and -II from bovine pituitary neurosecretory granules separated on a large scale from other subcellular organelles, *Biochem. J.* **106**:565–573 (1968).

108. E. Koenig, Synthetic mechanisms in the axon- IV, in vitro incorporation ^{3}H-precursors into axonal protein and RNA, *J. Neurochem.* **14**:437–446 (1967).

109. L. Austin and I. G. Morgan, Incorporation of ^{14}C-Labelled Leucine into Synaptosomes from Rat Cerebral Cortex *in Vitro*, *J. Neurochem.* **14**:377–387 (1967).

110. M. Vogt, Vasopressor, antidiuretic, and oxytocic activities of extracts of the dog's hypothalamus, *Brit. J. Pharm.* **8**:193–196 (1953).

111. E. de Robertis, Ultrastructure and function in some neurosecretory systems, *Mem. Soc. Endocrinol.* **12**:3–20 (1962).

112. R. Lasek, Axoplasmic transport in cat dorsal root ganglion cells: as studied with [^{3}H]-L-leucine, *Brain Res.* (In press).

113. H. Sachs, in *Regional Neurochemistry* (S. S. Kety, V. Elkes, eds.) pp. 264–273, Pergamon Press, New York (1960).

114. Y. Takabatake and H. Sachs, Vasopressin in biosynthesis III. *In vitro* studies, *Endocrinology* **75**:934–942 (1964).

115. S. Barondes, On the site of synthesis of the mitochondrial protein of nerve endings, *J. Neurochem.* **13**:721–727 (1966).

116. H. Sachs, Vasopressin biosynthesis II. Incorporation of [^{35}S] cysteine into vasopressin and protein associated with cell fractions, *J. Neurochem.* **10**:299–311 (1963).

117. E. Ito and J. Strominger, Enzymatic synthesis of the peptide in a uridine nucleotide from *Staphylococcus aureus*, *J. Biol. Chem.* **235**:PC5-6 (1960).

118. N. V. Bhagavan, The biosynthesis of gramicidin S. A restudy, *Biochemistry* **5**:3844–3850 (1966).

119. S. Otani, T. Yamanoi, Y. Saito, and S. Otani, Fractionation of an enzyme system responsible gramicidin S biosynthesis, *Biochem. Biophys. Res. Commun.* **25**:590–596 (1966).

120. B. Mach, E. Rich, and E. L. Tatum, Separation of the antibiotic polypeptide tyrocidine from protein synthesis, *Proc. Nat. Acad. Sci., U.S.A.* **50**:175–181 (1963).

121. H. Sachs and Y. Takabatake, Evidence for a precursor in vasopressin biosynthesis, *Endocrinology* **75**:943–948 (1964).

122. H. Sachs, R. Portanova, E. W. Haller, and L. Share, in Neurosecretion (F. Stutinsky, ed.) pp. 146–154, Springer-Verlag, Berlin, Heidelberg, New York (1967).

123. E. Scharrer and S. Brown, Electron-microscopic studies of neurosecretory cells in Lumbricus terrestris, Mem. Soc. Endocrinol. 12:103–108 (1962).

124. J. Osinchak, A fine structure and cytochemical study of neurosecretory cells in the rat, Int. Cong. Cell Biol. 77:33–34 (1964).

125. D. Zambrano and E. de Robertis, The secretory cycle of supraoptic neurons in the rat, Z. Zellforsch. 73:414–431 (1966).

126. J. Edstrom, D. Eichner, and N. Schor, in Regional Neurochemistry (S. Kety and J. Elkes, eds.) pp. 274–278, Pergamon Press, New York (1961).

127. D. F. Steiner, D. Cunningham, L. Spieglman, and B. Aten, Insulin biosynthesis: evidence for a precursor, Science 157:697–700 (1967).

128. H. Valtin, Hereditary hypothalamic diabetes insipidus in rats (Brattleboro strain). A useful experimental model, Am. J. Med. 42:814–827 (1967).

129. H. Valtin, W. H. Sawyer, and H. W. Sokol, Neurohypophysial principles in rats homozygous and heterozygous for hypothalamic diabetes insipidus (Brattleboro strain), Endocrinology 77:701–706 (1965).

130. P. M. Orkand and S. L. Palay, Effects of treatment with exogenous vasopressin on the structural alterations in the hypothalamo-neurohypophysial system of rats with hereditary diabetes insipidus, Anat. Rec. (In press.)

131. J. A. Shannon, The control of the renal excretion of water. II. The rate of liberation of the posterior pituitary antidiuretic hormones in the dog, J. Exp. Med. 76:387–399 (1942).

132. B. A. Cross and I. A. Silver, Electrophysiological studies on the hypothalamus, Brit. Med. Bull. 22:254 (1966).

133. C. McC. Brooks, K. Koizumi and G. A. Zeballos, A study of factors controlling activity of neurons within the paraventricular, supraoptic and ventromedian nuclei of the hypothalamus, Acta Physiol. Latinoam 16:83–96 (1966).

134. C. Brooks, T. Ishikawa, K. Koizumi, and H. Lu, Activity of neurones in the paraventricular nucleus of the hypothalamus and its control, J. Physiol. (Lond.) 182:217–231 (1966).

135. T. Ishikawa, K. Koizumi, and C. McC. Brooks, Electrical activity recorded from the pituitary stalk of the cat, Am. J. Physiol. 210:427–431 (1966).

136. K. Koizumi, T. Ishikawa, and C. McC. Brooks, Control of activity of neurons in the supraoptic nucleus, J. Neurophys. 27:878–891 (1964).

137. M. Pickford, The inhibitory effect of acetylcholine on water diuresis in the dog, and its pituitary transmission, J. Physiol. (Lond.) 95:226–238 (1939).

138. H. Mizuno, P. K. Talwalker and J. Meites, Central inhibition by serotonin of reflex release of oxytocin in response to suckling stimulus in the rat, Neuroendocrinol. 2:222–231 (1967).

139. J. W. Woods, P. Bard, and R. Bleier, Functional capacity of the deafferented hypothalamus: water balance and responses to osmotic stimuli in the decerebrate cat and rat, J. Neurophysiol. 29:751–767 (1966).

140. W. Douglas and A. Poisner, Stimulus–secretion coupling in a neurosecretory organ. The role of calcium in the release of vasopressin from the neurohypophysis, J. Physiol. (Lond.) 172:1–30 (1964).

141. V. Abrahams, G. Koelle, and P. Smart, Histochemical demonstration of cholinesterases in the hypothalamus of the dog, J. Physiol. (Lond.) 139:137–144 (1957).

142. G. B. Koelle and C. N. Geesey, Localization of acetylcholinesterase in the neurohypophysis and its functional implications, Proc. Soc. Exp. Biol. and Med. 106:625–628 (1961).

143. H. Gerschenfeld, J. Tramezzani, and E. de Robertis, Ultrastructure and function in neurohypophysis of the toad, Endrocrinology 66:741–762 (1960).

144. S. L. Palay, in Ultrastructure and Cellular Chemistry of Neural Tissue (H. Waelsch, ed.), pp. 31–44, Hoeber-Harper, New York (1957).

145. J. H. Burn and M. J. Rand, Sympathetic postganglionic mechanism, *Nature* **184**:163–165 (1959).
146. W. Feldberg and M. Vogt, Acetylcholine synthesis in different regions of the central nervous system, *J. Physiol. (Lond.)* **107**:372–381 (1948).
147. K. Lederis and A. Livingston, Acetylcholine content in the rabbit neurohypophysis, *J. Physiol. (Lond.)* **185**:37P–38P (1966).
148. E. Bindler, F. S. LaBella, and M. Sanwal, Isolated nerve endings (neurosecretosomes) from the posterior pituitary, *J. Cell Biol.* **34**:185–205 (1967).
149. E. W. Haller, *In vitro* studies of vasopressin secretion from the posterior pituitary. (Ph.D. dissertation, Western Reserve University, Cleveland, Ohio, 1967.)
150. H. A. Bern, in *The General Physiology of Cell Specialization* (D. Mazia and A. Tyler, eds.), McGraw-Hill, New York (1963).
151. K. Lederis, Fine structure and hormone content of the hypothalamo–neurohypophysial system of the rainbow trout (*Salmo iridens*) exposed to sea water, *Gen. and Comp. Endocrinol.* **4**:638–661 (1964).
152. W. W. Douglas, in *Neurosecretion* (F. Stutinsky, ed.) pp. 178–190, Springer-Verlag, Berlin, Heidelberg, New York (1967).
153. W. Vale, R. Burgus, and R. Guillemin, Presence of calcium ions as a requisite for the *in vitro* stimulation of TSH-release by hypothalamic TRF, *Experimentia* **23**:853–855 (1967).
154. M. Schramm, Secretion of enzymes and other macromolecules, *Ann. Rev. Biochem.* **36**:307–320 (1967).
155. B. Katz, The transmission of impulses from nerve to muscle, and the subcellular unit of synaptic action, *Proc. Roy. Soc. Ser. B.* **155**:455–477 (1962).
156. W. W. Douglas, T. Kanno, and S. R. Sampson, Influence of the ionic environment on the membrane potential of adrenal chromaffin cells and on the depolarizing effect of acetylcholine, *J. Physiol. (Lond.)* **191**:107–121 (1967).
157. W. Douglas, A. Ishida and A. Poisner, The effect of metabolic inhibitors on the release of vasopressin from the isolated neurohypophysis, *J. Physiol. (Lond.)* **181**:753–759 (1965).
158. H. Sachs and E. W. Haller, Further studies on the capacity of the neurohypophysis to release vasopressin, *Endocrinology* **83**:251–262 (1968).
159. M. E. Krass and F. S. LaBella, Hexosemonophosphate shunt in endocrine tissues. Quantitative estimation of the pathway in bovine pineal body, anterior pituitary, posterior pituitary and brain, *Biochimica et Biophysica Acta* **148**:384–391 (1967).
160. N. A. Thorn, *In vitro* studies of the release mechanism for vasopressin in rats, *Acta Endocrinol.* **53**:644–654 (1966).
161. E. de Robertis and A. Vaz Ferreira, Electron microscope study of the excretion of catechol-containing droplets in the adrenal medulla, *Exp. Cell. Res.* **12**:568–574 (1957).
162. W. W. Douglas, A. Poisner, and R. P. Rubin, The role of calcium in the secretory response of the adrenal medulla to acetylcholine, *J. Physiol. (Lond.)* **159**:40–57 (1961).
163. P. Banks, The release of adenosine triphosphate catabolites during the secretion of catecholamines by bovine adrenal medulla, *Biochem. J.* **101**:536–541 (1966).
164. F. H. Schneider, A. D. Smith, and H. Winkler, Secretion from the adrenal Medulla: Biochemical evidence for exocytosis, *Brit. J. of Pharm.* **31**:94–104 (1967).
165. G. E. Palade, P. Siekevitz, and L. G. Caro, in *The Exocrine Pancreas* (A. V. S. De Reuck and M. P. Cameron, eds.) pp. 23–49, Churchill, London (1962).
166. C. De Duve, in *Lysosomes* (A. V. S. De Reuck and M. P. Cameron, eds.) Churchill, London (1963).
167. E. R. Kandel and W. A. Spencer, Cellular neurophysiological approaches in the study of learning, *Physiol. Rev.* **48**(1):65–134 (1968).
168. H. Grundfest, in *The Neurosciences* (G. C. Quarton, T. Melnechuk, and F. O. Schmitt, eds.) pp. 353–372, Rockefeller University Press, New York (1967).

169. J. C. Eccles, *in The Neurosciences* (G. C. Quarton, T. Melnechuk, and F. O. Schmitt, eds.) pp. 408–427, Rockefeller University Press (1967).

170. H. Weinstein, R. M. Berne, and H. Sachs, Vasopressin in blood: effect of hemorrhage, *Endocrinology* **66**:712 (1960).

171. J. C. Eccles, *The Physiology of Synapses*, Academic Press, New York (1964).

172. H. Hydén, *in The Neuron* (H. Hydén, ed.) pp. 179–219, Elsevier, New York (1967).

173. J. E. Azcoaga and F. E. D'Amelio, Jr., Neuronal and neuroglial alterations induced experimentally in the magnocellular hypothalamic nuclei, *Intern. Congr. Neuropathol., 5th, Zurich, 1965,* 950–952 (1965).

174. L. W. Duchen, Effects of ingestion of hypertonic saline on the pituitary gland of the rat: a morphological study of the pars intermedia and posterior lobe, *J. Endocrinal.* **25**:161–168 (1963).

175. T. F. Leveque and M. Small, The relationship of the pituicyte to the posterior lobe hormones, *Endocrinology* **65**:909–915 (1959).

176. H. Sachs and J. Osinchak, Incorporation of tritiated uridine into RNA of pituicytes of rat neural lobes, *J. Cell. Biol.* **35**:186A (1967).

Chapter 18

ACTION OF GONADAL HORMONES

Roger A. Gorski and Lynwood G. Clemens*

Department of Anatomy and Brain Research Institute
UCLA School of Medicine, Los Angeles, California

I. INTRODUCTION

Only recently has significant progress been made in elucidating the mechanisms of action of gonadal steroids on their peripheral target organs. It is not surprising, therefore, that the mechanisms of action of these hormones on the brain are poorly understood. In this chapter we will consider the nature of the functional interrelationships between brain and gonads, and then attempt to relate existing data on the chemical effects of these hormones on peripheral tissue or on the brain, to these functional interactions. Broad understanding can arise from a discussion of specific examples only if an overview of the entire system is available to place any example in its proper perspective. Therefore, a brief summary of the fundamental concepts of reproductive neuroendocrinology will be presented first.

II. FUNCTIONAL INTERACTIONS BETWEEN BRAIN AND GONADS

Four general points deserve special emphasis as they form the basis of the subsequent discussion of functional interactions between brain and gonads.

a. Sex Differences. The extent to which processes in the male and female differ, and the causes of these differences, have received considerable attention. The factors involved in differentiation and control of these processes are fundamental to the study of brain chemistry.

b. Functional Cerebral Localization. Neurochemical analyses of the CNS are of limited value when performed at a gross level. Although the

* Present address: Dept. of Zoology, Michigan State University, East Lansing, Michigan.

division of the brain into many centers of totally independent function is unlikely, it is probable that different functional systems exist but continuously interact. Therefore, the activity of limited regions of the brain may be specifically facilitated or inhibited. The likelihood that polar neurochemical changes will cancel increases markedly as the size of a tissue sample is increased.

c. Temporal Pattern of Hormone Action. The action of a hormone changes over time, either because the condition of the animal changes, or because the hormone is metabolized. Since it is unlikely that any one hormone acts alone, the interaction of hormones, each with its own specific temporal pattern, must be considered in the design and interpretation of experiments.

d. Animal Age. The age of an animal strongly influences the nature of the interaction between the brain and gonadal hormones. These steroids exert a permanent effect on the brain during its rapid neonatal development in the rat, whereas in the adult, their effects are transient.

A. The Adult Animal

1. Basic Principles of Neuroendocrinology

Specialized hypothalamic neurons of the tuberoinfundibular system terminate in close association with the capillary loops of the median eminence, and there secrete factors into the hypophyseal portal blood. These neurochemical factors are carried to the adenohypophysis where they regulate the secretion of adenohypophyseal hormones. In response to the specific neurochemical "releasing factors" for luteinizing hormone (LH) and follicle stimulating hormone (FSH), these gonadotrophins are secreted; the gonad responds to LH and FSH with the secretion of steroid hormones. These steroids exert characteristic actions on peripheral target tissues, and also on the CNS where they modify the activity of the tuberal neurons, either by direct action or through a modification in their afferent input. Gonadal hormones also produce conditions within the CNS which result in the display of sexual behavior upon appropriate stimulation. General reviews of neuroendocrinology are available.[1,2]

2. Reproductive Neuroendocrinology in the Female

a. Regulation of Pituitary Activity. Ovulation characterizes reproductive function in the female and is produced by the abrupt release of LH in response to LH-releasing factor from neurons in the median eminence. In a group of mammals represented by the rabbit and cat, ovulation is a neuroendocrine reflex. The neural processes which result in the discharge of LH from the pituitary are initiated by copulation or mechanical stimulation of the cervix. In a second group of mammals, including the rat and the human female, ovulation occurs spontaneously and in cycles. In these spontaneous ovulators, the CNS is capable of the specific rhythmic activity which can induce ovulation in absence of cervical stimulation.

Whether the neural component of the ovulation process occurs in reflex to sexual stimuli, or spontaneously, gonadal hormones play a critical role. Vaginal stimulation induces ovulation only in the estrogen-treated rabbit. In the spontaneously ovulating rat, the importance of the temporal pattern of hormone action and functional neuroanatomical localization are also well documented. Under controlled environmental conditions (14 hr of light per day) the release of the ovulating burst of LH occurs during an experimentally defined period (2–4 P.M. colony time on the day of vaginal proestrus). The administration of barbiturates or other neural blocking agents during this 2-hr period blocks a neural stimulus and ovulation is prevented. The action of gonadal hormones on the CNS determines the effectiveness of this neural stimulus for ovulation. Exogenous estrogen or progesterone may advance or delay ovulation, depending on the time in an animal's estrous cycle the hormones are administered.[3] Thus, the potent influence of the temporal pattern of endogenous hormone activity may even reverse the overt action of an injected hormone.

The regulation of ovulation (and perhaps of all adenohypophyseal activity) is achieved by two neural systems which are apparently anatomically distinct. The tissues of the median eminence region can induce and partly regulate the discharge of hypophyseal gonadotrophin (GTH) even in the absence of neural afferent input from other regions of the hypothalamus or brain. This region, however, cannot independently initiate ovulation. Neural isolation of this region results in the continuous or tonic discharge of GTH, the continuous growth of generation after generation of ovarian follicles, and the continuous secretion of estrogen—the persistent vaginal estrous syndrome. Under normal circumstances another neural system apparently induces ovulation by activating this same median eminence region. This ovulation control system may include extrahypothalamic structures which eventually transmit an ovulatory stimulus via a diffuse septo-preoptico-antero-hypothalamo-median eminence pathway. It is probable that ovarian hormones exert different effects on these two neuroanatomical systems which may well have individual temporal characteristics.

In summary, the process of ovulation is a complex temporal interaction between at least two functional neural systems and the gonadal hormones. These concepts of the anatomical localization of regulatory systems, and the probable sites of hormonal feedback have been reviewed recently.[2–7]

b. Regulation of Sexual Behavior. In the normal female, ovulation is temporally coordinated with sexual receptivity, and the probability of fertilization thereby increased. In those species in which ovulation does not occur until coital stimuli are provided by the male, the coordination of ovulation with the presence of sperm is achieved via peripheral stimulation. In the spontaneously ovulating rat or guinea pig this coordination is achieved by the action of estrogen and progesterone. Estrogen appears to prime the neural tissues which mediate female mating behavior. A rise in progesterone secretion at proestrus induces high levels of sexual receptivity just a few hours before freshly ovulated ova reach the oviduct. The significance of the temporal pattern of hormone action is emphasized by the observation that in the guinea pig, the continued action of progesterone inhibits sexual receptivity.[8]

The neural substrate which mediates female mating behavior appears to overlap with that which regulates ovulation. Insofar as the functional integrity of the median eminence is required for estrogen secretion, female mating behavior is indirectly dependent upon this region. However, if exogenous hormones are supplied, lesions in this area do not disrupt sexual behavior. Lesions in the anterior hypothalamus, which appears to be essential for ovulation, do interfere with female receptivity even when exogenous estrogen and progesterone are supplied.[9] The direct implantation of estrogen into this region will induce sexual receptivity in the gonadectomized rat.[9] Although ovulation and sexual receptivity are temporally coordinated in the normal animal, it is readily possible to alter these processes independently. It should not be assumed that the degree of anatomical overlap between the two regulatory systems is necessarily extensive.

For quantitative analysis of sexual behavior, test conditions must be rigidly defined. The detection of sperm in the vagina of a female rat following a night of cohabitation with males is evidence that mating has occurred, but not that normal sexual behavior has occurred. Meaningful data are obtained only if the actual behavior of test animals is observed. The display of a postural lordosis reflex by the female rat when she is mounted by a male is used as an index of sexual receptivity. When a record is kept of the female's response to successive mounts, one female may be found to display the lordosis posture once every 10 times she is mounted, another animal may display 9 or 10 lordoses every 10 mounts. Obviously these two females are behaving differently. These differences reflect significant variation in the degree of sexual receptivity, and probably relevant neurochemical status. Differences in hormonal treatment, reactivity to hormones, or behavioral thresholds may individually or collectively determine the level of sexual receptivity. A detailed discussion of the problems of measuring sexual behavior can be found elsewhere.[10]

3. Reproductive Neuroendocrinology in the Male

Although minor fluctuations in androgen and GTH secretion have been recorded in males,[11] these "cycles" do not appear to be functionally important and are not equivalent to the dramatic cycles in GTH secretion or sexual behavior of the female. While males of certain species may have a definite seasonal cycle, during the breeding season the male represents a neuro-endocrine steady state; spermatogenesis occurs continually, androgen secretion by the testes is functionally constant, and the intact male is essentially "on call" for sexual behavior. In the male rat the neuronally isolated median eminence region is capable of supporting the GTH secretion necessary for testicular function without afferent input.[2,6]

Male sexual behavior is quantified by factors such as the latency to ejaculation, the number of intromissions per ejaculation, and the refractory interval following ejaculation. This behavior is heavily dependent upon androgen, although the experienced male often continues to exhibit male sexual behavior for a long period after castration.[10] The central substrate for male sexual behavior is extensive.[12] In the rat, for example, the neocortex is essential, and subcortical structures such as the preoptic area play significant roles in the regulation of male sexual behavior. To what extent the neural substrate of masculine sexual behavior is anatomically distinct from that of female behavior is not known. Regulation of sexual behavior in the two sexes, therefore, involves the action of different hormones on possibly different neuroanatomical loci.

It is interesting to note that both behavioral systems can function in the same adult animal, and can even be initiated by the same hormone, i.e., exogenous androgen in the female rat may induce mounting activity or lordosis behavior.[10] Whether such events reflect upon neurochemical or neurophysiological processes is unknown.

B. Influence of Animal Development on Hormone Action

The preceding section stressed that the functional effect of gonadal hormones is influenced by previous hormonal conditions and by the temporal sequence of hormone action. Similarly, the precise neural substrate upon which hormones act, and its reactivity at any point in time, will significantly influence steroid action. Also critical in the action of hormones on the brain is the age or stage of development of the animal. In the case of reproductive neuroendocrinology, four distinct stages can be described: paranatal, pubertal, adult, senile.

1. Puberty

Puberty represents the establishment of the complex functional interactions between brain and gonad characteristic of the adult which have just been discussed. A currently popular theory of puberty assumes that the CNS of the prepubertal animal is extremely sensitive to gonadal hormones.[13] Levels of gonadal steroids which would be considered low in the adult markedly suppress pituitary activity in the prepubertal animal. At puberty the hypothalamus inexplicably changes in its sensitivity to steroids. Sensitivity is decreased and gonadal hormone titers can reach peripherally effective levels before GTH secretion is inhibited by negative feedback. Extrahypothalamic systems which involve the amygdala have been implicated in the inhibition of prepubertal activation of the reproductive system.[14] Since exogenous estrogen can initiate precocious puberty, an alteration in neurochemical processes induced by endogenous gonadal hormones may play a critical role in normal puberty.[15]

2. The Paranatal Animal

The major difficulty in the study of the mechanisms of gonadal hormone action on the CNS is that alterations in neural function, and presumably related neurochemical changes, are predominantly transient in nature. During one period in the development of the mammalian organism, however, gonadal hormones exert a permanent influence on the CNS as judged by a permanent alteration in the functional capacity of the CNS to regulate genetically appropriate pituitary GTH secretion or sexual behavior. From study of the effects of gonadal steroids during fetal development or shortly after birth, a concept of sexual differentiation of the mammalian brain has developed.[5,6,8,16,17] According to this concept these general statements are valid: (1) The brain of the male and female, although different in the adult, pass through a phase of similar functional potential during development. (2) During this period the brain is competent to react to endogenous or exogenous gonadal steroid and can be permanently influenced. (3) Androgen appears to be the physiological determinant of this process. The presence of androgen during the competent period leads to the development of a pattern of brain function that is typically male. In the absence of androgen feminine neuroregulatory patterns are maintained or firmly established. (4) Sexual differentiation of the brain clearly applies to both the regulation of pituitary secretory activity and sexual behavior.

a. Sexual Differentiation of the Brain. The major difference in GTH secretory patterns of the male and female is that GTH secretion is cyclic in the female, but tonic or continuous in the male. Since a spurt of LH is required for ovulation, the consequence of this tonic secretory pattern is that ovarian tissue transplanted subcutaneously in male rats castrated as adults will not develop corpora lutea. However, ovulation does occur in ovarian grafts placed subcutaneously in male rats that have been castrated within the first 3 days of life. Therefore, the male rat is born with the potential to secrete GTH in a cyclic pattern, and requires the presence of his testes for more than the first 3 days of life to ensure development of what is considered to be the normal masculine pattern of tonic GTH release. In the male that is castrated at birth and, therefore, can support ovulation when ovarian grafts are present, the neural stimulus for ovulation appears to occupy the same anatomical substrate as in the female.[18] The administration of exogenous androgen to newborn female rats during this period of sexual differentiation leads to a noncyclic pattern of GTH secretion in the adult. These females develop the syndrome of persistent vaginal estrus, and never ovulate.

The neural substrate for the release of GTH, represented conceptually by the preoptic area, appears to be the site of androgen action; androgen renders this substrate refractory to its normal, and even exogenous stimuli.[5,17] The median eminence region which regulates the tonic discharge of GTH is comparable in both the male and female. Sexual differentiation involves the higher level of neural regulation, presumably the level at which integration of numerous sensory modalities (hormone titers, environmental lighting) and a "biological clock" occur. The precise level of this integration, and of the action of neonatally administered androgen are unknown.

The sexual differentiation of behavioral mechanisms is perhaps more difficult to study since environmental stimuli play a great role in the regulation of sexual behavior. In general terms, however, androgen acting in the paranatal animal appears to diminish the capacity of the adult male or female to display lordosis, and in some cases to increase the capacity of the female to display masculine behavior.[8,19,20] In the female rat, and perhaps the male, exogenous androgen appears to render the lordosis behavioral system relatively insensitive to progesterone.[21] Following such treatment androgenized females are significantly less receptive as judged by low lordosis frequencies, but in these animals the interplay between environmental stimuli and the expression of behavior becomes especially evident. These females display higher levels of sexual receptivity following adaptation to the novel environment of the test arena.[21]

This discussion has stressed the effect of neonatally administered androgen because experimental studies with this hormone reflect, as accurately as is currently possible, a physiological process. Other steroids, most notably estradiol benzoate, can also permanently alter neuroendocrine processes. Estrogen when administered to the neonatal female or to the male shortly after neonatal castration, will permanently prevent ovulation, and possibly masculinize the hypothalamus.[22,23] It is likely that this response represents a pharmacological action of relatively high levels of this potent steroid. It has been suggested that very small amounts of estrogen may actually prevent sexual differentiation in the normal male rat.[24]

b. Relation to the Adult. A significant contribution of these developmental studies has been a further elucidation of the regulation of both ovulation and sexual behavior in the adult. The concept of two fundamental levels of hypothalamic control of pituitary function has emerged in large part from the comparative study of normal and androgenized females. Similarly, study of the differentiation of the sexual behavioral substrate permits functional dissection of this system in the adult animal. Unfortunately, gonadal hormones need not exert their widely different actions in the developing and adult animal through the same or even similar mechanisms. Nevertheless, when appropriate physiological and chemical tools become available, the elucidation of the permanent effects of hormones in the developing animal could yield significant insight into the problem of hormone action in general. Although study of the processes which produce, or follow, the cessation of normal reproductive activity in the "senile" animal may prove similarly valuable, this subject will not be considered in this chapter.

III. GONADAL HORMONES AND NEUROBIOCHEMICAL PROCESSES

A. The Initial Association

Since the first demonstration[25] that the rat uterus incorporates and retains tritiated estradiol, the concept that steroid target organs contain specific trapping or binding sites has flourished. These sites appear to exist in finite numbers, and to have a saturation point.[26] They bind steroids quite firmly since cold estradiol is not able to displace already bound labeled estradiol; similarly, pretreatment of the mouse with cold estradiol reduced labeled estradiol accumulation. The uptake of steroids need not identify a receptor site, but may only indicate a metabolically active locus. The retention of a labeled hormone may also represent a trapping mechanism which ensures continual exposure of some functional substrate to the steroid long after it has disappeared from the circulation.[27] While binding may well serve as an indicant of the site of hormone action, the possibility that steroid hormones act without becoming associated with a cell in a detectable amount, or for a measurable time period, should not be overlooked. The biochemistry of steroid hormone action has been reviewed recently.[28]

Recent studies of the uptake of tritiated estradiol,[29,30] progesterone,[31] and testosterone[32] in the brain indicate that the hypothalamus concentrates more labeled steroid, or retains it longer than other cerebral tissues. This response of the hypothalamus resembles that of the classical target tissues such as the uterus, and suggests the existence of receptor or binding sites in the CNS. The apparent validity of this basic principle has been suggested by the effectiveness of intracerebral implants of hormones in some, but not all regions of the brain. In addition, the absence of an effect of exogenous hormone following placement of a lesion in the brain is circumstantial evidence for a local effect of the hormone. The complexity of the brain, however, precludes direct application of the concepts derived from the study of steroid binding within peripheral organs to the topic of gonadal-neural interactions. The brain is not an organ that is uniformly responsive to steroids. It is possible that steroid hormones can exert a profound action on the CNS without accumulating in specific areas, or conversely, the lipid solubility of steroids might lead to the solution of a steroid within myelin sheaths of nonresponsive neurons. Finally, the fact that the hypothalamus has many functions makes it very difficult to identify the component parts of any one functional system. A neuron which accumulates estrogen may not participate in reproductive processes.

In evaluating studies on the uptake of radioactive steroid hormones, it is important to consider the active form of the hormone. Hormones may be present in both bound and free forms in the plasma. In the case of thyroid hormone, it is the free form which is active in the feedback control of the CNS.[33] Comparison of cerebral levels of labeled steroid with plasma levels may be misleading unless possible alterations in the ratio of free to bound hormone are taken into account. Similarly, simultaneous analysis of several metabolic forms of an administered hormone may provide useful data. Resko, Goy, and Phoenix,[32] for example, have reported significant differences in the uptake and retention of free and conjugated testosterone between the hypothalamus, prostate, liver, and plasma. The time of analysis dramatically affects the relative distribution of labeled hormone. Selective distribution may reflect a "physical" blood–brain barrier or a limited capacity of neurons to metabolize these hormones.[34,35]

The recent studies of Eisenfeld and Axelrod[29] indicate that various steroids and drugs compete for estradiol binding sites in the hypothalamus, which suggests one possible mechanism for the interaction of various steroid hormones. However, steroid interaction may occur at many levels. In the ring dove, progesterone can antagonize the effect of androgen on one behavioral pattern (the bow and coo), while not interfering with the androgen-induced nest call.[36] The mechanisms underlying this functionally specific interaction between androgen and progesterone are unknown, and not limited to receptor inhibition. Although exogenous progesterone effectively antagonizes the stimulatory effect of estrogen on the uterus of the immature mouse, for example, Terenius[26] did not detect an effect of progesterone on estradiol uptake by the uterus. The complex processes which follow the initial association of steroid and neuron provide many additional opportunities for interaction. For purposes of discussion only, it is convenient to divide the possible mechanisms of hormone action into three arbitrary classes: morphological, metabolic, and molecular processes.

B. Morphological Processes

1. Membrane Integrity

Because of the functional significance of the neural membrane, an hormonally induced structural or functional alteration in the cell membrane would exert a profound influence on neural activity. Hyper- or depolarization of various neuronal elements could readily diminish or increase efferent output or the effectiveness of incoming stimuli from other neurons. Changes in electrical sensitivity of the CNS following hormone administration are suggestive evidence that the neuronal membrane has been altered (see Section IV). Actual study of the influence of gonadal hormones on membrane physiology has been restricted in large measure to peripheral target tissues. The experiments of Szego and her collaborators on estrogen-induced histamine release and the subsequent alteration in permeability of uterine cell membranes, document a possible indirect role of hormones on membrane physiology.[39] This and other possible mechanisms of the peripheral action of steroids have been critically reviewed.[38]

2. Myelination

The neonatal process of active myelination of the brain is accelerated by exogenous estradiol in the rat.[39,40] Correlated with estrogen-induced myelination of the CNS is a maturation of the electroconvulsive seizure pattern in the immature rat.[41] Unfortunately, an acceleration in the rate of cerebral maturation, although interesting from the developmental point of view, may not correlate with the action of these hormones in the adult brain. Nevertheless, these results are consistent with a potentially significant effect of estrogens on myelin, perhaps as a functional component of the neuronal membrane.

3. Neuronal Synaptic Configuration

The microanatomy of neuronal processes, thus the synaptic configuration of a neuron, is exceedingly complex and labile.[42] Changes in synaptic contact would significantly alter neuronal function. Although studies of the influence of gonadal hormones on the synaptic configuration of neurons have not been reported, such an influence could well underlie the permanent effect of gonadal steroids on the developing brain.

4. Nuclear Size

Several investigators have demonstrated and quantified morphological changes in nuclei or nucleoli of neurons in correlation with hormonal manipulations.[2] Following the implantation of estradiol in the region of the median eminence in the rat, there is gonadal atrophy and a decrease in neuronal nuclear size in the arcuate nucleus. The reactivity of hypothalamic neuronal nuclei following marked changes in the titers of several hormones has been extensively studied in Hungary.[2] Neuronal nuclear size change is only a qualitative index of changes in biochemical activity of the neurons. An increase in nuclear size does not necessarily signify an increase in biochemical function, just that a change has occurred. Thus, although these morphological changes may parallel basic processes, they may not have predictive value in relation to the direct mechanism of action of gonadal hormones.

5. The Glia

Glial cells actively incorporate DNA precursors, have considerable amounts of RNA, and actively incorporate amino acids.[43] Intimate metabolic relations between glial cells and neurons have been described.[35,44] Glia may participate in electrolyte, fluid, and respiratory exchange with neurons, as well as supply them with energy. Until the function of these cells which constitute the bulk of the brain is better understood, it may be premature to exclude the possibility that gonadal hormones may act in part on glial function.

C. Metabolic Processes

Since neurons are metabolically very active, neural metabolism represents a likely level of action of gonadal hormones. The possible action of steroids on cellular metabolism has been extensively investigated with peripheral tissues. In contrast to muscle, or actively secreting tissues, neurons perform very minimal external work, yet their activity is interrupted after only a few seconds without their normal blood supply. This large energy requirement is needed to maintain membrane potentials and the complex metabolic machinery of neurons. Gonadal hormones could exert major effects on nerve cells by altering their respiratory, protein, or electrolyte metabolism.

1. Respiratory Metabolism

Treatment of the rat with gonadal steroids affects the respiratory rate of brain suspensions[45]: progesterone and testosterone inhibit oxygen uptake while castration and estradiol propionate elevate the respiratory rate. Moguilevsky and his collaborators have studied in detail the respiratory quotient (RQ) of specific areas of the rat brain in vitro. Castration of the male depressed the RQ in the anterior and posterior hypothalamus, while that of the cortex and pituitary was increased.[46] The RQ of the anterior and posterior hypothalamus was highest when these tissues were removed from females sacrificed in the estrous stage of the vaginal cycle.[47] Since estrogen added to this in vitro system did not increase the RQ in hypothalamic tissue from the diestrous female, elevated respiration during estrus may not represent a direct effect of estrogen.[48] The RQ may reflect only the general level of activity of the neurons. Increased respiratory activity of hypothalamic cells could be the result of the activation of these neurons by other regions of the brain. The increase in estrogen secretion might as likely be the consequence of increased hypothalamic metabolism as the cause of the latter.

In peripheral tissues the effect of gonadal hormones on cellular respiration has been studied in detail. Progesterone inhibits electron transport in submitochondrial particles of beef heart.[49] When the rat is castrated, the observed decrease in prostate respiration is due, at least in part, to a decrease in the number of mitochondria. Doeg[50] has detected an increase in mitochondrial synthesis within the prostate following androgen injection to the castrate rat. These studies of nonnervous tissue suggest a possible mechanism whereby steroid hormones might act on neurons, which are profoundly dependent upon a very active respiratory metabolism.

2. Protein Metabolism

While a vast literature is rapidly accumulating on the influence of gonadal hormones on protein synthetic mechanisms in cells of sex accessory tissues, there is a dearth of knowledge relating to the CNS. The former studies have been extensively reviewed,[38,51] and it seems evident that gonadal hormones stimulate cytoplasmic and nuclear RNA and protein synthesis.[52,53] The response of target tissues such as the uterus or prostate is characterized by cell growth, cell division, or active secretion. These are all processes which require active protein synthesis, but which do not characterize the response of the adult brain to hormones. Therefore, the mechanisms proposed for the action of estrogen in the uterus, for example, may have little relationship to its action on the brain. Nevertheless, neurons do actively metabolize RNA, and the complex machinery for the maintenance of normal neural function undoubtedly requires active protein synthesis. The action of gonadal hormones on uterine RNA appears to be specific for certain RNA molecules, particularly nuclear RNA. A specific action of gonadal steroids on functionally distinct protein synthetic mechanisms within nerve cells would readily account for altered sensitivity or altered performance. If certain neurons are particularly sensitive to hormones, because of specialized receptor or binding mechanisms, alteration in their function may have profound functional or behavioral consequences.

3. Electrolyte Metabolism

In spite of the importance of electrolyte concentration and distribution for normal neural activity, few studies have attempted to clarify the effect of gonadal steroids on this system. Estrogen treatment of developing rats increases both electroconvulsive sensitivity and the amount of chloride in the cerebellum.[54] Wooley and Timaras[55] observed a significant increase in brain chloride following administration of varying doses of gonadal steroids to gonadectomized adult male and female rats. Such an alteration in electrolyte concentration might underlie the action of gonadal steroids on brain excitability. Because of the high chloride content of glial cells, these authors raise the possibility that gonadal hormones may act on these structural elements.

Gonadal steroids when applied directly to nervous tissue may also act by altering local ionic concentration. The possibility that these hormones may act as chelating agents has been reviewed.[56] Gonadal atrophy, inhibition of postcastration changes in GTH secretion, precocious puberty, and complex sexual and maternal behavior patterns are induced by hormonal application directly to specific neural tissue. These dramatic effects could result from a significant alteration in electrolyte concentration in the region of a steroid implant. If a hormonal implant acts as a chelating agent, one might expect an action on a functional system anywhere along its anatomical pathway. This concept would not support the common view that a steroid implant is effective only when placed amidst neurons sensitive to the hormone. Precise localization of steroid-sensitive neurons by means of the implantation of crystalline hormones is subject to another limitation. Steroids may be carried by diffusion within lipid membranes back to neuronal soma remote from an implant, or similarly, steroid molecules may enter a neuron and be carried by axoplasmic flow to nerve terminals scattered over a great distance. Radiochemical techniques[57] currently available to study the spread of hormones depend on the specific activity of the labeled hormone. These techniques may lack the sensitivity required to detect selective distribution of hormone although such distribution might be critical for biological activity.

In the case of endogenous gonadal hormones, their concentration by binding or trapping mechanisms in certain neurons might permit the action of the hormones on the local electrolyte environment. Therefore, changes in cerebral electrolytes following systemic injection of steroids and the possible chelating action of intracerebrally administered hormones offer a model for one mechanism of gonadal hormone action.

4. Hormone Metabolism by the Brain

The metabolism of gonadal hormones by cerebral tissues may modify the influence of gonadal hormones on cerebral function. Significant differences exist in the ability of brain and sexual accessory tissue to accumulate, metabolize, or retain testosterone and its metabolites.[30,32] Individual hormones are also handled differently. Laumas[31] postulates that progesterone acts via rapid uptake and metabolism within the brain, while estrogen is retained and incorporated into the neural tissue. The possible metabolic fate of gonadal steroids is of critical importance in studies where labeled hormones are injected or implanted and the radioactive label traced over time. Identification of the actual labeled component is essential for accurate interpretation of experiments of this type.

Experiments which utilize the direct application of crystalline hormones to the brain face yet another important challenge. The functional response to gonadal steroids, and presumably their mechanisms of action, is dependent upon the concentration of the hormone. Low levels of estrogen may exert a positive feedback action on certain cells, at higher concentrations these cells may be inhibited, and finally the effect of estrogen may become entirely nonspecific. Since hormonal concentration is critical, and since nerve cells including those possibly specialized as steroid receptors may not be physiologically exposed to huge local concentrations of crystalline hormone, the results of hormone implantation studies should not be interpreted without caution.

D. Molecular Processes

For this discussion of the possible levels of action of gonadal hormones on the CNS, we have arbitrarily classified the processes of neurotransmission and neurosecretion together. Individual neurons communicate across synaptic junctions through the release and reception of neurotransmitter molecules. Neurons also produce neurohormones such as oxytocin, and neuroendocrine mediators, such as the hypothalamic releasing factors. Gonadal hormone could readily modify cerebral function by an action at the level of these molecular processes.

There is an extensive and growing literature which correlates reproductive neuroendocrinology with the biogenic amines. The median eminence region, which is of profound importance for the regulation of pituitary function, is rich in neurons which contain monoamines or serotonin.[58–60] Changes in cerebral amine levels have been correlated with the estrous cycle.[61] Castration leads to an increased turnover,[62] and increased concentration of amines within the whole brain or specific regions of the hypothalamus.[63] Drug administration frequently alters both GTH secretion and monoamine levels.[64] Biogenic amines have also been implicated in the regulation of sexual behavior to the extent that progesterone may be thought to facilitate sexual behavior by monoamine depletion.[65]

Because of the concentration of aminergic and serotonergic neurons within the hypothalamus, and the many reported alterations in monoamine levels following hormonal manipulation, one can postulate that steroid-induced changes in the complex biogenic amine system represent a fundamental mechanism of action of gonadal hormones. Although it is difficult to deny some association, it is not always possible to correlate amine levels with hypothalamic function.[66,67] These negative findings may reflect the insensitivity of the assay system, or the complexity of the possible alterations. Biogenic amines may act at numerous levels within functional neuroendocrine systems, and gonadal steroids need not act directly upon the biogenic amine system. Removal of gonadal hormone by castration, for example, could free the metabolic machinery of a neuron from negative feedback. As a consequence of increased general neuronal activity, hypothalamic amine levels might rise along with hypophyseal GTH levels. The amines, therefore, may only mediate the transfer of information from a neuron that has been more directly altered by gonadal hormones to another neuron or to the pituitary. Although the relationship between gonadal hormones, neuronal activity, and the biogenic amines may be significant, there is no assurance that this relationship is direct.

IV. ELECTRICAL CORRELATES OF GONADAL HORMONE ACTION

Study of the electrical correlates of gonadal hormone action on the brain may yield insight into the functional alterations induced by these hormones. Gonadal steroids have been shown to alter the excitability of certain neurons as reviewed by Sawyer.[68] Progesterone can exert a biphasic effect on the threshold for electrical activation of certain EEG patterns which have been correlated with behavioral (EEG-arousal) and endocrinological (EEG-after-reaction) processes in the female rabbit. It is possible to elevate the threshold of the after-reaction selectively with the appropriate choice of gonadal steroid. Apparently a specific component of a complex system of steroid-sensitive neurons may be selectively altered. At the present time no explanation can be offered for the functional specificity of hormone action on the CNS.

Intravenous progesterone can modify the electrical activity of single cells within the hypothalamus, and also change cortical activity. If the amount of progesterone is sufficiently high, behavioral sleep and ultimately anesthesia is produced. Although it is possible that hypothalamic neurons are directly sensitive to progesterone, microanalysis of electrical changes in the nervous system is open to many interpretations. Sawyer and his colleagues have postulated that hypothalamic units follow general changes in the EEG which may be induced by changes in blood pressure, and need not represent a direct action of progesterone on the unit cells. Cortical synchronization induced by progesterone in the rat under urethane anesthesia appears to be the result of an activation of baroreceptors by an increase in blood pressure. An implication of these studies is that gonadal hormones may alter cerebral function through the modification of fundamental, but nonneural processes, such as blood pressure, local blood flow, and brain temperature. It is similarly possible that neurochemical modifications apparently induced by gonadal hormones directly may accompany these more generalized processes.

V. THE TEMPORAL PATTERN OF HORMONE ACTION

Following the administration of a hormone, a finite period of time elapses before a functional effect is observed. Of what significance is the changing hormonal environment during this period? What is the temporal pattern of action of individual hormones? In order to consider adequately this aspect of hormonal action on the CNS, the nature of the functional response is critical. The varied processes which result in an altered discharge of releasing factor and eventually an endocrine event such as ovulation, may differ markedly from the processes which permit a complex behavioral response.

A. Regulation of Hypophyseal Activity

Two examples illustrate the relevance of the temporal pattern of hormone action: the induction of ovulation by coitus in the rabbit, and the permanent androgenization of the hypothalamus of the newborn rat. In the rabbit, ovarian estrogen promotes the state of sexual receptivity, and in addition, creates the appropriate environment so that upon the initiation of mating, the afferent stimuli received by the female reflexly cause the release of ovulating hormone. Analysis of the response of the rabbit ovary to this reflex release of LH has revealed that ovulation in the rabbit is a complex process of dynamic hormonal-neural interaction.[69] The quantity of LH released in response to the mating stimuli is not sufficient to induce ovulation. Instead, this hormone initiates the active secretion of 20-α-hydroxy-pregn-4-en-3-one within 15 min of coitus. This steroid, in turn, acts back upon the hypothalamus to promote the continued release of LH which is essential for ovulation.

Reference has been made to the permanent effect of androgen on the developing hypothalamus. In the newborn male rat, testicular androgen induces the development of the masculine pattern of the tonic discharge of GTH. Similarly, exogenous androgen masculinizes the control of GTH secretion in the newborn female. Since the experimentor can initiate at will hypothalamic exposure to androgen by injecting this steroid into the female, studies of the dynamics of the action of androgen have been conducted in neonatal female rats. Initially we confirmed that reserpine, chlorpromazine, or progesterone could protect the newborn female from this permanent effect of androgen.[70]

Barbiturate or the antiandrogen, cyproterone acetate, also protects the 5-day-old female rat from a simultaneous injection of androgen. When these inhibitors of androgen action were injected at specific intervals after androgen, an estimate of the duration of barbiturate and antiandrogen-sensitive components of androgen action was determined.[71] Cyproterone acetate or phenobarbital inhibits the action of androgen only if given within the first 3 hr after androgen injection, while pentobarbital interferes with androgen as long as 12 hr after injection of the steroid. It has been postulated that during the first several hours after subcutaneous injection of androgen, the steroid is taken up by hypothalamic neurons. The more prolonged pentobarbital-labile phase may represent a more intimate association between androgen and the neuron, perhaps firm membrane binding, intracellular action, or the initiation of complex biochemical changes. After this pentobarbital-sensitive phase, many additional changes may still occur; perhaps additional protective agents will be discovered which will elucidate the nature of androgen action.

The elucidation of an induced permanent change in neuronal function could contribute significantly to an understanding of the mechanisms of action of gonadal hormones on the brain, and a detailed time analysis of this action of androgen should permit more productive research. If the temporal pattern as postulated above is realistic, it would be unnecessary, for example, to test possible inhibitors of cerebral uptake more than about 2 hr after androgen is administered. On the other hand, administration of labeled substances to monitor various metabolic processes, or the administration or depletion of brain monoamines, may be most effective if carried out between 2–9 hours after an injection of androgen. It must be emphasized that processes induced in the CNS by gonadal hormones at any age are clearly time dependent. The temporal domain of hormone action must be added to the problem of the spatial localization of hormone effects within the brain.

The liver may play a more significant role in the temporal characteristics of hormone action than is commonly assumed. If cerebrally active steroid is in equilibrium with plasma titers, the more rapidly the latter are removed, the smaller the amount available for action on the CNS. Experiments in adult neuroendocrinology are commonly based on the assumption that experimental procedures do not alter peripheral metabolism of hormones. The administration of barbiturate, however, is one treatment which routinely alters liver metabolism of steroid hormones. Although a single injection of phenobarbital is apparently without a significant functional effect, repeated injections of this drug stimulate liver microsomal enzymes which hydroxylate steroid hormones.[72,73] This stimulation can be detected functionally; an injection of estradiol benzoate which induces a uterine weight increase in the spayed female is without effect when administered to spayed females also treated for 4 days with phenobarbital. Whenever barbiturate is administered, especially repetitively (but not necessarily daily), the utilization of endogenous or exogenous steroids may be significantly altered. The potential consequences of this phenomenon must be considered during experimental design or data interpretation.

B. Regulation of Sexual Behavior

Steroid-induced alterations in neuronal function which eventually lead to overt behavior are difficult to assess since a behavioral response is subject to many influences, including the test condition imposed by the investigator. Nevertheless, certain general statements can be put forth in relation to the dynamics of hormone action. The intrahypothalamic implantation of crystalline estrogen initiates sexual receptivity after a minimum latency of 2–3 days.[9] Systemic injection of estrogen yields data which suggest that continual exposure of the brain to the hormone for that total period is not obligatory. Sexual behavior is induced following the administration of a radioactively tagged estrogen at a time when the tracer is no longer detectable in the hypothalamus.[74] Similarly, the use of an antiestrogen suggests that estrogen triggers a process which requires approximately 48 hr to complete in the rat. During this process continued exposure of neurons to estrogen beyond the first 24 hr is unnecessary.[75] Analysis of the temporal neurobiochemical changes induced by estrogen would be an obvious next step, if the appropriate neurons could be identified.

It is interesting to note that intrahypothalamic estrogen implants can induce sexual receptivity without any progesterone treatment. Progesterone, however, is dependent upon previous estrogen action. Perhaps because of this dependence, sexual behavior follows progesterone administration after only a short delay. Sexual receptivity can occur within 15 min following intraventricular or intravenous injection of progesterone, and we have observed a similar latency following progesterone implantation in the reticular formation (unpublished observations).

A fundamental question about the mechanisms of action of estrogen and progesterone is whether they act on the same neural substrate. Does progesterone act on neurons which have been prepared chemically by estrogen, or does progesterone act independently on different neurons which are functionally related to estrogen-sensitive cells? These questions cannot be answered at this time. In the general economy of the organism, the regulation of sexual behavior is closely related to yet another system, the control of hypophyseal activity. Although the neural substrates involved in these two processes may differ to a great extent, one common and uniting feature is the action of gonadal hormones. It is likely that the neurochemical changes induced by gonadal hormones are basically comparable; the functional expression of these changes, however, depends on the capacity of the neural substrate acted upon.

VI. CONCLUSION

The readers of this chapter will have been impressed with the lack of factual knowledge concerning many fundamental points, the elucidation of which must precede a clear understanding of the action of gonadal hormones on the CNS. Rather than being able to summarize published data which support uniformly popular concepts, we have been forced to consider very generally only the possible levels of hormone action. Elegant biochemical techniques which have led to considerable progress in the study of peripheral hormone action are less useful in studies involving the CNS. The brain is not uniformly affected by steroid hormones and is capable of many simultaneous functions. The identification of proposed steroid-sensitive neurons and the elucidation of their role in various neuroregulatory processes may be essential before specific neurochemical changes can be recognized and evaluated.

We have reviewed briefly the reciprocal functional relationship between gonadal hormone secretion and brain function. It is apparent that the simultaneous temporal and spatial characteristics of action of several hormones determines the functional interaction between hormones and the brain. Further progress in the elucidation of both the mechanisms of action of gonadal hormones on peripheral tissues and the mechanisms of neuroendocrine regulation will contribute to the continued progress in our understanding of the action of gonadal steroids on the brain.

VII. REFERENCES

1. L. Martini and W. F. Ganong, eds. *Neuroendocrinology*, Academic Press, New York (1966).
2. J. Szentágothai, B. Flerkó, B. Mess, and B. Halász, "Hypothalamic control of the anterior pituitary," *Akadémiai Kiadó*, Budapest (1968).
3. J. W. Everett, Central neural control of reproductive functions of the adenohypophysis, *Physiol. Rev.* **44**:373–431 (1964).
4. J. W. Everett, in *Major Problems in Neuroendocrinology* (E. Bajusz and G. Jasmin, eds.) pp. 346–366, Williams & Wilkins, Baltimore (1964).
5. C. A. Barraclough, Modifications in the CNS regulation of reproduction after exposure of prepubertal rats to steroid hormones, *Recent Prog. Hormone Res.* **22**:503–539 (1966).
6. R. A. Gorski, in *Biology of Gestation* (N. S. Assali, ed.) Vol. 1, pp. 1–66, Academic Press, New York (1968).
7. B. Halász and R. A. Gorski, Gonadotrophic hormone secretion in female rats after partial or total interruption of neural afferents to the medial basal hypothalamus, *Endocrinology* **80**:608–622 (1967).
8. C. H. Phoenix, R. W. Goy, and W. C. Young, in *Neuroendocrinology* (L. Martini and W. F. Ganong, eds.) Vol. 2, pp. 163–196, Academic Press, New York (1966).
9. R. D. Lisk, in *Neuroendocrinology* (L. Martini and W. F. Ganong, eds.) Vol. 2, pp. 197–239, Academic Press, New York (1966).
10. W. C. Young, in *Sex and Internal Secretions* (W. C. Young, ed.) Vol. 2, pp. 1173–1239, Williams & Wilkins, Baltimore (1961).
11. J. M. Davidson, in *Neuroendocrinology* (L. Martini and W. F. Ganong, eds.) Vol. 1, pp. 565–611, Academic Press, New York (1966).
12. F. A. Beach, Cerebral and hormonal control of reflexive mechanisms involved in copulatory behavior, *Physiol. Rev.* **47**:289–316 (1967).
13. B. T. Donovan and J. J. Van Der Werff Ten Bosch, *Physiology of Puberty*, E. Arnold, Ltd., London (1965).
14. V. Critchlow and M. E. Bar-Sela, in *Neuroendocrinology* (L. Martini and W. F. Ganong, eds.) Vol. 2, pp. 101–162, Academic Press, New York (1966).

15. V. D. Ramirez and C. H. Sawyer, Advancement of puberty in the female rat by estrogen, *Endocrinology* 76:1158–1168 (1965).

16. G. W. Harris, Sex hormones, brain development and brain function, *Endocrinology* 75:627–648 (1964).

17. R. A. Gorski, Localization and sexual differentiation of the nervous structures which regulate ovulation, *J. Reprod. Fert.*, Suppl. 1:67–88 (1966).

18. R. A. Gorski, Localization of the neural control of luteinization in the feminine male rat, *Anat. Rec.* 157:63–69 (1967).

19. K. L. Grady, C. H. Phoenix, and W. C. Young, Role of the developing rat testis in differentiation of the neural tissues mediating mating behavior, *J. Comp. Physiol. Psychol.* 59:176–182 (1965).

20. R. E. Whalen and D. A. Edwards, Hormonal determinants of the development of masculine and feminine behavior in male and female rats, *Anat. Rec.* 157:173–180 (1967).

21. L. G. Clemens, M. Hiroi, and R. A. Gorski, Induction and facilitation of female mating behavior in rats treated neonatally with low doses of testosterone propionate, *Endocrinology* 84:1430–1438 (1969).

22. R. A. Gorski, Modification of ovulatory mechanisms by postnatal administration of estrogen to the rat, *Am. J. Physiol.* 205:842–844 (1963).

23. S. Levine and R. Mullins, Estrogen administered neonatally affects adult sexual behavior in male and female rats, *Science* 144:185–187 (1964).

24. B. Flerkó, P. Petrusz, and L. Tima, On the mechanism of sexual differentiation of the hypothalamus: Factors influencing the "critical period" of the rat, *Acta Biol. Acad. Sci. Hung.* 18:27–36 (1967).

25. E. V. Jensen and H. I. Jacobson, Basic guides to the mechanism of estrogen action, *Recent Prog. Hormone Res.* 18:387–414 (1962).

26. L. Terenius, Uptake of radioactive oestradiol in some organs of immature mice, *Acta Endocrinol.* 50:584–596 (1965).

27. R. J. Wurtman, Estrogen receptor: Ambiguities in the use of this term, *Science* 159:1261 (1968).

28. H. G. Williams-Ashman, New facets of the biochemistry of steroid hormone action, *Cancer Res.* 25:1096–1120 (1965).

29. A. J. Eisenfeld and J. Axelrod, Evidence for estradiol binding sites in the hypothalamus—effect of drugs, *Biochem. Pharm.* 16:1781–1785 (1967).

30. J. Kato and C. A. Villee, Preferential uptake of estradiol by the anterior hypothalamus of the rat, *Endocrinology* 80:567–575 (1967).

31. K. R. Laumas and A. Farooq, The uptake *in vivo* $(1,2-^3\mathrm{H})$-progesterone by the brain and genital tract of the rat. *J. Endocrinol.* 36:95–96 (1966).

32. J. A. Resko, R. W. Goy, and C. H. Phoenix, Uptake and distribution of exogenous testosterone-1,2,^{3}H in neural and genital tissues of the castrate guinea pig, *Endocrinology* 80:490–498 (1967).

33. L. K. Christensen, Pituitary regulation of thyroid activity, *Acta Endocrinol.* 33:111–116 (1960).

34. J. Dobbing, The blood–brain barrier: Some recent developments, *Guys Hosp. Rep.* 112:267–286 (1963).

35. S. W. Kuffler and J. G. Nicholls, The physiology of neuroglial cells, *Ergebnisse Der Physiologie* 57:1–90 (1966).

36. C. J. Erickson, R. H. Bruder, B. R. Komisaruk, and D. S. Lehrman, Selective inhibition by progesterone of androgen-induced behavior in male ring doves (*Streptopella risoria*), *Endocrinology* 81:39–44 (1967).

37. C. M. Szego, Role of histamine in mediation of hormone action, *Fed. Proc.* 24:1343–1352 (1965).

38. O. Hechter and I. D. K. Halkerstron, Effects of steroid hormones on gene regulation and cell metabolism, *Ann. Rev. Physiol.* **27**:133–162 (1965).
39. J. Curry and L. M. Heim, Brain myelination after neonatal administration of oestradiol, *Nature* (*London*) **209**:915–916 (1966).
40. R. Casper, A. Vernadakis, and P. S. Timiras, Influence of estradiol and cortisol on lipids and cerebrosides in the developing brain and spinal cord of the rat, *Brain Res.* **5**:524–526 (1967).
41. L. M. Heim and P. S. Timiras, Gonad-brain relationship: Precocious brain maturation after estradiol in rats, *Endocrinology* **72**:598–606 (1963).
42. A. Globus and A. B. Scheibel, The effect of visual deprivation on cortical neurons: A Golgi study, *Exptl. Neur.* **19**:331–345 (1967).
43. H. Koenig, in *Morphological and Biochemical Correlates of Neural Activity* (M. M. Cohen and R. S. Snider, eds.) pp. 39–56, Hoeber, New York (1964).
44. E. Giacobini, in *Morphological and Biochemical Correlates of Neural Activity* (M. M. Cohen and R. S. Snider, eds.) pp. 15–38, Hoeber, New York (1964).
45. G. S. Gordan, R. C. Bentinck, and E. Eisenberg, The influence of steroids on cerebral metabolism, *Ann. N.Y. Acad. Sci.* **54**:575–607 (1951).
46. J. A. Moguilevsky, O. Schiaffini, and V. Foglia, Effect of castration on the oxygen uptake of different parts of hypothalamus, *Life Sci.* **5**:447–452 (1966).
47. J. A. Moguilevsky, Oxidative activity of different hypothalamic areas during sexual cycle in rats, *Acta Physiol. Latin Am.* **15**:423–424 (1965).
48. J. A. Moguilevsky and M. R. Malinow, Endogenous oxygen uptake of the hypothalamus in female rats, *Am. J. Physiol.* **206**:855–857 (1964).
49. F. Varricchio and D. R. Sanadi, Inhibition of mitochondrial respiration by progesterone and an azasteroid, *Arch. Biochem. Biophys.* **121**:187–193 (1967).
50. K. A. Doeg, Control of mitochondrial lipid biosynthesis by testosterone in male sex accessory gland tissue and liver of castrate rats, *Endocrinology* **82**:535–539 (1968).
51. P. Karlson, *Mechanism of Hormone Action*, Academic Press, New York (1965).
52. T. Fujii and C. A. Villee, Effect of testosterone on ribonucleic acid metabolism in the prostate, seminal vesicle, liver and thymus of immature rats, *Endocrinology* **82**:463–467 (1968).
53. R. W. Barton and S. Liao, A similarity in the effect of estrogen and androgen on the synthesis of ribonucleic acid in the cell nuclei of gonadohormone-sensitive tissues, *Endocrinology* **81**:409–412 (1968).
54. T. Valcana, A. Vernadakis, and P. S. Timaras, Influence of estradiol and cortisol on electrolytes in the central nervous system of developing rats, *Neuroendocrinology* **2**:326–329 (1967).
55. D. E. Wooley and P. S. Timaras, Water and electrolyte alterations in plasma, brain and liver of rats after castration and sex hormone administration, *Acta Endocrinol.* **46**:12–24 (1964).
56. A. E. Fisher, in *Current Trends in Psychological Theory*, pp. 70–86, Univ. Pittsburgh Press, Pittsburgh (1961).
57. Y. Palka, V. D. Ramirez, and C. H. Sawyer, Distribution and biological effects of tritiated estradiol implanted in the hypothalamus-hypophysial region of female rats, *Endocrinology* **78**:487–499 (1966).
58. K. Fuxe, Cellular localization of monoamines in the median eminence and infundibular stem of some mammals, *Z. Zellforsch.* **61**:710–724 (1964).
59. H. Kobayashi, Y. Oota, H. Vemura, and T. Hirano, Electron microscopic and pharmacological studies on the rat median eminence, *Z. Zellforsch.* **71**:387–404 (1966).
60. T. Hokfelt, The possible ultrastructural identification of tubero-infundibular dopamine-containing nerve endings in the median eminence of the rat, *Brain Res.* **5**:118–121 (1967).
61. F. J. E. Stefano and A. O. Donoso, Norepinephrine levels in the rat hypothalamus during the estrous cycle, *Endocrinology* **81**:1405–1406 (1967).
62. F. Anton-Tay and R. J. Wurtman, Norepinephrine-turnover in rat brains after gonadectomy, *Science* **159**:1245 (1968).

63. A. E. Donoso, F. J. E. Stephano, A. Biscardi, and J. Cukier, Effects of castration on hypothalamic catecholamines, *Am. J. Physiol.* **212**:737–739 (1967).
64. W. Lippman, R. Leonardi, J. Ball, and J. A. Coppola, Relationship between hypothalamic catecholamines and gonadatrophin secretion in rats. *J. Pharm. Exp. Therap.* **156**:258–266 (1967).
65. B. J. Meyerson, Central nervous monoamines and hormone induced estrous behaviour in the spayed rat, *Acta Physiol. Scand.* **63**(Suppl. **241**), 1–32 (1964).
66. K. Fuxe, T. Hökfelt, and O. Nilsson, Activity changes in the tubero-infundibular dopamine neurons of the rat during various states of the reproductive cycle, *Life Sci.* **6**:2057–2061 (1967).
67. K. M. Knigge, Catecholamines of rat median eminence: Lack of response in altered hypothalamo-pituitary-endocrine states, *Anat. Rec.* **160**:377 (1968).
68. C. H. Sawyer, Some endocrine aspects of forebrain inhibition, *Brain Res.* **6**:48–59 (1967).
69. J. Hilliard, R. Penardi, and C. H. Sawyer, A functional role of 20α-hydroxypregn-4-en-3-one in the rabbit, *Endocrinology* **80**:901–909 (1967).
70. Y. Arai and R. A. Gorski, Protection against the neural organizing effect of exogenous androgen in the neonatal female rat, *Endocrinology* **82**:1005–1009 (1968).
71. Y. Arai and R. A. Gorski, Critical exposure time for androgenization of the developing hypothalamus in the female rat. *Endocrinology* **82**:1010–1014 (1968).
72. W. Levin, R. M. Welch, and A. H. Conney, Effect of chronic phenobarbital treatment on the liver microsomal metabolism and uterotropic action of 17β-estradiol, *Endocrinology* **80**:135–140 (1967).
73. S. Burstein, Determination of initial rates of cortisol 2α and 6β-hydroxylation by hepatic microsomal preparations in guinea pigs: Effect of phenobarbital in two genetic types, *Endocrinology* **82**:547–554 (1968).
74. R. P. Michael, *in Proc. 22nd Intern. Physiol Congr.* Leiden, Vol. 1 Part 2, pp. 650–652, Exerpta Medica Foundation, Amsterdam (1962).
75. Y. Arai and R. A. Gorski, Effect of anti-estrogen on steroid induced sexual receptivity in ovariectomized rats, *Physiol. Behav.* **3**:351–354 (1968).

Chapter 19

PINEAL HORMONES

R. J. Wurtman

Department of Nutrition and Food Science
Massachusetts Institute of Technology
Cambridge, Massachusetts

I. INTRODUCTION

After a long period of disgrace in the limbo reserved for vestigial organs, the mammalian pineal has recently become the object of vigorous experimental analysis. A large body of information about the structure and biochemical function of the pineal has been accumulated, and important, if preliminary, observations have been made which begin to define the place of this organ in the economy of the body. Investigators not primarily concerned with the pineal per se have begun using it in their studies because some of its properties make it favorable for experimentation (e.g., the ease with which it is separated from the brain; its high concentration of sympathetic nerve endings). In this sense, the mammalian pineal is probably of special interest to the neurochemist for the following reasons:

1. It contains relatively large amounts of the biogenic amines serotonin and norepinephrine and the enzymes which are responsible for their synthesis and metabolism. Moreover, pineal serotonin undergoes metabolic transformations which appear to occur only within this structure.

2. The contents of serotonin and norepinephrine in the pineal vary as a function of environmental lighting and time of day. Nature thus performs daily "experiments" on amines in the pineal; these experiments provide excellent models for studying the physiological factors normally regulating tissue amine levels.

3. The synthesis, storage, release, and metabolism of certain compounds in pineal glandular cells appear to be under the direct control of sympathetic nerve endings, many of which terminate on these cells. This use of sympathetic nerves for metabolic regulation is unusual in the body and provides a relatively simple system for studying how nerve impulses affect enzymes.

II. THE EVOLUTION OF THE MAMMALIAN PINEAL GLAND

The pineal appears to have undergone extraordinary changes as verte-brates evolved from amphibians to mammals. An appreciation of these changes is essential to an understanding of the operation of the mammalian organ.

In the frog the pineal or epiphysis is not a gland; it is part of a photo-receptive complex on the roof of the brain.[1,2] The cells of the amphibian pineal contain organelles which are characteristic of all known biological units that convert light waves to nerve impulses.[1,3] These cells are oriented around a lumen; typical "outer segments," which may contain photopigments, project from their apices. Nerves leaving the frog pineal carry information directly to the brain[4]; along these nerves it is possible to record action potentials whose frequencies are altered when a light is shined on the pineal.[5] In all mammals examined to date, the pineal appears to have lost all of its photoreceptor cells; it is no longer capable of directly transducing a photic input to a neural or any other kind of signal.[1,2,6] In mammalian organs, the pineal photoreceptive cell has been replaced by the pinealocyte, a new kind of cell with many elongated processes which terminate adjacent to or near capillaries.[6,7] The mammalian pineal no longer sends nerve fibers to the epithalamus or elsewhere in the brain. It receives no input from the central nervous system, which is somewhat surprising in view of the fact that the pineal continues to originate embryologically as an outpouching of the roof of the third ventricle.[4,8] In the rat, the cat, the primate, and all other mam-mals examined,[4,9] the pineal receives an extensive sympathetic innervation which originates in the superior cervical ganglia and terminates at least in part directly on the pineal glandular cells.[6,10] As described below, informa-tion transmitted along these sympathetic nerves appears to regulate the metabolic function of the pinealocytes.

III. MEDIATION OF THE PHOTIC CONTROL OF THE MAMMALIAN PINEAL BY SYMPATHETIC NERVES

Although the mammalian pineal is no longer capable of a direct response to environmental lighting, its level of function continues to be influenced by light via an indirect pathway: Light impinges upon the retinas, generating nerve impulses which are carried along the optic nerves to the optic chiasm. At this point many of the nerve fibers carrying the primary retinal input decussate, and most of the crossed and uncrossed fibers continue as part of the classic visual system, terminating in the lateral geniculate bodies, the superior colliculus, and elsewhere. At least two additional nerve bundles which carry photic inputs leave the main visual system just beyond the optic chiasm: These are the inferior accessory optic tract and the superior accessory optic tract.[11] The former tract joins the medial forebrain bundle in the lateral hypothalamus and terminates in the medial terminal nucleus of the inferior accessory optic tract, which lies near the junction of the subthalamus and the mesencephalon. When the inferior accessory optic tracts of the rat are transected (either by cutting both within the medial forebrain bundle or by cutting one and removing the ipsilateral eye), the pineal of the animal is no longer influenced by light or darkness, even though the animal can still see.[12–15] On the other hand, when the classic visual system is transected by making large lesions in the lateral region of the geniculate bodies, the animal is functionally blind but its pineal can still respond to a change in environmental illumination. Hence, it seems likely that the inferior accessory optic tract mediates the photic control of the pineal. Almost nothing is known about the route taken by this photic input between the medial terminal nucleus of the inferior accessory optic tract and the preganglionic fibers to the superior cervical ganglia. However, if these or the postganglionic fibers are damaged, the pineal again becomes incapable of a metabolic response to environmental light.[16,17]

In summary, in lower vertebrates the pineal responds directly to light by generating nerve impulses which are transmitted to the brain; in mammals the pineal is no longer capable of a direct response to light, but is influenced indirectly via a neural pathway involving the eyes, the inferior accessory optic tract, and phylogenetically new pineal sympathetic nerves.

IV. METABOLISM OF SEROTONIN IN THE MAMMALIAN PINEAL

The pineal glands of primates[18] and certain rodents[19–21] contain very high concentrations of serotonin (Table I). If properly oxygenated, rat pineal glands grown in organ culture are capable of synthesizing this amine (as well as melatonin and 5-hydroxyindoleacetic acid) from tryptophan without the addition of cofactors.[22,23] The synthesis of serotonin from its parent amino

TABLE I

Concentrations of Indoles in Mammalian Pineal Glands

Species	Time of day	Melatonin (μg/g)	Serotonin (μg/g)
Cow	—	0.2	0.3
Rat	—	0.4	—
	Day	—	60–100
	Night	—	10–20
Human	—	0.3^a	0.5–20
Kangaroo	—	0.5–0.7	0.2

^{a} Pinealoma.

acid requires the action of two enzymes, tryptophan hydroxylase and aromatic *l*-amino acid decarboxylase; both have been identified in the rat pineal at very high levels of activity.[24,25] Pineal serotonin can undergo a variety of metabolic fates, two of which (conversion to methoxyindoles and binding within sympathetic nerve endings) appear to be unique to this organ. Probably the greatest fraction of the serotonin is destroyed within the gland by oxidative deamination; this process is catalyzed by the enzyme monoamine oxidase.[26] The product, 5-hydroxyindoleacetaldehyde, can then be oxidized to 5-hydroxyindoleacetic acid or reduced to 5-hydroxytryptophol. Both of these compounds are substrates for hydroxyindole-*O*-methyltransferase (HIOMT).[27,28] Some pineal serotonin is probably released into the circulation and metabolized elsewhere. A portion of the serotonin produced in pinealocytes can be taken up and stored within the adjacent sympathetic nerve endings. It is possible that this material is released following nerve stimulation; however, this has not yet been demonstrated. The ability of pineal sympathetic nerves to take up serotonin probably results not from any peculiar properties of these nerves, but from the very high concentration of the amine in their milieu.[29] Finally, a small but very important fraction of the pineal serotonin is *N*-acetylated within the gland to form *N*-acetyl-serotonin, and this intermediate is then *O*-methylated to form 5-methoxy *N*-acetyltryptamine (melatonin)[30,31] (Fig. 1). The formation of melatonin is catalyzed by the enzyme HIOMT; the methyl donor *S*-adenosylmethionine serves as a cofactor in this reaction.[27] The concentration of serotonin in the rat pineal varies with environmental lighting and the time of day.[19,32] Pharmacological studies have demonstrated that this phenomenon results from rhythmic changes in the rate at which the serotonin stored in the pineal is liberated from its binding sites.[33] Throughout the 24-hour day the serotonin is synthesized in the pineal at a more or less constant rate: During the daylight hours most of the amine is apparently bound within the pinealocyte in such a way as to make it invulnerable to destruction by monoamine oxidase; with the onset of darkness this bound serotonin is liberated. Various fractions are then probably either secreted unchanged, deaminated, or converted to melatonin and other methoxyindoles. The physiological mechanism responsible for this serotonin rhythm is described below.

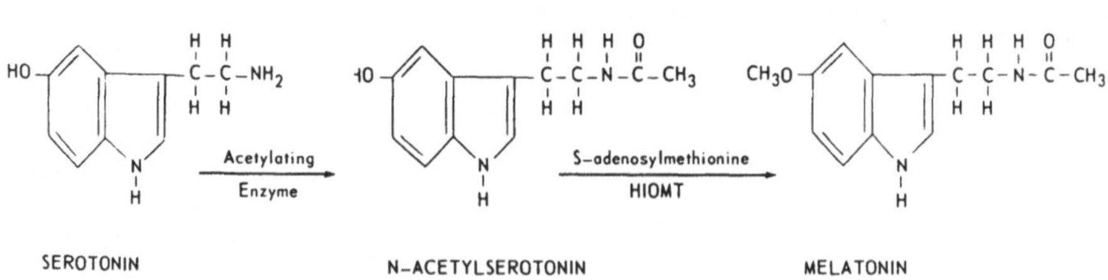

Fig. 1. Biosynthesis of melatonin. A portion of the serotonin in the pineal is *N*-acetylated to form *N*-acetylserotonin. This compound is then *O*-methylated, through the activity of the enzyme HIOMT, to form melatonin. The methyl donor is *S*-adenosylmethionine.

V. SYNTHESIS AND FATE OF MELATONIN

The chemical structure of melatonin isolated from bovine pineal glands was identified by Lerner and his collaborators in 1958.[34–36] These investigators were seeking the pineal compound, first described by McCord and Allen four decades earlier,[37] which blanched amphibian skin when fed to tadpoles or when added to the media in which they were swimming. The identification of melatonin was of crucial importance in the subsequent accumulation of knowledge about the pineal: For the first time, a unique biochemical property could be ascribed to the pineal and could be examined following the experimental modification of pineal function. Soon after the structure of melatonin was reported, Axelrod and Weissbach discovered HIOMT and showed that the pineal gland was the unique locus of this enzyme in mammals[38] (Table II). Melatonin and HIOMT have since been found in the pineal glands of all species studied,[2] and the methoxyindole has also been identified by bioassay techniques in peripheral nerve and human urine[39] (indicating that it is actually secreted from the pineal). However, no simple chemical assay sufficiently sensitive to measure the concentration of melatonin in individual rodent pineal glands has been described; hence, most studies which have used melatonin synthesis as an index of the pineal's functional activity have followed *in vitro* changes in HIOMT activity and assumed that such changes actually mirror alterations in the *in vivo* production of the indole (e.g., Refs. 26 and 40).

The extreme localization of melatonin synthesis to the pineal gland is characteristic only of mammals (Table II). Thus, in the frog, HIOMT activity can be demonstrated in the brain and the eye as well as in the pineal.[2,41] The evolutionary process through which melatonin synthesis became localized to the pineal paralleled other changes described above (i.e., the loss of pineal

TABLE II

Distribution of HIOMT Activity in Various Species

Species	Pineal ($\mu\mu$m/g/hr melatonin-^{14}C formed)	Other
Monkey	5800	—
Cow	2100	—
Rat	10–100[a]	—
Human	700	—
Hen	9000	—
Frog	36	Brain : 11
		Eye : 111
Toad	63	Eye : 55
Turtle	423	Eye : 29

[a] HIOMT activity in the rat varies with time of day and environmental lighting.

photoreception and the development of a pineal sympathetic innervation). It is possible that the functions of pineal methoxyindoles also changed with evolution: In mammals it seems likely that melatonin is released as a hormone[42] and acts at the level of the brain[43] and elsewhere[44] to influence the development and functional activity of the gonads, the pituitary, the thyroid, and perhaps other organs. The fact that several types of nervous tissue besides that in the pineal area can produce melatonin in frogs raises the possibility that this compound may function as a neurotransmitter in some lower vertebrates.

The fate of circulating melatonin has been studied in the rat and the cat.[44-46] In both species the indole disappears rapidly from the circulation; most of it is destroyed in the liver by 6-hydroxylation followed by conjugation with sulfuric or glucuronic acid. Circulating melatonin is taken up by all tissues, including the brain, and is highly concentrated in the pineal and certain endocrine glands.[44] The presence of endogenous melatonin in urine and in peripheral nerve,[39] a tissue which lacks HIOMT activity,[27,42] indicates that some of the indole normally gains access to the bloodstream. It is not known whether melatonin is secreted by the pineal directly into the blood or whether it first enters the cerebrospinal fluid. When radioactively labeled melatonin is placed in the lateral ventricles of rats, the indole concentrates in the midbrain and the hypothalamus.[47]

The high degree of localization of HIOMT to the pineal allows this enzyme to serve as a useful biochemical "marker" for pineal tissue. Significant HIOMT activity has been identified in two pineal tumors, a pinealoma located within the gland itself[48] and an ectopic pinealoma of the hypothalamus.[49] Pineal cancers can be induced by infecting hamster pineal cells in tissue culture with SV 40 or polyoma viruses and implanting this material in other animals; such tumors also contain melatonin-forming activity.[50]

VI. PINEAL BIOCHEMICAL RHYTHMS

When rats are maintained for several days under continuous darkness, their pineals contain several times as much HIOMT activity as those of animals kept under continuous light.[26] This effect of illumination can be observed in animals exposed to the 12-hr periods of light and darkness which occur in nature: Under these conditions pineal HIOMT activity at the end of the light period is only half as great as it is 5 hr after the onset of darkness[51] (Fig. 2). The mechanism responsible for this rhythm in pineal biochemical function appears to differ from those responsible for most daily rhythms (e.g., adrenocortical secretion, eating behavior, physical activity, and body temperature) in its absolute dependence upon cyclic changes in environmental lighting. If the experimental animals are deprived of a single period of light or darkness, this rhythm is rapidly extinguished.[51,52] Cyclic changes in pineal HIOMT activity are also lost if the eyes are removed or if the neural pathway which transmits photic impulses to the pineal is disturbed (e.g., by destroying the superior cervical ganglia).

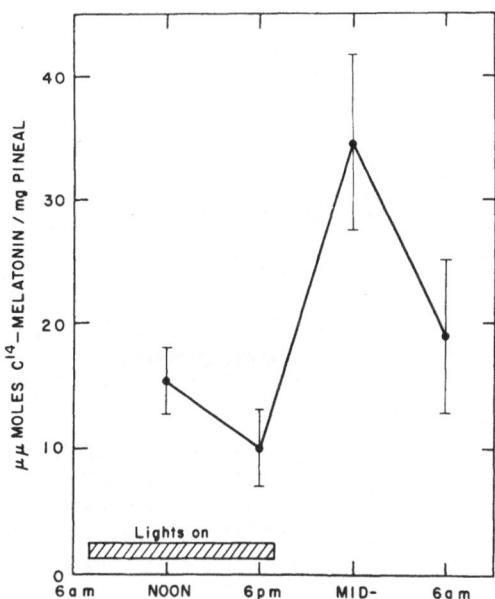

Fig. 2. Daily rhythm in the activity of HIOMT in the rat pineal. Animals were kept under light from 7 AM to 7 PM and were killed at 6-hr intervals.

The demonstration that this pineal rhythm was generated by impulses delivered to the gland through its sympathetic nerves suggested that the turnover of pineal norepinephrine might also undergo cyclic changes. It was subsequently found that the content of norepinephrine in the rat pineal also varies severalfold each day: it rises throughout the daily dark period and falls during the light period[53] (Fig. 3). This rhythm is also lost if animals are blinded or are kept under continuous darkness.[15] It is not known whether the nocturnal rise in pineal norepinephrine content results from increased synthesis, decreased release, or decreased metabolism of the catecholamine. However, the first hypothesis is supported by the observation that the activity of tyrosine hydroxylase in the rat pineal also varies diurnally.[54]

The content of serotonin in the rat pineal is also a function of the time of day; however, this rhythm is almost 180° out of phase with the rhythms in HIOMT activity or norepinephrine content. Serotonin levels are high during the daylight period and fall sharply with the onset of darkness.[19,32] This phase difference suggests that the key event in the daily changes in pineal metabolism is, in fact, the onset of darkness; this change in lighting causes serotonin to be liberated from its bound state and to become available for conversion to melatonin and other methoxyindoles.[33] At the same time of day the activity of the pineal enzyme HIOMT, which changes N-acetyl-serotonin to melatonin, increases, possibly as a result of enzyme induction.[51] At least two explanations can be offered for the photic inhibition of HIOMT activity: (1) Light may cause *increased* sympathetic nervous activity, with the

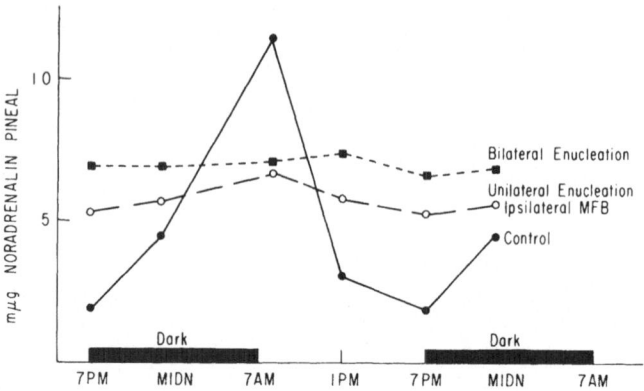

Fig. 3. Daily rhythm in norepinephrine content of rat pineals. Animals were kept under darkness for 12 hr daily and were killed at intervals. No rhythm was observed in animals whose pineals were deprived of a cyclic photic input (i.e., by blinding or by unilateral blinding and ipsilateral transection of the inferior accessory optic tract in the medial forebrain bundle.)

liberation of *more* of a neurotransmitter that *inhibits* HIOMT synthesis or activity. (2) Light may *depress* the activity of the pineal sympathetic nerves, causing the release of *less* of a neurotransmitter which normally *stimulates* HIOMT synthesis or activity. The choice of the proper explanation cannot be made until better data are available as to (1) whether light increases or decreases the tone of the sympathetic nerves which run to the pineal; (2) the nature of the neurotransmitter which is released by these nerves (i.e., norepinephrine or serotonin); and (3) whether this neurotransmitter stimulates or inhibits HIOMT activity.

The weight of the rat pineal is also subject to diurnal fluctuations[51]; this suggests that many biochemical processes in the pineal in addition to those described above depend upon environmental lighting and time of day.

The precise role of the pineal in mammalian physiology is not known; however, it has been suggested that this organ functions as a "neuroendocrine transducer."[2,52] It receives an input of sympathetic nervous impulses (ultimately generated by the retinal response to environmental lighting) and converts these to an output of the hormone melatonin. It is not clear whether melatonin is secreted into the blood or the cerebrospinal fluid, or precisely where the methoxyindole acts to produce its reported endocrine effects. However, it does seem highly likely that, in its role as transducer, the pineal provides the body with a circulating "biologic clock" which is absolutely geared to the environment. Whether or not any other organs normally watch the pineal clock awaits further investigation.

VII. SUMMARY

The pineal has undergone marked changes as vertebrates have evolved from amphibians to mammals: The amphibian organ is a photoreceptor, which sends information about lighting to the brain; the mammalian pineal is a gland which has no direct connection to the central nervous system and whose metabolism is controlled by light via an indirect pathway involving its peripheral sympathetic nerves. In mammals only the pineal contains significant amounts of the enzyme HIOMT, which is necessary to convert N-acetylserotonin to melatonin; in lower vertebrates this enzyme is widely distributed. Pineal serotonin, like the indoleamine in other organs, can be destroyed by oxidative deamination. However, it also can undergo two unique fates: it can be converted to methoxyindoles such as melatonin or it can be taken up and bound within pineal sympathetic nerve endings. Both the ability of the rat pineal to make melatonin and the contents of serotonin and norepinephrine in this organ are subject to wide daily fluctuations. With the onset of darkness serotonin is liberated from binding sites in pineal parenchymal cells, and an increasing fraction of the amine is probably converted to melatonin. Pineal norepinephrine levels also rise during the hours of darkness. The mammalian pineal functions as a "neuroendocrine transducer," which converts a neural signal (generated by the retinal response to environmental lighting) to a hormonal output (melatonin). The pineal thus provides the body with a "biologic clock," which is absolutely geared to the lighting environment.

ACKNOWLEDGMENT

The editorial assistance of Mrs. Arlene Siegel is gratefully acknowledged.

VIII. REFERENCES

1. D. E. Kelly, Pineal organs: photoreception, secretion, and development, *Am. Scientist* **50**:597–625 (1962).
2. R. J. Wurtman, D. E. Kelly, and J. Axelrod, *The Pineal*, Academic Press, New York (1968).
3. D. E. Kelly, in *Progress in Brain Research* (J. A. Kappers and J. P. Shadé, eds.) Vol. 10, pp. 270–287, Elsevier, New York (1965).
4. J. A. Kappers, in *Progress in Brain Research* (J. A. Kappers and J. P. Shadé, eds.) Vol. 10, pp. 87–153, Elsevier, New York (1965).
5. E. Dodt and E. Heerd, Mode of action of pineal nerve fibers in frogs, *J. Neurophysiol.* **25**:405–429 (1962).
6. D. E. Wolfe, in *Progress in Brain Research* (J. A. Kappers and J. P. Shadé, eds.) Vol. 10, pp. 332–388, Elsevier, New York (1965).
7. K. Scharenberg and L. Liss, in *Progress in Brain Research* (J. A. Kappers and J. P. Shadé, eds.) Vol. 10, pp. 193–217, Elsevier, New York (1965).

8. J. A. Kappers, The development, topographical relations and innervation of the epiphysis cerebri in the albino rat, *Z. Zell. Mikroskopisches Anatomie* **52**:163–215 (1965).
9. G. C. T. Kenny, The innervation of the mammalian pineal body, *Proc. of the Australian Association of Neurologists* **3**:133–140 (1965).
10. D. E. Wolfe, L. T. Potter, K. C. Richardson, and J. Axelrod, Localizing tritiated norepinephrine in sympathetic axons by electron microscopic autoradiography, *Science* **138**:440–442 (1962).
11. W. R. Hayhow, C. Webb, and A. Jervie, The accessory optic fiber system in the rat, *J. Comp. Neurol.* **115**:187–200 (1960).
12. R. Y. Moore, A. Heller, R. J. Wurtman, and J. Axelrod, Visual pathway mediating pineal response to environmental light, *Science* **155**:220–223 (1967).
13. R. Y. Moore, A. Heller, R. K. Bhatnagar, R. J. Wurtman, and J. Axelrod, Central control of the pineal gland: visual pathways, *Arch. Neurol.* **18**:208–218 (1968).
14. R. J. Wurtman, J. Axelrod, E. W. Chu, A. Heller, and R. Y. Moore, Medial forebrain bundle lesions: blockade of effects of light on rat gonads and pineal, *Endocrinology* **81**:509–514 (1967).
15. R. J. Wurtman, J. Axelrod, G. Sedvall, and R. Y. Moore, Photic and neural control of the 24-hour norepinephrine rhythm in the rat pineal gland, *J. Pharm. Exp. Therap.* **157**:487–492 (1967).
16. R. J. Wurtman, J. Axelrod, E. W. Chu, and J. E. Fischer, Mediation of some effects of illumination on the rat estrous cycle by the sympathetic nervous system, *Endocrinology* **75**:238–242 (1964).
17. R. J. Wurtman, *in Neuroendocrinology* (L. Martini and W. F. Ganong, eds.) Vol. 2, pp. 20–59, Academic Press, New York (1967).
18. N. J. Giarman, D. X. Freedman, and L. Picard-Ami, Serotonin content of the pineal gland of man and monkey, *Nature* **186**:480–481 (1960).
19. S. H. Snyder, M. Zweig, and J. Axelrod, Control of circadian rhythm in the serotonin content of the rat pineal gland, *Life Sci.* **3**:1175–1179 (1964).
20. C. Owman, *in Progress in Brain Research* (J. A. Kappers and J. P. Shadé, eds.) Vol. 10, pp. 423–453, Elsevier, New York (1965).
21. P. Miline, P. Stern, and S. Hukovic, Sur la presénce de la serotonin dans la glande pinéale, *Bull. Sci., Conseil Acad. RPF Yougoslavie* **4**:75 (1959).
22. R. J. Wurtman, F. Larin, J. Axelrod, H. M. Shein, and K. Rosasco, Formation of melatonin and 5-hydroxyindoleacetic acid from ^{14}C-tryptophan by rat pineal glands in organ culture, *Nature* **217**:953–954 (1968).
23. H. M. Shein, R. J. Wurtman, and J. Axelrod, Synthesis of serotonin by pineal glands of the rat in organ culture, *Nature* **213**:730–731 (1967).
24. W. Lovenberg, E. Jequier, and A. Sjoerdsma, Tryptophan hydroxylation: measurement in pineal gland, brain stem, and carcinoid tumor, *Science* **155**:217 (1967).
25. S. H. Snyder and J. Axelrod, A sensitive assay for 5-hydroxytryptophan decarboxylase, *Biochem. Pharm.* **13**:805–806 (1964).
26. R. J. Wurtman, J. Axelrod, and L. S. Phillips, Melatonin synthesis in the pineal gland: control by light, *Science* **142**:1071–1073 (1963).
27. J. Axelrod and H. Weissbach, Purification and properties of hydroxyindole-*O*-methyl transferase, *J. Biol. Chem.* **236**:211–213 (1961).
28. W. M. McIsaac, G. Farrell, R. G. Taborsky, and A. N. Tayer, Indole compounds: isolation from pineal tissues, *Science* **148**:102–103 (1965).
29. B. Falck, C. Owman, and E. Rosengren, Changes in rat pineal stores of 5-hydroxytryptamine after inhibition of its synthesis or breakdown, *Acta Physiol. Scand.* **67**:300–305 (1966).
30. H. Weissbach, B. G. Redfield, and J. Axelrod, Biosynthesis of melatonin: enzymic conversion of serotonin to *N*-acetylserotonin, *Biochim. Biophys. Acta* **43**:352–353 (1960).

31. J. Axelrod and H. Weissbach, Enzymatic *O*-methylation of *N*-acetylserotonin to melatonin, *Science* **131**:1312 (1960).
32. W. B. Quay, Circadian rhythm in rat pineal serotonin and its modifications by the estrous cycle and photoperiods, *Gen. Comp. Endocrinol.* **3**:473–479 (1963).
33. S. H. Snyder and J. Axelrod, Circadian rhythm in pineal serotonin: effect of monoamine oxidase inhibition and reserpine, *Science* **149**:542–544 (1965).
34. A. B. Lerner, J. D. Case, Y. Takahashi, T. H. Lee, and W. Mori, Isolation of melatonin, the pineal gland factor that lightens melanocytes, *J. Am. Chem. Soc.* **80**:2587 (1958).
35. A. B. Lerner, J. D. Case, and R. V. Heinzelman, Structure of melatonin, *J. Am. Chem. Soc.* **81**:6084 (1959).
36. A. B. Lerner, J. D. Case, and Y. Takahashi, Isolation of melatonin and 5-methoxyindole-3-acetic acid, *J. Biol. Chem.* **235**:1992–1997 (1960).
37. C. P. McCord and F. P. Allen, Evidences associating pineal gland function with alterations in pigmentation, *J. Exp. Zool.* **23**:207–224 (1917).
38. J. Axelrod, P. D. MacLean, R. W. Albers, and H. Weissbach, *in Regional Neurochemistry* (S. S. Kety and J. Elkes, eds.) pp. 307–311, Pergamon Press, Oxford (1962).
39. J. D. Barchas and A. B. Lerner, Localization of melatonin in the nervous system, *J. Neurochem.* **11**:489–491 (1964).
40. R. J. Wurtman, J. Axelrod, and S. H. Snyder, Changes in the enzymatic synthesis of melatonin in the pineal during the estrous cycle, *Endocrinology* **76**:798–800 (1965).
41. J. Axelrod, W. B. Quay, and P. C. Baker, Enzymatic synthesis of the skin-lightening agent, melatonin, in amphibians, *Nature* **208**:386 (1965).
42. R. J. Wurtman, J. Axelrod, and E. W. Chu, Melatonin, a pineal substance: effect on rat ovary, *Science* **141**:277–278 (1963).
43. L. Martini, F. Fraschini, and M. Motta, Neural control of anterior pituitary functions, *Recent Prog. Hormone Res.* **24**:439–496 (1968).
44. R. J. Wurtman, J. Axelrod, and L. T. Potter, The uptake of H³-melatonin in endocrine and nervous tissues and the effects of constant light exposure, *J. Pharm. Exp. Therap.* **143**:314–318 (1964).
45. I. J. Kopin, C. M. B. Pare, J. Axelrod, and H. Weissbach, The fate of melatonin in animals, *J. Biol. Chem.* **236**:3072–3075 (1961).
46. R. G. Taborsky, P. Deloigs, and I. H. Page, 6-Hydroxyindole and the metabolism of melatonin, *J. of Med. Chem.* **8**:855–861 (1965).
47. F. Anton-Tay and R. J. Wurtman, Unpublished observations (1968).
48. R. J. Wurtman, J. Axelrod, and R. Toch, Demonstration of hydroxyindole-*O*-methyl transferase, melatonin, and serotonin in a metastatic parenchymatous pinealoma, *Nature* **204**:1323–1324 (1964).
49. R. J. Wurtman and H. Kammer, Melatonin synthesis by an ectopic pinealoma, *New Engl. J. Med.* **274**:1233–1237 (1966).
50. S. A. Wells, R. J. Wurtman, and H. S. Rabson, Viral neoplastic transformation of hamster pineal cells *in vitro*: retention of enzymatic function, *Science* **154**:278–279 (1966).
51. J. Axelrod, R. J. Wurtman, and S. H. Snyder, Control of hydroxyindole-*O*-methyl transferase activity in the rat pineal gland by environmental lighting. *J. Biol. Chem.* **240**:949–955 (1965).
52. R. J. Wurtman and J. Axelrod, The pineal gland, *Sci. Am.* **213**:50–60 (1965).
53. R. J. Wurtman and J. Axelrod, A 24-hour rhythm in the content of norepinephrine in the pineal and salivary glands of the rat, *Life Sci.* **5**:665–669 (1966).
54. E. G. McGeer and P. L. McGeer, Circadian rhythm in tyrosine hydroxylase, *Science* **153**:73–74 (1966).

Chapter 20

THE HYPOTHALAMO-HYPOPHYSEAL COMPLEX

Max Reiss

Neuroendocrine Research Unit
Willowbrook State School
Staten Island, New York

I. INTRODUCTION

The hypothalamo–hypophyseal complex forms the center piece of a regulatory system that, as far as is known at present, plays a decisive role in maintaining homeostasis and equilibrium of many physiological functions. The ability of the body to comply with the various demands of everyday life depends on the normal functioning of this neuroendocrine system. The physiological actions of the system are either set off by some environmental stimuli or by feedback systems existing between the hormones and the metabolism of the body periphery, receptor compartments in the brain and their influence on the pituitary hypothalamic system.

It has recently become clearer that the hypothalamo–hypophyseal system has an influence on practically all phases of the body's metabolism. Whatever biological process has been thoroughly investigated, including molecular reactions, the primary role of the hypothalamo–hypophyseal system has been established. At present less is known about its influence on the brain than on the other organs of the body.

II. THE SIGNIFICANCE OF THE HYPOTHALAMO– HYPOPHYSEAL COMPLEX IN THE HOMEOSTASIS OF THE BODY

Figure 1 might be a suitable illustration of the general principles concerned.

The hypothalamus acts as a receptor organ for stimuli originating in higher brain parts, from the body periphery, and therefore also from various environmental demands.

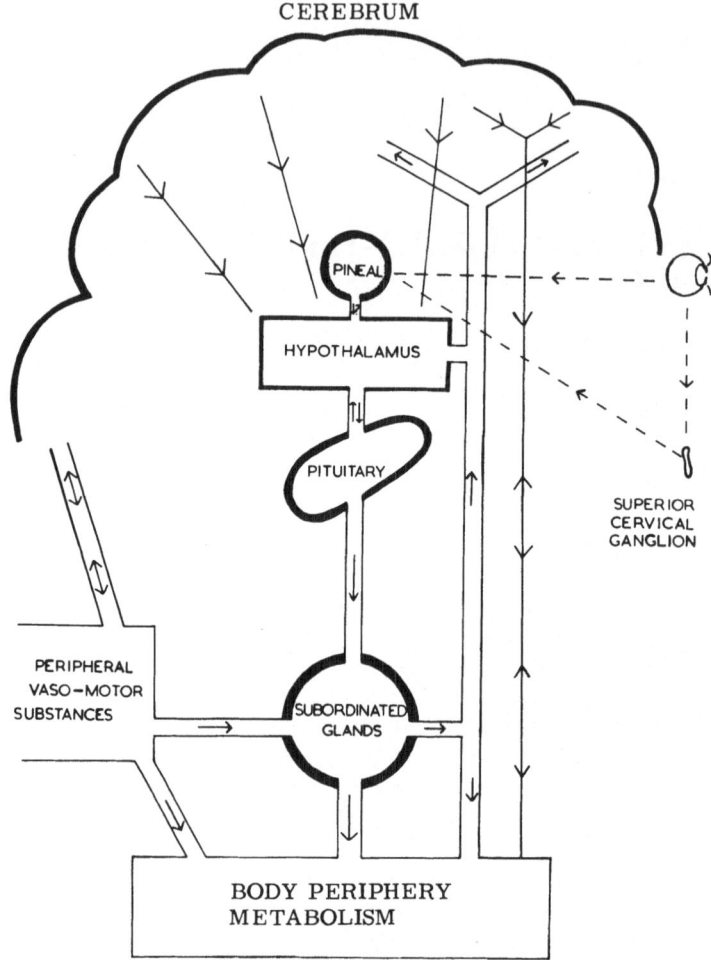

CEREBRUM

Fig. 1. (For explanation see text.)

At the same time the hypothalamus regulates the function of the pituitary gland in maintaining the homeostasis of the body. This regulation takes place either by means of various release factors produced in the hypothalamus, which reach the pituitary anterior lobe via the portal vessels, or by nerve tracks, which, similar to the portal vessels, descend with the pituitary stalk to the posterior lobe.

The pituitary anterior lobe produces several hormones which stimulate increased production in the subordinated endocrine glands such as thyroid, adrenals, and gonads. They also have some side effects not directly connected with the target gland. It also produces growth hormone which acts directly on practically all organs of the body and lactogenic hormone which stimulates mainly the mammary glands.

The posterior lobe is mainly a storage organ for hormones produced in the hypothalamus, and various somatic conditions cause their release.

One of the most interesting features of the neuroendocrine interrelation is the multitude of feedback systems which exist. The most characteristic examples can be seen when insufficient amounts of sex hormones, corticoids, or thyroid hormones are circulating in the body and the hypothalamus produces more release factors to stimulate the production of whichever tropic hormone is needed. On the other hand, when there is an overproduction of hormone from one of these peripheral glands, or the hormone is applied externally, the circulation of the relevant pituitary tropic hormone is reduced, or suppressed.

The influence of the various vasomotor substances, produced in the periphery, on the hypothalamus and higher brain centers is still not sufficiently circumscribed.

There is no doubt that the pineal gland which is now definitely recognized as an endocrine organ innervated by sympathetic (myelin free) nerve fibers only, interferes with the hypothalamo-hypophyseal system by being an antagonist of the pituitary. The pineal plays an essential role in homeostasis, reacting to light stimuli which reach it via the ganglion cervicale superior (see Chapter 19 of this handbook). It also produces hormones which are antagonistic to some of the pituitary anterior lobe hormones, to the melanocyte-stimulating hormone of the intermediate lobe and oxytocin of the posterior lobe.[9]

III. BIOLOGICAL CRITERIA AND PROOF FOR THE PRODUCTION OF RELEASE FACTORS IN THE HYPOTHALAMUS

For relevant review articles and surveys see Refs. 1–16.

A. Interrelation Between the Hypothalamus and Pituitary Posterior Lobe

1. Hormone Production

The idea that the hormones originally found only in the posterior lobe of the pituitary might also be produced in the hypothalamus was first proposed by Trendelenburg and Sato[17] in 1928. They made the observation that in dogs after hypophysectomy the tuber cinereum hypertrophied and contained large amounts of pituitary posterior lobe hormones (oxytocic pressor and antidiuretic activity). For the next 20 years these statements did not find any response among research workers. Only many years later, after bioassay methods were considerably improved, was it found that the hypothalamus under normal conditions contains only 0.5 to 2 % (Heller)[6] of the amount of active substances found in the posterior lobe. This was the basis for the assumption that oxytocin and vasopressin are produced in the hypothalamus and stored in the pituitary posterior lobe, the release of these substances from the posterior lobe depending apparently on the demands of the body and its environmental conditions. The first relevant example was presented by Gersh[18] who found a decrease of the antidiuretic factor of the neuro-hypophysis and an increase in its concentration in the body fluids in states of water deprivation. Verney[19–20] showed that the osmotic pressure of blood of the internal carotid artery is responsible for the release of vasopressin that contains the antidiuretic principle.

Heller[21] infused dogs with large volumes of 3 % NaCl solution, 3 % of the body weight, and found that after this change in osmotic conditions there was maximal vasopressin concentration in the blood. Lederis[22] found increased vasopressin release after dehydration (feeding animals with dry food). Vasopressin is also released after pain, shock, and hemorrhage.

Oxytocin release seems to be connected with parturition or lactation. In suckling, for instance, there is a considerable release of oxytocin,[23–24] which has the ability to contract the mammary myoepithelium, thus causing the ejection of milk.

Sawyer[27] and Heller[7] have studied the phylogenesis of the posterior lobe hormones. Vasotocin, a hormone that is produced mainly for the regulation of water and mineral metabolism, seems to be phylogenetically the oldest hormone isolated in all vertebrates except the elasmobranchs and the mammals. It is mainly antidiuretic and differs from vasopressin (see Table I) only inasfar as it contained Ile in position 3 instead of the Phe as in arginine vasopressin. Heller also isolated chromatographically a further hormone, the ichthyotocin that was originally regarded as oxytocin but was proved, on the basis of quantitative bioassays and in its chemistry, to differ from oxytocin. Heller and his co-workers isolated a further hormone in extracts of Neo-ceratodus, 8-isoleucin-oxytocin which in its quantitative biological test differs from ichthyotocin and oxytocin. At the moment one can only assume that the differing environmental demands under which the various species live produce biochemical mutations which are in accordance with the changing homeostatic needs of the various species. It is not understandable at present why arginine vasopressin occurs in many species, including man, while lysine vasopressin is found only in pituitaries of pigs and hippopotami.[8]

Rothballer[14] has carried out extensive investigations of the effect of stimulation of various brain parts on the release of vasopressin and oxytocin. Heller[8] surveyed the afferent pathways to the hypothalamus of rabbits, rats, cats, and goats which stimulate the production of oxytocin and vasopressin, the supraoptic nucleus being mainly responsible for the production of vaso-pressin, and the paraventricular nucleus for that of oxytocin.

2. Biological Actions of the Hormones

Oxytocin produces uterine contractions in vivo and in vitro. The response depends on the status of the uterus musculature. After castration, for instance, rats or rabbits show only a small contraction after treatment with oxytocin, while treatment with estrogen increases considerably its muscular thickness and contractility. The oxytocin sensitivity of rats in estrus is higher than during diestrus. Treatment with progesterone reduces the responsitivity of the uterus to oxytocin. That is also the case during pregnancy when progesterone is dominant. Only at the end of pregnancy when the circulation rate of progesterone is reduced does the uterus become more sensitive to the action of oxytocin. The latter intensifies the uterine contractions during delivery of the fetus.

In lactating animals oxytocin causes milk ejection by contraction of the myoephithelial cells around the alveoli; suckling can evoke this response by stimulation of endogenous production of oxytocin in the hypothalamus.

The rise in blood pressure by vasopressin comes about by stimulation of smooth muscle elements in the blood vessels. Owing to the contraction of the coronary heart vessels, one can often, in animal experiments, observe a slight and transient decrease in blood pressure after the initial rise.

The antidiuretic effect in birds, reptiles, and amphibians is due to vasotocin, which differs chemically from vasopressin, in position 3. Vasotocin also increases water absorption from the bladder and water movement through the skin. Frogs become heavier after injection of pituitary posterior lobe extracts (vasotocin?) owing to water retention while pitressin does not, apparently, have this effect.[7] Cyclostomes, elasmobranchs, or bony fishes do not show any antidiuretic reaction to pituitary posterior lobe extracts.

In higher vertebrates the antidiuretic effect is produced by vasopressin. The mechanism of this action, and the area of the kidney on which the vasopressin acts, is not yet completely clarified.

All these biological actions are now used as the basis for biological assay of the hormones. An instructive table on the quantitative sensitivity of the various test methods has been compiled by Zarrow et al.[25]

A number of authors are of the opinion that oxytocin and vasopressin are release factors for pituitary anterior lobe hormones. This question is discussed on page 474.

As far as the posterior lobe hormones in humans are concerned, it has been known for many years that the blood of women in labor contains considerably increased amounts of oxytocin. It was found at this time that the oxytocin content in blood started to rise about the third month of pregnancy, but its action on the uterus was inhibited by the presence of progestin. It was also reported that immediately after birth the oxytocin level decreased sharply.

A number of authors have suggested that states of eclampsia are due to a pathological increase in oxytocin, vasopressin, and antidiuretic components in the blood of pregnant women.

Diabetes insipidus does not occur after hypophysectomy, only after traumatic or other damage to the hypothalamus. This again suggests strongly that the antidiuretic substance is produced in the hypothalamus and only stored in the posterior lobe.

TABLE 1

Amino Acid Sequences in Neurohypophyseal Hormones

	Oxytocin	Arginine Vasopressin	Lysine Vasopressin	Arginine Vasotocin	8-Ile-Oxytocin (Mesotocin)	4-Ser-8-Ile-Oxytocin (Ichthyotocin)
1.	Cys	Cys	Cys	Cys	Cys	Cys
2.	Tyr	Tyr	Tyr	Tyr	Tyr	Tyr
3.	Ile	Phe	Phe	Ile	Ile	Ile
4.	Glu(NH₂)	Glu(NH₂)	Glu(NH₂)	Glu(NH₂)	Glu(NH₂)	Ser
5.	Asp(NH₂)	Asp(NH₂)	Asp(NH₂)	Asp(NH₂)	Asp(NH₂)	Asp(NH₂)
6.	CyS	CyS	CyS	CyS	CyS	CyS
7.	Pro	Pro	Pro	Pro	Pro	Pro
8.	Leu	Arg	Lys	Arg	Ile	Ile
9.	Gly(NH₂)	Gly(NH₂)	Gly(NH₂)	Gly(NH₂)	Gly(NH₂)	Gly(NH₂)

3. Chemistry of the Hormones

The hormones were also prepared from the posterior lobes by various simple extraction methods. For the purification, various combinations of chromatographic methods, countercurrent distribution, and zone electrophoresis were used.

Vigneaud and his co-workers[26–29] have isolated and successfully synthesized the pituitary posterior lobe hormones from the hypothalamus and the pituitary posterior lobe.

The hormones are octapeptides and all have a pentapeptide ring closed by a disulfide link. From this ring branches off a tripeptide side chain. Changes in the composition of these hormones occur mainly in positions 3 and 8 (see Table I). Substitution of the various amino acids in these positions are responsible for changes in the biological action. The intactness of the pentapeptide ring is essential for biological action; any chemical change of this ring is connected with a loss of biological activity. The strongest biological action is dependent on the structural formations described in Table I. Changes produced, for instance, by acetylation of the free amino groups or prolongation of the end chain diminish biological activity considerably.

B. Interrelation between the Hypothalamus and Pituitary Anterior Lobe

1. Historical Remarks

Increased environmental demands, various psychological and physiological stresses, states of mental and physical deprivation, variations in light and temperature, can change the production and concentration in the body fluids of the pituitary anterior lobe hormones. These facts led to the conclusion that the various receptor areas in the nervous system play a decisive role in the pituitary anterior lobe function.

There was much controversy concerning the mechanics of stimulation of the pituitary anterior lobe. This was mainly due to the many discrepancies in the results seen after pituitary stalk transsection. The majority of the workers assumed that stimulation of the anterior lobe took place via the nervous pathways coming in the stalk from the hypothalamus to the pituitary anterior lobe. Considerable time was needed before the full significance of the portal system was realized. The portal system starts as capillaries in the median eminence of the hypothalamus and the tuber cinereum, forms vascular trunks on the pituitary stalk, and ends in the sinusoids of the anterior lobe. These blood vessels can easily regenerate after stalk transsection, so that some authors assumed that stalk transsection had no influence on the function of the anterior lobe. The whole question was finally clarified by experiments carried out by Harris[30–31] who, by implantation of small unabsorbable paper plates after stalk transsection, prevented the regeneration of the portal system.

2. Basic Evidence for the Production of Hypothalamic Release Factors for the Secretion of the Pituitary Anterior Lobe Hormones: Investigational Approaches

Destruction of the median eminence of the hypothalamus produces changes comparable to those of hypophysectomy. After transplantation of the pituitary anterior lobe into other parts of the body, its action is very much reduced but not if it is transplanted in the near vicinity of the median eminence. These facts made the hypothalamus–pituitary anterior lobe interrelation much more probable. Final proof was reached when the role of the portal vessels became established. It was found possible to stimulate the pituitary anterior lobe activity with blood taken from the portal vessels or by extracts produced from median eminences themselves. The latter extracts could also, when injected into the portal vessel system, or in the pituitaries themselves, stimulate the pituitary anterior lobe hormone production.

It was found later that such extracts could not only increase the hormone concentrations of pituitary anterior lobe hormones in the blood, but also of the target glands. These actions could not be observed after hypophysectomy.

Finally, it was found that addition of median eminence or their extracts to pituitaries *in vitro* could stimulate considerably the production of the pituitary anterior lobe hormones.

3. Growth Hormone (Somatotropic Hormone) Release Factor (SRF)

The main proof for the existence of a growth hormone release factor is based on experiments in which it was found that extracts of the median eminence increased the release of growth hormone from pituitaries incubated *in vitro*.[32–35] Further, a long time after hypophysectomy of rats considerable amounts of SRF can be found in the hypothalamus[36] which are similar to the increased posterior lobe hormone content in the hypothalamus of dogs after hypophysectomy.[17] A depletion of pituitary growth hormone was also seen after intraperitoneal injection of crude bovine hypothalamus extracts.

Since the new radioimmunoassay for human growth hormone has become routine, it has been shown how a number of treatments can increase the growth hormone concentration in the blood. This increase is apparently due to stimulation of the hypothalamic release factor. The insulin response test, which produces, 30–60 min after insulin injection, an increase in the growth hormone concentration of sometimes several hundreds percent[37–39] is obviously due to a stimulation of the hypothalamic production of SRF. Such an acute increase is sometimes produced by injection of chorionic gonadotropin which also maintains a high growth hormone content in the blood 3–4 weeks after treatment.[40] This action also is most probably due to stimulation of the hypothalamus.

Hypoglycemia, as such, produced by insulin treatment does not seem to be the only stimulus for increased growth hormone production. There is no hypoglycemia connected with the stimulation seen after chorionic gonadotropin.[40]

The best example of the significance of environmental stimulation of the SRF is shown in the recently published paper on the state of "deprivation dwarfism." This state is characterized by extremely short stature and voracious appetite. The condition was apparently connected with the experience of severe emotional and psychological stress in the home. After change in environment, such as placing the patients in other surroundings (foster homes or hospital) the children started to grow at an increased rate. The growth hormone concentration in their blood also increased considerably.[41-42]

Numerous attempts were already made to purify the growth hormone release factor from hypothalamus extracts.[43-47]

4. Thyrotropic Release Factor (TRF)

A TSH hypothalamic interrelation was claimed many years ago but the TSH release factor was only prepared and purified after suitable biological assay methods were elaborated. Houssay,[48] for instance, observed a decrease in the height of the thyroid epithelium after hypothalamic injury which was similar to the atrophy seen after hypophysectomy, while Uotila[49] showed the absence of cold hyperplasia in the thyroid after dissection of the pituitary stalk.

The most convincing evidence comes from investigations carried out *in vitro* which showed how the TSH production of anterior pituitaries can be stimulated to a degree that permits the measurement of TSH concentration by bioassay.[50-51] Particularly instructive is a report[52] that the TSH release seen after incubation with hypothalamic extracts can be inhibited by addition of thyroxine (feedback action).

The existence of the TSH release factor will explain why patients with severe mental stress can sometimes produce in a very short time well-developed symptoms of hyperthyroidism which can be treated successfully with thyroxine. The inhibitory feedback action of thyroxine is well known in the clinical therapy of hyperthyroidism, and is apparently based on the inhibition of the production of the hypothalamic release factor.[52]

See Refs. 53–58 and Section III, B.8 for the chemistry of the TRF.

5. Gonadotropin Release Factors (LRF and FSHRF)

A central regulation of the anterior pituitary lobe's production of gonadotropin was assumed after the observations that follicle growth in the rabbit ovary only started after copulation, that a state of pseudopregnancy could be produced by artificial stimulation of the uterus orifice, and that the ovulation could be prevented by hypophysectomy.

Sexual prematurity, as often observed after encephalitis, was explained by stimuli coming from the basis of the brain to the pituitary.[59] The frequent occurrence of menstrual disturbances in women after psychic trauma also points to an important regulatory function of the central nervous system in the production of gonadotropins (see also Ref. 60).

The decisive proof for the existence and production of gonadotropin release factors was supplied by Harris[61] who produced follicular rupture in rabbits with 3 min of electric stimulation of the tuber cinereum and hypothalamus, while stimulation of the pituitary anterior lobe, even when lasting several hours, was without result.[61] More recently it was shown that electric stimulation of the median eminence of the hypothalamus for 10 sec could even produce changes in ovaries which had been transplanted into male castrated rats.[62–63] The importance of the portal vascular system leading to the pituitary was underlined by the observations that after transsection of the pituitary stalk the gonadal disturbance was only transient, and that the ovaries started to function normally again as soon as the portal vessels regenerated. However, if the regeneration of the portal vessels is prevented by putting a nonabsorbable little paper plate on the place where the transsection was performed, the ovaries do not regain their functional ability.[61,63]

Progress in methods of bioassay and chemical separation of polypeptides in hypothalamic extracts has, during the past years, supplied us with more evidence of the existence of separate release factors for the luteinizing hormone (LRF) and for the follicular-stimulating hormone (FSHRF) produced in the anterior pituitary lobe.

McCann and Dhariwal[13] were successful in showing the action of the luteinizing hormone release factor (LRF) by using the ovarian ascorbic acid depletion method of Parlow.[64] The fact that LRF is thermostable while luteinizing hormone is destroyed by heat was a very helpful factor in their physiological investigations. Using the Parlow method they were able to achieve a very good dose-response curve of LRF concentration in extracts of the median eminence of the hypothalamus.[65]

A further proof for the existence of LRF was shown by experiments in which the spontaneous production in rats of LH was prevented by hypothalamic lesion. Replacement therapy, using extracts of sheep hypothalamic tissue, was able to repair the disturbance.[66]

A good parallel to the above-mentioned increase in neurohypophyseal hormone content of the hypothalamus after hypophysectomy was seen when plasma from long-term hypophysectomized rats was investigated and found to contain considerable amounts of LRF, while the gonads were atrophic. As soon as the hypothalamus of these animals, however, was destroyed by electrolytic lesions, the increased LRF content of the plasma disappeared.[67] Finally, it was also shown that the addition of hypothalamic extracts to incubated rat's pituitary anterior lobe tissue, in vitro, increased the production of gonadotropin.[68]

The known biological assay methods for follicle-stimulating hormone were not sensitive enough and progress in this field was slow. Igarashi and McCann[69] developed a mouse uterine weight augmentation method which was claimed to be ten to twenty times more sensitive than previously developed methods. They were able to show with this method that crude acidic extracts of rat or bovine median eminences elevated plasma FSH within 10 min after intravenous injection, while plasma of hypophysectomized rats treated with the same extracts was inactive. Further proof for the production of a FSHRE in the hypothalamus was seen in the *in vitro* experimental results of Mittler and Meites[70] who showed increased release of FSH into the medium of pituitary incubates after adding hypothalamic extracts.

6. Prolactin Inhibiting Factor (PIF)

The growth and lactation-promoting action of estrogen, progesterone, and prolactin on the mammary gland is well known. In the section on hormones of the neurohypophysis the milk-ejecting activity of oxytocin was described.

In the literature a great number of experiments are described in which the milk production is increased. It was therefore very interesting to find that the hypothalamus produces a hormone that decreases the prolactin content of the pituitary and also the prolactin production of the pituitary. After stalk transsection milk production is increased. It can be assumed that this inhibitory factor is produced in order to prevent certain exhaustive states that can occur with too much suckling and overproduction of milk.

The hypothalamic production of a prolactin inhibitor could be demonstrated in both in *in vitro* and *in vivo* experiments.[13,71-73]

7. Corticotropin Release Factor (CRF)

The alarm reaction and adaptation theory of Selye has triggered off thousands of investigations into the actions of corticotropin, and the various environmental and endogenous conditions stimulating its production. This whole field, however, appears at present to be very chaotic. It is, for instance, possible that an autonomic transmitter such as adrenaline stimulates production of ACTH. Large series of experiments[74] pointed to the possibility of adrenaline being the essential release factor since it acted even on pituitaries implanted into the anterior chamber of the eye. The adrenaline action was, however, found to differ in various species and to be absent in man. Significantly increased corticosteroid content of blood is, for instance, usually seen in patients after major operations but need not be accompanied by increased adrenaline concentration.[75]

It seems advisable to restrict this survey to the hypothalamic pituitary interrelation, leaving the possibility open that there might perhaps exist also some other regulatory mechanisms which control ACTH release from the pituitary anterior lobe. There is, of course, little doubt that the inhibition of ACTH output by increased corticosteroid concentrations in blood might always be due to the same feedback mechanism.

The first definite evidence of hypothalamic control of ACTH secretion was supplied by Groot and Harris[76] who observed, after electric stimulation of the tuber cinereum, lymphopenia which was regarded as the result of increased ACTH excretion. Slusher and Hyde[77] investigated the effect of electrical stimulation on ACTH release by measuring ACTH levels in the jugular vein with a very sensitive biological assay method, namely the change in ascorbic acid concentration in the venous adrenal blood of hypophysecto-mized rats. They observed increased ACTH excretion after stimulation of various sites in the dorsal midbrain and of several areas in the diencephalon. They also found that the midbrain contains centers which, after electrical stimulation, inhibit the release of ACTH normally induced by stimulation of the hypothalamus. A number of authors found that after hypothalamic lesion in the tuber cinereum various stress conditions no longer stimulate increased ACTH secretion.[78-81]

Further proof for regulation of the corticotropin secretion has been supplied by testing hypothalamic extracts which stimulated the ACTH secretion rate. These extracts, however, were not fully purified and a new problem arose on observation that vasopressin can stimulate ACTH release. Many investigators assumed vasopressin was actually the corticotropin release factor, particularly when it was shown that it could produce adrenal ascorbic acid depletion in rats with chronic lesions in the median eminence.[82] However, in other experiments it was shown that the release of antidiuretic hormone could be inhibited by an overdose of water while ACTH release is stimulated at the same time.[83-84] Furthermore, it was shown that in hypo-thalamic extracts, the CRF can be inactivated by enzymes which leave the antidiuretic activity unchanged.[85] Further evidence for the existence of a separate CRF was supplied by experiments[86] in which hypotonic saline infused in the carotid of dogs produced antidiuresis, while the blood cortisone concentration was unchanged.

Saffron et al.[87] have studied the action of CRF on isolated pituitary tissue in vitro and established the existence of a CRF that does not contain vasopressin. The same results were also achieved after using pituitary tissue culture.[88] In all these experiments highly purified vasopressin was without effect. The authors were also able to separate a pure CRF from crude vaso-pressin preparations by paper chromatography.

McCann and Dhariwal[13] are of the opinion that vasopressin activity in an ACTH release is not a biological but rather a pharmacological effect of high vasopressin dose, and that vasopressin in a sufficiently high dosage can stimulate almost any endocrine gland directly (adrenal cortex, ovary, and thyroid).

8. Chemistry of the Release Factors

The establishment of the production of pituitary release factors in the hypothalamus has opened a new field of brain chemistry. At the moment the presence of these substances is being investigated in the hypothalamus, but the possibility of their presence in other brain compartments has been investigated very rarely. All of the factors investigated so far are heat stable, which is, with the exception of ACTH, contrary to the qualities of the pituitary anterior lobe hormones. The advantage of this is that the release factors can be easily separated from the latter so that the question of con-tamination with pituitary anterior lobe does not arise in biological testing.

However, most of these release factors can be inactivated by pepsin or trypsin. Most of the authors assume that they are small polypeptides which differ from vasopressin and oxytocin by the absence of the disulfide link which has been found between position 1 and 6 in the neurohypophyseal hormones. This is concluded from the fact that thioglycollic acid treatment, which inactivates vasopressin and oxytocin, has no influence on anterior lobe release factors, particularly CRF and LRF. It is therefore assumed that, in the anterior lobe release factor, we are dealing with a new family of poly-peptides produced in the hypothalamus.

The greatest difficulty in the chemical characterization of the release factors is their very low concentration in the hypothalamus. Guillemin et al.[89] for instance, had to work up 55 kg of sheep hypothalami in order to isolate about 0.5 mg of TRF. They used many steps for the fractionation, leading finally to gel filtration, countercurrent distribution, ion-exchange chromatography, and thin-layer chromatography on cellulose. The result was that the purified material contained nine components, of which only one had TRF activity. The TRF had the following amino acid sequence: Lys, His, Thre, Ser, Glu, Pro, Gly, Ala, Met, Leu, Try. Schally et al.[90] who carried out similar purification steps but used pig hypothalami found another amino acid sequence for TRF, namely, Gly, His, Pro, Thre, Leu, Ser, Lys. Thus, unless one assumes far-reaching differences between the TRF in various species, it must be obvious that the chemistry of TRF is not yet established.

The amino acid sequences found in bovine LRF[91] and the ovine LRF are in better agreement.

Bovine LRF: Asp Glu Gly Ala Lys His Arg Thre Ser Pro Leu

Ovine LRF: Asp Glu Gly Ala Lys His Arg Thre Ser — —

It is probable that the final chemical structure of these release factors will be established only after successful synthesis. Schally and Bowers[92] describe procedures (carried out partly with Guillemin) for separation of vasopressin from CRF. These were chromatography on carboxymethyl cellulose or IRC-50, column electrophoresis, countercurrent distribution, and molecular siev-ing on Sephadex. They found CRF to have a larger molecular weight than vaso-pressin and describe the following amino acid sequences for the two hormones.

Vasopressin   Cys-Phe-Tyr-Asp-Glu-Pro-Lys(or Arg)-Gly

Corticotropin RF Cys-Phe-Tyr-Asp-Glu-Pro-Lys(or Arg)-Gly-Ser-His-Ala
    CRF.

The CRF prepared in this way did not show any vasopressin activity in the biological tests and the Lys vasopressin was devoid of CRF activity. The amino acid sequence of the vasopressin differs in several points from that described in Table I. It would be interesting to see how addition of the last three amino acids of CRF to the vasopressin would change its biological activity. Many such investigations are still going on in the laboratories of Guillemin, Schally, and Saffron, and cannot be described here in detail.

In a recent communication, however, Guillemin[93] questioned the classically accepted concept of the simple polypeptide nature of hypothalamic releasing factors. He did this on the basis of quantitative studies during purification of TRF, while trying to establish the stability of TRF to incubation with proteolytic enzymes. According to Guillemin, it seems to be very probable that, in the case of TRF at least, we are not dealing with a simple polypeptide. "Studies with high resolution nuclear magnetic resonance spectra in particular seem to indicate that we are dealing with a highly saturated heterocyclic or alicyclic molecule with peripheral CH_3 groups."

The last results in particular confirm the opinion that the biochemistry of the hypothalamic release factors is still far from being established. The main shortcomings at the moment are (1) the release factors are present in only minute amounts in the hypothalamic region and thousands of hypothalami have to be collected in order to obtain a small quantity of the substance—just enough to carry out one or two biological tests and attempts to define the structure. (2) Even the most sensitive biological tests necessitate the use of relatively considerable amounts of "purified" extract. It is therefore hardly possible to also investigate the existence of, or contamination with, another of the release factors. The most ideal investigation would consist of *in vitro* assays in which pituitary anterior lobes are stimulated by one of the assumedly purified release factors, and the content of the pituitary anterior lobe hormones in the suspension fluid tested simultaneously for their content of growth hormone, thyrotropic hormone, corticotropin, and gonadotropic hormone. (3) The individual chemical structure of the various release factors will only be established when synthetically produced substances have the same action as the biologically produced release factor. (4) It is still possible at this juncture to speculate that all the activities of the release factors might be contained in one large molecule, the composition of which is variable. Such a consideration could be supported by the fact that in emergency situations, and also during everyday life, simultaneous production of all the release factors can become necessary. It might be relevant to mention here the recent investigations of Mason et al.[94] They measured the hormonal response in male Rhesus monkeys following their initial placement in a restraining chair. The responses appear to be organized in a reproducible pattern comprising two major response subgroups defined on the basis of response direction. Epinephrine, corticosteroid, growth hormone, thyroxine (BEI), and norepinephrine levels rise initially while insulin, testosterone, androsterone, etiocholanolane, estrone, estradiol, and estriol levels fall initially. It is concluded that endocrine regulations are not revealed by the study of one endocrine system at a time.

C. Pituitary Anterior Lobe Hormones

1. Historical Remark

A decade ago we regarded the pituitary anterior lobe as conductor of a large orchestra, each player being represented by a number of target glands and metabolic functions in the periphery. The target glands were stimulated to produce their specific hormone which in turn influenced a multitude of metabolic functions. The other metabolic functions were influenced directly from the anterior pituitary lobe by stimulation or inhibition of certain enzyme systems which, as can be seen as our knowledge increases, influence and regulate the greater number of our vital processes.

With the progressing knowledge of the hypothalamic release factors, however, we see that the pituitary anterior lobe is controlled by higher centers situated in the brain.

If anything is disturbed in the peripheral endocrine function, signals (feedback) go out to the original regulators of the pituitary anterior lobe which, in due course, either increase or reduce the relevant pituitary function so that the homeostasis necessary for the normal life process is maintained.

2. Growth Hormone

Growth hormone is a polypeptide which stimulates longitudinal bone growth and also causes growth in many other tissues of the body. It is generally assumed to be produced in the eosinophil cells and the regulation of its excretion rate was discussed in connection with the growth hormone release factor. These actions are mainly due to the anabolic function of growth hormone and its role in the protein metabolism. It is even being claimed that the action of various anabolic steroids, and particularly of those mobilized by chorionic gonadotropin, act via stimulation of growth hormone production.[40,95]

After hypophysectomy, many parameters of the metabolism of various organs are considerably decreased. The growth hormone influences and regulates many phases of the protein carbohydrate and fat metabolism. These actions can be seen *in vivo* and *in vitro*.

Growth hormone, for instance, stimulates considerably the accumulation of α-aminoisobutyric acid in the diaphragm of hypophysectomized rats[96] both when the hormone is injected *in vivo*, prior to incubation, or added *in vitro*. At the moment there is no explanation of why this stimulation of an amino acid uptake takes place in the organs of hypophysectomized animals but not in those of normal animals. Similar experiments were carried out by a number of other authors (see Ref. 97). Such results illustrate mainly the influence of growth hormone on the transfer across cell membrane to various organs but do not supply information about its role in metabolism (protein synthesis). Growth hormone can stimulate, however, the uptake of glycine-^{14}C, *l*-aminocyclopentanecarboxylic acid-^{14}C, L-alanine, and L-histidine by diaphragms of hypophysectomized rats.[98–102] The incorporation of *l*-phenylalanine-^{14}C in rat liver and brain protein was found *in vivo* to be considerably decreased 7 days after hypophysectomy and could be increased again by treatment with growth hormone. Investigation of the protein fraction showed an increase in the tyrosine phenylalanine ratio.[103] In normal equally treated rats these changes either did not occur or took place to a much smaller degree than in the hypophysectomized animals.

The influence of growth hormone was also shown in cell-free systems. The uptake of leucine-^{14}C into cell-free liver systems of rats was considerably reduced after hypophysectomy and could be increased again by treatment of the hypophysectomized rats with growth hormone.[104] These observed changes were due to the effect of growth hormone on the ribosomes.[105]

Our increased knowledge of the action of growth hormone upon RNA seems to be of great importance to the understanding of the mechanism of protein metabolism. It is well known that nucleic acids participate in several stages in the synthesis of proteins. A few facts are already known, namely, that growth hormone increases the synthesis of messenger and ribosomal nucleic acid which, after hypophysectomy, is decreased.[106–112] In preliminary investigations of brain tissue, the RNA was found to be diminished after hypophysectomy and increased after treatment with growth hormone, while the DNA concentration remained unchanged (in preparation). Extensive investigation of RNA metabolism in brain and brain parts is of greatest importance since in recent years much data have been accumulated on the relation between RNA, learning, memory, and conditioned behavior.[113–118] It has also been shown that prenatal treatment with growth hormone stimulates the proliferation of cortical neurons.[119]

One result of the increased uptake of amino acids in various tissues of the body is the considerable decrease in their concentration in plasma after treatment with growth hormone. In this case, however, there is no difference between normal and hypophysectomized rats.[120–124] In dogs, too, a decrease in the plasma concentration of the essential amino acids can be observed after growth hormone treatment. In rabbits the considerable increase of l-phenylalanine in the serum seen after intraperitoneal injection of 1 g of l-phenylalanine can, by treatment with growth hormone, be reduced to normal levels after 1–2 hr while the same process takes 6–7 hr in untreated animals.[95]

In past years the testing of the "specific dynamic action" of proteins (the change in basal metabolism after feeding fasting patients with a measured amount of protein) was an important clinical test. The significance of this was much debated. Some connected the results with the clinical status of the thyroid function, others with hypophyseal pathology. Careful investigations on rats showed that when 1–2 ml egg albumin was given orally to the animals, their oxygen consumption rose after 2 hr by 20–40%. Fifteen days after hypophysectomy, however, not only was their oxygen consumption reduced but they showed no rise after administration of egg albumin. However, treatment of the same animal with small amounts of growth hormone repaired the specific dynamic action of egg albumin. Thyrotropic hormone did not have such an action.[125] It is possible that growth hormone influences the protein metabolism in various phases, not only in the transport of amino acid and synthesis of protein, but also in splitting administered protein into peptides and amino acids that can be utilized in the body.

Growth hormone's anabolic action and ability for protein saving and accumulation is reflected in the urinary nitrogen retention, leading to a positive nitrogen balance.[126–127]

While growth hormone increases the protein content of the body tissues, the fat stores are decreased. Growth hormone plays an important role in fat mobilization in all animal species investigated so far.[128-134] It remains to be seen how far this action of growth hormone on the fat metabolism is due to its primary function in the protein storage and saving. The most important observation in this direction was made originally by Raben and Hollenberg[135-136] who found half an hour to 1 hr after intravenous injection of growth hormone there was a rise in the concentration of free fatty acids in blood.

This increase in the concentration of free fatty acids in the blood is due to the ability of growth hormone to stimulate the transfer of hydrolyzed fat from adipose tissue to other organs,[136-137] where it might (after transformation to sugar) serve as a source of energy.

In vitro investigations showed that small doses of growth hormone do not produce fat mobilization, but the addition of large doses to adipose tissue increases the lipogenesis.[138-139] A depression in synthesis of fatty acids from acetate was also observed *in vitro* after addition of growth hormone.[140]

The relation of growth hormone to carbohydrate metabolism, and diabetes was a matter of discussion for many years, but now seems to be clarified. Originally, Young[141] produced permanent experimental diabetes by injection of pituitary anterior lobe extracts into dogs. It was afterwards assumed that there was a special diabetogenic hormone accompanying the growth hormone in the extract which was responsible for the development of diabetes. After further purification procedures it was shown, however, that it was the purified growth hormone itself that was diabetogenic.

Growth hormone inhibits the utilization of glucose. This can be shown, for instance, *in vivo* where after addition of growth hormone, the utilization of glucose is diminished but the oxidation of fatty acids increased.[142] This effect can also be seen *in vitro* when the diaphragm of hypophysectomized rats is investigated. Growth hormone will (as mentioned before) stimulate the incorporation of radioactive glycine into the protein but it will inhibit the utilization of carbohydrate. It will also act antagonistically to insulin, which increases the uptake of glucose.[143]

Three to six weeks after hypophysectomy of rats, the oxygen consumption of liver is decreased by about 45% while the aneróbic glycolysis is increased by over 100%. The oxygen consumption of cerebral cortex, in the same animals, is barely changed, but the anerobic glycolysis is increased by about 30%. The hexokinase activity of the brain gray matter was, 5 to 8 days after hypophysectomy, found increased by about 72%. At the same time the ^{32}P content of the acid-soluble brain fraction is increased by over 150%.[144]

Growth hormone appears to have an important effect upon the ^{35}S metabolism of cartilage. After hypophysectomy the disappearance of ^{35}S is delayed but normalized again by administration of growth hormone.[145] The incorporation of ^{35}S into rat cartilage is considerably decreased and can be stimulated by growth hormone. It was assumed that the growth hormone quantity that was used to stimulate the ^{35}S incorporation was equivalent to that necessary for widening the tibial epiphysis of hypophysectomized rats.[146–149]

Changes in the electrolyte metabolism are mentioned here mainly because they might, similar to changes in protein and carbohydrate metabolism, contribute in the future to the explanation of some physiological and pathogenetic changes in the nervous system. These investigations were mainly carried out in humans since human growth hormone (HGH) was at the disposal of the investigators. After treatment with HGH the calcium excretion rate in the urine was increased but the intestinal absorption of calcium was also increased. At the same time, the fecal calcium was decreased so that the calcium balance did not become negative, and in most cases it even became positive.[150–156] Frequently the urinary excretion rate of potassium and phosphorus was decreased during HGH treatment.[157–160] There was also a sodium retention during HGH treatment which stopped soon after cessation of treatment, sometimes being followed by an increase in the sodium excretion rate.[153,156–157]

See page 470 on the stimulation of growth hormone secretion by insulin.

3. Thyrotropic Hormone

Growth hormone has a direct influence on a multitude of metabolic parameters and various enzymatic processes in the body. Thyrotropic hormone has as its main action an effect on the thyroid. There again, however, we deal with a complexity of changes in the thyroid morphology and biochemistry, and in addition there are some thyrotropic actions which apparently are not directly transmitted via thyroid stimulation.

As far as an increased growth of thyroid tissue is concerned, it is obvious that thyrotropic hormone produces similar general changes in the protein metabolism and energy consumption of the thyroid, as observed under the influence of growth hormone in other tissues.

As far as functional stimulation of the thyroid is concerned, it has to be borne in mind that thyrotropic hormone acts separately on the iodine uptake of the thyroid and the iodinization of various proteins and amino acids. Either of these functions can be disturbed separately and subsequently alter the quality of the substances produced. One can, for instance, observe in some patients a normal figure for the blood protein-bound iodine (PBI) while the active thyroid hormones such as thyroxine and triiodothyronine are very much reduced or even missing; the butanol-extractable iodine (BEI) can be considerably decreased. The mechanism of this anomaly is equally unknown as is that of the disturbance observed when the thyroid is insensitive to treatment with thyrotropic hormone. The mechanism of the TSH action on the thyroid is not yet known. The biological action of thyroid hormones in metabolism are discussed by Sokoloff in Vol. VI.

There are many unanswered questions concerning the action of TSH on the brain. It is known that thyroid extracts do not act on the brain oxygen consumption, yet 5–6 days' treatment with TSH produces a moderate but significant increase in the oxygen consumption of rat brain. The most significant change after TSH treatment can be seen in rats, 3–6 weeks after hypophysectomy, where the brain oxygen consumption is increased by almost 100 % after 5–6 days' treatment with TSH. In neither case does corticotropic hormone have an influence on brain oxygen consumption. The anaerobic glycolysis which is usually increased after hypophysectomy is not influenced by TSH treatment.[144]

The exophthalmos-producing action of thyrotropic hormone in rabbits and guinea pigs was, some time ago, regarded as an extrathyroidal action of thyrotropic hormone. More recently it has been found that this action is due to a substance chemically different from TSH.[161] While the thyroid-stimulating activity of TSH is not in uenced by methylation, the exophthalmos-producing activity is considerably reduced.[162]

Studies on the secretion rates of thyrotropic hormone in patients are now considerably simplified since the elaboration of an apparently very reliable immunoassay.[163]

4. Corticotropic Hormone

The influence of corticotropic hormone on size structure of the adrenals and the production of various corticoids is generally known and the subject of a number of handbook contributions. Corticotropic hormone was the first anterior lobe pituitary hormone to be synthetically produced. The corticotropic release factor was discussed (see page 473).

The action of ACTH on adrenal glands is not yet understood. Hechter and Halkerston[164] reviewed critically the various attempts made to explain the mechanism of the ACTH action on the adrenals, but did not reach any final conclusions. This might be because, as pointed out in the discussion of thyrotropin, in the action of corticotropin two different biological functions are concerned. One of these may consist of stimulation of growth of adrenal cortex tissue; the other, in stimulation of the production of the multitude of adrenal cortical hormones known.

The existence of a number of extraadrenal actions of ACTH is still under discussion. A list of established facts, and some possibilities, has been compiled.[165] Particularly interesting is the adipokinetic action and the effect on the melanocytes.

5. Gonadotropins

The great progress made in this field consisted in the proof of the existence of gonadotropin release factors, and their partial chemical identification (see page 472). Fundamental for this progress was the elaboration of more sensitive biological assay methods of the hormones *in vivo* and *in vitro*. This development and the possibilities offered to us now by the immunoassay methods in humans will do much to clarify the still complicated relevant feedback mechanisms between periphery and central nervous system.

Very little is known about the mechanism of the action of the various gonadotropins on the gonads. Investigations into the influence of chorionic gonadotropin on energy consumption of gonads are few in number. The oxygen consumption of immature rat ovaries, carried out with the Warburg methods in serum as suspension fluid, showed that the oxygen consumption of the ovaries starts to rise 1 hr after injection of 20 units of chorionic gonadotropin. This rise in the oxygen consumption reaches a maximum after 15–48 hr. Six hours after the injection the anaerobic glycolysis also starts to rise, and it was also interesting that these immature ovaries, which did not show any aerobic glycolysis before treatment, showed a considerable rise in the aerobic glycolysis at the same time. These changes occurred much sooner than any change could be observed in the growth or blood circulation rate in the ovaries.[166] The same changes in oxygen consumption and glycolysis were also observed in the testicles of immature rats after injection with chorionic gonadotropin.[167] After hypophysectomy, oxygen consumption and glycolysis of testes of adult rats were considerably decreased.[168] Hypophysectomy does not influence the uptake of α-aminoisobutyric-^{14}C acid of ovaries, but its transport is increased 4 hr after the immature rats were treated with FSH. In rabbit testes investigated *in vitro* the α-aminoisobutyric-^{14}C is not influenced by FSH or LH while tryptophan-^{14}C and valine-^{14}C are taken up.[169–170] The influence of the sex hormones on brain chemistry is discussed in Chapter 18.

Extensive observations have been made on the action of chorionic gonadotropin on behavior of mentally retarded children,[171] but the influence of any of the gonadotropic hormones on brain metabolism is still unknown. See Chapter 18 for a discussion of the various gonadotropins and their action.

6. Prolactin

Contrary to the other pituitary anterior lobe hormones, prolactin release is not stimulated by the hypothalamus which, however, does produce a prolactin inhibiting factor, PIF. After transection of the pituitary stalk, production of prolactin is increased.[172]

Prolactin has various biological actions, the physiological connections of which are not yet clarified. It stimulates a proliferation of the epithelium lining of the crop sac of pigeons. It produces milk secretion in the mammary glands of rabbits which were pretreated with esterone and progesterone. It has a luteotropic action which prolongs the life of the corpus luteum in rats and inhibits the estrous cycle. Finally, it has an important effect on behavior, increasing the nesting instinct and maternal behavior; it produces the broody instinct in fowl (for reviews see Refs. 173–175).

7. Chemistry of Pituitary Anterior Lobe Hormones

The history of this field is fraught with difficulties. Initially, large quantities of glandular extract had to be prepared and purified in order to obtain enough active material to inject the test animals. These crude extracts had to pe purified sufficiently to remove, at least, the toxic impurities, and to be able to exert a specific biological action it was necessary to inject large quantities. In order to proceed successfully with the purification, elaborate biological tests had to be designed to allow measurement of the hormonal content in the extract after each step in the purification process. Successful and unsuccessful attempts made in this direction during the last four decades can be counted in their hundreds. Today their interest is mainly historical.

More recently, it was necessary to prepare all the pituitary anterior lobe hormones from the same batch of human pituitary glands. The latter's relative scarcity demanded the selection and combination of the best extraction and purification methods in order to obtain the highest yield, with a relatively low degree of contamination of one hormone with the other. Wilhelmi[176] and later Parlow, Wilhelmi, and Reichert[177] succeeded in an admirable way in fractionating human pituitary glands by combining various extraction and purification methods previously used when working up cattle, pig, or sheep pituitaries. The degree of purity that they achieved with their fractionation methods was sufficient for the preparation to be used as a standard for the immunoassays carried out now in human plasma. They are distributed by the endocrinology section of the National Institutes of Health.

Dixon[178] and Butt[179] surveyed the various methods for further purification of the hormones, and the attempts for their synthesis. Lately purification methods have consisted mainly of chromatographic, electrocataphoretic procedures, in countercurrent distribution and gel filtration. Great progress has been made in recent years, particularly with the use of a molecular sieve provided by Sephadex gels of different densities.

There are a number of criteria for the purity of the hormones used. These consist not only in the finding that the preparation in question contains the highest hormone content per weight but also in determining the homogeneity of the hormone; for instance, single bands in starch-gel electrophoresis, zone electrophoresis, and lately also immunological methods of assessing homogeneity are used.

In the structural analyses of the hormones, great advances came from new methods of end group analyses as used by Sanger[180] and Edman[181] and their modifications.

Successful synthesis of the polypeptides and their biological activity usually confirms the structure found by analysis. It was also found that a synthetic product, even when sometimes incomplete, but containing just a group of the essential amino acids, showed the expected biological activity even if in diminished concentration per milligram. Since the original classical attempts for protein synthesis by Emil Fischer, considerable progress was made in the methodology of peptide and protein synthesis.

a. Growth Hormone. Multiple attempts have been made to purify the growth hormone found in the pituitaries of various species. The latter show differences between the purified products but they all are simple proteins and

TABLE II

Amino Acid Sequence of Human Pituitary Growth Hormone[a]

H-Phe-Pro-Thr-Ile-Pro-Leu-Ser-Arg-Leu-Phe-Asp-Asn-Ala-Met-Leu-Arg-Ile-Leu-Ser-Leu-Glu-Leu-Ile-Ser-Try-Leu-Glu-Pro-Val-Glu-
1   10   20   30

Phe-Ala-His-Arg-Leu-His-Gln-Leu-Ala-Phe-Asp-Thr-Tyr-Glu-Glu-Phe-Glu-Glu-Ala-Tyr-Ile-Pro-Lys-Glu-Gln-Lys-Tyr-Ser-Phe-Leu-
                     40   50   60

Gln-Asp-Pro-Glu-Thr-Ser-Leu-CyS-Phe-Ser-Ser-Ile-Glu-Ser-(Asp,Pro,Pro,Thr)-Arg-Glu-Gln-Thr-Gln-Lys-Ser-Asp-Leu-Glu-Leu-Leu-
                     70   80   90

Arg-Ser-Val-Phe-Ala-Asn-Ser-Leu-Val-Tyr-Gly-Ala-Ser-Asn-Ser-Asp-Val-Tyr-Asp-Leu-Leu-Lys-Asp-Leu-Glu-Glu-Gly-Ile-Glu-Thr-
                     100   110   120

Leu-Met-Gly-Arg-Leu-Glu-Asp-Pro-Ser-Gly-Arg-Thr-Gly-Gln-Ile-Phe-Lys-Glu-Thr-Tyr-Ser-Lys-Phe-Asp-Thr-Asn-Ser-His-Asn-Asp-
                     130   140   150

Asp-Ala-Leu-Leu-Lys-Asp-Tyr-Gly-Leu-Leu-Tyr-CyS-Phe-Arg-Lys-Asp-Met-Asp-Lys-Val-Glu-Thr-Phe-Leu-Arg-Ile-Val-Gln-CyS-Arg-
                     160   170   180

Ser-Val-Glu-Gly-Ser-CyS-Gly-Phe-OH
                     188

[a] From Li, Liu, and Dixon.[191]

in some the composition and amino acid sequences of the polypeptide chains are already analyzed.[182-201] The most recent analysis of the amino acid sequence determined in human pituitary growth hormone comes from the laboratory of Li. It consists of 188 amino acids and has, like some other growth hormone preparations, phenylalanine on both ends[191] (see Table II).

The molecular weights of the purified growth hormones prepared from pituitaries of various species differ. It is interesting that the molecular weight of human growth hormone is 21,500[191] and of simian growth hormone 25,400,[184] while the molecular weight of growth hormone prepared from ox, sheep, and pig are approximately twice as high (42–48,000).[202] This difference makes it possible to consider that bovine growth hormone might possibly consist of two molecules, one of which could be identical with that of human growth hormone. In fact, Laron et al.[203] digested bovine pituitary growth hormone with pepsin and found that during electrophoresis one component moved toward the anode, similar to the reaction of human growth hormone. This fraction also behaved immunologically like HGH. For a discussion of immunoassays of growth hormone, see Refs. 37 and 39.

b. *Corticotropic Hormone.* Progress in extraction and purification of ACTH took place faster than in other pituitary anterior lobe hormones owing to its greater solubility in acid alcohol and glacial acetic acid, and greater resistance to heat. Great progress in purification was made as soon as it was found that pig pituitaries contain much more ACTH than bovine pituitaries, and when the Sawyer assay based on the reduction of ascorbic acid in the adrenals of hypophysectomized animals was introduced. It was possible to extract ACTH in 80 % acetone which was acidified to 0.2 N HCl.[204] It was further possible to extract it with glacial acetic acid.[205] Using these methods, comparatively clean raw extracts were achieved which were then further purified by absorption on oxycellulose.[206-207] Essential for further purification was countercurrent distribution.[208-209] The various workers followed these procedures with starch electrophoresis, chromatography, Amberlite (IRC-50) and, later on, gel filtration and Sephadex G-50 led to purified homogeneous preparations.

The amino acid sequences of bovine, ovine, and porcine ACTH were finally determined and they all proved to be straight-chain polypeptides containing 39 amino acids with molecular weights of about 4500. The highest biological potency was 140 units/mg. The amino acid sequences of the corticotropins of these three species are practically identical. There are only some differences between the amino acids in position 25 to 33[210-213] (see Table III). A short time afterward, the amino acid sequence of human corticotropin was published and was found similar to corticotropins from other species, but the polypeptide chain was apparently shorter[214] (see Table IV).

ACTH is the first anterior pituitary lobe hormone that was synthesized, proving the validity of the structural analyses.[215]

Later on, parts of the ACTH polypeptide chains and various analogues were synthesized which also had considerable ACTH activity in the biological

TABLE III

Amino Acid Sequence of Bovine Corticotropin[a]

Ser-Tyr-Ser-Met-Glu-His-Phe-Arg-Try-Gly-Lys-Pro-Val-Gly-Lys-Lys-Arg-Arg-Pro-Val-Lys-Val-Tyr-Pro-Asp-

Gly-Glu-Ala-Glu-Asp-Ser-Ala-Glu-(NH_2)-Ala-Phe-Pro-Leu-Glu-Phe

[a] From Li, Dixon, and Chung.[213]

TABLE IV

Amino Acid Sequence of Human Corticotropin (ACTH)

Ser-Tyr-Ser-Met-Glu-(His-Phe-Arg)-(Try-Gly-Lys$_2$-Pro-Val)-(Lys$_2$-Arg$_2$-Pro-Val)-(Val-Tyr-Pro$_2$-Asp$_2$-Ser-

Gly-Glu$_3$-Ala$_3$-Phe)-Leu-Glu-Phe

[a] From Lee, Lerner, and Buettner-Janusch.[214]

test. Synthesis of the first 26 amino acid residues only of bovine ACTH had a considerable biological ACTH activity,[216] while synthesis of a peptide containing the 20 units with a structure of the amino end of ACTH showed only a very slight activity.[217] A 23-unit peptide synthesized by Hofman et al.[218] had the full activity of ACTH as a natural hormone, while a 16-unit terminal peptide[219] has no activity and a 19-unit terminal peptide has 50% of the activity[220] of corticotropin in vitro and in vivo.

One of the most important side activities of ACTH was its lipolytic activities in adrenolectomized animals. They were characterized mainly by increased release of free fatty acids from adipose tissue, increase in plasma free fatty acids, lipolytic activation of lipase in vitro, and decreased incorporation of acetate-1-^{14}C into protein and fatty acids in vitro. While this activity was originally regarded as due to contaminants, the fact that it was maintained in synthetically produced peptide chains proved it to be an intrinsic property of ACTH. The quantitative range of this activity depends, as in corticotropic hormone, mainly on the number of amino acid residues. The smallest synthetic peptide that can stimulate lipolysis in the rabbit perirenal adipose tissue is the pentapeptide (His-Phe-Arg-Try-Gly). Addition of glutamic acid residue of the peptide increases the lipolytic activity by about 10 times and a decapeptide containing the first 10 amino acids of the ACTH contains already one-tenth of the potency of the corticotropic hormone, while the lipolytic activity of a heptadecapeptide is higher than that of the natural ACTH. The corticotropic activity, however, of this synthetic product (as mentioned before) is low.[221] The peptides with the 24 first amino acid residues already have immunological properties.[222] For a discussion of melanophore-stimulating activity, see page 491.

c. Thyrotropic Hormone. There are a great number of extraction methods available for thyrotropic hormone. Crude extracts are usually made by water extraction, removing impurities by 50% alcohol extraction, precipitating the hormone by increasing the alcohol concentration to 80%, or precipitating the water extracts with saturated picric acid, eluting the hormone with ammonia water at pH 10 and precipitating the eluate with acetone, etc. Purification of the hormone can be carried out on Amberlite IRC-50 and later by chromatography on carboxymethyl cellulose. Great progress in the purification was made after introduction of gel filtration on Sephadex.

There are a number of sensitive bioassay methods for TSH at our disposal which are based either on the thyroid uptake of ^{131}I in rats, which is particularly sensitive in the newborn,[144] or the thyroid iodine loss in various in vitro methods. Some of these biological methods were made so sensitive that TSH in blood could be determined. Even the most sensitive biological methods, however, will be abandoned as soon as the much more sensitive immunoassay methods become routine.[223-225]

A purified homogeneous preparation of TSH has not yet been prepared. The most active preparations are glycoproteins with a molecular weight of about 28,000. The purified preparations contain 8% carbohydrate, consisting of mannose, fucose, glucosamine, and galactosamine with traces of glucose and galactose but no uronic acid.[226–227] The active preparations also contain all the common amino acids with the exception of tryptophan. Recently, a report on crystallization of bovine TSH was published.[228]

d. Gonadotropins. Two of the gonadotropins are of pituitary origin and secreted by the anterior pituitary lobe (luteinizing (LH) and follicle-stimulating hormone (FSH)). The other two are of extrapituitary origin. One is chorionic gonadotropin (HCG) which is excreted in large amounts in the urine of pregnant women after production in the placenta, acting biologically similarly to LH; the other is the hormone prepared from the serum of pregnant mares (PMS). The gonadotropic hormone that can be extracted from the urine of menopausal women is most probably produced in the pituitary and its biological activity is most likely the same as that of FSH. The biological action of PMS too, is similar to that of FSH.

The method for the preparation of crude extracts of all these hormones is simplified by their low solubility in alcohol and acetone. Glandular aqueous extracts of urine and serum alike can first be brought to an alcohol concentration of 40–50 %, the precipitated impurities discarded, while the hormone can be precipitated by increasing the alcohol concentration to 70–80 %. The acetone-washed and dried powders are then a suitable starting material for the preparation of purified extracts.

Hormones excreted in urine can also be directly absorbed with substances like kaolin. The alkaline eluate of the latter can be further purified after acetone precipitation.

i. FSH and LH. Separation of LH and FSH in pituitary extracts can be carried out by fractional precipitation with ammonium sulfate, LH being insoluble in halfsaturated ammonium sulfate. Since neither of the hormones is dialyzable, the salt accumulated during the repeated fractionation can be dialyzed and the separated hormones dried either by lyophylization or after final acetone precipitation. Another property useful for separation of LH from FSH is that FSH is not adsorbed to carboxymethyl cellulose while LH is. A separation of extracts from human pituitaries can also be achieved by DEAE cellulose.[229] Gel filtration,[230–231] chromatography on calcium filtrate,[232] and electrophoresis on starch[233–234] were also used. A particularly pure LH preparation from human pituitaries was made by Reichert and Parlow[235] which contained 900 i.u./mg. The purest FSH prepared contained 200 i.u./mg.[236]

So far no one has succeeded in preparing a homogeneous pure preparation of FSH or LH. The publications at present show the molecular weights varying between 26,000 for human LH, 45,000 for porcine LH, and 67,000 for porcine FSH.

The only fact that is definitely established is that FSH and LH are glycoproteins, similar to thyrotropin. Estimates of the carbohydrate content of the preparations differ. One highly purified human FSH preparation with a molecular weight of 25,000–30,000 was reported to contain 8 % carbohydrate.[236]

In other investigations, the carbohydrate content of bovine and ovine FSH and LH was measured and found to vary between 2.1 and 5.5 %,[237–239] the hexose being a mixture of galactose and mannose. Butt[179] found a considerably higher carbohydrate content in FSH and LH of human pituitaries. The total carbohydrate was found to be 22 % of the FSH molecule, consisting of 10.7 % mannose, 1.18 % galactose, 1.7 % fucose, 4.5 % glucosamine, and 3.37 % N-acetylneuraminic acid. Human and ovine LH contained more mannose than galactose. Human FSH and LH contain approximately the same amount of sialic acid, while ovine FSH contains at least twice as much sialic acid as ovine LH. Sialic acid seems to be a very important part of the molecule because inactivation takes place with purified neuraminidase. FSH can easily be inactivated by amylase and glucuronidase.

Butt[179] analyzed the amino acids in FSH and LH and found no methionine in FSH and no tryptophan in LH, while all the other amino acids were present (see also Ref. 240). A considerable cysteine content was reported to be present in LH.[241]

All these factors show us how difficult it will be to find the final structure of LH and FSH, a task which will be made still more complicated by the great similarity of these hormones with TSH and the difficulty of separation. It is understandable why homogeneous purified substances have not yet been found of either of these hormones, and that it is futile to consider their molecular weight at present.

ii. Menopausal urinary gonadotropin. Precipitation and purification of urinary gonadotropin is much simpler than the preparation of glandular gonadotropins since their extracts contain many more contaminants than does urinary gonadotropin. For purification, the same adsorption, electrocataphoretic, and countercurrent methods as mentioned for TSH, LH, and FSH can be used. The menopausal urine contains mainly follicular-stimulating hormones but there are also traces of luteinizing hormone. Most preparations were carried out with adsorption methods.[238] A very good purification, believed to be homogeneous, was carried out with electrophoresis and had a carbohydrate content of 30 %.[242]

iii. Human chorionic gonadotropin. The same purification and separation methods already mentioned can be used. A combination of various adsorption methods and lately particularly Sephadex with electrocataphoretic and countercurrent methods, should lead to very purified preparations. One preparation was described by Got and Bourrillon[243] which was claimed to be homogeneous and to have a molecular weight of 30,000, an isolectric point of 2.9, a total carbohydrate content of 30 % and a biological activity of 12,000 units/mg. In a more recent purification procedure,[244] the material, after ethanol fractionation, was purified on a Sephadex 25 column and later through a Sephadex 200 column. Preparations with an activity of 14–16,000 i.u./mg were reached and this activity was increased further by gel filtration with Sephadex G-100 to a potency of 16–21,000 i.u./mg. Continuing this purification with starch-gel immunoelectrophoresis, the authors obtained a preparation with an activity of 30,000 i.u./mg, but even this was not yet fully homogeneous.

Analytical investigations were carried out with six preparations in various stages of purification. Amino acid analysis showed a slight increase in the content of proline and a decrease in the content of aspartic acid, glutamine, and glycine with increasing potency. The sialic acid portion contained only N-acetyl neuraminic acid. Investigation of the carbohydrate did not show significant differences in galactose and mannose content (ratio 2 : 1) and in glucosamine and galactosamine (ratio 4 : 1). The sialic acid content increased and fucose decreased with increasing potency.[244]

For a discussion of immunochemistry and immunoassay of HCG see References 245–257). LH was found to interfere with the determination of HCG,[258–259] which might indicate a chemical identity between HCG and LH. It is interesting that the immunological qualities are not destroyed by heating, while the biological activity disappears.[260]

    iv. Gonadotropin of pregnant mare serum (PMS). The crude preparation is usually made by a fractional precipitation by alcohol or acetone. Further purification is usually carried out by adsorption on kaolin or benzoic acid or permutite ion-exchange resin after zone electrophoresis on starch[261] at different pH values. No separation between FSH and LH content of the serum was reached even if the purest fraction contained 30,000 i.u./mg.[261] The carbohydrate content of the PMS is an intrinsic quality of, and important for, its biological activity. Treatment with carbohydrate-splitting enzymes destroys the biological activity, similar to HCG. The carbohydrate content of PMS is also high (hexose 18.6 %, hexosamine 17.5 %, fucose 1.4 %, l-rhamnose 0.7 %, and sialic acid 10.4 %).[179]

    e. Prolactin. For a review of the chemistry of prolactin see References 262–263, 179.

Efforts to purify and isolate prolactin were made mainly by using sheep, cattle, and human pituitaries as raw material. Originally the main principles of purification were based on precipitation at an isoelectric point of 5.6 and countercurrent distribution in a system of s-butanol and dichloracetic acid.

The preparations prepared in this way were regarded as homogeneous after electrophoresis and ultracentrifugation. Their molecular weight determinations varied between 24,100 and 26,000, and they contained 33 i.u./mg of crop sac stimulating activity.[264–265]

In later purification attempts, chromatography and DEAE cellulose in 0.01 M tris buffer at pH 8.2, and stepwise elution with increasing concentrations of NaCl were used instead of countercurrent distribution.[266] In this way three active fractions were isolated, which all had practically identical amino acid composition.

Further purification procedures in which the previous approaches were combined with or replaced by filtration on Sephadex G-75 did not produce purer preparations of prolactin.[267–268] They all had similar amino acid compositions. It is interesting that they seem to contain 4 Cys in their composition and there are approximately 205 amino acid residues.[263]

In more recent investigations molecular weights of prolactin and pituitary growth hormone were determined by gel filtration.[269]

Investigations about the species specifity of prolactin[271] and the immunological properties of human prolactin were made.[270]

Dixon and Li[263] and Li[221] report that the recently purified human growth hormone possesses the various activities demonstrated in ovine lactogenic hormone: promotion of pigeon crop sac growth, stimulation of functional hypertrophy of corpora lutea, alveoli mammary growth, and induction of localized milk secretion. Moreover, if the growth-promoting activity is destroyed, the lactogenic activity disappears as well. The HGH has per milligram about 10 to 20% activity of the purified ovine or bovine prolactin. Li assumed that the HGH has intrinsic lactogenic properties, and hypothesized that certain amino acid sequences may exist in common when the complete primary structure of the two hormones is elucidated.

D. Pars Intermedia Melanocyte Stimulation and the Hypothalamus

The size of the intermediate lobe varies considerably in different species and under various environmental conditions. We are, however, still in the dark concerning the biological significance of the melanophore-stimulating hormone (MSH) produced in this region. An extensive review on color change and the environment in invertebrates and vertebrates and the physiological role that hormones play in these changes has been written by Barrington.[272] The best-known physiological action of MSH consists in expanding the melanophores in frog skin *in vivo* and *in vitro*. Certain authors assume that MSH, and particularly MSH activity contained in the corticotropin, increased the pigmentation of the skin of treated patients. It has, for instance, been shown that ACTH given in doses of 2400 units per day[273] darkens the skin of men. For clinical consideration of the role of MSH, see Reference[165]. Another theory is that MSH improves visual acuity[274] in normal subjects. It is interesting that MSH is present in the human fetal pituitary (together with ACTH). It was found in one fetus of only 11 weeks, gestational age.[275]

Recently, the existence of an MSH release inhibiting factor[275] and of an MSH release stimulating factor have been claimed. The latter also stimulated the synthesis of the hormones in the intermedia cells after transection of the pituitary stalk.[276]

It will be interesting in the future to define biologically the relation between MSH and melatonin. At present their actions seem to be antagonistic but the physiological significance is not clarified.

Early observations on the chemical extraction and purification of MSH were made by Landgrebe *et al.*[277] It was originally shown that the MSH activity is based on two substances which are α-MSH which had an isoelectric point between 10.5 and 11, while the other, β-MSH, had an isoelectric point at *p*H 5. It was later found that the amino acid sequences of these two MSH forms differ.

TABLE V

Comparison of Amino Acid Sequence of β-MSH and ACTH[a]

	1	2	3	4	5	6	7	8	9	10	11	12	13	14	15	16	17	18	
Corticotropin	Ser	Tyr	Ser	Met	Glu	His	Phe	Arg	Try	Gly	Lys	Pro	Val	Gly					...Phe
	1	2	3	4	5	6	7	8	9	10	11	12	13	14					39
β-MSH (pig)	Asp	Glu	Gly	Pro	Tyr	Lys	Met	Glu	His	Phe	Arg	Try	Gly	Ser	Pro	Pro	Lys	Asp	
	1	2	3	4	5	6	7	8	9	10	11	12	13	14	15	16	17	18	

[a] From Harris and Roos.[280]

MSH was prepared mainly from pituitary posterior lobe extracts in acid–acetone mixture followed by salt precipitation and adsorption on oxycellulose. The eluate was purified by countercurrent distribution and further fractionated by electrophoresis.[278] Later a combination of most of the purification methods, as used for pituitary anterior lobe extracts, were used in order to achieve an homogeneous preparation. α- and β-MSH were separated by countercurrent distribution, the partition coefficient for α was 1.5 and for β, 0.05.[279]

The first amino acid sequence determination of β-MSH was made by Harris and Roos.[280] The knowledge of the amino acid sequences of MSH and corticotropin clarified previous assumptions that corticotropin was contaminated by MSH. This was due to the fact that even synthetic ACTH was found to show some MSH activity and further, that the amino acid sequence of β MSH is in the position 5 to 15, practically identical to the amino acid position 2 to 12 in the corticotropin (see Table V).

This similarity of the initial amino acid sequence in ACTH and MSH indicates that the biological MSH activity is an intrinsic factor of the ACTH molecule, and also explains why ACTH has much less activity in the biological melanocyte-stimulating assay than the purified MSH.

IV. CONCLUDING REMARKS

It can be seen from this article that the hypothalamo-pituitary complex is well founded and its place in the homeostasis of the body beginning to be defined. There is, however, a great deal to be done before the physiological circle between brain, hypothalamus, pituitary, target organs, and various feedback systems is completely understood.

Too little is known about the actions of the various participants in the hypothalamo-pituitary complex on biochemical parameters of the brain. If only our knowledge in this direction could be increased, and the significance of various brain chemical changes on mentation and behavior be analyzed, we would be much further ahead in the task of understanding the mechanism of action, and role played by the endocrine factors in normal and pathological behavior. The therapeutic possibilities in this case would be enormous.

V. REFERENCES

1. I. Assenmacher, Le role de l'hypothalamus dans les régulations hypophysaires. *Presse Méd.* **65**:1612–1614, 1670–1671 (1957).
2. M. Saffran, *in Ergebnisse der inneren Medizin und Kinderheilkunde* (L. Heilmeyer *et al*, eds.) pp. 594–612, Springer, Berlin (1958).
3. W. J. Schindler, Hypothalamic neurohumoral control of pituitary function, *Proc. Roy. Soc. Med.* **55**:125–130 (1962).
4. G. W. Harris, The development of ideas regarding hypothalamic-releasing factors, *Metabolism* **13**:1171–1176 (1964).

5. G. W. Harris, M. Reed, and C. P. Fawcett, Hypothalamic releasing factors and the control of anterior pituitary function, *Brit. Med. Bull.* **22**:266–272 (1966).

6. H. Heller, Hormone content of hypothalamo–neurohypophysial system, *Brit. Med. Bull.* **22**:227–231 (1966).

7. H. Heller, Biochemie und phylogenese der hypophysenhinterlappenhormone, *Mitteilungen der naturforschenden gesellschaft in Bern*, **23**:63–81 (1966).

8. H. Heller, Neural control of hormone secretion, *J. Scientific and Industrial Res.* **25**:298–302 (1966).

9. M. Reiss, Neuroendocrinology and Psychiatry (A critical assessment of the present status), *Int. J. Neuropsychiatry* **3**:441–463 (1967).

10. R. Guillemin, Hypothalamic factors releasing pituitary hormones, *Rec. Prog. Hormone Res.* **20**:89–130 (1964).

11. R. Gullemin, The adenohypophysis and its hypothalamic control, *Ann. Rev. Physiol.* **29**:313–348 (1967).

12. A. V. Schally, E. E. Muller, A. Arimura, C. Y. Bowers, T. Saito, T. W. Redding, S. Sawano, and P. Pizzolato, Releasing factors in human hypothalamic and neurohypophysial Extracts, *J. Clin. Endo. Metab.* **27**:755–762 (1967).

13. S. M. McCann and A. P. S. Dhariwal, in *Neuroendocrinology* (L. Martini and W. F. Ganong, eds.) Vol. 1, pp. 261–296, Academic Press, New York (1966).

14. A. B. Rothballer, in *Endocrines and the Central Nervous System*, Proc. of the Assoc. for Res. in Nerv. and Ment. Dis. (R. Levine, ed.) pp. 86–131, Williams and Wilkins, Baltimore (1966).

15. A. V. Nalbandov, *Advances in Neuroendocrinology*, University of Illinois Press, Urbana, Ill. (1963).

16. V. H. T. James and J. Landon, *The Investigation of Hypothalamic Pituitary Adrenal Function*, Memoirs for the Society of Endocrinology, No. 17, Cambridge University Press, Cambridge, England (1968).

17. P. Trendelenburg, *Die Hormone*, p. 119, Julius Springer, Berlin (1929).

18. I. Gersh, Water metabolism: endocrine factors, *Res. Publ., A Nerv. Ment. Dis.* **20**:436–448 (1940).

19. E. B. Verney, The antidiuretic hormone and the factors which determine its release, *Proc. Roy. Soc. Ser. B.* **135**:25–106 (1947).

20. E. B. Verney, Agents determining and influencing the function of the pars nervosa of the pituitary, *Brit. Med. J.* **2**:119 (1948).

21. J. Heller, Evidence for the conditioned reflex secretion of antidiuretic hormone, *Cs. Fysiol.* **5**:373 (1956).

22. K. Lederis, *Mem. Soc. Endocr.* (12):227 (1962).

23. G. W. Theobald, The separate release of oxytocin and antidiuretic hormone, *J. Physiol.* (*Lond.*) **149**:443 (1959).

24. M. Pickford, Factors affecting milk release in the dog and the quantity of oxytocin liberated by suckling, *J. Physiol.* (*Lond.*) **152**:515 (1960).

25. M. X. Zarrow, J. M. Yochin, and J. L. McCarthy, *Experimental Endocrinology, A Source Book of Basic Techniques*, p. 350, Academic Press, New York, London (1966).

26. A. H. Livermore and V. Du Vigneaud, Preparation of high potency oxytocic material by the use of counter-current distribution, *J. Biol. Chem.* **180**:365 (1949).

27. R. A. Turner, J. G. Pierce, and V. Du Vigneaud, The purification and the amino acid content of vasopressin preparations, *J. Biol. Chem.* **191**:21 (1951).

28. R. Acher, A. Light, and V. Du Vigneaud, Purification of oxytocin and vasopressin by way of a protein complex, *J. Biol. Chem.* **233**:116–120 (1958).

29. V. Du Vigneaud, G. Winestock, V. V. S. Murti, D. B. Hope and R. D. Kimbrough Jr., Synthesis of 1-β-mercaptopropionic acid oxytocin (desaminooxytocin), a highly potent analogue of oxytocin, *J. Biol. Chem.* **235**:PC64–PC66 (1960).

30. G. W. Harris, Oestrous rhythm, pseudopregnancy and the pituitary stalk in the rat, *J. Physiol.(Lond.)* **111**:347–360(1950).

31. G. W. Harris, *Neural control of the pituitary gland*, p. 87, Edward Arnold, Ltd., London (1955).

32. J. Franz, C. H. Haselbach, and O. Libert, Studies on the effect of hypothalamic extracts on somatotrophic pituitary function, *Acta Endocrin.* **41**:336–350(1962).

33. R. R. Deuben and J. Meites, Stimulation of pituitary growth hormone release by a hypothalamic extract *in vitro*, *Endocrinology* **74**:408–414 (1964).

34. A. V. Schally, S. L. Steelman, and C. Y. Bowers, *Stimulation of Release of Growth Hormone In Vitro by a Hypothalamic Factor*, p. 143, 46th Annual Meeting Endocrine Society, San Francisco (1964).

35. S. Symchowica, W. D. Peckham, R. Oneri, C. A. Korduba, and P. L. Perlman, The effect *in vitro* of purified hypothalamic extract on the growth hormone content of the rat pituitary, *J. Endocr.* **35**:379–383(1966).

36. G. Falconi, E. Muller, and A. Pecile, Maintenance of growth hormone releasing activity in the hypothalamus of long-term hypophysectomized rats, *Experientia* **22**:333(1966).

37. J. Roth, S. M. Glick, R. S. Yalow, and S. A. Berson, Secretion of human growth hormone: physiologic and experimental modification, *Metabolism* **12**:577(1963).

38. F. C. Greenwood, J. Landon, and T. C. B. Stamp, The plasma sugar, free fatty acid, cortisol and growth hormone response to insulin. I. In control subjects, *J. Clin. Invest.* **45**:429 (1966).

39. A. W. Root, R. L. Rosenfield, A. M. Bongiovanni, and W. R. Eberlein, The plasma growth hormone response to insulin-induced hypoglycemia in children with retardation of growth, *Pediatrics* **39**:844(1967).

40. J. Hillman, J. Hammond, J. Sokola, and M. Reiss, Changes in plasma growth hormone levels in retarded children, *J. Ment. Def. Res.* **12**:294 (1968).

41. H. K. Silver and M. Finkelstein, Deprivation dwarfism, *J. Pediatrics* **70**:317(1967).

42. G. F. Powell, J. A. Brasel, S. Raiti, and R. M. Blizzard, Emotional deprivation and growth retardation stimulating iodiopathic hypopituitarism, *New Eng. J. Med.* **276**:1279(1967).

43. A. P. S. Dhariwal, L. Krulich, S. H. Katz, and S. M. McCann, Purification of growth hormone-releasing factor, *Endocrinology* **77**:932-936(1965).

44. L. Krulich, A. P. S. Dhariwal, and S. M. McCann, Growth hormone-releasing activity of crude ovine hypothalamic extracts, *Proc. Soc. Exp. Biol. Med.* **120**:180–184(1965).

45. A. Pecile, E. Muller, G. Falconi, and L. Martini, Growth hormone-releasing activity of hypothalamic extracts at different ages, *Endocrinology* **77**:241–246(1965).

46. A. P. S. Dhariwal, J. Antunes-Rodrigues, L. Krulich, and S. M. McCann, Separation of growth hormone-releasing factor (GHRF) from corticotrophin-releasing factor (CRF). *Neuroendocrinology* **1**:341–349(1966).

47. A. V. Schally, A. Kuroshima, Y. Ishida, A. Arimura, T. Saito, C. Y. Bowers, and S. L. Steelman, Purification of growth hormone-releasing factor from beef hypothalamus, *Proc. Soc. Exp. Biol. Med.* **122**:821–823(1966).

48. B. A. Houssay, A. Biasotti, and R. Sammartino, Modificaciones producidas por las lesiones Infundibulo-tube-rianas en el sapo, *Rev. Soc. Argent. Biol.* **11**:318(1935).

49. U. U. Uotila, Role of pituitary stalk in regulation of thyrotropic and thyroid activity, *Proc. Soc. Exp. Biol. Med.* **41**:106(1939).

49a. U. U. Uotila, On the role of the pituitary stalk in the regulation of the anterior pituitary, with special reference to the thyrotropic hormone, *Endocrinology* **25**:605(1939).

50. J. M. McKenzie, The thyroid-activating hormones and hypothalamic control, *Res. Publ. Ass. Res. Nerv. Ment. Dis.* **43**:47–58(1966).

51. S. H. Solomon and J. M. McKenzie, Release of thyrotropin by the rat pituitary gland *in vitro*, *Endocrinology* **78**:699–706(1966).

52. D. K. Sinha and J. Meites, Stimulation of pituitary thyrotropin synthesis and release by hypothalamic extract, *Endocrinology* **78**:1002–1006 (1966).

53. P. Ducommun, E. Sakiz, and R. Guillemin, Increase of plasma TSH concentration following injection of purified hypothalamic TRF, *Endocrinology* **77**:792–796 (1965).

54. R. Guillemin, D. N. Ward, and E. Sakiz, Filtration chromatography on substituted sephadex in the purification of the hypothalamic hormone TRF (TSH-Releasing Factor), *Proc. Soc. Exp. Biol. Med.* **120**:256–258 (1965).

55. R. Guillemin, R. Burgus, E. Sakiz, and D. N. Ward, Nouvelles données sur la Purification de l'hormone hypothalamique TSH-hypophysiotrope, TRF, *C. R. Acad. Sci.* (D) **262**:2278–2280 (1966).

56. A. V. Schally, C. Y. Bowers, T. W. Redding, and J. F. Barrett, Isolation of thyrotropin releasing factor (TRF) from procine hypothalamus, *Biochem. Biophys. Res. Commun.* **25**:165 (1966).

57. A. V. Schally, C. Y. Bowers, and T. W. Redding, Purification of thyrotropic hormone-releasing factor from bovine hypothalamus, *Endocrinology* **78**:726–732 (1966).

58. A. V. Schally, E. E. Muller, A. Arimura, C. Y. Bowers, T. Saito, T. W. Redding, S. Sawano, and P. Pizzolato, Releasing factors in human hypothalamic and neurohypophysial extracts, *J. Clin. Endocr. Met.* **27**:755–762 (1967).

59. W. Hohlweg, Der Mechanismus der Wirkung von Gonadotropen Substanzen auf das Ovar der infantilen Ratte, *Klin. Wschr.* **15**:1832–1835 (1936).

59a. W. Hohlweg and K. Junkmann, Die hormonal-nervose Regulierung der Funktion des Hypophysenborderlappens, *Klin. Wschr.* **11**:321–323 (1932).

60. I. Rothchild, The central nervous system and disorders of ovulation in women, *Am. J. Obstet. Gynec.* **98**:719–747 (1967). (Review)

61. G. W. Harris, The induction of ovulation in the rabbit, by electrical stimulation of the hypothalamo–hypophysial mechanism, *Proc. Roy. Soc. Ser. B.* **122**:374–394 (1937).

61a. G. W. Harris, Electrical stimulation of the hypothalamus and the mechanism of neural control of the adenohypophysis, *J. Physiol.* **107**:418–429 (1948).

62. J. Moll and G. H. Zeilmaker, Ovulatory discharge of gonadotrophins induced by hypothalamic stimulation in castrated male rats bearing a transplanted ovary, *Acta Endocr.* **51**:281–289 (1966).

63. G. W. Harris, Sex hormones, brain development and brain function, *Endocrinology* **75**:627–648 (1964).

64. A. F. Parlow, in *Human Pituitary Gonadotropins* (A. Albert and Charles C. Thomas, eds.) p. 300, Charles C. Thomas, Springfield, Illinois (1961) and *Fed. Proc.* **17**:402 (1958).

65. S. M. McCann, A hypothalamic luteinizing hormone-releasing factor, *Am. J. Physiol.* **202**:395–400 (1962).

66. R. Schiavi, M. Jutisz, E. Sakiz, and R. Guillemin, Stimulation of ovulation by purified LH-releasing factor (LRF) in animals rendered anovulatory by hypothalamic lesion, *Proc. Soc. Exp. Biol. Med.* **114**:426 (1963).

67. R. Nallar and S. M. McCann, Luteinizing hormone-releasing activity in plasma of hypophysectomized rats, *Endocrinology* **76**:272–275 (1965).

68. T. Kobayashi, T. Kobayashi, T. Kigawa, M. Mizuno, and Y. Amenomori, *Endocr. Jap.* **10**:16 (1963).

69. M. Igarashi and S. M. McCann, A new sensitive bioassay for follicle stimulating hormone, *Endocrinology* **74**:440–445 (1964).

70. J. C. Mittler and J. Meites, In vitro stimulation of pituitary follicle stimulating hormone release by hypothalamic extract, *Proc. Soc. Exp. Biol. Med.* **117**:309–313 (1964).

71. C. E. Grosvenor, S. M. McCann, and R. Nallar, Inhibition of nursing induced and stress induced fall in pituitary prolactin concentration in lactating rats following injection of acid extracts of bovine and rats hypothalamus, *Endocrinology* **76**:883–889 (1965).

72. J. Pasteels, Administration d' extraits hypothalamiques a' l'hypophyse de Rat *in vitro* dans le but d'en controler la sécrétion de prolactine, *Compt. Rend.* **254**:2664–2666 (1963).

73. P. K. Talwalker, A. Ratner, and J. Meites, *In vitro* inhibition of pituitary prolactin synthesis and release by hypothalamic extract, *Am. J. Physiol.* **205**:213–218 (1963).

74. W. V. McDermott, E. G. Fry, J. R. Brobeck, and C. N. H. Long, Mechanism of control of adrenocorticotrophic hormone, *Yale. J. Biol. Med.* **23**:52–65 (1951).

75. C. Franksson, C. A. Gemzell, and U. S. von Euler, Cortical and medullary adrenal activity in surgical and allied conditions, *J. Clin. Endocr.* **14**:608–621 (1954).

76. J. de Groot and G. W. Harris, Hypothalamic control of the anterior pituitary gland and blood lymphocytes, *J. Physiol.* **111**:335–346 (1950).

77. M. A. Slusher and J. E. Hyde, Effect of diencephalic and midbrain stimulation on ACTH levels in unrestrained cats, *Am. J. Physiol.* **210**:103–108 (1966).

78. S. M. McCann, Effect of hypothalamic lesions on the adrenal cortical response to stress in the rat, *Am. J. Physiol.* **175**:13–20 (1953).

79. R. W. Porter, Hypothalamic involvement in the pituitary adrenocortical response to stress stimuli, *Am. J. Physiol.* **172**:515–519 (1953).

80. D. M. Hume, The neuro-endocrine response to injury: present status of the problem, *Ann. Surg.* **138**:548–557 (1953).

81. C. Fortier, Dual control of adrenocorticotrophin release, *Endocrinology* **49**:782–788 (1951).

82. S. M. McCann, ACTH releasing activity of extracts of the posterior lobe of the pituitary *in vivo*, *Endocrinology* **60**:664–676 (1957).

83. C. S. Nagareda and R. Gaunt, Functional relationship between the adrenal cortex and posterior pituitary, *Endocrinology* **48**:560 (1951).

84. R. K. McDonald, E. N. Wagner Jr., and V. K. Weise, Relationship between endogenous antidiuretic hormone activity and ACTH release in Man, *Proc. Soc. Exp. Biol. Med.* **96**:652 (1957).

85. P. C. Royce and G. Sayers, Corticotropin releasing activity of a pepsin labile factor in the hypothalamus, *Proc. Soc. Exp. Biol. Med.* **98**:677 (1958).

86. B. L. Nichols and R. Guillemin, Endogenous and exogenous vasopressin on ACTH release, *Endocrinology* **64**:914–920 (1959).

87. M. Saffron, A. V. Schally, and B. G. Benfey, Stimulation of the release of corticotropin from the adrenohypophysis by a neurohypophysial factor, *Endocrinology* **57**:439 (1955).

88. R. Guillemin, W. R. Hearn, W. R. Cheek, and D. E. Householder, Control of corticotropin release: further studies with *in vitro* methods, *Endocrinology* **60**:488 (1957).

89. R. Guillemin, E. Sakiz, and D. N. Ward, Further purification of TSH releasing factor (TRF) from sheep hypothalamic tissues with observations on the amino acid composition, *Proc. Soc. Exp. Biol. Med.* **118**:1132–1137 (1965).

90. A. V. Schally, T. W. Redding, J. F. Barrett, and C. Y. Bowers, Purification of porcine thyrotrophin releasing factor (TRF), *Fed. Proc.* **25**:Ref. 895 (1966).

91. A. V. Schally and C. Y. Bowers, Purification of luteinizing hormone releasing factor from bovine hypothalamus, *Endocrinology* **75**:608 (1964).

92. A. V. Schally and C. Y. Bowers, Corticotropin releasing factor and other hypothalamic peptides, *Metabolism* **13**:1190–1205 (1964).

93. R. Guillemin, *in The Investigation of Hypothalamic Pituitary Adrenal Function* (V. H. T. James and J. Landon, eds.) p. 19, Memoirs of the Society of Endocrinology, Cambridge University Press, Cambridge, England (1968).

94. J. W. Mason, E. Mougey, F. Wherry, D. Collins, E. Taylor, and M. Wool, Report of the Forty-ninth Annual Meeting of the Endocrine Society, Abstract 124, p. 90 (1967).

95. M. Reiss, M. B. Sideman, and E. S. Plichta, Influence of anabolic hormones on phenylalanine metabolism: II. Studies in animals, *J. Ment. Def. Res.* **10**:130–140 (1966).

96. A. Hjalmarson, and K. Ahren, Sensitivity of the rat diaphragm to growth hormone, *Acta Endocr.* **54**:645–662 (1967).

97. E. Knobil and J. Hotchkiss, in *Annual Review of Physiology* (V. E. Hall, ed.) p. 47 (1964).

98. P. F. Brande and E. Knobil, Further evidence for amino-acid transport as a site of action of growth hormone, *Proc. Soc. Exp. Biol. Med.* **110**:5–6 (1962).

99. J. L. Kostyo, J. Hotchkiss, and E. Knobil, Stimulation of amino acid transport in isolated diaphragm by growth hormone added *in vitro, Science* **130**:1653–1654 (1959).

100. J. L. Kostyo and F. L. Engel, *In vitro* effects of growth hormone and corticotropin preparations on amino acid transport by isolated rat diaphragms, *Endocrinology* **67**:708–716 (1960).

101. J. L. Kostyo, C. A. Snipes, and J. E. Schmidt, Effects of hypophysectomy and STH on the uptake of non-utilizable and utilizable amino acids by muscle, *Fed. Proc.* **21**:198 (1962).

102. C. A. Snipes and J. L. Kostyo, Effects of hypophysectomy and growth hormone on utilizable amino acid accumulation, *Am. J. Physiol.* **203**:933–938 (1962).

103. S. Takahashi, A. Lajtha, N. Penn, and M. Reiss, The influence of growth hormone on phenylalanine incorporation into brain and liver protein, *in Protein Metabolism of the Nervous System*, A. Lajtha (ed.), Plenum Press, New York (1970).

104. A. Korner, The effect of hypophysectomy of the rat and of treatment with growth hormone on the incorporation of amino acids into liver proteins in a cell-free system, *Biochem. J.* **73**:61 (1959).

105. A. Korner, The effect of hypophysectomy and growth-hormone treatment of the rat on the incorporation of amino acids into isolated liver ribosomes, *Biochem. J.* **81**:292 (1961).

106. G. P. Talwar, N. C. Panda, G. S. Sarin, and A. I. Tolani, Effect of growth hormone on ribonucleic acid metabolism, I. *Biochem. J.* **82**:173 (1962).

107. G. P. Talwar, S. L. Gupta, and F. Gros, Effect of growth hormone on ribonucleic acid metabolism, *Biochem. J.* **91**:565–572 (1964).

108. I. Geschwind, C. H. Li, and H. M. Evans, The partition of liver nucleic acids after hypophysectomy and growth hormone treatment, *Arch. Biochem.* **28**:73 (1950).

109. A. Korner, Regulation of the rate of synthesis of messenger ribonucleic acid by growth hormone, *Biochem. J.* **92**:449 (1964).

110. A. Korner and A. E. Pegg, Growth hormone action on rat liver RNA polymerase, *Nature* **205**:904 (1965).

111. K. G. Dawson, P. Patey, D. Rubinstein, and J. C. Beck, Growth hormone and protein synthesis, *Mol. Pharmacol.* **2**:269–274 (1966).

112. C. D. Jackson and B. H. Sells, The effect of bovine growth hormone on formulation of RNA by rat liver slices, *Biochim. Biophys. Acta* **142**:419–429 (1967).

113. F. O. Schmitt, *in Horizon in Biochemistry* (N. Kasha and B. Pullman, eds.) p. 437, Academic Press, New York (1962).

114. H. Hyden and E. Egyhazi, Glial RNA Changes During a Learning Experiment in Rats, *Proc. Nat. Acad. Sci.* **49**:618 (1963).

114a. H. Hyden and E. Egyhazi, Changes in RNA content and base composition in cortical neurons of rats in a learning experiment involving transfer of handedness, *Proc. Nat. Acad. Sci.* **52**:1030 (1964).

115. S. H. Appel, Effects of inhibition of RNA synthesis on neural information storage, *Nature* **207**:1163 (1965).

116. D. E. Cameron, S. Sved, L. Solyom, B. Wainrib, and H. Barik, Effects of ribonucleic acid on memory defect in the aged, *Am. J. Psychiat.* **120**:320 (1964).

117. T. I. Chamberlain, G. H. Rotschild, and R. W. Gerard, Drugs affecting RNA and learning, *Proc. Nat. Acad. Sci.* **49**:918 (1963).

118. L. Cook, A. B. Davidson, D. J. Davis, H. Green and E. J. Fellows, Ribonucleic acid: effect on conditioned behavior in rats, *Science* **141**:268 (1963).

119. S. Zamenhof, J. Mosley, and E. Schuller, Stimulation of the proliferation of cortical neurons by prenatal treatment with growth hormone, *Science* **152**:1396 (1966).

120. E. G. Frame and J. A. Russell, The effects of insulin and anterior pituitary extract on the blood amino nitrogen in eviscerated rats, *Endocrinology* **39**:420–429 (1946).

121. C. Griffen, J. M. Luck, V. Kulakoff, and M. Mills, Further observations on the endocrine regulation of blood amino acids, *J. Biol. Chem.* **209**:387–393 (1954).

122. C. H. Li, M. E. Simpson, and H. M. Evans, The influence of growth and adrenocorticotropic hormones on the fat content of the liver, *Arch. Biochem.* **23**:51–54 (1949).

123. J. A. Russel, in *Hypophyseal Growth Hormone, Nature and Actions* (R. W. Smith, O. H. Gaebler, and C. N. H. Long, eds.) pp. 213–224, McGraw-Hill, New York (1955).

124. W. D. Lotspeich, Relations between insulin and pituitary hormones in amino acid metabolism, *J. Biol. Chem.* **185**:221–229 (1950).

125. M. Reiss, Influence of the pituitary anterior lobe upon the specific dynamic action of protein, *J. Endocr.* **2**:329–338 (1940).

126. M. S. Raben, Growth hormone, *New Engl. J. Med.* **266**:31–35 (1962).

127. G. A. Brown, L. Stimmler, and J. G. Lines, Growth hormone induced nitrogen retention in children of short stature, *Arch. Dis. Childh.* **42**:239–244 (1967).

128. B. Ketterer, P. J. Randle, and F. G. Young, The pituitary growth hormone and metabolic processes, *Ergeb. Physiol.* **49**:127 (1957).

129. R. C. deBodo and N. Altszuler, *Vitamins Hormones* **15**:205 (1957).

130. H. R. Engel, L. Hallman, S. Siegel, and D. M. Bergenstal, Effect of growth hormone on plasma unesterified fatty acid levels of hypophysectomized rats, *Proc. Soc. Exp. Biol. Med.* **98**:753 (1958).

131. H. M. Goodman and E. Knobil, Effect of pH and route of administration on the fatty acid mobilizing activity of growth hormone solutions, *Endocrinology* **65**:977 (1959).

132. H. M. Goodman and E. Knobil, Growth hormone and fatty acid mobilization: the role of the pituitary, adrenal and thyroid, *Endocrinology* **69**:187 (1961).

133. E. Knobil, in *Growth in Living Systems* (M. X. Zarrow, ed.) p. 758, Basic Books, New York (1961).

134. H. M. Goodman and E. Knobil, The effects of fasting and of growth hormone administration on plasma fatty acid concentration in normal and hypophysectomized rhesus monkeys, *Endocrinology* **65**:451 (1959).

135. M. S. Raben and C. H. Hollenberg, Effect of growth hormone on plasma fatty acids, *J. Clin. Invest.* **38**:484–488 (1959).

136. M. S. Raben, in *Recent Progress in Hormone Research* (G. Pincus, ed.) Vol. 15, p. 71, Academic Press, New York (1959).

137. M. S. Raben and C. H. Hollenberg, in *Ciba Foundation Colloquia on Endocrinology* (G. E. W. Wolstenholme and C. M. O'Connor, eds.) Vol. 13, *Human Pituitary Hormones*, pp. 89–105, Little, Brown, Boston (1960).

138. A. I. Winegrad, W. N. Shaw, F. D. W. Lukens, and W. C. Stadie, Lipogenesis in adipose tissue, *Am. J. Clin. Nutr.* **8**:651 (1960).

139. A. I. Winegrad, W. N. Shaw, F. D. W. Lukens, W. C. Stadie, and A. E. Renold, Effects of growth hormone in vitro on the metabolism of glucose in rat adipose tissue, *J. Biol. Chem.* **234**:1922 (1959).

140. R. D. Orth, W. D. Odell, and R. H. Williams, Some hormonal effects on the metabolism of acetate-I-C by rat adipose tissue, *Am. J. Physiol.* **198**:640 (1960).

141. F. G. Young, Permanent experimental diabetes produced by pituitary (anterior lobe) injections, *Lancet* **2**:372–374 (1937).

142. J. C. Shipp, L. H. Opie, and D. Challoner, Fatty acid and glucose metabolism in perfused heart, *Nature (London)* **189**:1018 (1961).

143. G. Bolodia and F. G. Young, Growth hormone and carbohydrate metabolism *in vitro*, *Nature (London)* **215**:960 (1967).

144. M. Reiss, in *Chemical Pathology of the Nervous System* (J. Folch, ed.) p. 437, Pergamon Press, New York (1961).

145. S. Ellis, J. Huble, and M. E. Simpson, Influence of hypophysectomy and growth hormone on cartilage sulfate metabolism, *Proc. Soc. Exp. Biol. Med.* **84**:603–605 (1953).
146. E. J. Collins and V. F. Baker, Growth hormone and radiosulfate incorporation: I. A new assay method for growth hormone, *Metabolism Clin. and Exptl.* **9**:556–560 (1960).
147. E. J. Collins, S. C. Lyster, and O. S. Carpenter, Growth hormone and radio-sulfate incorporation in costal cartilage, *Acta Endocr.* **36**:51–56 (1961).
148. C. W. Denko and D. M. Bergenstal, The effect of hypophysectomy and growth hormone on S^{35} fixation in cartilage, *Endocrinology* **57**:76–86 (1955).
149. W. R. Murphy, W. H. Daughaday, and C. Hartnett, The effect of hypophysectomy and growth hormone on the incorporation of labeled sulfate into tibial epiphyseal and nasal cartilage of the rat, *J. Lab. Clin. Med.* **47**:715–722 (1956).
150. P. H. Henneman, A. P. Forbes, M. Moldawer, E. F. Dempsey, and E. L. Carroll, Effects of human growth hormone in man, *J. Clin. Invest.* **39**:1223–1238 (1960).
151. J. C. Beck, E. E. McGarry, I. Dyrenfurth, and E. H. Venning, The metabolic effects of human and monkey growth hormone in man, *Ann. Internal. Med.* **49**:1090–1105 (1958).
152. D. M. Bergenstal and M. B. Lipsett, Metabolic effects of human growth hormone and growth hormone of other species in man, *J. Clin. Endocr. and Met.* **20**:1427–1436 (1960).
153. J. J. Hutchings, R. F. Escamilla, W. C. Deamer, and C. H. Li, Metabolic changes produced by human Growth hormone (LI) in a pituitary dwarf, *J. Clin. Endocr. Met.* **19**:759–769 (1959).
154. D. Ikkos, R. Luft, and C. A. Gemzell, The effect of human growth hormone in man, *Acta Endocr.* **32**:341–361 (1959).
155. M. B. Lipsett, D. M. Bergenstal, and F. G. Dhyse, Metabolic studies with human growth hormone in dwarfism and acromegaly, *J. Clin. Endocr. Met.* **21**:119–128 (1961).
156. D. Ikkos, R. Luft, and C. A. Gemzell, The effect of human growth hormone in man, *Lancet* i:720–721 (1958).
157. T. H. Shepard, R. L. Nielsen, M. L. Johnson, and N. Bernstein, Human growth hormone, *A. M. A. J. Dis. Child.* **99**:90–96 (1960).
158. D. M. Bergenstal, H. A. Lubs, L. F. Hallman, J. Patten, H. J. Levine, and C. H. Li, *J. Lab. Clin. Med.* **50**:791–792 (1957).
159. H. Gershberg, Metabolic and renotropic effects of human growth hormone in disease, *J. Clin. Endocr. Met.* **20**:1107–1119 (1960).
160. A. Korner, P. Randle, F. G. Young, A. C. Crooke, R. F. Fletcher, H. G. Sammons, R. Fraser, K. Ibbeston, I. D. F. Wooton, F. G. Prunty, R. R. McSwiney, R. Vaughn Jones, A. Stuart-Mason, and C. E. King, *Lancet* i:7–12 (1959).
161. B. M. Dobyns and S. L. Steelman, Thyroid stimulating hormone of anterior pituitary as distinct from exophthalamos producing substance, *Endocrinology* **52**:705–711 (1953).
162. T. P. Haynie, R. J. Winzler, J. Matovinovic, E. A. Carr Jr., and W. H. Beirwaltes, Thyroid-stimulating and exophthalamos-producing activity of biochemically altered thyrotropin, *Endocrinology* **71**:782–790 (1962).
163. W. D. Odell, P. L. Rayford, and G. T. Ross, Simplified partially automated method for radioimmunoassay of human thyroid-stimulating, growth, luteinizing and follicle stimulating hormones, *J. Lab. Clin. Med.* **70**:937–980 (1967).
164. O. Hechter and I. D. K. Halkerston, *in The Hormones* (G. Pincus, K. V. Thimann, and E. B. Astwood, eds.) pp. 697–825, Academic Press, New York (1964).
165. H. Friesen and E. B. Astwood, Hormones of the anterior pituitary body, *New Eng. J. Med.* **272**:1276 (1965).
166. M. Reiss, H. Druckrey, and F. Fischl, Die Stoffwechselreaktion des Ovars unter dem Einflusse des Hypophysenvorderlappensexualhormons, *Endokrinologie* **10**:241–250 (1932).
167. M. Reiss, H. Druckrey, and F. Fischl, Die Stoffwechselreaktion des Hodens uter dem Einflusse des Hypophysenvorderlappensexualhormons, *Endokrinologie* **10**:329–335 (1932).

168. M. Reiss, H. Druckrey, and A. Hochwald, Der Einfluss der Hypophysektomie auf den Hodenstoffwechse, *Endokrinologie* **12**:243–250 (1933).
169. P. F. Hall and K. B. Eik-nes, Interstitial Cell-Stimulating Hormone and Penetration of D-Xylose-1-C^{14} and α-aminoisobutyric Acid-1-C^{14} into Slices of Testis, *Proc. Soc. Exp. Biol. Med.* **110**:148–151 (1962).
170. P. F. Hall and K. B. Eik-nes, The action of gonadotropic hormones upon rabbit testis *in vitro Biochim. Biophys. Acta* **63**:411–422 (1962).
171. M. Reiss, H. H. Berman, J. J. Pearse, K. Albert-Gasorek, and J. C. Hillman, Investigations into the interrelation of physical and mental retardation, *J. Neuropsychiatry* **2**:109–137 (1961).
172. J. Meites, C. S. Nicoll, and P. K. Talwalker, in *Advances in Neuroendocrinology* (A. V. Nalbandov, ed.) pp. 238–288, University of Illinois Press, Urbana, Ill. (1963).
173. O. Riddle and R. W. Bates, in *Sex and Internal Secretions* (Allen, Baliere, Tindall and Cox, eds.) pp. 1088–1117 (1939).
174. O. J. Riddle, *Natn. Canc. Inst.* **31**:1039 (1963).
175. C. W. Turner, in *Sex and Internal Secretions* (Allen, Baliere, Tindall and Cox, eds.) pp. 740–803 (1939).
176. A. E. Wilhelmi, Fractionation of human pituitary glands, *Can. J. Biochem. Physiol.* **39**:1659–1668 (1961).
177. A. F. Parlow, A. E. Wilhelmi, and I. E. Reichert Jr., Further studies on the fractionation of human pituitary glands, *Endocrinology* **77**:1126–1134 (1965).
178. H. B. F. Dixon, in *The Hormones* (G. Pincus, K. V. Thimann, and E. B. Astwood, eds.) pp. 1–68, Academic Press, New York (1964).
179. W. R. Butt, *Hormone Chemistry*, Van Nostrand, London (1967).
180. F. Sanger, The free amino groups of insulin, *Biochem. J.* **39**:507 (1945).
181. P. Edman, Method for determination of the amino acid sequence in peptides, *Acta Chem. Scand.* **4**:283 (1950).
182. J. M. Dellachs and A. V. Fontanive, Quantitative *N*-terminal amino acid analysis by thin layer chromatography, *Experientia* **21**:351–352 (1965).
183. P. Andrews, Molecular weights of prolactins and pituitary growth hormone estimated by gel filtration, *Nature* **209**:155–157 (1966).
184. C. H. Li and H. Papkoff, Preparation and properties of growth hormone from human and monkey glands, *Science* **124**:1293 (1956).
185. C. H. Li and B. Starman, Human pituitary growth hormone. IX. Molecular weight of the monomer, *Biochim. Biophys. Acta* **86**:175 (1964).
186. W. -K. Liu, J. S. Dixon and C. H. Li, Human pituitary growth hormone. X. Isolation and amino acid sequence of the NH_2-terminal octapeptide, *Biochim. Biophys. Acta* **93**:428 (1964).
187. C. H. Li, A. J. Parcells, and H. Papkoff, The C-terminal amino acid sequence of growth hormones from human, monkey, whale, and sheep pituitary glands, *J. Biol. Chem.* **233**:1143 (1958).
188. C. H. Li, W. -K. Liu, and J. S. Dixon, Human pituitary growth hormone. VI. Modified procedure of isolation and NH_2-terminal amino acid sequence, *Arch. Biochem. Biophys.* Suppl. **1**:327 (1962).
189. C. H. Li and G. Samuelsson, Human pituitary growth hormone. XII. Rate of hydrolysis by trypsin, chymotrypsin and pepsin: Effect of trypsin on the biological activity, *Mol. Pharmacol.* **1**:47 (1965).
190. W. A. Shroeder, R. T. Jones, J. Cormick, and K. McCalla, Chromatographic separation of peptides on ion exchange resins, *Anal. Chem.* **43**:1570 (1962).
191. C. H. Li, W. -K. Liu, and J. S. Dixon, Human pituitary growth hormone. XII. The amino acid sequence of the hormone, *J. Am. Chem. Soc.* **88**:2050–2051 (1966).

192. J. M. Dellacha, M. A. Enero, and I. Faiferman, Molecular weight of bovine growth hormone, *Experientia* 22:16–17 (1966).

193. F. Reusser and H. Ko, Fractionation of highly purified bovine growth hormone on Sephadex G-25 Gel, *Experientia* 22:310–312 (1966).

194. B. B. Saxena and P. H. Henneman, Isolation and properties of the electrophoretic components of human growth hormone by Sephadex-gel filtration and preparative polyacrylamide-gel electrophoresis, *Biochem. J.* 100:711–717 (1966).

195. M. Wallis and H. B. F. Dixon, A chromatographic preparation of ox growth hormone, *Biochem. J.* 100:593–600 (1966).

196. C. A. Free and M. Sonenberg, Separation and properties of multiple components of bovine growth hormone, *J. Biol. Chem.* 241:5076–5082 (1966).

197. C. E. Wolfenstein, J. A. Santome, and A. C. Paladini, Amino acid composition of purified bovine growth hormone, *Acta Physiol. Lat. Amer.* 16:194–199 (1966).

198. W. D. Peckham, The preparation of homogenous monkey and human pituitary growth hormone, *J. Biol. Chem.* 242:190–196 (1967).

199. J. B. Mills, C-terminal sequence of pig growth hormone, *Nature* 213:631–632 (1967).

200. K. J. Catt, B. Moffat, and H. D. Niall, Human growth hormone and placental lactogen: structural similarity, *Science* 157:321 (1967).

201. L. M. Sherwood, Similarities in the chemical structure of human placental lactogen and pituitary growth hormone, *Proc. Nat. Acad. Sci.* 58:2307–2314 (1967).

202. C. H. Li, Comparative biochemical endocrinology of pituitary growth hormone, *Proc. First Int. Congr. of Endocrinology*, p. 75, Copenhagen (1960).

203. Z. Laron, A. Yed-Lekach, S. Assa, and A. Kowadlo-Silbergeld, Immunochemical properties of bovine and human pituitary growth hormone after pepsin digestion, *Endocrinology* 74:532–537 (1964).

204. W. R. Lyons, Preparation and assay of mammotropic hormone, *Proc. Soc. Exp. Biol. Med.* 35:645 (1937).

205. R. W. Payne, M. S. Raben, and E. B. Astwood, Extraction and purification of corticotropin, *J. Biol. Chem.* 187:719 (1950).

206. E. B. Astwood, M. S. Raben, R. W. Payne, and A. B. Grady, Purification of corticotropin with oxycellulose, *J. Am. Chem. Soc.* 73:2969 (1951).

207. E. B. Astwood, in *The Hormones: Physiology, Chemistry and Applications* (G. Pincus and K. V. Thimann, eds.) Vol. 3, pp. 235-308, Academic Press, New York (1955).

208. W. F. White, Studies on adrenocorticotropin, V. The isolation of corticotropin-A, *J. Am. Chem. Soc.* 75:503 (1953).

209. R. G. Shepherd, K. S. Howard, P. H. Bell, A. R. Cacciola, R. G. Child, M. C. Davies, J. P. English, B. M. Finn, J. H. Meisenhelder, A. W. Moyer, and J. van der Scheer. Studies with corticotropin. I. Isolation, purification and properties of β-corticotropin, *J. Am. Chem. Soc.* 78:5051 (1956).

210. R. G. Shepherd, S. D. Willson, K. S. Howard, P. H. Bell, D. S. Davies, E. A. Eigner, and N. E. Shakespeare, Studies with corticotropin. III. Determination of structure of β-corticotropin and its active degradation products, *J. Am. Chem. Soc.* 78:5067–5076 (1956).

211. C. H. Li, I. Geschwind, D. Cole, I. D. Raacke, J. I. Harris and J. S. Dixon, Amino acid sequence of alpha corticotropin, *Nature* 176:687–689 (1955).

212. C. H. Li, Hormones of the anterior pituitary gland. Part I. Growth and adrenocorticotropic hormones, *Advance Protein Chem.* 11:101–190 (1956).

213. C. H. Li, J. S. Dixon, and D. Chung, Structure of bovine corticotropin, *J. Am. Chem. Soc.* 80:2587–2588 (1958).

214. T. H. Lee, A. B. Lerner, and V. Buettner-Janusch, Isolation and structure of human corticotropin (ACTH), *J. Am. Chem. Soc.* 81:6084 (1959).

215. R. Schwyzer and P. Sieber, Total synthesis of adrenocorticotrophic hormone, *Nature (London)* **199**:172–174 (1963).

216. J. Ramachandran and C. H. Li, Adrenocorticotropins. XXXIII. Synthesis of a biologically active hexacosapeptide corresponding to the first 26 residues of bovine ACTH, *J. Amer. Chem. Soc.* **87**:2691–2696 (1965).

217. R. A. Boissonnas, St. Guttmann, J. -P. Waller, and P. -A. Jaquenoud, Synthesis of a polypeptide with ACTH-like structure, *Experientia* **12**:446 (1956).

218. K. Hofmann, H. Yajima, T. Liu, and N. Yanaihara, Studies on polypeptides. XXIV. Synthesis and biological evaluation of a tricosapeptide possessing essentially the full biological activity of ACTH, *J. Am. Chem. Soc.* **84**:4475 (1962).

219. K. Hofmann, N. Yanaihara, S. Lande, and H. Yajima, Studies on polypeptides. XXIII. Synthesis and biological activity of a hexadecapeptide corresponding to the N-terminal sequence of the corticotropins, *J. Am. Chem. Soc.* **84**:4470 (1962).

220. C. H. Li, J. Meienhofer, E. Schnabel, D. Chung, T. -b. Lo and J. Ramachandran, Synthesis of a biologically active nonadecapeptide corresponding to the first nineteen amino acid residues of adrenocorticotropins, *J. Am. Chem. Soc.* **83**:4449 (1961).

221. C. H. Li, Perspectives in the biochemical endrocrinology of adenohypophyseal hormones, *Bull. N.Y. Acad. Med.* **39**:143–155 (1963).

222. J. Gelzer, Immunochemical study of β-corticotropin-(1-24)-tetracosapeptide, *Immunochemistry* **5**:23–31 (1968).

223. R. D. Utiger, W. D. Odell, and P. G. Condliffe, Immunologic studies of purified human and bovine thyrotropin, *Endocrinology* **73**:359–365 (1963).

224. W. D. Odell, J. F. Wilber, and W. E. Paul, Radioimmunoassay of human thyrotropin in serum, *Metabolism* **14**:465–467 (1965).

225. S. Kusaka, Immunological studies on thyroid stimulating hormone (TSH). I. Immunological and biological properties of thyrotropin antiserum. II. Basic investigation for application of haemagglutination inhibition reaction to human TSH assay. III. Basic investigation for application of radioimmunoprecipitation reaction to human TSH assay, *Jap. Arch. Intern. Med.* **13**:473–483, 485–493, 539–548 (1966).

226. J. G. Pierce, M. E. Carsten, and L. K. Wynston, Chemistry and physiology of the thyroid-stimulating hormone, *Ann. N.Y. Acad. Sci.* **86**:612 (1960).

227. M. E. Carsten and J. G. Pierce, Chemical studies on thyrotropin preparations and related glycoproteins, *J. Biol. Chem.* **238**:1724 (1963).

228. P. Condliffe and W. B. Jakoby, Crystallization of bovine thyroid stimulating hormone, *Endocrinology* **80**:203–204 (1967).

229. S. L. Steelman, A. Segaloff, and R. N. Anderson, Purification of human pituitary follicle stimulating and luteinizing hormone, *Proc. Soc. Exp. Biol. Med.* **101**:452 (1959).

230. G. Bettendorf, M. Apostolakis, and K. D. Voigt, Darstellung von Gonadotropin aus Menschlichen Hypophysen, *Acta Endocr. Copenh.* **41**:1 (1962).

231. L. E. Reichert and A. F. Parlow, Preparation of ovine luteinizing hormone (LH, ICSH) having high biological activity, *Endocrinology* **73**:285 (1963).

232. W. R. Butt, A. C. Crooke, and F. J. Cunningham, Studies on human urinary and pituitary gonadotrophins, *Biochem. J.* **81**:596 (1961).

233. W. R. Butt, A. C. Crooke, F. J. Cunningham, and A. Wolf, Biological and immunological properties of human pituitary follicle stimulating hormone obtained by starch gel electrophoresis, *J. Endcr.* **25**:541 (1963).

234. P. Roos and C. A. Gemzell, The isolation of human pituitary follicle stimulating hormone, *Biochim. Biophys. Acta* **82**:218 (1964).

235. L. E. Reichert and A. F. Parlow, Further studies on purification of human pituitary luteinizing hormone, *Endocrinology* **75**:815–817 (1964).

236. S. L. Steelman, A. Segaloff, and R. N. Andersen, Purification of human pituitary follicle stimulating (FSH) and luteinizing (LH) hormones, *Proc. Soc. Exp. Biol. Med.* **101**:452–454 (1959).
237. U. Gröschel and C. H. Li, On the carbohydrate moiety of ovine and human pituitary gonadotropins, *Biochim. Biophys. Acta* **37**:375 (1960).
238. S. L. Steelman and A. Segaloff, Recent studies on the purification of the pituitary gonadotropins, *Recent. Prog. Horm. Res.* **15**:115 (1959).
239. E. F. Walborg and D. Ward, The carbohydrate components of ovine luteinizing hormone, *Biochim. Biophys. Acta* **78**:304 (1963).
240. G. B. Pierce and L. K. Wynston, On the composition of sheep thyrotropic hormone, *Biochim. Biophys. Acta* **43**:538 (1960).
241. D. N. Ward, E. F. Walborg, and M. Adams-Mayne, Amino acid composition and electrophoretic properties of ovine luteinizing hormone, *Biochim. Biophys. Acta* **50**:224 (1961).
242. R. Bourrillon, R. Got, and R. Marcy, Isolement et quelques caracteristiques d'une gonadotrophine urinaire de ménopause homogéne á l'électrophorése, *Acta Endocr.* **35**:225–234 (1960).
243. R. Got and R. Bourrillon, New method of purification of human chorionic gonadotropin, *Biochim. Biophys. Acta* **42**:505–512 (1960).
244. Goverde B. C. Van Hell, A. H. W. M. Schuurs, E. De Jager, R. Matthijsen and J. D. H. Homan, Purification, characterization and immunochemical properties of human chorionic gonadotropin, *Nature* **212**:261–262 (1966).
245. J. Lewis, S. Dray, S. Genuth, and H. S. Schwartz, Demonstration of immunological similarities of human pregnancy, gonadotropin and choriocarcinoma gonadotropin with antisera prepared in rabbits and monkeys, *J. Clin. Endocr. and Met.* **24**:197 (1964).
246. C. M. McKean, Preparation and use of antisera to human chorionic gonadotrophin, *Am. J. Obst. Gynaec.* **80**:596 (1960).
247. S. Kaivola, U. Kiistala, and E. Axelson, Studies on a serological pregnancy test: III, *Acta Endocr. Copenh.* **42**:395 (1963).
248. S. Brody and G. Carlström, Immuno-assay of human chorionic gonadotropin in normal and pathologic pregnancy, *J. Clin. Endocr. Met.* **22**:564 (1962).
249. M. G. Carlsson, The use of haemagglutination inhibition reaction for qualitative and quantitative determinations of HCG in normal and pathological pregnancies, *Acta Endocr. Copenh.* **46**:142 (1964).
250. A. R. Midgley, G. B. Pierce, and W. O. Weigle, Immunobiological identification of human chorionic gonadotropin, *Proc. Soc. Exp. Biol. Med.* **108**:85 (1961).
251. B. Lunenfeld, C. Isersky, and M. C. Shelesnyak, Immunologic studies on gonadotropins. I. Immunogenic properties and immunologic characterization of human chorionic gonadotropin preparations (HCG) and their homologous antisera, *J. Clin. Endocr. Met.* **22**:555 (1962).
252. H. S. Schwartz and N. Mantel, A rapid bio-assay for chorionic gonadotropin (HCG) and gonadotropin inhibition, *J. Clin. Endocr. Met.* **22**:393 (1962).
253. D. K. Keele, J. Remple, J. Bean, and J. Webster, Immunologic reactions to human chorionic gonadotropin, *J. Clin. Endocr. Met.* **22**:287 (1962).
254. S. Hamashige and E. R. Arquilla, Immunological studies with a commercial preparation of human chorionic gonadotropin, *J. Clin. Invest.* **42**:546 (1963).
255. W. E. Paul and W. D. Odell, Radiation inactivation of the immunological and biological activities of human chorionic gonadotropin, *Nature, (London)* **203**:979 (1964).
256. A. J. Fulthorpe, J. E. Tovey, J. A. C. Parke, and J. C. Monckton, Pregnancy diagnosis by a none-stage passive haemagglutination inhibition method, *Brit. Med. J.* **i**:1049 (1963).
257. W. K. Whitten, *Aust. J. Scient. Res.* Series B **3**:346 (1950).
258. W. R. Butt, A. C. Crooke, and F. J. Cunningham, Immunological study of human gonadotrophins, *Proc. Roy. Soc. Med.* **54**:647 (1961).

259. L. Wide, R. Roos, and C. Gemzell, Immunological determination of human pituitary luteinizing Hormone (LH), *Acta Endocr. Copenh.* **37**:445–449 (1961).
260. L. Wide and C. A. Gemzell, *Ciba Fdn. Colloq. Endocr.* **14**:296 (1962).
261. I. D. Raacke, A. J. Lostroh, J. M. Boda, and C. H. Li, Some aspects of the characterization of pregnant mare serum gonadotrophin, *Acta Endocr. Copenh.* **26**:377 (1957).
262. A. White, The chemistry and physiology of adenohypophyseal luteotropin (prolactin), *Vitamins and Hormones* **7**:253 (1949).
263. J. S. Dixon and C. H. Li, Chemistry of prolactin, *Metabolism* **13**:1093–1101 (1964).
264. R. D. Cole, I. I. Geschwind, and C. H. Li, Studies on pituitary lactogenic hormone. XV. N-terminal residue analysis and N-terminal sequence analysis, *J. Biol. Chem.* **224**:399 (1957).
265. C. H. Li, Hormones of the anterior pituitary gland. Part II, *Advances in Protein Chem.* **12**:269 (1957).
266. M. Sluyser and C. H. Li, Studies in pituitary lactogenic hormones. XXIII. Isolation of the monomer of ovine prolactin, *Arch. Biochem. Biophys.* **104**:50 (1964).
267. P. G. Squire, B. Starman, and C. H. Li, Studies of pituitary lactogenic hormone, *J. Biol. Chem.* **238**:1389 (1963).
268. R. A. Reisfeld, D. E. Williams, V. J. Cirillo, G. L. Tong and N. G. Brink, Characterization of sheep prolactin, *J. Biol. Chem.* **239**:1777 (1964).
269. P. Andrews, Molecular weights of prolactins and pituitary growth hormone estimated by gel filtration, *Nature* **209**:155–157 (1966).
270. Z. Laron and M. Apostolakis, Immunological properties of a human pituitary prolactin preparation, *J. Endocr.* **35**:117–118 (1966).
271. P. Baranyai, I. Nagy, M. Kurcz and A. Orosz, Species specificity of prolactin, *Nature* **212**:1255–1256 (1966).
272. E. J. W. Barrington, *in The Hormones* (G. Pincus, K. V. Thimann, and E. B. Astwood, eds.) Vol. IV, pp. 299–363, Academic Press, New York (1964).
273. A. B. Lerner and J. S. McGuire, Melanocyte stimulating hormone and adrenocorticotrophic hormone: relation to pigmentation, *New Eng. J. Med.* **270**:539–546 (1964).
274. R. Motias, *Bull. Inst. Oceanog.* No. 1198 (1961). (quoted from Barrington)
275. A. J. Kastin, G. Genser, A. Arimura, M. Clinton Miller, and A. V. Schally, Melanocyte stimulating and corticotrophic activities in human foetal pituitary glands, *Acta Endocr.* **58**:6–10 (1968).
276. T. Ito, Experimental studies on the hypothalamic control of the pars intermedia activity of the frog rana nigromaculata, *Neuroendocrinology* **3**:25–33 (1968).
277. F. W. Landgrebe, E. Reid, and H. Waring, Further observations on the intermediate lobe pituitary hormones, *Quart. J. of Exp. Physiol.* **32**:121–141 (1943).
278. T. H. Lee and A. B. Lerner, Isolation of melanocyte-stimulating hormone from hog pituitary gland, *J. Biol. Chem.* **221**:943 (1956).
279. J. S. Dixon and C. H. Li, The isolation and structure of β-melanocyte stimulating hormone from horse pituitary glands, *Gen. Comp. Endocr.* **1**:161 (1961).
280. J. I. Harris and P. Roos, Studies on pituitary polypeptide hormones. 1. The structure of β-melanocyte stimulating hormone from pig pituitary glands, *Biochem. J.* **71**:434–445 (1959).

SUBJECT INDEX

Acetylcholine
assay methods, 267
binding to AChE, 278–279
chromatography, 267
distribution in tissue, 264
receptors, 276
solubility, 266
stability, 266–267
structure, 265–266
as transmitter, 263–265
uptake by brain, 349
Acetylcholinesterase
assay methods, 276–277
inhibitors, 280
kinetics, 280–281
levels
during innervation, 276
in regenerating nerve, 275
localization
cellular, 274–275
subcellular, 275
physical properties, 278
purification, 277–278
substrates, 278–280
binding forces, 278–279
N-Acetylhistamine, 233
N-Acetylnormetanephrine, 206
Actin, interaction with ATP, 18–19
Active transport
of ammonium compounds, 349
and ATPases
and cation transport, 28–29
and metabolic regulation, 4, 28–29
sodium-potassium ATPase, 22–27
and sodium pump, 27–28
and catecholamines, 341–342
and ion flux, 322–324
of serotonin, 344–345
Adenosine diphosphate, hexokinase
inhibition, 8
Adenosine monophosphate
effect of histamine, 239
and phosphorylase activity, 5
role in glycogen breakdown, 5
Adenosine triphosphate
inhibition of phosphofructokinase, 7
and muscle contraction, 19–20
Adenochromes, and catecholamine
oxidation, 214
Aldolase, activity in brain, 2
Aliphatic amines, and monoamine
oxidase, 293

γ-Aminobutyric acid, as neurotransmitter
in invertebrates, 120–122
in vertebrates, 122–124
Amine oxidases
benzylamine oxidase, 303
classification, 288
cofactors, 303
competitive inhibition, 287
effect of cyanide, 286
diamine oxidase, 301
dopamine-β-oxidase, 301
monoamine oxidase
in adrenergic transmission, 213–214
in catecholamine metabolism, 55
characteristics, 209–210
effect of cortisone, 299
detection, 288–289
distribution
cellular, 210–211, 290
subcellular, 290–292
in developing brain, 292
estimation, 289–290
in heart, 215
inhibitors, 211–212, 294–298
effects, 303–304
isozymes, 301
metabolic functions, 299–301
effect of oxygen tension, 292
effect of riboflavin, 299
substrates, 210, 292–294
effect of thyroxine, 299
spermine oxidase, 302
substrates, 285–287
competition, 286–287
transformation, 286
Amino acids
uptake
effect of growth hormone, 478
effect of insulin, 365, 366–367
Amphetamine
and amine oxidases, 287
inhibition, 296
structure, 295
L-Arabinose, and blood-brain barrier,
331, 332
Aromatic L-amino acid decarboxylase
assay methods, 170–171
substrate specificity, 200
ATPase
actomyosin ATPase, 19
activity
in active transport, 22–33

507